AF697804

BIMOLECULAR COLLISIONS

Advances in Gas-Phase Photochemistry and Kinetics

Edited by M. N. R. Ashfold and J. E. Baggott

Each title in this series aims to provide critical reviews on a specific area of current interest at the interface of molecular spectroscopy, photochemistry, and chemical kinetics.

Molecular Photodissociation Dynamics
Bimolecular Collisions

How to obtain future titles on publication

'Bimolecular Collisions' is the second title in the series 'Advances in Gas-Phase Photochemistry and Kinetics'. The first title in the series dealt with gas-phase molecular photodissociation. Future titles may be obtained as soon as they are published by sending a standing order to:

The Royal Society of Chemistry
Distribution Centre
Blackhorse Road
Letchworth
Herts. SG6 1HN

Telephone: Letchworth (0462) 672555
Telex: 825372

Advances in
Gas-Phase Photochemistry and Kinetics

Bimolecular Collisions

Edited by
M. N. R. Ashfold
School of Chemistry
University of Bristol

J. E. Baggott
Department of Chemistry
University of Reading

ROYAL
SOCIETY OF
CHEMISTRY

QC
173
.B336
1989

British Library Cataloguing in Publication Data
Bimolecular collisions
1. Molecules. Collisions
I. Ashfold, M. N. R. (Michael Norman Royston), *1954–*
II. Baggott, J. E. (James Edward), *1957–*
539.6

ISBN 0-85186-393-0

Front Cover Illustration
Two quasiclassical trajectories for $H + H_2 \rightarrow H_2 + H$ as functions of mass-scaled coordinates. The underlying dashed contours are the potential energy surface. The two trajectories begin in the lower right with the same vibrational energy but different phases, they experience a long-lived collision and exhibit features corresponding to the classical analogue of a Feshbach internal-excitation resonance, and finally they evolve upwards to the product region with slightly different final vibrational energies.
Adapted, with permission, from J. W. Duff and D. G. Truhlar, *Chem. Phys.*, 1974, **4**, 1.

© The Royal Society of Chemistry, 1989

All Rights Reserved
No part of this book may be reproduced or transmitted in any form or by any means—graphic, electronic, including photocopying, recording, taping, or information storage and retrieval systems—without written permission from The Royal Society of Chemistry

Published by The Royal Society of Chemistry
Burlington House, Piccadilly, London W1V 0BN

Typeset by KEYTEC, Bridport, Dorset

Foreword

This volume is the second in an informal series designed to provide authoritative and timely reviews of advances in the fields of gas-phase photochemistry and kinetics. We have continued with the philosophy, described in Volume 1 ('Molecular Photodissociation Dynamics'), of using the series to provide a forum for a selection of internationally acknowledged experts to present an overview of the latest exciting developments in their field of study. In this volume, we have collected together reviews which deal with the general area of bimolecular collision processes. These include quantum state resolved studies of reactive processes, experimental and theoretical studies of the association reactions of atoms and free radicals, experimental studies of collisional energy transfer, of bimolecular collision processes involving alkaline earth atoms, of the reactions of bare metal clusters, and of collision processes involving ions in drift fields.

For many years the hydrogen exchange reaction, $H + H_2 \rightarrow H_2 + H$, and its isotopic variants, has been the prototypical bimolecular reaction used by theoretical chemists to develop an understanding of chemical reactivity. The opening chapter, by Valentini and Phillips, provides a survey of the latest experimental and theoretical developments in the study of this reaction. The authors describe how modern spectroscopic techniques have enabled the quantitative measurement of the population distributions of the various quantum states of the H_2 product following reaction at a number of well-defined collision energies. Such measurements provide a searching test of the quality of the presently available *ab initio* potential energy surfaces for this reaction, and probe the accuracy of our treatment of the nuclear motions on such surfaces. The agreement between measured and calculated product quantum state distributions is impressive, and encourages us to believe that measurements of the fully resolved state-to-state reaction cross-sections, and perhaps even the observation of the predicted dynamical resonances associated with the H_3 transition state itself, lie just around the corner.

The study of reactions between species more complicated than H atoms and H_2 molecules presents us with some obvious difficulties. The available experimental data for more complicated systems tend to be less state-specific and the associated theoretical treatments are, of necessity, more empirically based. In Chapter 2, Smith considers the effects of reagent vibrational excitation on bimolecular collision processes, and describes results for a number of benchmark systems, including the hydrogen exchange reaction. The effects of reagent vibrational excitation are most manifest in reactions in which the vibrational excitation assists in surmounting an activation barrier along the reaction co-ordinate, producing dramatic rate enhancements. Vibrationally adiabatic transition state theory provides a sound theoretical framework within which many of the experimentally observed effects may be understood.

The association reactions of atoms and free radicals proceed across a potential energy surface which is often characterized by a well, rather than a barrier. Reactions which proceed through a transitory complex intermediate are relatively insensitive to reagent vibrational excitation. However, the possibility of collisional stabilization of the intermediate makes these reactions sensitive to factors which control the collision rate and the rate of energy transfer. In Chapter 3, Davies and Pilling present an overview of recent experimental studies of association reactions, choosing, like Smith, to present material in the form of case histories. Reactions whose rates vary as functions of temperature and pressure need to be studied with a precision which places considerable demands on current technology. Data of high precision are needed, however, if it is the intention to extrapolate the rate measurements outside the regions of temperature and pressure which are accessible in the experiments. Data fitting also requires considerable attention, and these authors make full use of the framework provided by transition state theory.

In Chapter 4, Hase and Wardlaw show exactly what kind of contribution transition state theory can make to our understanding of the rates of association reactions. With no barrier to pinpoint the location of the transition state, it is necessary to turn to alternative definitions. These authors provide a comprehensive survey of recent developments in variational transition state theory, and include a useful introduction to the subject for the newcomer. The application of theory is illustrated by reference to two systems, $H + CH_3$ and $CH_3 + CH_3$, which are also discussed at length from the experimentalists' viewpoint in Chapter 3. Closer collaboration between the theoretical and experimental chemist of the type exemplified in the reviews of Davies and Pilling and Hase and Wardlaw surely holds much promise for the future.

Non-reactive deactivating collisions play an important role in the kinetics of gas-phase atom and free-radical reactions, but are of special significance in association reactions of the type described in Chapters 3 and 4. Our understanding of energy-transferring collisions has improved as increasingly elaborate experimental techniques have been developed for their study. In Chapter 5, by Hippler and Troe, the results of recent direct studies of collisional energy transfer are reviewed. These authors identify a number of general behavioural trends exhibited by molecules which range in size from simple triatomics (SO_2, CS_2) to quite large polyatomics (azulene, substituted cycloheptatrienes). These studies have focused on the experimental measurement of the average amount of energy transferred per collision, $\langle \Delta E \rangle$, and the factors which influence the size of $\langle \Delta E \rangle$. As with most frontier research, the study of energy transfer is not without controversy, and Hippler and Troe provide us with an opportunity to form conclusions on the basis of the most recent experimental developments.

Bimolecular collision processes involving alkaline earth atoms represent a fairly mature field of study. Husain and Roberts describe in Chapter 6 how presently available pulsed dye laser techniques have been used to prepare excited atoms in a state-specific manner and thence to monitor their decay in real time. In reactions of the excited atoms with oxidants, the dominant product channels can often be deduced from an analysis of the intense chemiluminescence which accompanies the reaction. Detailed reaction mechanisms can sometimes be inferred using *ab initio* theory or through a consideration of adiabatic correlation diagrams.

In contrast, the experimental study of the reactions of metal clusters represents a relatively new field. In Chapter 7, Jarrold describes the various experimental methods for generating such clusters (both neutrals and ions), together with the techniques which have been developed to study some of their simpler reactions. This survey clearly illustrates the impressive progress which has been made in this new field, and highlights the many questions which remain to be answered. For example, factors which might control the reactivity of clusters include electronic structure and geometrical structure; but it is not always clear which exerts the dominant influence. Interestingly, some cluster sizes appear to be particularly inert compared to others. There are obvious connections to be made here between the properties of clusters and the properties of bulk metals, and certainly the study of clusters may be expected to provide important information on aspects of chemisorption and catalysis.

We close in Chapter 8 with a review, by Leone, of the experimental study and theoretical interpretation of bimolecular collision processes involving ions in drift fields. This account begins with a survey of theories of drift motion, and goes on to explain how tunable dye lasers are used to probe important ion drift parameters, ion internal excitation, and ion alignment. The ability to garner such detailed information concerning ion quantum state distributions and ion alignment holds much promise for the study of simple ion–molecule reactions with the resolution now available in the study of the reactions of neutrals.

In the opening paragraph of the foreword for 'Molecular Photodissociation Dynamics', we commented on the fact that the ready availability of tunable laser sources has initiated a revolution in the study of molecular photodissociation. Of course, any experimental science which makes regular use of sources of electromagnetic radiation will have benefitted from the development of these remarkable devices, and this is certainly true for the study of bimolecular collisions. Lasers are used to prepare high concentrations of atoms, free radicals, and ions, often in selected quantum states and they are used to probe the progress of the collision process (whether reactive or unreactive). We are now in a position to examine the collision process in unprecedented detail, striking right at the heart of the act of chemical reaction. We hope that in this volume we have been able to convey something of the flavour of the excellent science now in progress around the world.

Bristol
July 1988

M. N. R. Ashfold
J. E. Baggott

Contributors

J. W. Davies, *University of Oxford*
W. L. Hase, *Wayne State University, Detroit, Michigan, U.S.A.*
H. Hippler, *Universität Göttingen, Göttingen, West Germany*
D. Husain, *University of Cambridge*
M. F. Jarrold, *AT & T Bell Laboratories, Murray Hill, New Jersey, U.S.A.*
S. R. Leone, *University of Colorado, Boulder, Colorado, U.S.A.*
D. L. Phillips, *University of California, Irvine, California, U.S.A.*
M. J. Pilling, *University of Oxford*
G. Roberts, *University of Cambridge*
I. W. M. Smith, *University of Birmingham*
J. Troe, *Universität Göttingen, Göttingen, West Germany*
J. J. Valentini, *University of California, Irvine, California, U.S.A.*
D. M. Wardlaw, *Queen's University, Kingston, Ontario, Canada*

Contents

Chapter 5 Recent Direct Studies of Collisional Energy Transfer in Vibrationally Highly Excited Molecules in the Ground Electronic State

H. HIPPLER and J. TROE

CHAPTER 1

Experimental and Theoretical Studies of the Dynamics of the Hydrogen Exchange Reaction: The Maturation of a Prototype

J. J. VALENTINI and D. L. PHILLIPS

1 Introduction

The hydrogen exchange reaction has been referred to so frequently as the 'prototype chemical reaction' that the impact of that appellation has by now diminished. This is ironic, because it is only recently that the $H + H_2 \rightarrow H_2 + H$ reaction and its isotopic variants have achieved the status of a prototype in experimental reaction dynamics. Although the reaction has held an unrivalled position as the most important and most frequently studied system in theoretical reaction dynamics for as long as this field has existed, that preeminence has not extended to experimental reaction dynamics. It is only in the past few years that experimentalists have produced studies of the hydrogen exchange reaction that are equal in detail and sophistication to those common in theoretical dynamics. This final realization of the full prototype status of the hydrogen exchange reaction has created a renaissance of interest in it, and has fostered an unprecedented interplay of theory and experiment. The comparison of the predictions of *ab initio* dynamical theories at the most detailed level of state-to-state cross-sections, state-resolved differential cross-sections, and state-selected kinetics opens up to our view the most intimate dynamics of the reaction. Perhaps most significant of all, the theoretical prediction and experimental observation of dynamical resonances provide a direct link to the very structure and dynamics of the transition state itself.

The role of the $H + H_2 \rightarrow H_2 + H$ reaction as a prototype derives from its uniqueness, and this uniqueness from simplicity: the hydrogen exchange reaction is the simplest chemical reaction, there being only a single electron associated with each atom and only a single adiabatic electronic surface involved. The simplicity of $H + H_2$ would confer a true, but insignificant, uniqueness were it not for the immense difficulty of generating an accurate, global, *ab initio* potential energy surface for any many-electron reaction.[1] Because of its relative simplicity in this regard, the hydrogen exchange reaction is the only chemical reaction for which an accurate, global, *ab initio* potential energy surface exists.[2–4]

The importance of the accuracy of the global H_3 surface is that theoretical descriptions of the $H + H_2$ reaction dynamics, both quantal and classical, which depend on the potential energy surface, are not significantly compromised by inadequacies of the surface. Comparison with experiment, the true test of the adequacy of the theory, then does indeed test the theoretical dynamics, not the potential energy surface. This is the truly valuable uniqueness of the $H + H_2$ reaction. For every other reaction the available potential surfaces have large enough uncertainties that comparison with empirical observation is more a test of the adequacy of the surface than a test of the theoretical chemical dynamics. To develop, test, and refine theoretical models of reaction dynamics requires a reaction system for which there is little question about the potential energy surface. The $H + H_2$ reaction provides the only example.

From a theoretical perspective the simplicity of the hydrogen exchange reaction extends beyond considerations about obtaining the potential energy surface. Each atom in the reaction not only has but one electron, but also a very light nucleus. Consequently, the reactant and product diatoms' moments of inertia are very small and the vibrational frequencies high. Thus, even though the reaction barrier is significant, almost 10 kcal mol^{-1}, and correspondingly high total energies are involved in the reaction, only a few rotational and vibrational states are typically accessible, and only a few partial waves (angular momentum states) need be considered. For quantum mechanical coupled-channel dynamics calculations this minimizes the number of reactant–product channels that need be included in the calculations. For example, at a total energy ($H + H_2$ collision energy plus H_2 internal energy) of 23 kcal mol^{-1} (1 eV) there are only 16 rotational and vibrational energy levels accessible, $j = 0$—9 in $v = 0$, and $j = 0$—5 in $v = 1$. However, this still represents a large number (134) of product states because of the $(2j + 1)$ rotational energy level degeneracy, so coupled-channel calculations are not so easy even for $H + H_2$. However, the simplicity of $H + H_2$ also makes all sorts of approximate quantum, semiclassical, and quasi-classical theoretical calculations possible. The symmetry of the

[1] T. H. Dunning, jun. and L. B. Harding, in 'Theory of Chemical Reaction Dynamics', ed. M Baer, CRC Press, Boca Raton, 1985, Vol. 1, p. 1.

[2] B. Liu, *J. Chem. Phys.*, 1973, **58**, 1925.

[3] P. Siegbahn and B. Liu, *J. Chem. Phys.*, 1978, **68**, 2457.

[4] D. G. Truhlar and C. J. Horowitz, *J. Chem. Phys.*, 1978, **68**, 2466; 1979, **71**, 1514.

reaction – the products and reactants being the same chemical species, at least for some of the isotopic variants – is also simplifying theoretically.

Unfortunately, the factors that make the hydrogen exchange reaction so appealing from the perspective of theoretical chemical dynamics are generally either of little consequence or actually of considerable disadvantage from an experimental perspective. Having a single 1*s* electron per atom makes the potential energy surface relatively easy to calculate, but it also makes bonds involving hydrogen atoms generally quite strong, so that production of H atoms for studies of the reaction is difficult. This is not of much concern if one is interested only in the thermal rate constant – for flow-tube kinetics studies, low-power microwave discharges are adequate sources of H. However, for chemical dynamics experiments, *i.e.* crossed molecular beam and time-resolved state-to-state dynamics experiments, hydrogen atoms are not easy to produce in sufficient quantities or with sufficient energies. And the energies needed are significant: the classical barrier height is equivalent to $\frac{1}{2}RT$ at 10 000 K. The easily obtainable isotopomers of molecular hydrogen – H_2, HD, and D_2 – are difficult to detect in molecular beam experiments because mass spectrometer backgrounds at these low masses are extremely high, and the situation for the atoms H and D is even worse. State-resolved detection of molecular hydrogen is similarly problematic. The homonuclear isotopomers lack a permanent dipole moment and therefore cannot be observed by infrared vibrational spectroscopy or by microwave rotational spectroscopy, and the dipole moments of the heteronuclear isotopomers, *e.g.* HD, are far too small to be useful spectroscopically. The first electronic absorptions of the hydrogen molecule lie in the vacuum ultraviolet (VUV), so electronic spectroscopy is also difficult, at least with the temporal resolution and sensitivity required for studies of the reaction dynamics.

As a result, the balance between the mass of theoretical work and that of experimental work has been tipped quite far towards the theoretical side. Although rate constant measurements were first carried out more than 50 years ago, by Farkas[5] and by Geib and Harteck[6] for $H + p\text{-}H_2 \rightarrow o\text{-}H_2 + H$, and by Farkas and Farkas[7] for $D + o\text{-}D_2 \rightarrow p\text{-}D_2 + D$, until very recently there was not a significant amount of experimental chemical dynamics research done on the hydrogen exchange reaction. However, that situation has changed considerably in the past five years, and there has been a flurry of experimental activity, including measurements for which the analogous theoretical computation is lacking.

This review is motivated principally by the flourishing of experimental

[5] A. Farkas, *Z. Phys. Chem.*, 1930, **B10**, 419.
[6] K. H. Geib and P. Harteck, *Z. Phys. Chem. Bodenstein Festband*, 1931, 849.
[7] A. Farkas and L. Farkas, *Proc. R. Soc. London, Ser. A*, 1935, **152**, 124.

studies of the dynamics of the hydrogen exchange reaction, primarily laser spectroscopic state-to-state experiments but also molecular beam experiments. However, in the past few years there have also been important developments in the theoretical chemical dynamics of $H + H_2$, including the implementation of new coupled-channel computational techniques, the application of approximate quantum methods, the understanding of resonances, and the application of variational transition-state theory. Along with all of this activity in theoretical and experimental chemical dynamics there have also been developments in the area of experimental rate constant measurements on the $H + H_2$ reaction, specifically the study of the reactions of vibrationally excited diatoms, and the opening up of muonium chemical kinetics.

This review will focus on these recent developments in the study of the hydrogen exchange reaction, with a particular emphasis on the new interplay of theory and experiment that the experimental methodological developments have made possible. The scope of this chapter is, consequently, something less than historically comprehensive, even if one limits history to just the past ten years. This seems justified because there are already several excellent reviews on the hydrogen exchange reaction. Rather than supplant them we aim to supplement them. Most recently an entire issue of the *International Journal of Chemical Kinetics*[8] was devoted to papers on the hydrogen exchange reaction, papers that resulted from the Farkas Memorial Symposium, 'Fifty Years of $H + H_2$ Kinetics', in Jerusalem in December 1985. At about the same time Schatz[9] presented an extensive review of recent theoretical work on the $H + H_2$ system. Although now more than a decade old, the two reviews by Truhlar and Wyatt,[10,11] remain valuable and provide a comprehensive discussion of the long and rich history of H_3 kinetics. Extensive discussion of the $H + H_2$ reaction is also presented in the encyclopaedic four-volume monograph, 'Theory of Chemical Reaction Dynamics', edited by Baer.[12] Although none of the 21 chapters therein deals specifically with the hydrogen exchange reaction, the H_3 system has been so important to the development of chemical dynamics theory that almost every chapter has a discussion of some aspect of it.

2 The H_3 Potential Energy Surface

The potential energy surface that so distinguishes the $H + H_2$ reaction is based on the configuration interaction (CI) calculations of Liu and Siegbahn.[2,3] The CI calculations were carried out for H_3 angles between

[8] *Int. J. Chem. Kinet.*, 1986, **18**.

[9] G. C. Schatz, in 'The Theory of Chemical Reaction Dynamics', ed. D. C. Clary, Reidel, Dordrecht, 1986, p. 1.

[10] D. G. Truhlar and R. E. Wyatt, *Annu. Rev. Phys. Chem.*, 1976, **27**, 1.

[11] D. G. Truhlar and R. E. Wyatt, *Adv. Chem. Phys.*, 1977, **36**, 141.

[12] M. Baer, ed. 'Theory of Chemical Reaction Dynamics', CRC Press, Boca Raton, 1985.

180 and 60° and bond lengths between 0.2 and 2.2 Å. This covers H—H—H geometries for which the interaction energy is less than ~80 kcal mol^{-1} (3.5 eV). This *ab initio* surface is globally accurate to within a fraction of a kcal mol^{-1}, and in the important region around the saddle point even more accurate, probably to within 0.1 or 0.2 kcal mol^{-1}. The saddle point occurs for collinear geometry, with two equal bond lengths of 0.929 Å (1.76 bohr), and gives a classical barrier height of 9.8 kcal mol^{-1} (0.42 eV).

The *ab initio* 'surface' is only a table of electronic energies for ~300 discrete H—H—H geometries, and therefore is not a complete description of the potential energy surface in terms of a continuous function of the three H—H distances involved. The parameterization of the Liu and Siegbahn *ab initio* surface in terms of such an analytic function was completed by Truhlar and Horowitz.[4] Since it is this analytic representation that provides the interaction potential used in classical and quantal dynamics calculations, it is appropriate, and has become customary, to refer to the *ab initio* H_3 surface as the Liu–Siegbahn–Truhlar–Horowitz surface, LSTH for short. This parameterization provides an exceptionally good fit to the *ab initio* points: it has a root mean square (r.m.s.) error of only 0.17 kcal mol^{-1}, and is within 0.05 kcal mol^{-1} in the saddle point region. Contour maps of this LSTH surface for various fixed values of the H—H—H angle are shown in Figure 1.

The LSTH surface is frequently referred to as being 'chemically accurate'. One can justify this in many ways, all of them comparative. The uncertainties in the $H + H_2$ interaction energies are comparable to $\frac{1}{2}RT$ at 300 K, and to the smallest energy level spacing in the H_2 reactant, which is the 0.34 kcal mol^{-1} energy difference between the two lowest energy rotational levels. The label 'chemically accurate' is also appropriate in a more pragmatic sense, for the potential energy of interaction at the saddle point is known to an accuracy rivalling or surpassing the 0.2 to 0.4 kcal mol^{-1} accuracy to which the closely related experimental quantity, the activation energy, can be determined. Even though highly accurate, the small remaining uncertainties are not completely insignificant. For example, the ±0.1 kcal mol^{-1} uncertainty in the barrier height produces an uncertainty in the thermal rate constant of approximately $\exp(\pm 0.1/RT)$, which at 300 K is +18%/−15%. This uncertainty is comparable to the accuracy with which actual measurements of the thermal rate constants for the $H + H_2$ reactions have been made.

Thus, even though the potential energy surface for the hydrogen exchange reaction is the most accurate available for any chemical reaction, reassessments and refinements of it continue. Additional *ab initio* calculations[13–15] have resulted in an adjustment of the classical

[13] B. Liu, *J. Chem. Phys.*, 1984, **80**, 581.
[14] D. M. Ceperly and B. J. Alder, *J. Chem. Phys.*, 1984, **81**, 5833.
[15] R. N. Bartlett, P. J. Reynolds, and W. A. Lester, jun., *J. Chem. Phys.*, 1985, **82**, 2700.

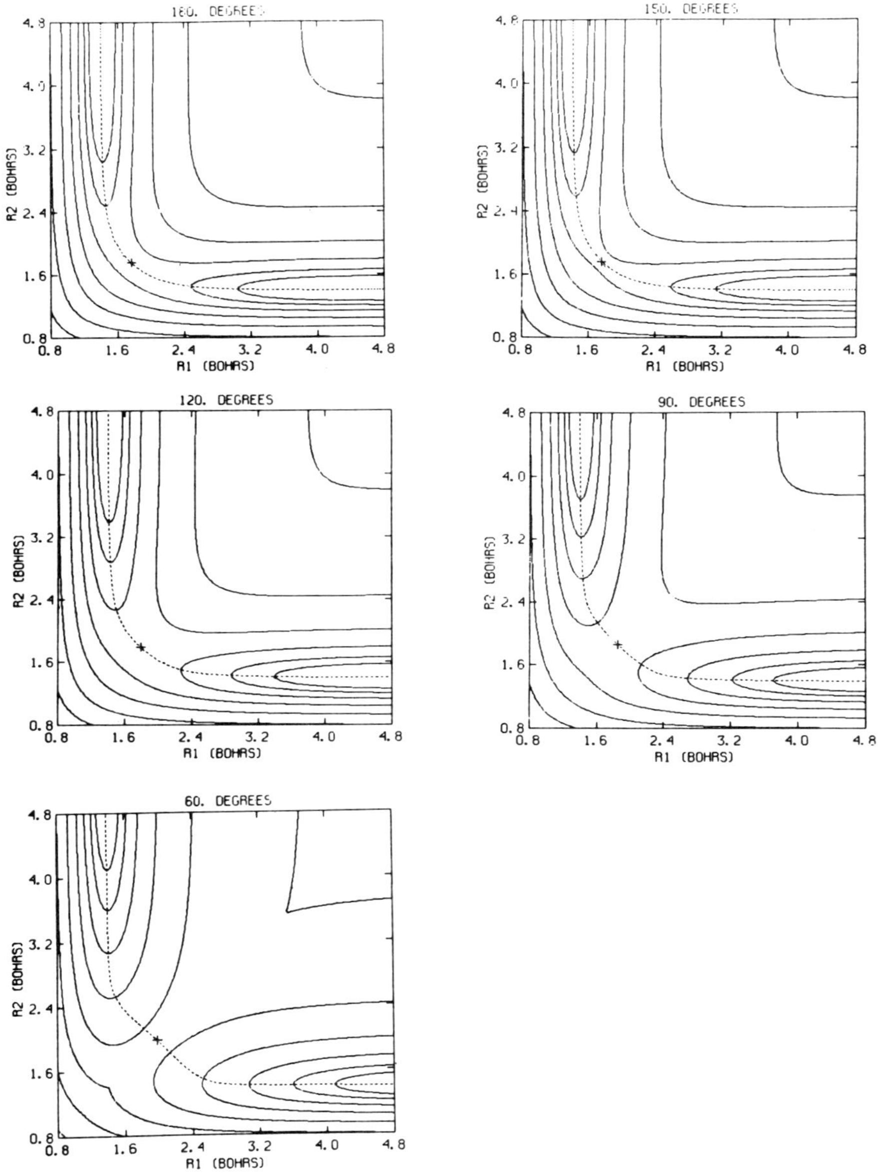

Figure 1 *Contour maps of the* H_3 *potential energy surface for fixed* H—H—H *angle. The solid lines give the contours corresponding to* 3, 6, 12, 24, 48, 96, 192, *and* 384 kcal mol^{-1}. *The dashed curves give the minimum energy paths and the crosses designate the saddle points*
(Reproduced with permission from *J. Chem. Phys.*, 1978, **68**, 2466)

barrier height from the 9.8 kcal mol^{-1} value of the LSTH surface to 9.65 ± 0.1 kcal mol^{-1}. In response to questions raised about the adequacy of the LSTH surface for representing the reaction of vibrationally excited $H_2(v = 1)$ and $D_2(v = 1)$ – questions arising from the discrepancies (discussed in Section 3) between theoretical and experimental values of the rate constants for the reactions with $v = 1$ reactants – Blomberg and Liu[16] carried out additional *ab initio* calculations near the $v = 1$ adiabatic barrier and other regions sampled in the reaction of the $v = 1$ species. These calculations were done for geometries not sampled in the original Liu and Siegbahn calculations. The results of Blomberg and Liu indicate that the analytic representation of the LSTH surface yields interaction energies within 0.25 kcal mol^{-1} of the true *ab initio* values in the region of the $v = 1$ reaction adiabatic barrier, and even for energies 46 kcal mol^{-1} (2.0 eV) above the $H + H_2$ asymptote the analytic representation is within 0.4 kcal mol^{-1}. That the discussion of and adjustments to the LSTH surface are at the 0.1—0.4 kcal mol^{-1} level further emphasizes the unique accuracy of the H_3 potential energy surface.

Partly to incorporate these new *ab initio* results, a new analytic representation[17] of the H_3 surface has been developed. Since this surface is based on a double many-body expansion it is referred to as DMBE. The new parameterization correctly describes the D_{3h} region of the surface, where a conical intersection with the surface for the first electronic excited state of H_3 occurs, at ~63 kcal mol^{-1} (2.7 eV), and also is more accurate in describing the van der Waals interaction[18,19] at large separation. However, the r.m.s. error in the new parameterization is not as small as the r.m.s. error in the LSTH representation, so the new form is not necessarily better than its antecedent for all purposes.

Table 1 *Saddle point properties for two* ab initio *potential energy surfaces of* H_3^a

Property	*Surface* *LSTH*	*DMBE*
H—H distance/Å	0.929	0.928
Classical barrier height/kcal mol^{-1}	9.80	9.65
Vibrationally adiabatic barrier height/kcal mol^{-1}	9.22	9.13
Harmonic vibrational frequencies/cm^{-1}		
Symmetric stretch	2059	2067
Bend	910	899
Antisymmetric stretch	1506i	1493i

^{a}All data from ref. 20

[16] M. R. A. Blomberg and B. Liu, *J. Chem. Phys.*, 1985, **82**, 1050.
[17] A. J. C. Varandas, F. B. Brown, C. A. Mead, D. G. Truhlar, and N. C. Blais, *J. Chem. Phys.*, 1987, **86**, 6258.
[18] F. Torello and M. G. Dondi, *Chem. Phys. Lett.*, 1978, **55**, 339.
[19] F. Torello and M. G. Dondi, *J. Chem. Phys.*, 1979, **70**, 1564.
[20] B. C. Garrett, D. G. Truhlar, A. J. C. Varandas, and N. C. Blais, *Int. J. Chem. Kinet.*, 1986, **18**, 1065.

Table 1 summarizes the saddle point properties of both the original LSTH and the newer DMBE potential energy surfaces.

Before the calculations of Liu and Siegbahn made possible an accurate *ab initio* potential energy surface for H_3, the surface most commonly used in dynamics calculations was the semi-empirical surface due to Porter and Karplus,[21] referred to as PK2. Because of its long history of use and the consequent large body of theoretical calculations done on it, new computations that use PK2 still occasionally appear, primarily to allow direct comparison with earlier calculations. The PK2 surface is only slightly different from LSTH. The PK2 classical barrier height, 9.18 kcal mol^{-1}, is 0.62 kcal mol^{-1} lower than that of the *ab initio* LSTH surface, and the vibrationally adiabatic barrier height on it, 8.81 kcal mol^{-1}, is 0.41 kcal mol^{-1} lower. This vibrationally adiabatic barrier height is what is important for quantum calculations, because it includes the required zero-point energy of the H_3 vibrations that are perpendicular to the reaction co-ordinate at the saddle point. The vibrationally adiabatic barrier height is the activation energy for the reaction at absolute zero. The PK2 barrier is thicker than the LSTH barrier, the imaginary frequency associated with the reaction co-ordinate being 1100i cm^{-1}. These differences are small but certainly significant, particularly in terms of calculated thermal rate constants; the rate constants obtained using the semi-empirical PK2 surface are significantly larger than those on LSTH.

3 Thermal Rate Constants

Despite being highly averaged quantities, thermal rate constants provide valuable and sensitive tests of theoretical descriptions of the hydrogen exchange reaction and the H_3 potential energy surface. The thermal rate constant is particularly sensitive to the barrier height and shape of the potential around the saddle point, and to tunnelling and zero-point energy dynamical effects. Because the more detailed and less averaged results of state-to-state and molecular beam experiments have only become available recently, thermal rate constants for a long time provided the only experimental data with which to test theory. Comparisons of theory and experiment for thermal rate constants are many. It is unfortunate, though, that no rate constants have yet been reported by 'exact' coupled-channel (CC) quantum dynamics using the accurate LSTH potential energy surface. Since the CC method is the only theoretical approach that makes no approximations, other than the Born–Oppenheimer separation of electronic and nuclear motions, the absence of coupled-channel rate constants is regrettable. However, there have been numerous rate constant calculations on LSTH using various approximate quantum methods, classical dynamics, and transition-state

[21] R. N. Porter and M. Karplus, *J. Chem. Phys.*, 1964, **40**, 1105.

theory. This review will deal with only those theoretical calculations of the rate constants that have used the LSTH surface. Thermal rate constants are quite sensitive to the surface used and only LSTH can be viewed as sufficiently accurate to allow the kind of comparison with theory that gauges the accuracy of the theoretical dynamics.

To illustrate the sensitivity of the thermal rate constant to the potential energy surface features at the saddle point we note the significant differences between theoretical rate constants calculated using the LSTH surface and the PK2 semi-empirical surface[21] for the $H + H_2$ reaction at 300 K. This reaction can be studied experimentally as $H + p\text{-}H_2 \rightarrow o\text{-}H_2 + H$, and the experiments of Schulz and LeRoy[22] give $k(300) = 2.7 \times 10^{-16}$ $cm^3\,molecule^{-1}\,s^{-1}$. Variational transition-state theory (VTST)[20] calculations yield a value only 15% lower, 2.3×10^{-16} $cm^3\,molecule^{-1}\,s^{-1}$, when carried out on LSTH. However, VTST gives 1.2×10^{-15} $cm^3\,molecule^{-1}\,s^{-1}$ on PK2,[23] more than a factor of 5 larger. On LSTH, quasi-classical trajectory (QCT) results[24] give a 300 K rate constant of 1.2×10^{-16} $cm^3\,molecule^{-1}\,s^{-1}$, but on PK2 a value a factor of two larger is obtained.[25] These are significant differences, both computationally and experimentally, that occur despite the fact that the PK2 surface differs only slightly from the LSTH surface around the saddle point.

The quantum coupled-channel rate constant for $H + H_2$ has been calculated only using the PK2 surface, and is 9.9×10^{-16} $cm^3\,molecule^{-1}\,s^{-1}$ at 300 K.[26] This CC rate is 18% lower than the corresponding VTST result, and we therefore estimate that on LSTH the CC rate will be somewhat less than 2×10^{-16} $cm^3\,molecule^{-1}\,s^{-1}$. In fact, the rate calculated using the coupled-states (CS) method, which should provide a very close approximation to the exact CC result, gives $k(300)$ of 1.7×10^{-16} $cm^3\,molecule^{-1}\,s^{-1}$.[27] Thus, we find that the CS and VTST methods, and probably the CC method as well, yield $H + H_2$ rate constants that are at 300 K within 10—40% of the experimental value. The poorest agreement with experiment occurs for the CS result, for which the estimated uncertainty is 25%. The combined statistical and random errors in the measured rate constant are probably about 15%. This figure is obtained by comparing the direct measurements of Schulz and LeRoy[22] with values for $k(T)$ obtained from a measurement of the rate constant for $H + D_2 \rightarrow HD + D$[28] and a measurement of the ratio of the rate constants for $H + p\text{-}H_2$ and $H + D_2$.[29] Actually, as Figure 2 shows, the CS rate constants are in better agreement with experiment at

[22] W. R. Schulz and D. J. LeRoy, *J. Chem. Phys.*, 1965, **42**, 3869.
[23] B. C. Garrett and D. G. Truhlar, *J. Chem. Phys.*, 1984, **81**, 309.
[24] H. R. Mayne and J. P. Toennies, *J. Chem. Phys.*, 1981, **75**, 1794.
[25] M. Karplus, R. N. Porter, and R. D. Sharma, *J. Chem. Phys.*, 1965, **43**, 3259.
[26] G. C. Schatz and A. Kuppermann, *J. Chem. Phys.*, 1976, **65**, 4668.
[27] M. C. Colton and G. C. Schatz, *Int. J. Chem. Kinet.*, 1986, **18**, 961.
[28] A. A. Westenberg and N. deHaas, *J. Chem. Phys.*, 1967, **47**, 1393.
[29] K. A. Quickert and D. J. LeRoy, *J. Chem. Phys.*, 1970, **53**, 1325; 1971, **54**, 5444.

all temperatures other than 300 K for which comparison can be made. Omitting this temperature, the largest difference between the CS results and either experimental determination is 24%, while the largest difference between the two experimental determinations is 28%. For the VTST results,[20] also shown in Figure 2, the agreement with experiment is as good or even better. Thus, the $H + H_2$ rate constants calculated using the best available theoretical calculations, variational transition-state theory, and coupled-states calculations, agree with experimental measurements within experimental error. The fact that the rate constants calculated by both VTST and CS methods are lower than the experimental values at low temperatures, and that recent *ab initio* calculations (discussed in Section 2) indicate that the barrier on the LSTH potential is 0.15 kcal mol^{-1} too high, suggests that closer agreement with experiment might be obtained using the newer DMBE potential, which incorporates the most recent *ab initio* data. However, VTST rates calculated[20] using the DMBE surface are actually about 10% smaller than those computed using LSTH. This comes about because the tunnelling contribution to the thermal rate constant on DMBE is smaller than on LSTH.

Theoretical calculations of the thermal rate constant using the *ab initio* LSTH potential energy surface are most abundant for the $D + H_2 \rightarrow HD + H$ variant of the hydrogen exchange reaction. Table 2 compares the experimental rate constants $k(300)$ and $k(600)$ with values

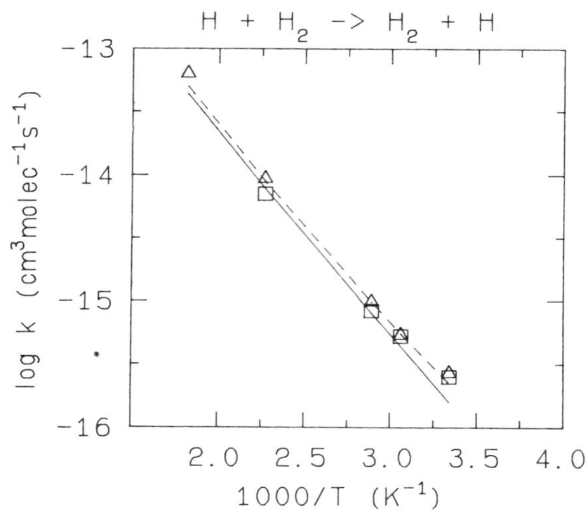

Figure 2 *Arrhenius plot of experimental and theoretical rate constants for the* $H + H_2 \rightarrow H_2 + H$ *reaction.* □ = *the experimental results of ref.* 22; △ = *results obtained by combining those of ref.* 28 *and* 29. *The solid line shows the rate constants calculated by the coupled-states method, ref.* 27, *and the dashed line those calculated by variational transition-state theory, ref.* 20

calculated using various approximate quantum methods[30–35] and quasi-classical trajectories[24] as well. Overall the agreement is very good. As for H + H_2, the 300 K rate constant calculated by VTST[34] is lower than the experimental value, by about 25%. At 600 K the transition-state theory calculations give a rate constant that is a little closer to the measured value, but still below it. At 300 K both the QCT rate constant[24] and that calculated by the infinite order sudden (IOS) approximation[30] are significantly larger than the measured value. The QCT rate cannot be expected to be exact since it allows reaction with energies below that of the lowest vibrational state of the transition state, and also neglects tunnelling. These two deficiences offset each other somewhat, so the QCT rate can be closer than its lack of fundamental correctness might suggest. The IOS approximation assumes that the reaction takes place on a time-scale that is short with respect to the H_2 rotational period. This approximation becomes more reasonable as the average translational energy of all collisions leading to reaction increases, *i.e.* as the temperature increases, so it is not surprising to find that it predicts a more accurate rate constant at 600 K than at 300 K.

Table 2 *Experimental and theoretical rate constants for the* D + H_2 *reaction at* 300 K *and* 600 K

Method	*Rate constant*/cm^3 molecule^{-1} s^{-1} 300 K ($\times 10^{16}$)	600 K ($\times 10^{13}$)	*Reference*
Infinite order sudden calculation (IOS)[a]	4.5	1.1	30
Quasi-classical trajectory calculation (QCT)	3.6	1.3	24
Reduced dimensionality quantum theory (BCRLM)	2.9	1.3	31
Adiabatic T-matrix calculation	2.9	1.0	32
Reduced dimensionality quantum theory (CEQB)	2.3	1.1	33
Variational transition-state theory (VTST)	2.2	1.0	34
Coupled-channel distorted wave calculation (CCDW)	1.3	1.0	35
Experiment	3.0	–	36
Experiment	2.9	1.2	28

[a] l_{in} values

[30] N. Abusalbi, D. J. Kouri, Y. Shima, and M. Baer, *J. Chem. Phys.*, 1985, **82**, 2650.
[31] R. B. Walker and E. Pollak, *J. Chem. Phys.*, 1985, **83**, 2851.
[32] J. C. Sun, B. H. Choi, R. T. Poe, and K. T. Tang, *J. Chem. Phys.*, 1983, **79**, 5376.
[33] J. M. Bowman, K. T. Lee, and R. B. Walker, *J. Chem. Phys.*, 1983, **79**, 3742.
[34] B. C. Garrett and D. G. Truhlar, *J. Chem. Phys.*, 1980, **72**, 3460.
[35] M. S. Bowers, B. H. Choi, and K. T. Tang, *Chem. Phys. Lett.*, 1987, **136**, 145.
[36] D. N. Mitchell and D. J. LeRoy, *J. Chem. Phys.*, 1973, **58**, 3449.

The coupled-channel distorted wave calculation (CCDW)[35] leads to a rate constant that is considerably smaller than that measured at 300 K, and somewhat smaller at 600 K.

While there are still small discrepancies between experiment and theory for the rate constants of the $H + H_2$, $D + H_2$, and $H + D_2$ reactions, there is generally good agreement between theory and experiment for the thermal rate constant and Arrhenius parameters of the hydrogen exchange reaction. Similarly, the agreement among different experimental determinations of the rate constants is very good. Thus, there has been no motivation for further studies of the purely thermal rates of the $H + H_2$, $D + D_2$, $H + D_2$, and $D + H_2$ reactions and no new thermal rate constant measurements for them since 1973. In the past three years, however, there has been a renewed interest in hydrogen exchange reaction thermal rate constants, driven by experimental developments that allow study of muonium isotopic variants of the reaction.

Fleming and co-workers[37] have constructed an apparatus that takes accelerator-generated muonium ions, charge neutralizes them, and monitors the reaction of the resultant muonium atoms with H_2 and D_2 using spin resonance techniques. What makes these studies interesting is the extreme kinematics of reactions involving muonium. Muonium (Mu) is a hydrogen-like atom in which a positive muon takes the place of the proton in H. The muon mass is only 1/9 the mass of the proton, so Mu serves as a very light 'isotope' of hydrogen, with an atomic mass of 0.113 amu. While the electronic potential energy surface is the same as for all other isotopic variants of the reaction, the light mass and zero-point energy restrictions make the thermochemistry and dynamics peculiar for the Mu reactions. For example, the $Mu + H_2$ and $Mu + D_2$ reactions are highly endothermic, ΔH°_0 of 7.6 and 9.2 $kcal\,mol^{-1}$, respectively, comparable to the classical barrier height, while for isotopic combinations involving only H and D ΔH°_0 is no more than 1 $kcal\,mol^{-1}$. The substantial endothermicity pushes the vibrationally adiabatic barrier into the exit valley. The reduced masses of the reactions, 0.231 and 0.332 amu, are very small, and the skew angles very large, 77.0° and 80.6°, for $Mu + H_2$ and $Mu + D_2$, respectively. The small reduced masses increase the importance of the tunnelling contribution to the thermal rate constant.

Experimental measurement of the $Mu + H_2$ and $Mu + D_2$ reaction rate constants have been made over the temperature range 473—853 K.[38] The reaction studies are carried out in a high-pressure gas cell into which a beam of muons is injected. The muons are neutralized and thermalized in collisions with inert diluent gas before reaction with H_2

[37] D. M. Garner, D. G. Fleming, and R. J. Mikula, *Chem. Phys. Lett.*, 1985, **121**, 80.

[38] I. D. Reid, D. M. Garner, L. Y. Lee, M. Senba, D. J. Arseneau, and D. G. Fleming, *J. Chem. Phys.*, 1987, **86**, 5578.

or D_2. A modest external magnetic field is applied to the high-pressure cell, leading to a spin precession frequency of free muonium of ~6 MHz. In the diamagnetic environment of MuH or MuD the precession frequency is much lower, ~60 kHz, so the reaction serves to damp exponentially the coherent precession of an ensemble of Mu atoms. This damping is revealed by the time dependence of the spatial distribution of muon decay events, in which a high-energy positron is ejected, preferentially along the muon spin direction. The use of high pressures of H_2 or D_2 is necessitated by the short muon decay lifetime, ~2.2 μs, and the small thermal rate of the reactions, 10^{-16} to 10^{-13} cm^3 $molecule^{-1}$ s^{-1}.

Figure 3 gives an Arrhenius plot of the measured rate constants for

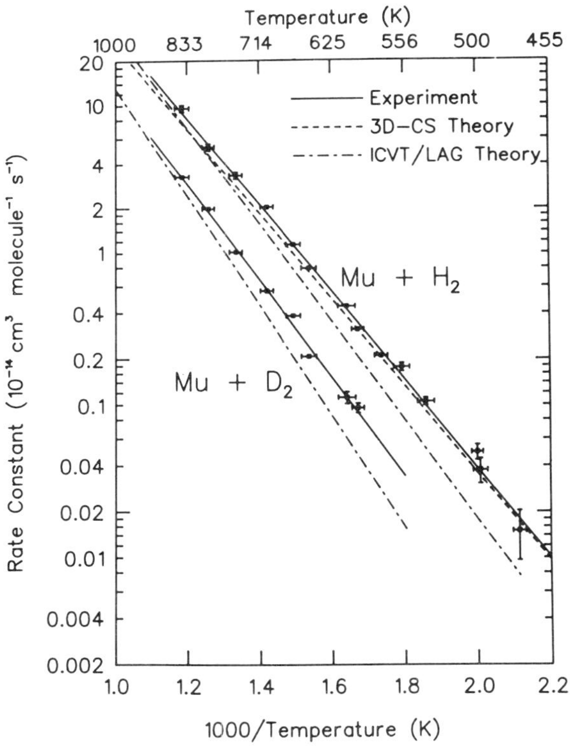

Figure 3 *Arrhenius plot of the experimental and theoretical rate constants for the* Mu + H_2 *and* Mu + D_2 *reactions. The points and solid lines are the experimental results of ref.* 38. *The dashed line represents the rate constants calculated via coupled states, ref.* 39, *and the dot–dash line gives the rate constants calculated via variational transition-state theory, ref.* 23

(Reproduced with permission from *J. Chem. Phys.*, 1987, **86**, 5578)

Mu + H_2 and Mu + D_2, along with theoretical predictions from accurate quantum coupled-states (CS) calculations[39] and variational transition-state theory (VTST).[23] The quantum dynamics calculations, available only for the reaction with H_2, are in excellent agreement with the measurements, while the transition-state results underestimate the rates and overestimate the activation energy. The experimental and theoretical Arrhenius parameters are compared in Table 3. In Table 4 are compared the experimental values for the rates and activation energies of all hydrogen exchange reaction isotopic variants that have been studied. Note that the activation energies for the thermoneutral or very nearly thermoneutral D + H_2,[28] H + H_2,[36] and H + D_2[28] reactions over a similar temperature range are much smaller than those for the endoergic Mu reactions, and the thermal rates are correspondingly much greater.

Coupled-channel distorted wave quantum calculations[35] give Mu + H_2 rate constants that are lower than the experimental values, but yield an activation energy of 13.7 kcal mol^{-1}, very close to the measured value. Quasi-classical trajectory calculations[40] give rate constants that are 10 to

Table 3 *Experimental and theoretical Arrhenius parameters for the* Mu + H_2 *and* Mu + D_2 *reactions*

	E_a/kcal mol^{-1}	$-\log(A/\text{cm}^3$ molecule^{-1} s$^{-1})$	*Reference*
Mu + H_2			
Experiment	13.29 ± 0.22	9.605 ± 0.074	38
CS theory	13.0	–	39
VTST theory	14.4	–	23
Mu + D_2			
Experiment	14.73 ± 0.40	9.67 ± 0.12	38
VTST theory	16.2	–	23

Table 4 *Activation energies and thermal rate constants for isotopic variants of the hydrogen exchange reaction*

Reaction	E_a/kcal mol^{-1}	$k(745\text{ K})/$ 10^{-14} cm^3 molecule^{-1} s^{-1}	*Reference*
D + H_2	7.6 ± 0.2	43.2	28
H + H_2	8.5 ± 0.5	40.0	28, 29
H + D_2	9.4 ± 0.3	14.4	28
Mu + H_2	13.3 ± 0.2	3.3	38
Mu + D_2	14.7 ± 0.4	1.0	38

100 times larger than the experimental values. However, it is probably wrong to interpret this result to mean that the QCT method is inapplicable for this system because it is more quantal than other variants of the hydrogen exchange reaction. The rate constants for $Mu + H_2 \rightarrow MuH + H$ obtained by application of detailed balance to the rate constants derived by QCT calculation for the reverse reaction are in reasonable agreement with the experimental values. This outcome may simply mean that QCT-calculated rate constants are unreliable if the dynamical bottleneck is located far towards the products, as it is in the $Mu + H_2$ reaction.

All of this discussion indicates that theory is capable of predicting thermal rate constants and Arrhenius parameters for all isotopic variants of the hydrogen exchange reaction with high accuracy. Using high-level theories like CS, VTST, reduced dimensionality quantum approaches, and T-matrix methods, *ab initio* prediction of the rate constants with an accuracy of ±25% is achievable. The accuracy of the calculations rivals that of the measurements themselves, an impressive accomplishment. Even with methods whose approximations are more severe, like IOS, QCT, and distorted wave methods, the calculated rates are usually within a factor of two of the experimental values. One could conclude then that a first-principles understanding of the thermal rate behaviour of the hydrogen exchange reaction had been achieved and that experiment had validated the theoretical descriptions of the reaction rate. Given this, one would be highly confident of theoretical predictions of the rates of the reaction under non-thermal-equilibrium conditions.

Such confidence in the theory is what has made the large discrepancy between experimental and theoretical values of the rate constant for the reactions $H + H_2(v = 1)$, $H + D_2(v = 1)$, and $D + H_2(v = 1)$ one of the most puzzling, and troubling, aspects of the hydrogen exchange reaction. The discrepancies are very large, and not easily explained away. For example, for the $H + H_2(v = 1)$ reaction quasi-classical trajectory calculations[24] (which for the $v = 0$ reactions give rate constants uniformly higher than the measured ones) give the 300 K rate constant as 1.3×10^{-13} cm^3 molecule^{-1} s^{-1}, while the reported[41] experimental rate constant is 52×10^{-13} cm^3 molecule^{-1} s^{-1}, 40 times larger. The rate constants calculated by approximate quantum methods are in no better agreement with this experiment. Variational transition-state theory[42] and a semi-classical adiabatic approach[43] both give 2.0×10^{-13} cm^3 molecule^{-1} s^{-1}. The reduced dimensionality bending corrected rotating linear model (BCRLM) yields[44] the largest calculated

[39] G. C. Schatz, *J. Chem. Phys.*, 1985, **83**, 3441.
[40] N. C. Blais, D. G. Truhlar, and B. C. Garrett, *J. Chem. Phys.*, 1983, **78**, 2363.
[41] E. B. Gordon, B. I. Ivanov, A. P. Perminov, V. E. Balalaev, A. V. Ponomarev, and V. V. Filatov, *Chem. Phys. Lett.*, 1978, **58**, 425.
[42] B. C. Garrett and D. G. Truhlar, *J. Phys. Chem.*, 1985, **89**, 2204.
[43] E. Pollak and R. E. Wyatt, *J. Chem. Phys.*, 1983, **78**, 4464.
[44] R. B. Walker and E. F. Hayes, *J. Phys. Chem.*, 1983, **87**, 1255.

rate constant, 4.8×10^{-13} cm^3 $molecule^{-1}$ s^{-1}, which is still more than an order of magnitude below the experimental value.

As Table 5 shows, the situation is even worse for the $D + H_2(v = 1) \rightarrow HD + H$ reaction. The experimental rate constants were determined by several different methods. Kneba *et al.*[45] produced D atoms by microwave discharge and $H_2(v = 1)$ by energy transfer from laser-excited HF, and followed the temporal evolution of the H atom product by VUV resonance absorption. Wellhausen and Wolfrum[46] used microwave discharges to generate both the D and the $H_2(v = 1)$ and Raman detection of the $H_2(v = 1)$, as did Dreier and Wolfrum.[47] Glass and Chaturvedi[48] also used microwave discharges to produce D and $H_2(v = 1)$, but detected the atomic product via e.p.r. Rozenshteyn *et al.*[49] employed e.p.r. detection of the product atoms, with microwave discharge production of D and simple thermal heating of H_2 to get the vibrationally excited reactant.

The largest, and therefore most troubling, experimental value in Table 5 is that reported[45] by the Wolfrum group in 1979, 1.2×10^{-11} cm^3 $molecule^{-1}$ s^{-1}. This value was so much larger than the value that any conceivable theoretical calculation could give that it was cause for significant concern. Subsequent measurements,[46,48,49] including

Table 5 *Experimental and theoretical rate constants for the* $D + H_2(v = 1)$ *reaction at* 300 K

Method	*Rate constant/* 10^{-13} cm^3 $molecule^{-1}$ s^{-1}	*Reference*
Infinite order sudden calculation (IOS)	6.7	50
Adiabatic distorted wave calculation (RADW)	5.0	51
Reduced dimensionality quantum calculation (BCRLM)	2.8	44
Infinite order sudden calculation (IOS)[a]	2.0	30
Variational transition-state theory (VTST)	2.1	42
Semi-classical adiabatic calculation (SCA)	2.1	43
Quasi-classical trajectory calculation (QCT)	1.7	24
Reduced dimensionality quantum calculation (BCRLM)	1.6	31
Reduced dimensionality quantum calculation (CEQB)	0.86	33
Experiment	120	45
Experiment	16	49
Experiment	9.8	48
Experiment	8.3	46
Experiment	1.0	47

[a] l_{in} values

another by the Wolfrum group in 1985,[46] gave values that were more than an order of magnitude smaller than this original rate constant. The accepted experimental value for $D + H_2(v = 1)$ seemed then to be converging at just under 1×10^{-12} cm^3 molecule^{-1} s^{-1}, still larger than all the theoretical calculations. However, Dreier and Wolfrum[47] recently reported yet a third measurement in their laboratory, one that gives a rate constant that is now significantly smaller than all but one of the calculated values.

Given the difficulty of measuring the rate of the $v = 1$ hydrogen exchange reaction, a variation in the experimental rate constants that is large compared with the variation in the measured values for the $v = 0$ reaction is expected. It is not possible to produce a perfectly clean source of $H_2(v = 1)$, that is $H_2(v = 1)$ without any contamination from higher vibrational states of H_2 or other energetic species, and relaxation of the reactant occurs in parallel with the reaction. Even with vibrational excitation the reaction is slow, about a hundredth of gas kinetic. Consequently, the study of the $H + H_2(v = 1)$ reaction and its isotopic analogues presents considerable experimental challenges. However, the two-orders-of-magnitude variation that is actually observed in the experimentally determined rate constants is disturbing, as is the fact that every new measurement but one has yielded a smaller value than all previous measurements. Despite the fact that several different methods have been used both to produce the $H_2(v = 1)$ species and to follow the reaction, it seems likely that some or all of the experimentally reported values are in error. Definitive experiments are certainly needed in order to determine conclusively the rate constants for $H + H_2(v = 1)$ and its isotopic variants. However, the concerns that the initial measurements of these rate constants raised about the adequacy of various theoretical approaches to $H + H_2$ dynamics and the accuracy of the LSTH potential energy surface appear to have been unsubstantiated.

The theoretical calculation that gives the largest $D + H_2(v = 1)$ rate constant is that of Pollak *et al.*[50] This is based on an infinite order sudden (IOS) calculation in which the IOS cross-sections are interpreted as the rotationally averaged cross-sections. This interpretation is different from that normally invoked, which treats the IOS cross-sections as those for the ground rotational state. This is an important distinction. Since in the normal IOS approach the cross-sections for higher rotational states are assumed to be the same as for the ground state at the same

[45] M. Kneba, U. Wellhausen, and J. Wolfrum, *Ber. Bunsenges. Phys. Chem.*, 1979, **83**, 940.
[46] U. Wellhausen and J. Wolfrum, *Ber. Bunsenges. Phys. Chem.*, 1985, **89**, 314.
[47] T. Dreier and J. Wolfrum, *Int. J. Chem. Kinet.*, 1986, **18**, 919.
[48] G. P. Glass and B. K. Chaturvedi, *J. Chem. Phys.*, 1982, **77**, 3478.
[49] V. B. Rozenshteyn, Yu. M. Gershenzon, A. V. Ivanov, and S. I. Kucheryavii, *Chem. Phys. Lett.*, 1984, **105**, 423.
[50] E. Pollak, N. Abusalbi, and D. J. Kouri, *Chem. Phys. Lett.*, 1985, **113**, 585.
[51] J. C. Sun, B. H. Choi, R. T. Poe, and K. T. Tang, *Phys. Rev. Lett.*, 1980, **44**, 1211.

translational energy, the cross-sections for higher rotational states are energy-shifted upward from the IOS value. If the IOS cross-sections are taken to be the rotationally averaged cross-sections, this energy shifting is absent. The result is that within the new IOS interpretation the rotationally averaged reaction cross-section is higher at a given temperature than in the conventional interpretation. The rate constants calculated for $D + H_2(v = 0)$ by the IOS method[30] (see Table 2) were obtained using the conventional interpretation of the IOS cross-sections. These thermal rate constants are already larger than the experimental values, so if the new IOS interpretation is applied here, the resultant rate constants will be in even poorer agreement with the experimental values.

The lowest temperature at which the rate constant of the hydrogen exchange reaction has been determined is 167 K.[36] Recently, however, two research groups[52–55] have succeeded in getting information about the rate of the reaction at temperatures as low as 1.9 K. These experiments are done in solid H_2 or one of its isotopomers, using e.s.r. to follow the time evolution of H or D atoms. The atoms are generated in the solid by condensation of hydrogen gas that has passed through a microwave discharge, or by γ-irradiation of the molecular solid itself. Since these experiments are done in an amorphous solid, direct comparison of the results with homogeneous gas-phase kinetics measurements is difficult. Similarly, interpretation of the rates of disappearance or appearance of the atoms solely in terms of the rate of the exchange reaction is problematic. Nonetheless, because of the very low temperatures at which they are done, these experiments provide important information on and insight about quantum mechanical tunnelling in chemical reactions.

Only for the $D + HD \rightarrow D_2 + H$ reaction has a quantitative experimental rate constant determination been made so far. For this reaction the pseudo-first-order rate constant in pure solid HD at 4.2 K is measured to be $1.1 \times 10^{-4}\ s^{-1}$.[54] At the density of solid HD this decay rate implies a second-order rate constant of 3.8×10^{-27} $cm^3\ molecule^{-1}\ s^{-1}$. This value is in good agreement with the rate constant calculated using transition-state theory (TST):[56] 1.7×10^{-27} $cm^3\ molecule^{-1}\ s^{-1}$. The TST calculations use an Eckart potential fit to the Marcus–Coltrin one-dimensional reaction path[57] on the LSTH H_3 surface to determine the tunnelling factor. Arrhenius plots of the

[52] A. V. Ivliev, A. S. Iskovskikh, A. Ya. Katunin, I. I. Lukashevich, V. V. Sklyarevskii, V. V. Suraev, V. V. Filippov, N. I. Filippov, and V. A. Shevtsov, *JTEP Lett.*, 1983, **38**, 379.

[53] H. Tsuruta, T. Miyazaki, K. Fueki, and A. Azuma, *J. Phys. Chem.*, 1983, **87**, 5422.

[54] T. Miyazaki, K.-P. Lee, K. Fueki, and A. Takeuchi, *J. Phys. Chem.*, 1984, **88**, 4959.

[55] T. Miyazaki and K.-P. Lee, *J. Phys. Chem.*, 1986, **90**, 400.

[56] T. Takayanagi, N. Masaki, K. Nakamura, M. Okamoto, S. Sato, and G. C. Schatz, *J. Chem. Phys.*, 1987, **86**, 6133.

[57] R. A. Marcus and M. E. Coltrin, *J. Chem. Phys.*, 1977, **67**, 2609.

calculated low-temperature rate constants for several isotopic variants of the hydrogen exchange reaction are shown in Figure 4. These plots show that for temperatures below ~5 K the calculated rate constants follow the expected Wigner threshold law,[58] which predicts a dependence of *k* on $T^{1/2}$ for thermoneutral reactions and T^0 for exothermic reactions. Similar plots for the endothermic reactions $H + HD \rightarrow H_2 + D$ and $H + D_2 \rightarrow HD + D$ have calculated rate constants that vary as $\exp(-\Delta H^\circ_0/RT)$, also in agreement with the Wigner threshold behaviour.

These TST results show clearly the very great sensitivity to isotopic composition that the hydrogen exchange reaction rate constants have at low temperature. For example, at 4.2 K the rate constants for $D + H_2$, $H + H_2$, $D + D_2$, and $H + D_2$, are 3.9×10^{-27}, 6.1×10^{-28}, 1.8×10^{-34}, and 6.8×10^{79} cm^3 molecule^{-1} s^{-1}, respectively.[56] The rate constant has not been quantitatively determined by experiment for any of these isotopic variants, but the available experimental observations

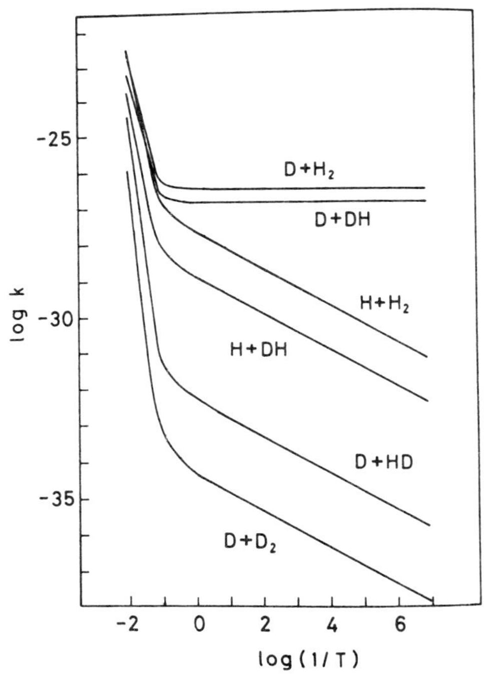

Figure 4 *Variational transition-state theory rate constants (cm^3 molecule^{-1} s^{-1}) for thermoneutral and exoergic isotopic variants of the hydrogen exchange reaction at low temperature (K)*
(Reproduced with permission from *J. Chem. Phys.*, 1987, **86**, 6133)

[58] E. P. Wigner, *Phys. Rev.*, 1948, **73**, 1002.

indicate that the first two of these must be many orders of magnitude faster than the last two. For example, a lower limit on the rate constant ratio $k(D + H_2 \rightarrow HD + H)/k(H + D_2 \rightarrow HD + D)$ at 4.2 K is 30 000.[53] This is understandable in terms of the endothermicity of the $H + D_2$ reaction, 0.98 kcal mol^{-1}, for which $\exp(-\Delta H_0^\circ/RT)$ is only 1.1×10^{-51} at 4.2 K. Consequently, the rate of this reaction is very much smaller than that of the 0.82 kcal mol^{-1} exothermic $D + H_2$ reaction at this temperature. In fact, in these cryogenic solids endothermic variants of the hydrogen exchange reaction are so slow that they cannot be directly observed.

4 Product Angular and Velocity Distributions

The crossed molecular beam technique has been the major tool for studies of the dynamics of chemical reactions. However, experimental difficulties associated with the production of beams of hydrogen atoms and the detection of hydrogen molecules have limited the quantity and quality of the information that molecular beam studies have provided about the hydrogen exchange reaction. The significant, ~0.3 eV, energy threshold for the reaction requires high-temperature thermal dissociation, arc discharge, or photolytic atomic beam sources, all of which are technically demanding. At the low mass of the product diatoms the electron-impact-ionization mass spectrometer detectors used in these experiments have high backgrounds, while the signal levels are low because of the small reactive cross-sections, ~1 Å^2, even at energies far above threshold.

Despite these difficulties, the hydrogen exchange reaction was one of the first reactions investigated in crossed molecular beam experiments. The first attempts[59,60] to study $D + H_2$ and $H + D_2$ were only partly successful, but in subsequent experiments Geddes *et al.*[61,62] were able to measure the laboratory angular distribution of the HD product of $D + H_2$ collisions, at an average collision energy of 0.48 eV. These measurements showed that the HD scatters backwards in the centre-of-mass system. Product velocity distributions were determined by the variation of product intensity with lock-in phase in the phase-sensitive detection system, and these indicated that the HD was formed with very little internal excitation. The experiments used a 2850 K thermal dissociation source to generate D atoms from D_2. While operation at this high temperature is necessary both to achieve appreciable dissociation of

[59] S. Datz and E. Taylor, *J. Chem. Phys.*, 1963, **39**, 1896.

[60] W. L. Fite and R. T. Brackman, in 'Proceedings of the Third International Conference on the Physics of Electronic and Atomic Collisions', ed. M. R. C. McDowell, Wiley, New York, 1964, p. 955.

[61] J. Geddes, H. F. Krause, and W. L. Fite, *J. Chem. Phys.*, 1970, **52**, 3296.

[62] J. Geddes, H. F. Krause, and W. L. Fite, *J. Chem. Phys.*, 1972, **56**, 3298; 1973, **59**, 566.

the D_2 and to produce collision energies above the reaction threshold, the high temperature Maxwell–Boltzmann velocity distribution in the effusive D beam causes a large spread in the D + H_2 collision energy, a FWHM spread of ~70% of the 0.48 eV average collision energy.

Kwei *et al.*[63,64] utilized a novel radiochemical approach in another crossed molecular beam study of the hydrogen exchange reaction. They chose the H + T_2 and D + T_2 reactions for study, employing an array of MoO_3-coated detectors to collect the T atoms reactively scattered into 10° angular intervals. The T atoms scattered into each interval were quantitatively measured by subsequent liquid scintillation counting of the MoO_3 coatings. In the case of the H + T_2 experiments the thermal dissociation atom source was operated at 2100 and 2600 K, to give average collision energies of 0.61 and 0.69 eV, while the D + T_2 experiments were done only at the lower source temperature, giving an average collision energy of 0.56 eV. Product velocity distributions were not measured. However, the laboratory angular distributions alone reveal the centre-of-mass angular distribution of the products, since the transformation between the laboratory and centre-of-mass co-ordinate systems iş single-valued as a consequence of the centre-of-mass T atom product recoil velocity being greater than the velocity of the centre-of-mass. These centre-of-mass angular distributions show that the T atoms are scattered forward, *i.e.* along the direction of the H(D) atom beam, so the HT(DT) molecular product must be backward scattered, indicative of a direct, rebound mechanism. The HT laboratory angular distributions increase in width significantly with only a small change in average collision energy, suggesting that the centre-of-mass angular distributions broaden with increasing collision energy. However, since the collision energy spread in the experiments, ~0.4—0.5 eV FWHM, is greater than the range over which the average energy was changed, 0.08 eV, this observation cannot be the basis of a strong conclusion.

With thermal dissociation sources of H and D atoms, the collision energies possible in a crossed molecular beam study of the hydrogen exchange reaction are limited to ~0.65 eV. To explore the dynamics at higher energies, different sources of atoms are needed. Götting *et al.*[65–67] have used an arc discharge source to investigate the D + H_2 reaction at collision energies of 1.0 and 1.5 eV. The arc discharge produces a D atom beam with a Maxwell–Boltzmann velocity distribution characterized by a temperature of 15 000 to 25 000 K. This makes possible the high average collision energies, but also gives rise to a broad distribution of energies, about ±1.3 eV at a most probable energy of 1.5 eV. Both angular and time-of-flight (TOF) velocity distributions

[63] G. H. Kwei, V. W. S. Lo, and E. A. Entemann, *J. Chem. Phys.*, 1973, **59**, 3421.
[64] G. H. Kwei and V. W. S. Lo, *J. Chem. Phys.*, 1980, **72**, 6265.
[65] R. Götting, H. R. Mayne, and J. P. Toennies, *J. Chem. Phys.*, 1984, **80**, 2230.
[66] R. Götting, J. P. Toennies, and M. Vodegel, *Int. J. Chem. Kinet.*, 1986, **18**, 949.
[67] R. Götting, H. R. Mayne, and J. P. Toennies, *J. Chem. Phys.*, 1986, **85**, 6396.

of the HD product were measured for a wide range of laboratory angles, and from these measurements centre-of-mass contour maps of scattered HD intensity have been constructed. The contour map for the higher collision energy is shown in Figure 5.[67]

The large amount of kinematic smearing that results from the broad velocity distribution of the arc-heated D atoms precludes an analysis of the data to yield a unique contour map deconvoluted of the averaging over experimental conditions. The map shown in Figure 5 is a simple velocity-weighted contour map, obtained by dividing the measured laboratory system product flux by the square of the product laboratory velocity, which yields a contour map that represents the centre-of-mass product flux divided by the square of the centre-of-mass velocity of the product. As such, this contour map must be interpreted with caution. However, the gross features of the contour map indicate sideways to backward peaking of the product, with modest internal energy. The angular distribution peaks at 125°, and the recoil velocity distribution indicates that ~50% of the available energy appears as product translation. The experimental contour map of Figure 5 is in quite good agreement with one calculated[67] for these experimental conditions using quasi-classical trajectories on the LSTH surface. However, the extensive averaging over the spread in experimental parameters makes this agreement less than conclusive.

The factor that obviously limits the information content, and therefore the value, of all these crossed molecular beam measurements of product angular and velocity distributions is the broad spread in the H or D

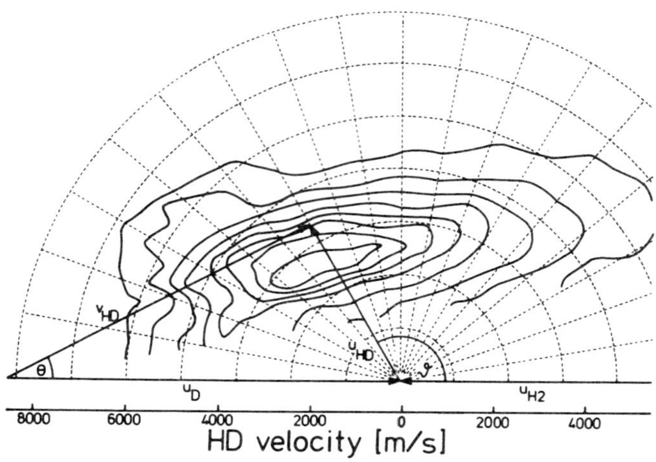

Figure 5 HD *product contour map for the* $D + H_2 \rightarrow HD + H$ *reaction at an average collision energy of* 1.5 eV. *The solid lines are the contours of product flux divided by the square of the product recoil velocity, and are given for relative intensities of* 1 *to* 8
(Reproduced with permission from *J. Chem. Phys.*, 1986, **85**, 6396)

atom velocity. The result is a collision energy that is poorly defined. This not only introduces ambiguities in the interpretation of the results of the experiments, but also makes the usual deconvolution of molecular beam measurements quite difficult, since the angular and velocity distributions of the products are sensitive functions of the collision energy. There is a clear need for velocity-and-angle-resolved differential cross-section measurements for the hydrogen exchange reaction made under conditions of well-defined collision energy. Such conditions are achievable using velocity selection of the beam produced by the arc discharge or thermal dissociation source, but it is not clear whether the resultant beams would have sufficient intensity.

Such conditions are also achievable using laser photolytic generation of the H or D atoms, an approach taken by Buntin *et al.*[68] in their crossed molecular beam study of the $D + H_2$ reaction. They used an ArF exciplex laser to photodissociate a D_2S beam at 193 nm, producing a fast D atom beam with well-defined velocity. The resultant collision energy, 0.95 eV, is comparable to that available with arc discharge sources, but the energy definition is much better, the spread in collision energy being only ±0.11 eV. Although a product angular distribution has not yet been measured, and a product velocity distribution has been reported at only one laboratory angle, the results of this experiment are both significant and encouraging. The limited energy spread and high time-of-flight temporal resolution achieved in the experiment permit a vibrational state distribution to be extracted from the product translational energy distribution. This is illustrated in Figure 6. The fast peak in the HD TOF distribution shown in the top of the figure is due to products scattering into angles between 60 and 120° in the centre-of-mass frame for which it is not possible to resolve different internal states. The two slower peaks in the TOF distribution are due to $v' = 1$ and $v' = 0$ HD products scattered near 180° in the centre-of-mass. Individual rotational states cannot be resolved, but the shape of the peaks does contain information about the rotational state distribution of the backward-scattered HD product. The $v' = 1/v' = 0$ ratio is 0.19, while the rotational state distributions implied by the data are fairly broad with peaks at $j' = 0$ in $v' = 1$ and $j' = 6$ in $v' = 0$. This vibrational state distribution is in good agreement with that from the quasi-classical trajectory calculations of Blais and Truhlar,[69] who find a $v' = 1/v' = 0$ ratio of 0.20 for products scattered into angles greater than 160°. However, the QCT-calculated HD rotational distributions at these large scattering angles are much more sharply peaked, and in $v' = 1$ peaked at larger j', 2—3, than are the experimental distributions. Further crossed-beam studies of this type on the hydrogen exchange reaction can be expected, and these will provide very detailed state-resolved differential cross-sections for comparison with theory.

[68] S. A. Buntin, C. F. Giese, and W. R. Gentry, *J. Chem. Phys.*, 1987, **87**, 1443.
[69] N. C. Blais and D. G. Truhlar, *J. Chem. Phys.*, 1988, **88**, 5457.

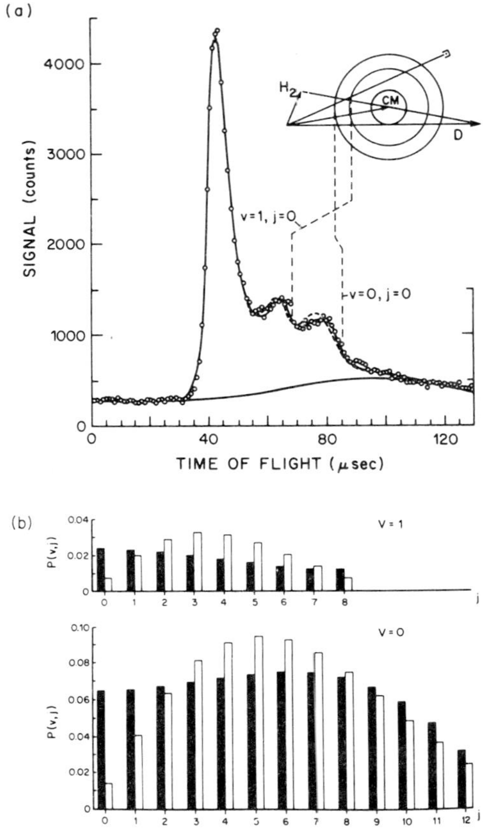

Figure 6 (a) *Time-of-flight velocity distribution for the* HD *product of the* $D + H_2 \rightarrow HD + H$ *reaction at a collision energy of* 0.95 eV. *The Newton diagram shows the laboratory angle at which this time-of-flight distribution was recorded. The dashed lines indicate the flight times corresponding to backward-peaked* HD *in* $v' = 1$, $j' = 0$, *and* $v' = 0$, $j' = 0$. (b) *The product quantum-state distribution for backward-peaked* HD *derived from the time-of-flight distribution in* (a). *The solid and open bars represent two different fits to the data: the solid bars yield the solid line through the data in* (a), *while the open bars give the dashed line in* (a). *The solid bars give a better fit*
(Reproduced with permission from *J. Chem. Phys.*, 1987, **87**, 1443)

Crossed-beam studies of reactive hydrogen atom–hydrogen molecule collisions are complemented by related studies of the elastic differential cross-sections, from measurements at collision energies below the thresholds for reaction and for rotational/vibrational excitation.[18,19] Measurements of the total elastic plus inelastic plus reactive collision cross-sections have also been made, at energies up to ~1.5 eV.[70-73] The elastic differential cross-section measurements[18,19] indicate a spherically averaged van der Waals potential well depth of 1.67 meV at 3.46 Å, and an r_0, the separation at which the potential energy of interaction is zero, of 2.99 Å. These values are close to those for the LSTH potential energy surface, for which the well occurs at 3.49 Å with a depth of 1.56 meV, and r_0 is 3.06 Å.[17]

Theoretical calculations of the product angular and velocity distributions in the hydrogen exchange reaction are much more numerous than the experimental measurements of these distributions. There are in particular extensive QCT results,[24,67,69,74-76] IOS calculations,[77] distorted wave calculations,[32,78-83] reduced dimensionality quantum calculations,[84] and exact coupled-channel quantum calculations.[26,85] All of these theoretical results are in at least semi-quantitative agreement with regard to the behaviour of the differential cross-sections. For all isotopic variants of the hydrogen exchange reaction the molecular product is back-scattered at collision energies not far above the reaction threshold, becoming less so as the collision energy is raised. This reflects the repulsive potential energy surface with collinear minimum energy path, leading to impulsive rebound dynamics.

The calculated differential cross-sections for $H + H_2(v = 0, j = 0) \rightarrow H_2(v' = 0, j' = 1) + H$ shown in Figures 7 and 8 illustrate the behaviour in the low collision energy regime. These angular distributions are from coupled-channel calculations on the PK2 potential energy surface.[26] The width at half maximum in these distributions is 48, 32, 33, 41, and 51°, at total energies (collision energies) of 0.30 (0.03), 0.40 (0.13), 0.50 (0.23), 0.60 (0.33), and 0.70 (0.43) eV, respectively. All but the highest of these energies is below the vibrationally adiabatic barrier on the PK2 surface, and the reaction cross-sections are correspondingly small, for example only ~10^{-14} Å^2 at $E_{rel} = 0.03$ eV. At these energies near threshold one might expect an isotropic angular distribution, since

[70] N. Hishinuma, *Chem. Phys. Lett.*, 1980, **70**, 361.
[71] W. Bauer, R. W. Bickes, jun., B. Lantzsch, J. P. Toennies, and K. Walaschewski, *Chem. Phys. Lett.*, 1975, **31**, 12.
[72] R. Gegenbach, C. Hahn, and J. P. Toennies, *J. Chem. Phys.*, 1975, **62**, 3620.
[73] W. C. Stwalley, A. Niehaus, and D. R. Herschbach, *J. Chem. Phys.*, 1969, **51**, 2287.
[74] H. R. Mayne, *J. Chem. Phys.*, 1980, **73**, 217.
[75] G.-D. Barg, H. R. Mayne, and J. P. Toennies, *J. Chem. Phys.*, 1981, **74**, 1017.
[76] N. C. Blais and D. G. Truhlar, *J. Chem. Phys.*, 1985, **83**, 2201.
[77] D. J. Kouri, V. Khare, and M. Baer, *J. Chem. Phys.*, 1981, **75**, 1179.
[78] B. H. Choi and K. T. Tang, *J. Chem. Phys.*, 1976, **65**, 5161.
[79] J. C. Sun, B. H. Choi, R. T. Poe, and K. T. Tang, *J. Chem. Phys.*, 1980, **73**, 6095.
[80] D. C. Clary and J. N. L. Connor, *Mol. Phys.*, 1981, **43**, 621.
[81] G. C. Schatz, L. M. Hubbard, P. S. Dardi, and W. H. Miller, *J. Chem. Phys.*, 1984, **81**, 231.
[82] B. H. Choi, R. T. Poe, and K. T. Tang, *J. Chem. Phys.*, 1984, **81**, 4979.
[83] S. H. Suck Salk and C. K. Lutrus, *J. Chem. Phys.*, 1985, **83**, 3965.
[84] R. B. Walker and E. F. Hayes, *J. Phys. Chem.*, 1984, **88**, 1194.
[85] A. Kuppermann and G. C. Schatz, *J. Chem. Phys.*, 1975, **62**, 2502.

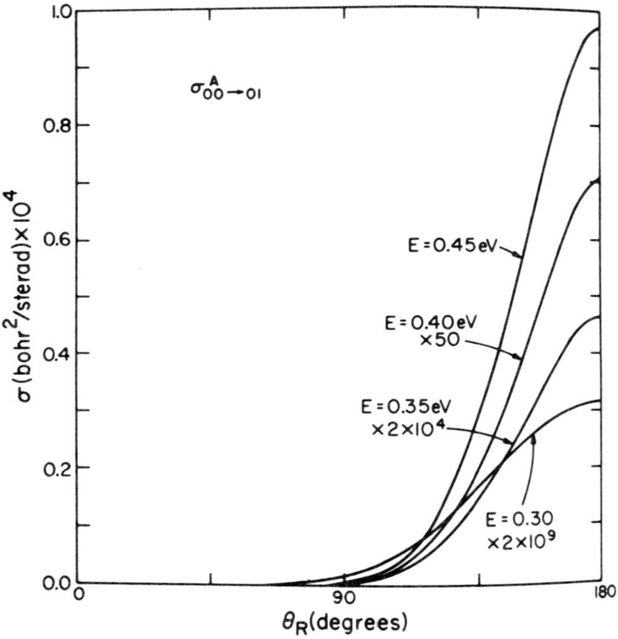

Figure 7 *Antisymmetrized differential cross-sections for the* $H + H_2(v = 0, j = 0) \rightarrow H_2(v' = 0, j' = 1) + H$ *reaction, from coupled-channel calculations at total energies (collision energy plus the* 0.27 eV *zero-point vibrational energy of* H_2) *between* 0.30 eV *and* 0.45 eV
(Reproduced with permission from *J. Chem. Phys.*, 1976, **65**, 4668)

in the limit of vanishing cross-section only the $l = 0$ partial wave would contribute to the reaction. However, even at the lowest energy, partial waves other than $l = 0$ contribute significantly to the reaction. Note, however, that the width of the angular distribution narrows with increase in collision energy from 0.03 to 0.13 eV, reflecting the increased contribution from larger impact parameters as the energy increases.

Angular distributions derived from quasi-classical trajectory calculations for $D + H_2 \rightarrow HD + H$ at higher collision energies are shown in Figure 9.[67] At energies considerably above the threshold, the differential cross-section becomes quite broad. At quite high energies, ~2 eV, the angular distribution is predicted to be nearly isotropic, characteristic of a hard-sphere interaction. Note that the LSTH surface, on which these QCT calculations were carried out, has a classical barrier height of 1.3 eV for 90° angle of approach, so the cone-of-acceptance[86] is very large for reaction at 2.0 eV.

[86] R. D. Levine and R. B. Bernstein, 'Molecular Reaction Dynamics and Chemical Reactivity', Oxford, New York, 1987.

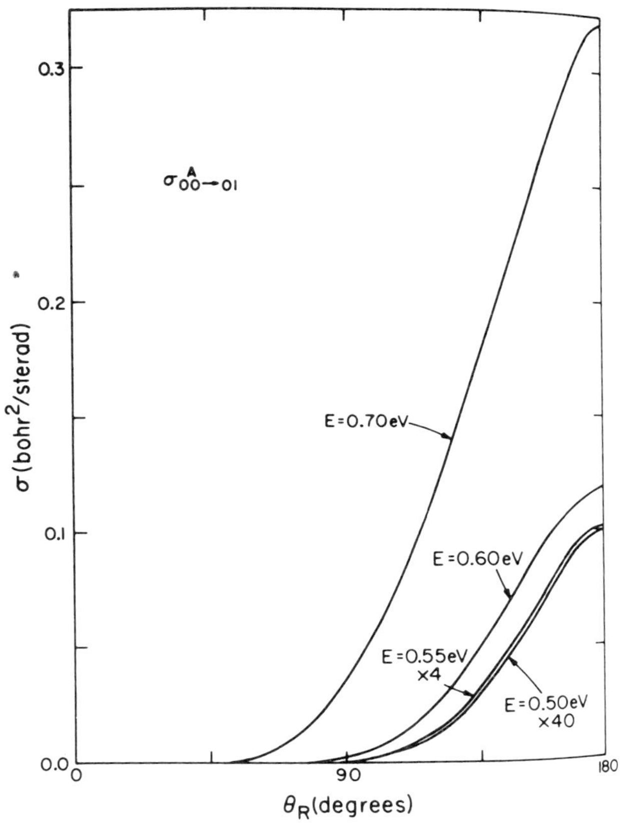

Figure 8 *Antisymmetrized differential cross-sections for the* $H + H_2(v = 0, j = 0) \rightarrow H_2(v' = 0, j' = 1) + H$ *reaction, from coupled-channel calculations at total energies (collision energy plus the* 0.27 eV *zero-point vibrational energy of* H_2*) between* 0.50 eV *and* 0.70 eV
(Reproduced with permission from *J. Chem. Phys.*, 1976, **65**, 4668)

5 Product Quantum-State Distributions

Product quantum-state distributions convey very detailed information about the dynamics of any reaction. The vibrational state distributions, and even more the rotational state distributions, are sensitive to the shape and location of the barrier to reaction, the anisotropy of the potential energy surface (*i.e.* 'steric' effects), and the opacity function. The product quantum-state distributions provide information that is complementary to, and in some cases superior to, that provided by recoil angle and velocity distributions. However, measuring product quantum-state distributions in the hydrogen exchange reaction has until recently proved even more difficult than obtaining product angle and velocity distributions.

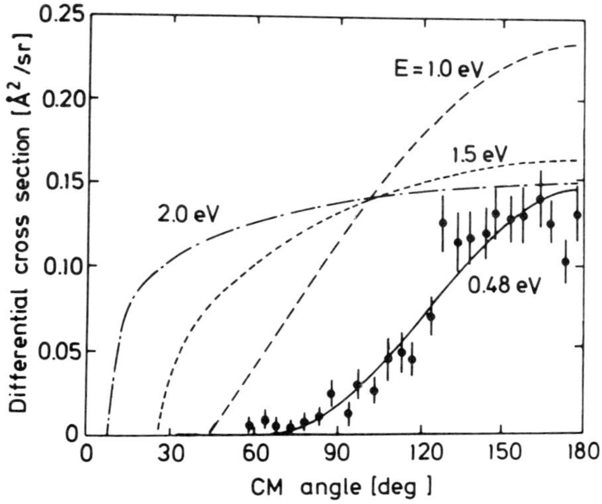

Figure 9 *Differential cross-sections for the* $D + H_2 \rightarrow HD + H$ *reaction at collision energies between* 0.48 *and* 2.0 eV, *from QCT calculations. The curves represent the fits to the trajectory data; the points give the actual trajectory results at* 0.48 eV
(Reproduced with permission from *J. Chem. Phys.*, 1986, **85**, 6396)

In principle, measurement of product recoil velocity can reveal the product internal state distribution, if the kinematics are favourable and if the velocity resolution is quite high. Indeed, as discussed in the previous section, Buntin *et al.*[68] have already demonstrated this for the $D + H_2 \rightarrow HD + H$ reaction, but only for a limited range of recoil angles near 180°, and even then the rotational distribution is not actually resolved. Since products with different quantum numbers may have the same recoil velocity, there is not in general a one-to-one correspondence between a particular product recoil velocity and a particular collection of product quantum numbers. In general, then, some spectroscopic technique is required to establish the rotational and vibrational state distributions. Therein lies the difficulty associated with state distribution measurement in the hydrogen exchange reaction. Hardly any other molecule is less amenable to spectroscopic detection than molecular hydrogen. The homonuclear variants have exactly zero dipole moments and the heteronuclear variants very nearly zero, and the first electronic absorption of molecular hydrogen lies in the vacuum ultraviolet. Thus, the infrared chemiluminescence and laser-induced fluorescence (LIF) techniques that have proved so valuable in studies of the dynamics of other reactions are at best problematic to implement and at worst inapplicable. These spectroscopic impediments are compounded by the difficulty of generating hydrogen atoms of sufficient energy to surmount the substantial barrier to reaction, a difficulty already discussed in regard to crossed molecular beam experiments.

The development of non-linear spectroscopic techniques, such as frequency tripling to generate tunable vacuum ultraviolet radiation from u.v. lasers, multi-photon ionization (MPI), two-photon fluorescence (TPF), and coherent anti-Stokes Raman scattering (CARS) has, however, greatly expanded experimental capabilities for state-resolved chemical dynamics studies. Further, the development of laser systems capable of producing intense, tunable u.v. radiation has made possible the generation of copious quantities of hydrogen atoms with high, well-defined, and selectable translational energy, using photolysis of hydrogen halides and hydrogen sulphide.

The first application of these techniques to the study of product-state distributions in the hydrogen exchange reaction was reported almost simultaneously by Marinero, Rettner, and Zare[87,88] and by Gerrity and Valentini.[89,90] These two groups both reported measurement of the rotational and vibrational state distributions in the $H + D_2 \rightarrow HD + D$ reaction at 1.3 eV collision energy, the Valentini group using CARS detection of HD, the Zare group using MPI. The state distribution measured by Gerrity and Valentini is shown in Figure 10. In the MPI

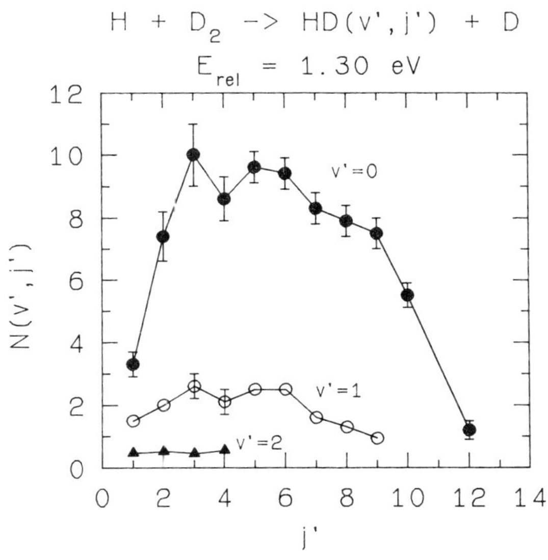

Figure 10 *Product quantum-state distribution for the* $H + D_2 \rightarrow HD(v', j') + D$ *reaction at a collision energy of* 1.30 eV, *from the CARS spectroscopic measurements of ref.* 90

[87] C. T. Rettner, E. E. Marinero, and R. N. Zare, in 'Physics of Electronic and Atomic Collisions', eds. J. Eichler, I. V. Hertel, and N. Stolterfoht, North-Holland, Amsterdam, 1984, p. 51.

[88] E. E. Marinero, C. T. Rettner, and R. N. Zare, *J. Chem. Phys.*, 1984, **80**, 4146.

[89] D. P. Gerrity and J. J. Valentini, *J. Chem. Phys.*, 1983, **79**, 5202.

[90] D. P. Gerrity and J. J. Valentini, *J. Chem. Phys.*, 1984, **81**, 1298.

experiments only a part of the product-state distribution was reported: the distributions could not be obtained for $v' = 0$ or for $j' > 6$, so a complete comparison with the CARS data was not possible. Over the range of v' and j' for which the comparison could be made there was good agreement for the vibrational state distribution, but the MPI-determined rotational distribution was somewhat colder than that from the CARS measurements. However, it now appears that the MPI measurements were in error, and recent re-measurement of the HD product-state distributions using MPI in the Zare laboratory[91] gives results in excellent agreement with the CARS data. Figure 11, which shows a comparison of the HD($v' = 1$) rotational distributions from the new MPI experiments and the original CARS experiments, illustrates this good agreement.

Since these first experiments, the Valentini group has carried out several additional measurements of the product-state distributions in the hydrogen exchange reaction, including the $H + D_2 \rightarrow HD + D$ reaction at lower collision energies (1.1 and 0.98 eV),[92,93] inelastic $H + D_2$ collisions at 1.1 and 1.3 eV,[93,94] inelastic plus reactive collisions in $H + n\text{-}H_2$ collisions at 1.1 eV,[95] the $H + p\text{-}H_2$ reaction at more than twenty collision energies between 0.7 and 1.1 eV,[96,97] and the $D + H_2$ reaction at 0.67 and 0.79 eV.[98]

These experiments adapt classic 'hot' atom techniques, which have been used in the past to study the hydrogen exchange reaction,[10] for use in single-collision, state-to-state dynamics experiments. The CARS experiments are carried out using the apparatus shown schematically in Figure 12. The sample is a flowing gas mixture containing H_2 or D_2 as well as HI or DI, at a total pressure of a few Torr. A pulsed, tunable u.v. beam at a wavelength between 250 and 310 nm, generated by frequency doubling a tunable dye laser, is used to generate H or D atoms by photolysis of the HI(DI). The products of the collisions of these H or D atoms with the H_2 or D_2 are detected by a second laser pulse that effects CARS spectroscopy after a single collision of the H(D) atom. The photolysis generates atoms with well-defined translational energies between 1.9 and 0.95 eV. Depending on the isotopes involved, these laboratory energies give hydrogen exchange reaction collision energies between 0.48 and 1.5 eV, with a spread in collision energy of

[91] R. S. Blake, K.-D. Rinnen, D. A. V. Kliner, and R. N. Zare, *Chem. Phys. Lett.*, 1988, **153**, 365

[92] D. P. Gerrity and J. J. Valentini, *J. Chem. Phys.*, 1985, **82**, 1323.

[93] J. J. Valentini and D. P. Gerrity, *Int. J. Chem. Kinet.*, 1986, **18**, 937.

[94] D. P. Gerrity and J. J. Valentini. *J. Chem. Phys.*, 1985, **83**, 2207.

[95] H. B. Levene, D. L. Phillips, J.-C. Nieh, D. P. Gerrity, and J. J. Valentini, *Chem. Phys. Lett.*, 1988, **143**, 317.

[96] J.-C. Nieh and J. J. Valentini, *Phys. Rev. Lett.*, 1988, **60**, 519.

[97] J.-C. Nieh and J. J. Valentini, *J. Chem. Phys.*, submitted for publication.

[98] D. L. Phillips, H. B. Levene, and J. J. Valentini, *J. Chem. Phys.*, 1989, **90**, 1600.

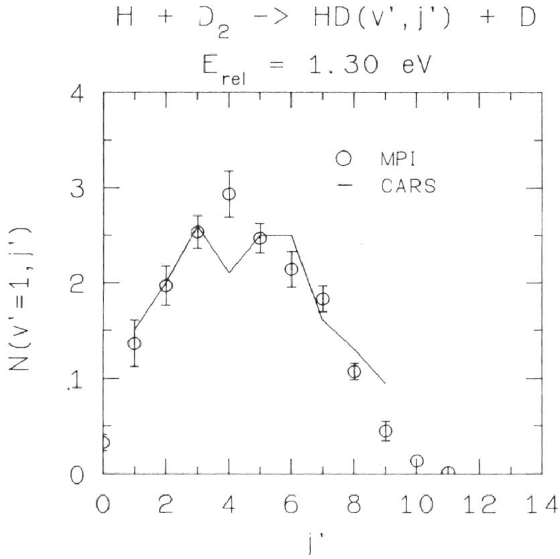

Figure 11 *Comparison of measured product quantum-state distributions for the* $H + D_2 \rightarrow HD(v', j') + D$ *reaction at a collision energy of* 1.30 eV. *The solid line gives the* $HD(v' = 1)$ *rotational distribution from the CARS results of ref.* 90 *that are also shown in Figure* 10. *The symbols give the results of the MPI measurements of the same quantity, from ref.* 91. *The two sets of data have been normalized such that* $\Sigma_{v'=1,\, j'=1 \text{ to } 9}$ *is the same for both*

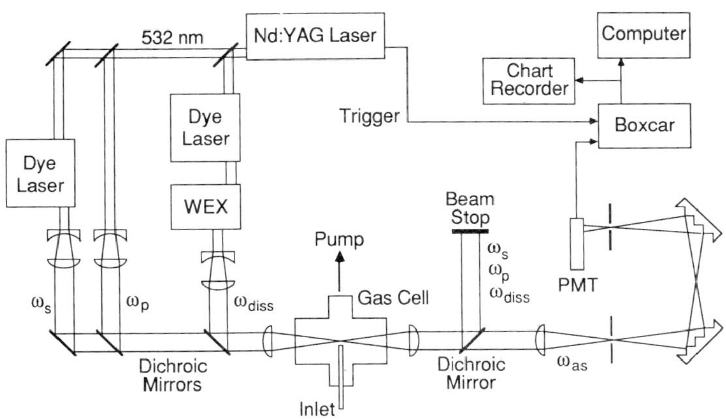

Figure 12 *Schematic diagram of the experimental apparatus used to determine product quantum-state distributions in the hydrogen exchange reaction by CARS spectroscopy. The gas cell contains a flowing mixture of* HI(DI) *and* $H_2(D_2)$. *The laser beams labelled* ω_p *and* ω_s *provide the CARS pulse that detects the reaction product, while the beam labelled* ω_{diss} *provides the u.v. pulse that photodissociates the* HI(DI) *to generate the* H(D) *reactant. See text for details*

0.1 to 0.2 eV. The MPI experiments of the Zare group, which have so far been done only for $H + D_2 \rightarrow HD + D$ at 1.3 eV relative energy, are similar in concept. They are carried out in an effusive flow containing HI and D_2, and photolysis of the HI is effected by the fourth harmonic of a Nd:YAG laser at 266 nm. The radiation near 200 nm needed for the two-photon-resonant, three-photon ionization of the HD product is obtained by mixing the fundamental output of a tunable dye laser with its second harmonic, or by Raman shifting a frequency-doubled dye laser beam.

The product-state distribution measurements that have resulted from the CARS spectroscopic measurements are very extensive – almost thirty different combinations of isotopes and collision energy for reactive collisions, and a similar number for inelastic collisions – and very detailed – complete product rotational and vibrational state distributions as well as absolute partial reaction cross-sections. As a result of the wealth of this new experimental information, some generalizations about the quantum-state distributions in the hydrogen exchange reaction can justifiably be made. First, the exchange reaction yields very little vibrationally excited product, even near Feshbach dynamical resonances (discussed in Section 7). Table 6 presents the vibrational state distributions and absolute partial cross-sections at a few selected collision energies between 0.6 and 1.3 eV for the three isotopic variants of the reaction that have been studied. In all cases at least 80% of the products are in the ground vibrational state, even at the highest energy yet studied, 1.3 eV in the $H + D_2$ reaction for which $v' = 3$ is energetically accessible. The dynamical bias against product vibrational excitation is conveyed by the vibrational surprisal parameters given in Table 6. These are obtained by an information theoretic analysis[99] of the data, fitting the observed distribution to a linear surprisal function. The surprisal parameters are λ_V and θ_R, whose magnitude and sign reflect the dynamical bias that characterizes the energy disposal in the reaction. The positive values of λ_V indicate a bias that disfavours vibrational excitation of the products. For all but the lowest energy $D + H_2$ and $H + p\text{-}H_2$ experiments λ_V is quite large, and the bias is correspondingly quite strong. Note that there is evidence for the influence of a $v' = 1$ enhancing Feshbach resonance in the $D + H_2$ reaction at 0.67 eV and in the $H + p\text{-}H_2$ reaction at 0.70 eV, and this may explain the reduction in λ_V for the 0.67 eV $D + H_2$ and 0.79 eV $H + p\text{-}H_2$ data in Table 6. This is discussed more fully in Section 7.

The product rotational state distributions are also characterized by large positive values of the rotational surprisal parameter, θ_R, as indicated by the data in Table 7, which summarizes the rotational energy distributions of the products for the same reactions, as in Table 6. The data show that $\langle j' \rangle$ and $j'_{\text{most probable}}$ are in all cases low, a small fraction

[99] R. D. Levine and R. B. Bernstein, *Acc. Chem. Res.*, 1974, **11**, 215.

Table 6 *Experimentally determined product vibrational state distributions and absolute partial cross-sections for three isotopic variants of the hydrogen exchange reaction*

Reaction	*Collision energy /eV*	$v' = 0$		$v' = 1$		$v' = 12$			
		P^a	$\sigma^b/Å^2$	P	$\sigma/Å^2$	P	$\sigma/Å^2$	$\lambda_V{}^c$	*Reference*
$D + H_2 \rightarrow HD + H$	0.67	0.85	0.68	0.15	0.12			0.6	98
$D + H_2 \rightarrow HD + H$	0.79	0.91	0.74	0.09	0.068			2.2	98
$H + p\text{-}H_2 \rightarrow o\text{-}H_2 + H$	0.79	0.94	0.66	0.06	0.044			1.5	96, 97
$H + p\text{-}H_2 \rightarrow o\text{-}H_2 + H$	1.1	0.96	1.25	0.04	0.050			4.2	96, 97
$H + D_2 \rightarrow HD + D$	0.98	0.87	0.76	0.13	0.11	0.005	0.004	2.2	92, 95
$H + D_2 \rightarrow HD + D$	1.1	0.89	0.54	0.11	0.065	0.006	0.004	3.3	92, 95
$H + D_2 \rightarrow HD + D$	1.3	0.80	0.96	0.18	0.21	0.025	0.030	2.6	90, 95

[a]The probability of forming the reaction product in state v', $[\Sigma P(v') = 1.0]$. [b]The partial reaction cross-section into the indicated product vibrational state. [c]The vibrational surprisal parameter computed by least-squares fit of the $P(v', j')$ distribution to a linear surprisal function

Table 7 *Experimentally determined product rotational state distributions for the $v' = 0$ product vibrational state in three isotopic variants of the hydrogen exchange reaction*

Reaction	*Collision energy*/eV	$\langle j' \rangle$ [a]	$j'_{\text{most probable}}$ [b]	j'_{limiting} [c]	θ_R [d]	*Reference*
$D+H_2 \rightarrow HD+H$	0.67	4.2	3	11	3.7	98
$D+H_2 \rightarrow HD+H$	0.79	5.3	5	12	2.5	98
$H+p\text{-}H_2 \rightarrow o\text{-}H_2+H$	0.79	2.7	2	10	7.6	96, 97
$H+p\text{-}H_2 \rightarrow o\text{-}H_2+H$	1.1	3.9	3	12	5.5	96, 97
$H+D_2 \rightarrow HD+D$	0.98	5.2	4	13	3.5	92
$H+D_2 \rightarrow HD+D$	1.1	5.6	4,5	14	3.4	92
$H+D_2 \rightarrow HD+D$	1.3	6.5	5	15	3.0	90

[a]The average value of j'. [b]The most probable value of j'. [c]The largest value of j' energetically accessible. [d]The rotational surprisal parameter computed by least-squares fit of the $P(v', j')$ distribution to a linear surprisal function

of j'_{limiting}, the highest energetically accessible rotational state. Large positive values of θ_R have been observed before for the rotational state distributions determined by coupled-channel calculations of the $H + H_2$ reaction at low collision energies.[26,100] These have been interpreted as a consequence of the coupling of the H_3 bending motion to product rotation, and the stiff bending vibration and collinear geometry of H_3.

Since the dynamics of the hydrogen exchange reaction strongly disfavour rotational and vibrational excitation of the products, the available energy in the reactions must appear primarily as product translation. Table 8, which provides quantitative data on the fraction of the available energy that appears as translation, rotation, and vibration for the same reactions as in Tables 6 and 7, indicates that this is indeed so. Between 69 and 84% of the available energy, which comes almost exclusively from reactant translation, remains as translational energy of the products. That is, translationally hot reactants beget translationally hot products.

The primary value of these experimentally measured, very detailed product quantum-state distributions is in providing critical tests of theoretical descriptions of reaction dynamics. The data have already served to do so, but not yet as extensively as will eventually be the case. Because of limitations imposed by available photolytic sources of monoenergetic hydrogen atoms, the experiments cannot be done easily at much lower energies than have already been explored. However, these are still quite high energies by the standards of coupled-channel calculations and other quantum mechanical dynamical methodologies for calculating product quantum-state distributions, since there are many channels open at these energies. Methodological developments in the quantum theoretical techniques promise to make calculations at these

[100] R. E. Wyatt, *Chem. Phys. Lett.*, 1975, **34**, 167.

Table 8 *Experimentally determined product energy disposal for three isotopic variants of the hydrogen exchange reaction*

Reaction	*Collision energy*	f'_{trans}[a]	f'_{rot}[b]	f'_{vib}[c]	*Reference*
$D+H_2 \rightarrow HD+H$	0.67	0.71	0.19	0.10	98
$D+H_2 \rightarrow HD+H$	0.79	0.70	0.25	0.05	98
$H+p\text{-}H_2 \rightarrow o\text{-}H_2+H$	0.79	0.84	0.12	0.04	96, 97
$H+p\text{-}H_2 \rightarrow o\text{-}H_2+H$	1.1	0.82	0.16	0.02	96, 97
$H+D_2 \rightarrow HD+D$	0.98	0.73	0.21	0.06	92
$H+D_2 \rightarrow HD+D$	1.1	0.74	0.22	0.04	92
$H+D_2 \rightarrow HD+D$	1.3	0.69	0.24	0.07	90

[a]The fraction of the total available energy in translation. [b]The fraction of the total available energy in rotation. [c]The fraction of the total available energy in vibration.

relatively high energies more tractable, but application of them is just now beginning. There are, however, methods for which collision energies of 1 eV or more are not difficult, specifically quasi-classical trajectory calculations and infinite order sudden calculations, and quite extensive comparison of the product-state distributions derived from such calculations with their experimental counterparts are possible and have been made. Comparison of observations with the results of distorted wave calculations has also been made.

In general one finds, perhaps somewhat surprisingly, that QCT-derived rotational and vibrational state distributions[67,76,101] are very close to those actually measured. Even the QCT-calculated absolute partial cross-sections are often in excellent agreement with the measurements. Figure 13 illustrates this good agreement for the $H + D_2 \rightarrow HD + D$ reaction at 0.98 eV relative energy. The agreement, even in terms of absolute partial cross-sections, is quite good, but tempered by the rather significant error bars on the experimental absolute cross-sections. The experimental partial cross-sections are known relative to one another to much greater accuracy than they are known absolutely, so plotted as a relative population distribution the experimental data of Figure 13 have error bars only a third as large as shown. Consequently, viewed as a comparison of relative cross-sections the QCT–experiment agreement shown is extremely good. There are cases for which the absolute partial cross-sections from the trajectory calculations are not in such good agreement with experiment, for example $H + D_2$ at 1.1 eV, and cases for which even the relative population distributions from QCT do not mimic the experimental results, as Figure 14 shows for $D + H_2$ at 0.67 eV relative energy. However, these are energies at which there is some evidence of dynamical resonances (discussed in Section 7), and classical mechanics will almost certainly fail when such quantum effects are evident. Even in the cases for which QCT results do not agree with

[101] N. C. Blais and D. G. Truhlar, *Chem. Phys. Lett.*, 1983, **102**, 120.

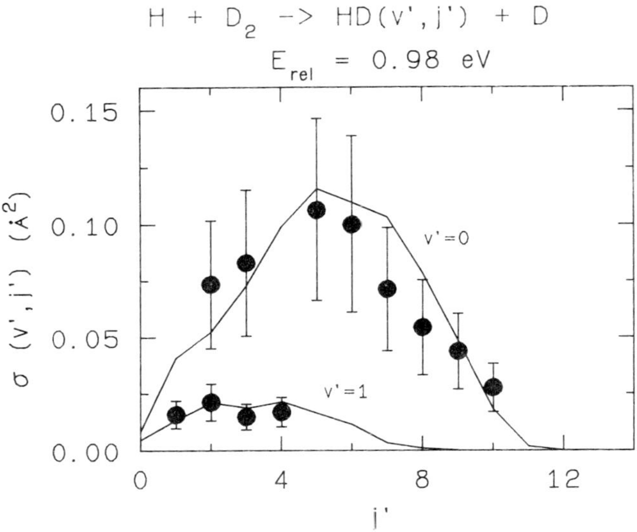

Figure 13 *Comparison of experimentally measured and theoretically predicted absolute partial cross-sections for the* H + $D_2 \rightarrow$ HD(v', j') + D *reaction at* 0.98 eV *collision energy. The solid circles with error bars give the experimental results of refs.* 92 *and* 95, *while the solid lines show the results of the QCT calculations of ref.* 76

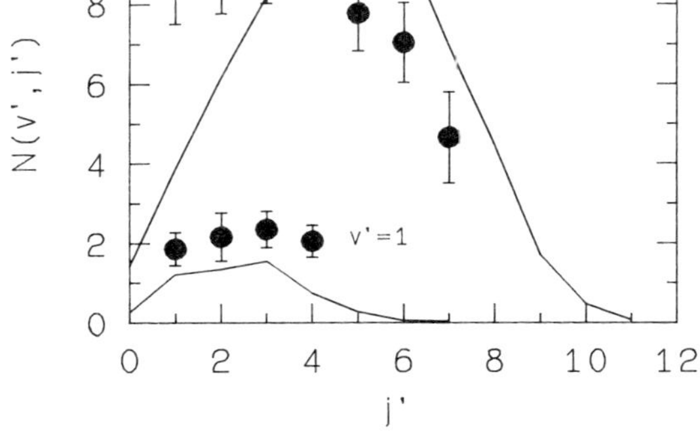

Figure 14 *Comparison of experimentally measured the theoretically predicted product quantum-state distributions for the* D + $H_2 \rightarrow$ HD(v', j') + H *reaction at* 0.67 eV *collision energy. The solid circles with error bars give the experimental results of ref.* 98, *while the solid lines show the results of the QCT calculations of ref.* 69

the experimental results within mutual uncertainties, they are not too far off. The apparent accuracy of (quasi-) classical mechanics in describing the product quantum-state distributions even when only a handful of states is accessible may be fortuitous. However, the agreement extends to distributions for both reactive and inelastic processes for most of the five $D + H_2$ and $H + D_2$ energies at which both the calculations and experiments have been done, so it is becoming hard to dismiss the correlations as accidental. The de Broglie wavelength for $D + H_2$ at 0.67 eV is 0.32 Å, and that for $H + D_2$ at 1.3 eV is 0.28 Å, while the range of the H_3 potential energy surface is 1—2 Å, so by the Ehrenfest criterion the reaction dynamics might be reasonably well approximated by classical mechanics.

Accompanying the first comparison of the results of these hot atom state-to-state dynamics experiments with quasi-classical trajectory calculations was an important comparison of QCT and accurate quantum results for product-state distributions and partial cross-sections. At 0.55 eV relative energy Schatz[102] used the coupled-states method to compute the $v' = 0$ and $v' = 1$ cross-sections for $H + D_2 \rightarrow HD + D$, and compared these with QCT results of Blais and Truhlar.[101] Overall, the agreement between the classical and quantal descriptions was remarkably good, considering the low collision energy involved and the resultant limited number of product channels (only eleven v', j' pairs are energetically accessible). For $v' = 0$, the CS and QCT rotational state distributions are very similar, although the QCT partial cross-sections are ~50% too large. The QCT calculations do predict orders of magnitude too much for $v' = 1$. This is understandable, since the 0.68 eV vibrationally adiabatic threshold for $v' = 1$ HD product is only a few meV below the total energies (collision energy plus reactant zero-point vibrational energy and rotational energy) available in the Boltzmann rotationally averaged, 0.55 eV relative energy collisions of these calculations. As a result, the CS cross-sections are very small, $<10^{-7}$ Å^2. The classical $v' = 1$ barrier, which does not include the zero-point energy of the H_3 transition state, is significantly lower than the vibrationally adiabatic barrier, and thus classical trajectories give much larger $v' = 1$ yields.

The experimentally observed product-state distributions in the $H + D_2 \rightarrow HD + D$ reaction at relative energies of 0.98, 1.1, and 1.3 eV can also be compared with distributions computed by distorted wave calculations. The distorted wave calculations of Connor and Southall[103] and those of Bowers *et al.*[104] for $H + D_2 \rightarrow HD + D$ give rotational state distributions and vibrational state distributions in reasonable agreement with the experimental results.[88,90,92] The variant of the

[102] G. C. Schatz, *Chem. Phys. Lett.*, 1984, **108**, 532.

[103] J. N. L. Connor and W. J. E. Southall, *Chem. Phys. Lett.*, 1986, **123**, 139.

[104] M. S. Bowers, B. H. Choi, R. T. Poe, and K. T. Tang, *Chem. Phys. Lett.*, 1985, **116**, 239.

distorted wave approach used is the so-called static–static distorted wave. The calculated vibrational distributions are a little colder than the experimentally measured ones. The computed rotational distributions are a little hotter than the experimental distributions in some vibrational states, but colder in others. Suck Salk *et al.*[105] have also carried out distorted wave calculations, using a slightly different method: they do not report vibrational state distributions, but do give rotational state distributions. At 1.30 eV relative energy, their calculated rotational distributions are significantly colder than the experimental distributions.[88,90,92]

The distorted-wave-computed rotational distributions agree (disagree) with one another to about the same extent that they agree (disagree) with the experimental results. Since the calculations use different distortion potentials, such differences are expected. There is considerable sensitivity to the distortion potential, since this potential is the source of the coupling between reactants and products. The absolute cross-sections are more sensitive to the distortion potential than are the product-state distributions and thus are not very accurate, and these are in poor agreement with the experimental values. At the high collision energies dealt with here, the underlying assumption on which the distorted wave method is based (that reaction is a small perturbation on the collision dynamics) may break down, as the reaction cross-sections are becoming appreciable, around 1 Å^2.

The only other theoretical calculation of product-state distributions over the energy range of the CARS experimental measurements are the infinite order sudden approximation results of Zhang *et al.*[106] While the vibrational state distributions from these IOS calculations are fairly close to the experimental values, the absolute partial and total cross-sections of Zhang *et al.* are uniformly too large.

6 Total Reaction Cross-sections

The first experimental values for the total cross-section for the hydrogen exchange reaction were derived from the crossed molecular beam experiments of Geddes *et al.*[62] on the $D + H_2 \rightarrow HD + H$ reaction. As in most crossed molecular beam experiments, the differential cross-sections measured in this experiment were relative, not absolute. However, absolute calibration of the relative measurements was done by comparison of the magnitude of the reactive scattering with the magnitude of the elastic scattering, for which the absolute cross-section was determined separately. The experimental data are consistent with a line-of-centres energy dependence for the cross-section, $\sigma(E) =$

[105] S. H. Suck Salk, C. R. Klein, and C. K. Lutrus, *Chem. Phys. Lett.*, 1984, **110**, 112.
[106] Y. C. Zhang, Z.-H. Zhang, D. J. Kouri, and M. Baer, *Chem. Phys.*, 1987, **114**, 267.

$\sigma_0(1 - E_0/E)$. Using the absolute elastic cross-section measurements of Gegenbach *et al.*,[72] one finds $\sigma_0 = 0.35 \pm 0.18$ Å^2, if a threshold energy, E_0, of 0.26 eV is assumed. This gives a total cross-section of 0.16 ± 0.08 Å^2 at the 0.48 eV average energy of the Geddes *et al.* experiments. Absolute calibration of the cross-sections measured in crossed molecular beam experiments is always difficult, and this difficulty is exacerbated for these $D + H_2 \rightarrow HD + H$ studies by the very large spread in the collision energy.

Theoretical calculations of the total reaction cross-sections for $D + H_2$ by classical trajectory,[24] infinite order sudden approximation,[107] and distorted wave approximation[79] methods, give values of ~0.5 Å^2 at 0.48 eV, with a threshold energy near 0.27 eV: in reasonable agreement with the crossed-beam measurements. However, the line-of-centres excitation function assumed in the analysis of the beam data rises much more rapidly and levels off at much lower energy than does the excitation function in any of the calculations. Because the crossed-beam experiments have such a large energy spread, ~0.35 eV FWHM, and were carried out at only one energy, one cannot expect to get a meaningful excitation function from them.

Accurate total cross-section measurements have been made at well-defined energies, but in low-pressure gas cell experiments using spectroscopic detection of the products, rather than in crossed molecular beams. Bersohn and co-workers[108,109] have determined the total cross-section for the $H + D_2 \rightarrow HD + H$ reaction at four collision energies between 0.87 and 2.70 eV and for the $H + HD \rightarrow H_2 + D$ reaction at three energies between 0.82 and 1.86 eV. They used exciplex laser photolysis of HCl of H_2S at 248, 193, or 157 nm to generate translationally hot hydrogen atoms in a gas sample containing H_2 or HD. The relative concentrations of the reactant H and product D atoms were monitored as a function of time after the photolysis pulse using VUV laser-induced fluorescence at Lyman-α, 121.6 nm. The Lyman-α radiation was generated by frequency tripling in krypton the output of a near-u.v. dye laser. Measurement of the concentrations of both reactant and product atoms obviates the need for absolute calibration of the LIF signals to determine the absolute cross-section. The same approach was used by Gerlach-Meyer *et al.*[110] to determine the $H + D_2$ cross-section at 1.54 and 2.11 eV. The combination of laser photolysis and single-collision product detection makes the collision energies in these experiments well defined. The spread in the energy, 0.1—0.2 eV FWHM, comes primarily from the thermal velocity of the molecular target.

[107] N. Abusalbi, D. J. Kouri, Y. Shima, and M. Baer, *Chem. Phys. Lett.*, 1984, **105**, 472.
[108] K. Tsukiyama, B. Katz, and R. Bersohn, *J. Chem. Phys.*, 1986, **84**, 1934.
[109] G. W. Johnston, B. Katz, K. Tsukiyama, and R. Bersohn, *J. Phys. Chem.*, 1987, **91**, 5445.
[110] U. Gerlach-Meyer, K. Kleinermanns, E. Linnebach, and J. Wolfrum, *J. Chem. Phys.*, 1987, **86**, 3047.

A comparison of the most recent measurements[109] of cross-sections for $H + D_2 \rightarrow HD + D$ and those computed from classical trajectory calculations[111,112] is shown in Figure 15. The measured cross-sections are in excellent agreement with the QCT results, both showing a broad maximum of ~1.3 Å^2 around 2 eV. The trajectory calculation results shown in Figure 15 are corroborated by other QCT calculations[24,76,101] over the same energy range. However, the total cross-sections computed by the IOS approximation[106] are ~40% higher than the QCT results shown in Figure 15 over the energy range 0.8—2.0 eV. For $H + HD \rightarrow H_2 + D$ the QCT-experiment agreement is equally good, both theory and experiment giving a cross-section that is approximately half that for $H + D_2$, ~0.65 Å^2 at 2 eV. It thus appears that both the LSTH surface on which the trajectories were run and the trajectory method itself provide an accurate representation of the cross-section excitation function for the hydrogen exchange reaction. This apparent validation of theory stands in contrast to the suspicions raised about the adequacy of the potential energy surface and the trajectory calculations when the first experimental measurements of total cross-sections for $H + D_2 \rightarrow HD + D$ using Lyman-α LIF detection of H and D atoms

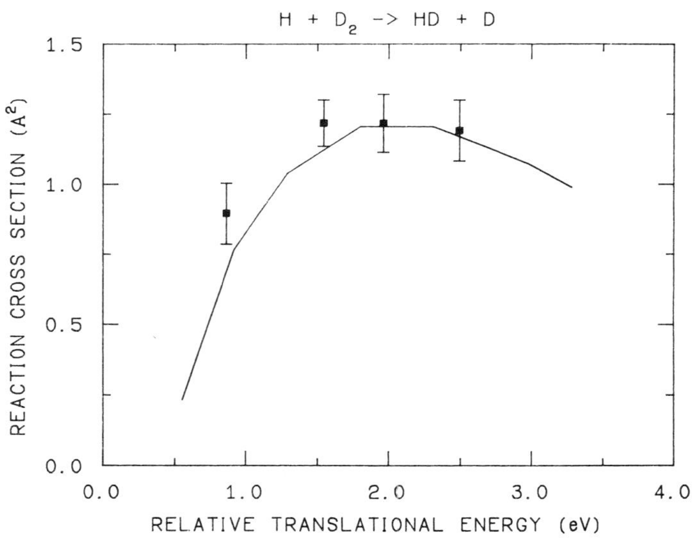

Figure 15 *Comparison of experimentally measured and theoretically predicted total reaction cross-sections for the* $H + D_2 \rightarrow HD + D$ *reaction. The solid squares with error bars give the experimental results of refs.* 109, *while the solid lines show the results of the QCT calculations of ref.* 111 *and* 112

[111] I. Schecter and R. D. Levine, *Int. J. Chem. Kinet.*, 1986, **18**, 1023.
[112] I. Schecter, R. Kosloff, and R. D. Levine, *J. Phys. Chem.*, 1986, **90**, 1006.

were reported. These experiments[108,110] gave total cross-sections around 2 eV that were 50% larger than the QCT-calculated values. However, it is now known[109] that one of these first experiments[108] was in error, and it seems likely that the other[110] was as well, for another direct measurement of $H + D_2 \rightarrow HD + D$ total cross-sections also confirms the QCT results.

This direct measurement comes from absolute calibration of product quantum-state distribution measurements to obtain absolute partial reaction cross-sections, *i.e.* cross-sections for production of specific product rotational and vibrational states, for several isotopic variants of the hydrogen exchange reaction.[90,92,95] Summation of these partial cross-sections over product states yields the total reaction cross-section. For $H + D_2 \rightarrow HD + D$ the total cross-sections obtained in this way are 0.88 ± 0.33 Å^2 at 0.98 eV relative energy, 0.61 ± 0.23 Å^2 at 1.1 eV, and 1.20 ± 0.43 Å^2 at 1.3 eV, in very good agreement with the experimental and theoretical values in Figure 15. The cross-section at 1.1 eV is somewhat lower than the QCT value. This may be the consequence of a dynamical resonance, and we note that the measured product-state distribution also differs slightly from the QCT-calculated state distribution at this relative energy. The absolute partial cross-section measurements from which these total cross-sections are derived are made by absolute calibration of the CARS signals from the HD product and D_2 reactant to establish their absolute densities, together with determination of the reactant H-atom absolute density using CARS signals from the HI precursor with and without photodissociation. CARS measurements[98] of the total cross-section for $D + H_2 \rightarrow HD + H$ at 0.67 and 0.79 eV are in reasonable agreement with corresponding QCT values.[69] The measured values are the same, 0.81 Å^2, with a ± 0.35 Å^2 experimental uncertainty, at both the lower and higher energy, while linear interpolation/extrapolation of the 0.70 and 0.85 eV trajectory results gives 0.99 Å^2 at 0.67 eV and 1.23 Å^2 at 0.79 eV.

The quasi-classical trajectory calculations that yield the theoretical total cross-sections in Figure 15 also suggest a surprising interpretation of the cross-section excitation function, involving a competition between abstraction and insertion. Levine and co-workers[111,112] make the distinction between abstraction and insertion on the basis of the angle of attack of the reactant atom and the response of the diatom to the approach of the atom. In the insertion mode the atom approaches the diatom sideways and the diatom bond elongates significantly, while for abstraction the approach is closer to collinear and the change in diatom bond length is smaller. The trajectories indicate that the abstraction mode is dominant at relative energies below 1 eV, while above this energy the insertion mode is more significant. This is illustrated in Figure 16, which plots the same QCT $H + D_2$ total reactive cross-section data as given in Figure 15, but also separates the insertion and abstraction contributions to this total cross-section. For reaction with a

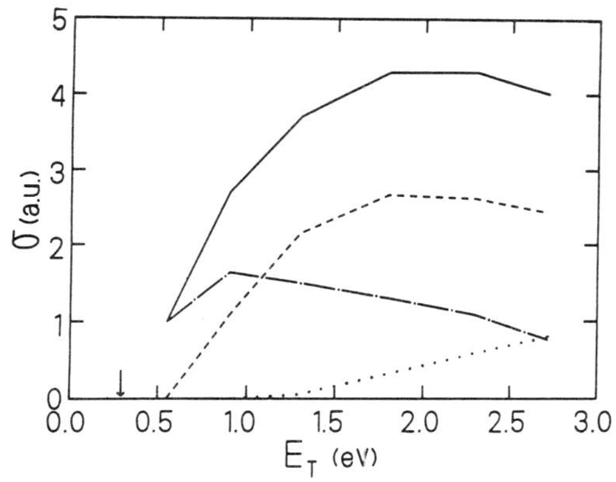

Figure 16 *Energy dependence of the reaction cross-section for the* $H + D_2 \rightarrow HD + D$ *reaction, from QCT calculations. The solid line shows the total cross-section. The dot–dash line represents the contribution to the total cross-section from those trajectories that proceed by abstraction. The dashed and dotted lines give the contribution from trajectories that proceed by insertion, the former through reaction with the* D *atom that is closer to the* H *at the moment of insertion, and the latter through reaction with the* D *atom that is farther away at the moment of insertion*
(Reproduced with permission from *J. Chem. Phys.*, 1986, **90**, 1006)

homonuclear hydrogen molecule, the trajectory calculations indicate that it is difficult to discern some experimentally measurable quantity that would distinguish insertion from abstraction. However, for reaction of H with HD the results show that the HD/H_2 branching ratio offers such a distinction, insertion leading to preferential production of HD in the reaction.

Most of the experimental determinations of the total reactive cross-section in the hydrogen exchange reaction have been done at energies well above the threshold. This is due to the fact that the cross-sections are very small near threshold, and so direct measurements at energies near threshold are the most difficult. Studies at these relatively low energies are difficult also because there are no photolytic sources of H atoms with well-defined energies below about 0.95 eV, which limits the range of accessible collision energies for $H + D_2$, $H + H_2$, and $D + H_2$ to values above 0.76, 0.63, and 0.48 eV, respectively. In contrast, essentially all theoretical calculations of the total cross-sections, except those from QCT calculations, are easiest at energies near threshold, and so most of such theoretical work has been at relative energies below 0.5 eV. The overlap between theory and experiment then is quite limited, and much of the theoretical work leading to calculation of total cross-sections has revolved around comparison of results from different theoretical methodologies. For the $H + H_2$ variant of the reaction, such comparisons invariably use the results of quantum coupled-channel calculations[26,85,113–116] as the exact reference against which the cross-sections from approximate computational methodologies are judged.

Quite good agreement is found among the cross-sections for relative energies below 0.5 eV computed by many, but not all, of the different methods. For example, using the LSTH potential energy surface for $H + H_2(v = 0,\ j = 0)$, exact quantum coupled-channel calculations,[116] coupled-states calculations,[9] adiabatic T-matrix calculations,[117] and coupled-channel distorted wave calculations[35] give cross-sections that agree within 40%. Quasi-classical trajectory calculations,[24,74] which are suspect at these low energies because of neglect of tunnelling and vibrational zero-point energy constraints, give cross-sections that are within 75% of the quantum values. Figure 17 summarizes all but the CS data, showing the good agreement. However, both rotationally adiabatic distorted wave (RADW) and vibrationally adiabatic distorted wave

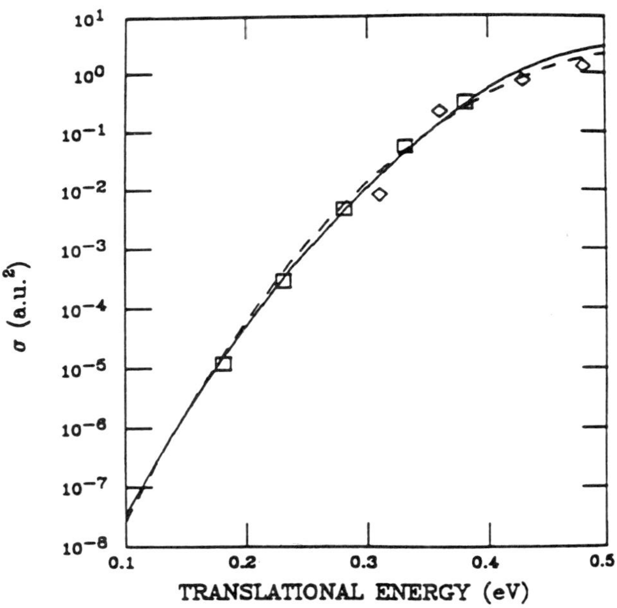

Figure 17 *Comparison of total reaction cross-sections for the* $H + H_2(v = 0,\ j = 0) \rightarrow H_2 + H$ *reaction as a function of collision energy, calculated by various theoretical methods. Solid line, CCDW calculations of ref.* 35; *dashed line, T-matrix results of ref.* 117; □, *CC results of ref.* 116; ◇, *QCT results of ref.* 24
(Reproduced with permission from *Chem. Phys. Lett.*, 1987, **136**, 145)

[113] G. C. Schatz and A. Kuppermann, *J. Chem. Phys.*, 1976, **65**, 4642.
[114] A. B. Elkowitz and R. E. Wyatt, *J. Chem. Phys.*, 1975, **62**, 2504.
[115] A. B. Elkowitz and R. E. Wyatt, *J. Chem. Phys.*, 1975, **63**, 702.
[116] R. B. Walker, E. B. Stechel, and J. C. Light. *J. Chem. Phys.*, 1978, **69**, 2922.
[117] J. C. Sun, B. H. Choi, R. T. Poe, and K. T. Tang, *J. Chem. Phys.*, 1983, **78**, 4523.

(VADW) calculations[80] give cross-sections that are significantly smaller than the coupled-channel results. The RADW total cross-sections are a factor of 2 to 10 smaller than the exact quantum values over the energy range of Figure 17, while the VADW cross-sections are an order of magnitude smaller than even the RADW cross-sections.

For $D + H_2 \rightarrow HD + H$, total cross-sections have been calculated on the accurate LSTH surface over the relative energy range 0.2–0.5 eV (and in a few cases above this) using the T-matrix approach,[32] the reduced dimensionality bending corrected rotating linear model (BCRLM),[31] adiabatic distorted wave methods,[79,80,118] the infinite order sudden approximation,[30,119] and quasi-classical trajectory calculations.[24,69,74] For $H + D_2 \rightarrow HD + D$, the only extensive theoretical computations of total cross-sections as a function of relative energy from threshold and above are trajectory calculations.[24,74,111,112] At a few high relative energies for which hot-atom experiments have been reported there are also theoretical calculations of the total reaction cross-section by coupled states,[102] distorted wave,[103,104] infinite order sudden approximation,[106] and additional trajectory calculations.[76,101]

The controversy surrounding the rate constant for the $H + H_2(v = 1)$ reaction (discussed in Section 3) has motivated several studies of the cross-section excitation function for this reaction and its isotopic variants. There have been theoretical calculations by exact quantum,[120] QCT,[24,75,121] IOS,[106,107] semi-classical adiabatic,[43] BCRLM,[122] and T-matrix[51] methods using the accurate LSTH surface. The results of these computations indicate that the energy threshold for the reaction with $H_2(v = 1)$ shifts by ~0.15 eV to lower energy from the threshold for the corresponding $H_2(v = 0)$ reaction. Note that this is considerably smaller than the $(v = 1) - (v = 0)$ energy difference, 0.51 eV, and in fact at low relative energies vibrational energy is less effective in promoting reaction than is translational energy. There is only one experimental measurement of the total reaction cross-section for vibrationally excited $H_2(v = 1)$ with which to compare these several theoretical calculations. A crossed molecular beam study[123] of the $D + H_2(v = 1) \rightarrow HD + H$ reaction at 0.33 eV relative energy yields a total cross-section of 1.14 ± 0.50 Å^2. The total cross-section at this energy computed by QCT[24] and IOS[106,107] methods is between 1.0 and 1.3 Å^2, in good agreement with the experimental value, but the T-matrix-computed cross-section[51] and that from semi-classical adiabatic theory[43] are more than a factor of three too large.

[118] Y. Y. Yung, B. H. Choi, and K. T. Tang, *J. Chem. Phys.*, 1980, **72**, 621.
[119] N. Abusalbi, D. J. Kouri, Y. Shima, and M. Baer, *Chem. Phys. Lett.*, 1984, **82**, 472.
[120] K. Haug, D. W. Schwenke, Y. Shima, D. G. Truhlar, J. Zhang, and D. J. Kouri, *J. Phys. Chem.*, 1986, **90**, 6757.
[121] I. Schecter, R. Kosloff, and R. D. Levine, *Chem. Phys. Lett.*, 1985, **121**, 297.
[122] R. B. Walker and E. F. Hayes, *J. Phys. Chem.*, 1983, **87**, 1255.
[123] R. Götting, V. Herrero, J. P. Toennies, and M. Vodegel, *Chem. Phys. Lett.*, 1987, **137**, 524.

7 Resonances and the Spectroscopy of the H_3 Transition State

One of the most intriguing features of the hydrogen exchange reaction is the existence of dynamical resonances: either shape or Feshbach. These manifest themselves as sharp changes in specific partial cross-sections, and sometimes total cross-sections as well, as the total energy changes. Figure 18 shows the results of quantum coupled-channel calculations[124] on the PK2 potential energy surface that give an example of this resonance behaviour, for $J = 0$, $H + H_2(v = 0,\ j = 0) \rightarrow H_2(v') + H$. These resonances were first observed for the hydrogen exchange reaction, in early one-dimensional quantum coupled-channel calculations by Truhlar and Kuppermann[125] and by Levine and Wu,[126] and later were shown to persist in three-dimensional coupled-channel calculations as well.[127] Since then, resonances have also been observed in theoretical calculations on $F + H_2 \rightarrow HF + H$, $Cl + HCl \rightarrow ClH + Cl$, $I + HI \rightarrow IH + I$, $Li + HF \rightarrow LiF + H$, and other reactions,[128,129] but the hydrogen exchange reaction remains the major system for which theoretical investigation of resonances takes place.

The sharp maxima and minima in the partial cross-sections are the result of a quantum mechanical interference between two scattering amplitudes: a direct one and a resonant one. The resonant scattering amplitude occurs through a quasi-bound excited state of the transition state, which forms because of non-adiabatic coupling between the relative motion (reaction co-ordinate) and the internal (vibrational) degrees of freedom that are perpendicular to the reaction co-ordinate motion. The resonance energies are the energies of the vibrational levels of the transition state relative to the energy of the separated reactants. For example, the resonances shown in Figure 18 at 0.971, 1.17, 1.38, and 1.56 eV correspond to the (1,0,0), $(1,1^1,0)$, (2,0,0), and $(2,2^2,0)$ excited vibrational levels of the linear H_3 transition state.[130–134] Here the conventional (v_1,v_2^{Ω},v_3) labelling of the vibrational levels of a linear triatomic molecule is used: v_1 designates the number of symmetric stretch vibrational quanta, v_2 gives the number of bend quanta with Ω

[124] P. G. Hipes and A. Kuppermann, *Chem. Phys. Lett.*, 1987, **133**, 1.
[125] D. G. Truhlar and A. Kuppermann, *J. Chem. Phys.*, 1970, **52**, 3841.
[126] R. D. Levine and S.-F. Wu, *Chem. Phys. Lett.*, 1971, **11**, 557.
[127] G. C. Schatz and A. Kuppermann, *Phys. Rev. Lett.*, 1975, **35**, 1266.
[128] D. G. Truhlar, ed. 'Resonances', ACS Symp. Ser. 263, American Chemical Society, Washington, 1984.
[129] A. Kuppermann, in 'Potential Energy Surfaces and Dynamics Calculations', ed. D. G. Truhlar, Plenum, New York, 1981, p. 375.
[130] M. C. Colton and G. C. Schatz, *Chem. Phys. Lett.*, 1986, **124**, 256.
[131] J. M. Bowman, *Chem. Phys. Lett.*, 1986, **124**, 260.
[132] B. C. Garrett, D. W. Schwenke, R. T. Skodje, D. Thirumalai, T. C. Thompson, and D. G. Truhlar, in 'Resonances', ed. D. G. Truhlar, ACS Symp. Ser. 263, American Chemical Society, Washington, 1984, p. 375.
[133] E. Pollak and R. E. Wyatt, *J. Chem. Phys.*, 1984, **81**, 1801.
[134] E. Pollak, *J. Chem. Phys.*, 1986, **90**, 3619.

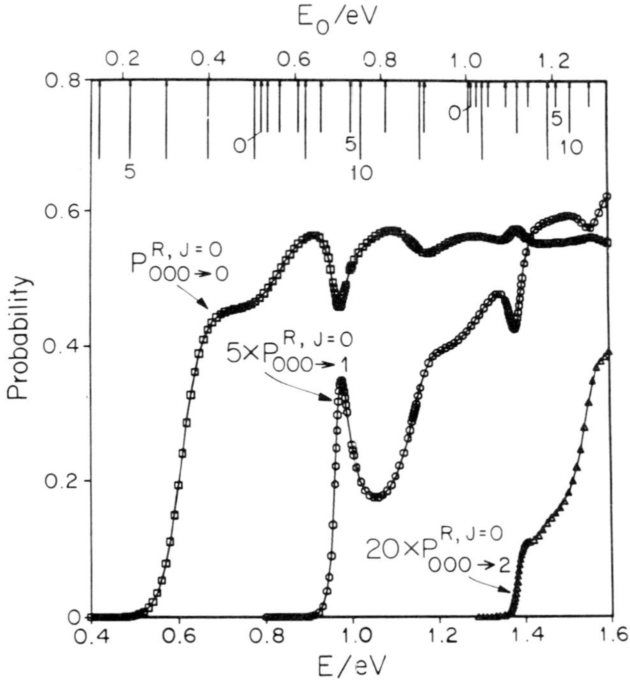

Figure 18 *Distinguishable atom reactive probabilities, $P_{v,j,m_j\to v'}$, as a function of collision energy, E_0, or total energy, E (collision energy plus reactant zero-point vibrational energy), for the $H + H_2(v = 0, j = 0, m_j = 0) \to H_2(v') + H$ reaction for $J = 0$, from coupled-channel calculations. The vertical lines at the top of the figure indicate the energies at which individual v', j' states open up*
(Reproduced with permission from *Chem. Phys. Lett.*, 1987, **133**, 1)

the component of vibrational angular momentum about the internuclear axis, and v_3 is the number of antisymmetric stretch quanta. The last of these is always zero as the antisymmetric stretch is the reaction co-ordinate itself.

It is not necessary to carry out coupled-channel quantum dynamics calculations in order to predict the resonance energies. In fact, the resonance energies for H_3 and its isotopic variants in three dimensions have been obtained from coupled states[130] and reduced dimensionality[84,122,131] approximate quantum calculations, and by semi-classical approaches,[132–135] as well as by several other coupled-channel calculations.[116,127,136,137] Table 9 summarizes the energies of all the reso-

[135] D. W. Schwenke and D. G. Truhlar, *J. Chem. Phys.*, 1987, **87**, 1095.
[136] F. Webster and J. C. Light, *J. Chem. Phys.*, 1986, **85**, 4744.
[137] R. T. Pack and G. A. Parker, *J. Chem. Phys.*, 1987, **87**, 3888.

nances with $v_1 = 1$ calculated by three of these methods using the LSTH *ab initio* potential energy surface. We do not include for comparison here the results of calculations that have been carried out on the semi-empirical PK2 surface, since the resonance energies are sensitive to the potential surface.

Table 9 H_3 *resonance energies*

Resonance[a]	*Energy*[b] /eV			
	Theory			*Experiment*[c]
	CS[d]	*CEQB*[e]	*SCSA*[f]	
1,0,0	0.71	0.70	0.71	0.71 ± 0.01
$1,1^1,0$	0.83	0.83	0.82	0.84 ± 0.01
$1,2^0,0$	0.93	0.93	0.94[g]	0.94 ± 0.01
$1,2^2,0$	0.96	0.96	0.94[g]	1.02 ± 0.01

[a]Labelling is v_1, v_2^{Ω}, v_3. [b]The relative energy at which the resonance occurs, for reactant in $v = 0$, $j = 0$. [c]Ref. 96, 97. [d]Coupled-states calculation, ref. 130. [e]Collinear exact quantum calculation with adiabatic bending corrections, ref. 131. [f]Small curvature semi-classical adiabatic calculation, ref. 132. [g]Harmonic approximation implies that $E(1,2^0,0) = E(1,2^2,0)$

Nieh and Valentini[96,97] recently reported an extensive set of experiments providing the first experimental evidence for the occurrence of these H_3 resonances. Their experiments, which are described more extensively in Section 5 above, use CARS spectroscopy to determine the rotational and vibrational state distributions and absolute partial cross-sections in the $H + p\text{-}H_2(v = 0, j = 0,2) \rightarrow o\text{-}H_2(v', j') + H$ reaction at 22 relative energies between 0.7 and 1.1 eV. The relative energy is selected by varying the wavelength of the u.v. laser used to photodissociate HI to produce H atoms. The use of pure *para*-hydrogen reactant allows separation of reactive from non-reactive processes in $H + H_2$ collisions, since only reaction can convert *para*-hydrogen into *ortho*-hydrogen. The resonances that Nieh and Valentini observed are the four $v_1 = 1$ resonances listed in Table 9. These particular resonances are predicted to produce an increase in both the relative and absolute amount of $v' = 1$ product. Owing to the dominant vibrational adiabaticity of the reaction, accessing a resonance with $v_1 = 1$ is strongly correlated with enhanced production of $v' = 1$. Figure 19 shows a plot of the behaviour of the $v' = 1$, j' absolute partial cross-sections as the relative energy is scanned over the resonances. Two partial cross-sections are plotted, that for $v' = 1$, $j' = 1$ and that for $v' = 1$, $j' = 3$. Together these constitute almost all the $v' = 1$ yield, so their sum is nearly equal to $\sigma(v' = 1)$. Both are plotted to show that the resonances affect different product rotational states differently. The $\sigma(v' = 0)$ partial cross-section, which is 4 to 15 times larger than $\sigma(v' = 1)$, does not show a change at the resonances, and so the yield of $v' = 1$ increases in a relative sense also.

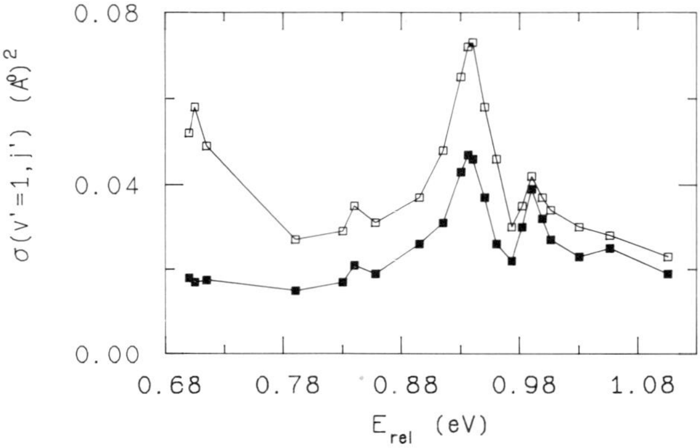

Figure 19 *Experimental measurements of the energy dependence of the absolute partial cross-sections for the* $H + p\text{-}H_2(v = 0,\ j = 0, 2) \rightarrow o\text{-}H_2(v' = 1, j') + H$ *reaction, from refs.* 96 *and* 97: □, $v' = 1$, $j' = 1$; ■, $v' = 1$, $j' = 3$

The energies of the maxima observed in $\sigma(v' = 1)$ are given in the last column of Table 9. These energies are in excellent agreement with the energies at which the (1,0,0), $(1,1^1,0)$, $(1,2^0,0)$, and $(1,2^2,0)$ H_3 resonances are predicted. It must be pointed out that the calculations were done for specific J, $J \leq \Omega$, while the experiments are perforce averaged over all possible partial waves and therefore all possible J. Since the total angular momentum, J, the vector sum of j and l in the $H + H_2$ collision, becomes the rotational angular momentum of the H_3 quasi-bound resonant state, the energy of the resonance should shift like $J(J + 1)B_{H_3}$. However, Nieh and Valentini argue that the partial wave sum must not include very large J, and even for $J = 5$ the $J(J + 1)B_{H_3}$ energy shift of the resonance is only 0.03 eV. Given the experimental energy spread of ~0.1 eV, a shift due to the average over J may not be discernable unless many J contribute.

Following the report of the experimental observation of these resonances, Mladenovic *et al.*[138] carried out a series of coupled-channel quantum calculations for $H + p\text{-}H_2(v = 0, j = 0, 2) \rightarrow o\text{-}H_2(v', j') + H$, for J up to 2, over the entire energy range sampled by the experiments, computing individual partial cross-sections. They did find resonance structure in the $v' = 1$ partial cross-sections, but it differed somewhat from that experimentally observed, and the partial cross-sections were much smaller than the experimental values. The computed cross-sections are not yet converged, and higher J must be included before a detailed comparison with experiment can be made.

The dynamical resonances are also predicted to influence the inelastic scattering in $H + H_2$.[124,139] The distinguishability of the H atoms in

[138] M. Mladenovic, M. Zhao, D. G. Truhlar, D. W. Schwenke, Y. Sun, and D. J. Kouri, *Chem. Phys. Lett.*, 1988, **146**, 358.
[139] F. A. Gianturco and U. T. Lamanna, *Chem. Phys.*, 1977, **25**, 401.

theoretical calculations makes it easy to separate inelastic from reactive processes, and thus distinguish the inelastic resonance effects from the reactive ones. In the experimental domain, the use of p-H_2 reactant makes it possible to identify quantum states (o-H_2) that can be produced only by $H + H_2$ reaction. However, there are no states that can be populated only by inelastic scattering; the p-H_2 product comes from both inelastic and reactive events, and the inelastic and reactive scattering amplitudes interfere with one another and are therefore inseparable. However, the inelastic scattering amplitude is much larger than the reactive amplitude, so the partial cross-sections for p-H_2 products are strongly dominated by the inelastic process. Analysis of the p-H_2 product partial cross-sections in the experiments of Nieh and Valentini[96,97] does show evidence of dynamical resonance effects in $H + p$-H_2 inelastic scattering. The same resonance energies are observed for both inelastic and reactive processes, but the magnitude of the effect of the resonances on the inelastic partial cross-sections is not identical to that on the reactive partial cross-sections. What is observed is in reasonable agreement with the predictions of Hipes and Kuppermann.[124]

From measurements of the product-state distributions and absolute partial cross-sections for $D + H_2 \rightarrow HD + H$ there is some evidence for a dynamical resonance in this system as well. The $D + H_2$ reaction has been studied[98] at relative energies of 0.67 and 0.79 eV, which are close to the $v' = 1$ vibrationally adiabatic threshold. The yield of $v' = 1$ product is larger, in both a relative and an absolute sense, at the lower collision energy than at the higher energy. The (1,0,0) Feshbach resonance should occur at ~0.65 eV in DH_2, so this behaviour may be a result of it. However, experiments will have to be carried out at many more energies around 0.7 eV before one can be more confident about ascribing the $v' = 1$ enhancement to the consequence of a resonance.

More work on these resonances, both theoretical and experimental, is clearly needed. Only a fully converged (with respect to J) quantum calculation will provide a meaningful comparison with the experimental results, and improvements in the state and energy definition of the experiments would make such a comparison more informative. The experiments are limited by the energy spread (~0.1 eV) in the collisions, which results from the 300 K thermal velocity spread of the p-H_2 target, and by the thermal rotational distribution of the p-H_2, which gives equal amounts of $j = 0$ and $j = 2$. Nonetheless, the experimental observation of dynamical resonances in $H + H_2$ is very exciting. It is not unfair to call the data of Figure 19 a vibrational spectrum of the transition state,[129] so the resonances provide the most detailed view of the transition state itself.

The use of resonances to probe the vibrational structure of the H_3 transition state is an indirect spectroscopy; that is, no direct optical interaction with the transition state itself is involved. Rather, the energy needed to effect a vibrational transition is carried by the colliding $H + H_2$, and the transition is detected through its effects on the $H_2 + H$ products. However, there has existed for some time the hope that direct optical spectroscopy of the H_3 transition could be accomplished, and recent developments suggest this possiblity may soon be realized. This optical spectroscopy would involve a continuum-to-bound electronic transition in H_3, from the ground $2pE'$ state on which the reaction takes place, to the stable excited electronic states $2sA'_1$ and $2pA''_2$. Electronic absorption spectra of these transitions have been theoretically computed.[140–145] The transitions lie in the u.v. or vacuum u.v. depending on the collision energy.

Recently a most important and exciting first step in making this direct H_3 transition-state optical spectroscopy an experimental reality was taken by Collings *et al.*,[146] who successfully observed the HD_2 transition state in the $H + D_2 \rightarrow HD + D$ reaction. They used a u.v. laser pulse to create the H atoms by H_2S photolysis in a molecular beam containing H_2S and D_2. They detected the HD_2 by two-photon ionization at 193 nm, using the $2pA''_2$ electronic state as the resonant intermediate. This was a fixed-frequency experiment, so spectra were not obtained, but it suggests that optical spectra of the H_3 transition state and its isotopic variants are now within the grasp of experimentalists.

8 Conclusion

In the past few years there has been a renaissance of interest in the dynamics of the hydrogen exchange reaction. This rebirth has been stimulated in large measure by experimental developments that have at last made this simplest of all reactions an archetype in experimental chemical dynamics as well as in theoretical chemical dynamics. This review has focused principally on these new experimental results, and the detailed comparison with theory that they have made possible. Technical and methodological improvements in experimental dynamics will most certainly continue, and the 'ideal' chemical dynamics experiment – a complete measurement of a fully quantum-state-resolved differential cross-section with quantum-state-prepared reactants – may soon be realized for the hydrogen exchange reaction. In addition, exploration of dynamical resonances and direct optical spectroscopy of

[140] H. R. Mayne, R. A. Poirier, and J. C. Polanyi, *J. Chem. Phys.*, 1984, **80**, 4025.
[141] H. R. Mayne, J. C. Polanyi, N. Sathyamurthy, and S. Raynor, *J. Phys. Chem.*, 1984, **88**, 4064.
[142] V. Engel, Z. Bacic, R. Schinke, and M. Shapiro, *J. Chem. Phys.*, 1985, **82**, 4844.
[143] V. Engel and R. Schinke, *Chem. Phys. Lett.*, 1985, **122**, 103.
[144] P. M. Agrawal, V. Mohan, and N. Sathyamurthy, *Chem. Phys. Lett.*, 1985, **114**, 343.
[145] T. Seideman and M. Shapiro, *J. Chem. Phys.*, 1988, **88**, 5525.
[146] B. A. Collings, J. C. Polanyi, M. A. Smith, A. Stolow, and A. W. Tarr, *Phys. Rev. Lett.*, 1987, **59**, 2551.

H_3 will open to view the transition state itself, the most fundamental construct in theory of chemical reactions.

Experimental information of this detail is most meaningful when comparison with the predictions of *ab initio* theory is possible. In fact, some of the experimental observations, for example resonances and transition-state spectroscopy, will be comprehensible only in the framework of a theoretical analysis of unprecedented detail. Almost certainly, such interpretation of the new experimental results will not suffer for lack of the necessary theoretical work. In parallel with the experimental developments we can expect the realization of the 'ideal' theoretical calculation – fully converged coupled-channel computation of a quantum-state-resolved differential cross-section as a function of collision energy and reactant quantum state at arbitrary energy. While every bit as difficult as the ideal experiments, the methodological developments in coupled-channel calculations that have been reported recently by several groups[120,124,137,138,147–152] put this ultimate *ab initio* calculation within reach.

Acknowledgement. Support for this research is provided by the US National Science Foundation.

[147] A. Kuppermann and P. G. Hipes, *J. Chem. Phys.*, 1986, **84**, 5962.
[148] G. A. Parker, R. T. Pack, B. J. Archer, and R. B. Walker, *Chem. Phys. Lett.*, 1987, **137**, 564.
[149] M. Baer, *J. Phys. Chem.*, 1987, **91**, 5846.
[150] L. Wolniewicz and J. Hinze, *J. Chem. Phys.*, 1986, **85**, 2012.
[151] M. Mishra, J. Linderberg, and Y. Ohrn, *Chem. Phys. Lett.*, 1984, **111**, 439.
[152] J. Linderberg and B. Vessal, *Int. J. Quantum Chem.*, 1987, **31**, 65.

CHAPTER 2

Vibrational Excitation in Bimolecular Reactions

I. W. M. SMITH

1 Introduction

In the mid 1950s, Norrish and his co-workers[1] were able to record spectra of vibrationally excited O_2 produced when the reactions of $O(^3P)$ atoms with NO_2 and ClO_2 were initiated by flash photolysis. Shortly afterwards, Cashion and Polanyi[2] recorded the first spectra of infrared chemiluminescence from vibrationally excited products of a bimolecular reaction. These were the original, pioneering experiments in state-to-state kinetics, a subject which forms an important part of the wider field of molecular reaction dynamics.[3,4]

Investigations involving vibrational excitation have remained central to the study and understanding of the dynamics of bimolecular reactions. There are two sides to this important role: first, there is the observation, characterization, and analysis of the vibrational product-state distributions from exothermic bimolecular reactions; second, there is the investigation of what influence specific reagent vibrational excitations exert on chemical reaction rates. As the examples already quoted indicate, the study of product-state distributions has by now quite a long history and has been reviewed many times.[5-13] The effects of reagent

[1] F. J. Lipscomb, R. G. W. Norrish, and B. A. Thrush, *Proc. R. Soc. London Ser., A*, 1956 **233,** 455.
[2] J. K. Cashion and J. C. Polanyi, *Proc. R. Soc. London Ser., A*, 1960, **258,** 529 and 570; *J. Chem. Phys.*, 1959, **30,** 317 and 1961, **35,** 600.
[3] I. W. M. Smith, 'Kinetics and Dynamics of Elementary Gas Reactions', Butterworths, London, 1980.
[4] R. D. Levine and R. B. Bernstein, 'Molecular Reaction Dynamics and Chemical Reactivity', Oxford University Press, Oxford, 1987.
[5] J. C. Polanyi, *Acc. Chem. Res.*, 1972, **5,** 161.
[6] I. W. M. Smith, *Adv. Chem. Phys.*, 1975, **28,** 1.
[7] M. J. Berry, in 'Molecular Energy Transfer', eds. R. D. Levine and J. Jortner, Wiley, New York, 1976, p. 114.
[8] J. D. McDonald, *Annu. Rev. Phys. Chem.*, 1979, **30**, 29.

vibrational excitations have, on the whole, been studied more recently.[14–19] Unfortunately, there are still very few successful experiments at the truly state-to-state level. Such experiments would provide a matrix of (vibrational states of reagents)-to-(vibrational states of products) rate coefficients rather akin to the Deslandres tables of band origins reported by molecular spectroscopists.[20] The present article attempts to construct a theoretical framework within which both kinds of vibrational effect, reagent and product, can be understood, but it lays slightly greater emphasis on the effects of exciting specific *reagent* vibrations.

The early spectroscopic experiments on product energy distributions required new theoretical concepts and methods if they were to be understood. Fortunately, help, in the shape of large high-speed computers, was at hand. In the 1960s, Bunker, Karplus, and Polanyi (with the aid of their co-workers) founded the quasi-classical trajectory method[21] which, by solving the classical equations of motion for representative samples of trajectories, made it possible to relate aspects of reactive collision dynamics (including the partitioning of energy between product degrees of freedom) to features of the potential energy surface and the kinematics for a reactive system.[5,22]

Several important generalizations were made by Polanyi.[5] Although these were based on the results of large, statistically meaningful samples of trajectories calculated in three dimensions,[22] it is possible to rationalize the conclusions by visualizing the motion of a single mass point across a potential energy surface for a collinear collision that has been properly scaled and skewed.[23] A surface of this kind is shown in Figure

9 B. E. Holmes and D. W. Setser, in 'Physical Chemistry of Fast Reactions. Vol. 2: Reaction Dynamics', ed. I. W. M. Smith, Plenum, New York, 1980, p. 83.
10 A. Ben-Shaul, Y. Haas, K. L. Kompa, and R. D. Levine, 'Lasers and Chemical Change', Springer-Verlag, New York, 1981.
11 J. C. Whitehead, in 'Chemical Kinetics', eds. C. H. Bamford and C. F. H. Tipper, Elsevier, Amsterdam, 1983, Vol. 24.
12 S. R. Leone, *Annu. Rev. Phys. Chem.*, 1984, **35,** 109.
13 C. E. Hamilton and S. R. Leone, in 'Gas-Phase Chemiluminescence and Chemi-Ionisation', ed. A. Fontijn, North-Holland, Amsterdam, 1985.
14 I. W. M. Smith, in 'Gas Kinetics and Energy Transfer', eds. P. G. Ashmore and R. J. Donovan, The Chemical Society, London, 1976, Vol. 2, p. 1.
15 S. H. Bauer, *Chem. Rev.*, 1978, **78,** 147.
16 C. B. Moore and I. W. M. Smith, *Faraday Discuss. Chem. Soc.*, 1979, **67,** 146.
17 M. Kneba and J. Wolfrum, *Annu. Rev. Phys. Chem.*, 1980, **31,** 47.
18 I. W. M. Smith, in 'Physical Chemistry of Fast Reactions. Vol. 2: Reaction Dynamics, ed. I. W. M. Smith, Plenum, New York, 1980, p. 1.
19 J. Wolfrum, *J. Phys. Chem.*, 1986, **90,** 375.
20 G. Herzberg, 'Molecular Spectra and Molecular Structure. I. Spectra of Diatomic Molecules', Van Nostrand, Princeton, 1950, p. 40.
21 L. M. Raff and D. L. Thompson, in 'Theory of Chemical Reaction Dynamics', ed. M. Baer, CRC Press, Boca Raton, 1985, Vol. 1, Chap. 1.
22 P. J. Kuntz, E. M. Nemeth, J. C. Polanyi, S. D. Rosner, and C. E. Young, *J. Chem. Phys.*, 1966, **44,** 1168.
23 Ref. 3, p. 99.

1a. The minimum energy path (MEP)* for reaction, *i.e.* leading from asymptotic A + BC to AB + C, is shown as a dashed line, and the maximum along this path, the potential energy barrier, is marked with a cross. The position of the barrier shown in Figure 1a is consistent with

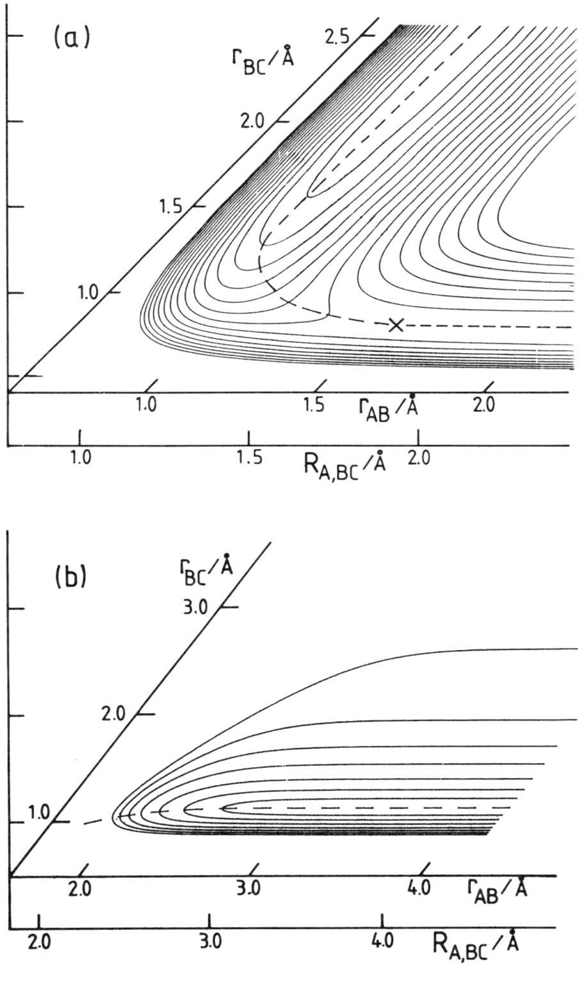

Figure 1 *Comparison of mass-scaled potential energy surfaces for collinear* A + BC *collisions where chemical reaction is and is not possible. Surface* (a) *is the Muckerman* 5 *surface*[72] *for the reaction* $F + H_2 \rightarrow HF + H$, *with the cross showing the position of the saddle-point. Surface* (b) *represents collision between* Ar *and* N_2 *with exponential repulsion between the* Ar—N *diatomic pairs*

*The term 'minimum energy path' is preferred in this article to 'reaction path', which is often used to mean the same thing, since realistic reactive trajectories will never actually follow the MEP from reagents to products.

Polanyi's first proposition: namely, that for exothermic reactions (as between A and BC) the barrier is 'early' whereas for endothermic reactions it is 'late'. For the 'early' barrier the BC bond, which is broken in the reaction, is only slightly stretched from its equilibrium length in the isolated BC molecule, while the A—B distance is much greater than its eventual equilibrium value in the AB product. The differences, Δr_{AB}^{sp} and Δr_{BC}^{sp}, between the inter-nuclear separations at the saddlepoint, r_{AB}^{sp} and r_{BC}^{sp}, and the equilibrium bond lengths, $r_{AB,e}$ and $r_{BC,e}$, provide a good means of characterizing the location of the saddlepoint along the MEP. The barrier locations predicted by the early/late rule are borne out by *ab initio* calculations of potential energy surfaces which have themselves improved greatly in scope and accuracy as computers have developed over the past 20 years.[24]

Polanyi's other generalizations were concerned with how (*a*) the yield of vibrationally excited products of bimolecular reactions and (*b*) the utilization of reagent vibrational energy are related to the position of the potential energy barrier for a direct, bimolecular reaction and the form of the profile of potential energy along the MEP. In general, the earlier the barrier, the greater the release of energy as the reagents approach 'attractively' and, as long as the reaction remains 'direct' without multiple collisions between the newly formed species, the greater the fraction of the reaction energy that is channelled into the vibration of AB. Kuntz *et al.*[22] showed that, for various mass combinations undergoing reactive collisions on semi-empirical, London–Eyring–Polanyi–Sato (LEPS) potential energy surfaces, there is a fair correlation between $\langle f_{v'} \rangle$, the fraction of the available energy which, on average, appears as product vibration, and the fraction of the energy that is released before typical trajectories first cross the line $r_{AB} = r_{AB,e}$ (*i.e.* before the AB separation first reaches its equilibrium value in AB).

The second general conclusion to emerge from the early quasi-classical trajectory studies of Polanyi's group concerned the utilization of reagent energy, or what Levine and Bernstein[4] term the selectivity of energy consumption. In a series of studies of model systems,[25] including reactions of the type $AB + CD \rightarrow AC + BD$, as well as of $A + BC \rightarrow AB + C$, it was shown that, for *early* barriers, energy in relative translational motion will generally be more effective in promoting reaction than vibrational excitation of the reagents. Conversely, when the potential barrier is *late*, excitation of the bond vibration or vibrations which are broken in the reaction will increase the reaction rate most.

Since the generalizations about vibrational effects were made by Polanyi,[3] a clearer understanding has been obtained with respect to

[24] T. H. Dunning, jun., L. B. Harding, R. A. Bair, R. A. Eades, and R. L. Shepard, *J. Phys. Chem.*, 1986, **90,** 344.

[25] (*a*) J. C. Polanyi and W. H. Wong, *J. Chem. Phys.*, 1969, **51,** 1439; (*b*) M. H. Mok and J. C. Polanyi, *ibid.*, 1970, **53,** 4588.

which features of the potential energy hypersurface for a reaction control, and are revealed by, particular dynamical effects. For example, the dependence of reaction rates on initial rotational state and the distribution of products over final rotational states provides evidence about the angular dependence of the reactive potential. This is perhaps best illustrated by considering the case of an A + BC reaction with a narrow 'cone-of-acceptance', *i.e.* a reaction that is very tightly constrained to collinearity as the system passes through the transition-state region. In these circumstances, the greater the rotational momentum of BC the more difficult it will be to maintain a favourable orientation as the trajectory traverses the dynamical bottleneck, so the rate coefficients for reaction will decrease as the rotational state of BC is increased. Furthermore, because low initial rotational and orbital angular momenta are favoured and the system will be nearly collinear as repulsive forces push the AB + C products apart, AB will on average be created with little rotational momentum.

In contrast, effects of vibrational excitation reveal the nature of the motion and the lowest energy path across the kind of potential energy surface (as distinct from hypersurface) shown in Figure 1a. The yield of vibrational energy in the reaction product can therefore be related, at least in classical terms, to the motion of a representative mass point in the region where the MEP is strongly curved. In the asymptotic limits, motion along the MEP corresponds to relative translation of the reagents and of the products. If the direction of an A + BC trajectory started directly along the MEP were to be changed sufficiently slowly, it could cling to the MEP and any energy released would appear entirely as relative motion of AB + C. However, this is not the case in real reactive collisions.

After a system crosses an early barrier and attractive forces set in, the trajectory leaves the MEP (the 'bobsled effect') and enters the exit valley on the surface with some considerable energy in motion perpendicular to the direction of that valley and of the MEP. In many cases, this energy is approximately maintained as the products separate and it therefore appears as vibrational energy in AB.

The selective effects of reagent energy can be visualized in a similar manner. At an early barrier, the direction of the MEP is almost parallel to the horizontal axis on the scaled and skewed potential energy surface and to the initial direction of an A + BC (collinear) trajectory in the absence of vibrational excitation. Consequently, there is an excellent correlation between the motion required to carry the system over the potential energy barrier and the initial, relative translational motion of A + BC. By contrast, vibrational motion in BC is orthogonal to the initial direction of the reaction path and approximately orthogonal to the crucial motion at the barrier. Arguments relating to the selective effects of energy on reaction in the endothermic direction are a little more complicated, but not greatly so. Now high translational energy will force the system to the head of the AB + C valley but will not help the trajectory turn the corner on the potential energy surface so as to approach the late barrier. Vibrational energy in AB can do this, and causes motion which is more nearly parallel to that required to carry the system over the barrier in the endothermic direction. However, it is important to recognize that AB + C collisions may also require some translational energy if the system is to reach the strong interaction region where the vibrational excitation can become effective. This point will be considered in greater detail and in a slightly different context later.

The arguments which have been presented in the last few paragraphs are not only oversimplified but also rest on the implicit assumption that nuclear motions can be considered adiabatic over certain sections of a given trajectory but that this assumption breaks down in the strong interaction zone where, for example, translational and vibrational motions (as far as this distinction remains) are strongly coupled. Theoretical treatments based on the notion of vibrational adiabaticity (and its breakdown) are amongst the most illuminating of recent advances in treating vibrational effects in bimolecular reactions, and they are given a strong emphasis in this article. In the remainder of this introduction, the concept of vibrational adiabaticity is discussed in general terms.

The Born–Oppenheimer approximation, which encapsulates the concept of *electronic* adiabaticity, is crucial to theories of chemical reactions. Because of it, one can frequently (though, of course, not always) discuss reactive collisions in terms of the motions of the nuclear particles over a single potential energy hypersurface which represents how the electronic energy depends on the nuclear geometry. The Born–Oppenheimer separation is justified because the period characteristic of electronic motions is much shorter than those associated with the more sluggish motions of the nuclei.

A similar separation is, of course, often made for different nuclear motions: for example, rotations and vibrations in isolated molecules are treated separately, and for similar reasons. In collisions, one must also consider relative translation. If this can be viewed as simply exerting a time-dependent force on the internal motions of one of the collision partners, one can use arguments based on the separation of time-scales to make qualitative predictions about the likelihood of inelastic collisions. Where the force results from an intermolecular motion which is intrinsically much slower than the internal motion, the amount of energy transferred, or in quantum mechanics the probability of a transition between states, is small.

Before we transfer our attention to reactive collisions, it is useful to consider inelastic collisions in a little more depth. Figure 1b shows a potential energy surface representing the interaction in a collinear collision of N_2 with Ar. It is assumed that atom-pair repulsive forces between the Ar atom and each of the N atoms dominate the intermolecular interaction. Consequently, approach of the Ar atom to the N_2 molecule tends to compress the N_2 bond, as is indicated by the slight curvature of the MEP towards the r_{Ar,N_2} axis as r_{Ar,N_2} is decreased. Examination of Figure 1b indicates how slight this curvature is until high energies are reached. The extent of curvature and its effect on the N_2 vibrational frequency can be analysed by calculating values of r_{N_2} along the MEP and of the frequency of the vibration orthogonal to the MEP. At an energy of 6.0 kcal mol^{-1}, the N—N distance on the MEP has been reduced by only 5×10^{-3} Å and the vibrational frequency raised by an amount equivalent to 58 cm^{-1}.

In simple classical terms, for energy to be transferred between the vibrational and translational motions it is necessary that the incoming and outgoing parts of the trajectory are different. For a classical collision between an atom and a non-vibrating diatomic molecule, the trajectory will start along the MEP and can only deviate significantly from it if it reaches the high-energy regions of the surface where the MEP curves appreciably and the bobsled or inertial effect can play a role.

In practice, the very low probabilities of vibrational energy transfer in collisions between noble gas atoms and diatomic molecules at low energies (and temperatures), and the increase in these probabilities at higher energies (and temperatures), confirm the essential correctness of this very simplified picture. The elements of this model are also used in quantum mechanical, distorted wave treatments of vibrational energy transfer in collinear collisions.[26,27] In the simplest such treatments, the vibrational and translational motions are separated and parallel potential energy curves are assumed to represent individual vibronic states. The probability of transfer between two levels is then related to the product of the squares of two matrix elements. The first involves the wavefunctions of the two vibrational states and that part of the intermolecular potential which depends on the vibrational co-ordinate. The second matrix element is related to the probability of the system tunnelling from one vibronic curve to the other. The analogy with the treatment of non-adiabatic electronic transitions is clear. In terms of distorted-wave theory, the probability of non-adiabatic transitions is increased by strong coupling between the translational and vibrational motions, which enhances both the value of the vibrational matrix element and the overlap of the wavefunction representing translational motion on the vibronic potential energy curves.

The concepts that have been used to describe collisions between species which do not interact chemically can be carried over to examine the concept of vibrational adiabaticity in reactive collisions. The potential energy surface in Figure 1a, and particularly the MEP, differ in two

[26] J. T. Yardley, 'Introduction to Molecular Energy Transfer', Academic Press, New York, 1980.
[27] D. Rapp and T. Kassal, *Chem. Rev.*, 1969, **69,** 61.

main respects from those in Figure 1b. First, the initial curvatures in the MEPs are in an opposite sense: in Figure 1b the approach of A tends to compress the B—C bond, whereas in the case of chemical interaction the B—C bond length starts to increase and the frequency of the periodic motion orthogonal to the reaction path is lowered. Second, in the reactive case represented by the typical potential energy surface in Figure 1a, there is a region where the MEP curves strongly between two other regions where it is essentially parallel to one or other of the two axes of the diagram. Analogy with the analysis of potentially inelastic, non-reactive collisions suggests that it is likely to be only in the intermediate strong-interaction zone that there are appreciable non-adiabatic effects. It is this idea which lies at the heart of vibrationally adiabatic transition-state theory (VA-TST),[28,29] as it is applied to the reactions of selectively excited species.

Because VA-TST is a central theme in this chapter, a brief summary will be given of its main features at this point. The next section will return to the theory and attempt to review its assumptions and accomplishments in more detail, along with those of other theoretical methods that have been used to explore the effects of reagent vibrational excitation on chemical reactivity. The remainder of the chapter is devoted to the examination of a number of test cases or case histories. These are chosen on two criteria: (*a*) the level of detailed information available about the effects of vibrational excitation, and (*b*) to provide a range of reactions of different type and complexity.

The fundamental basis of VA-TST is that potential curves can be constructed which correspond to vibrationally adiabatic MEPs leading from reagents to products. The energies along these adiabatic channels can be represented by the equation:

$$V_{ad}(v,x) = V_{el}(x) + E_{vib}(v,x) \tag{1}$$

where $V_{el}(x)$ is the electronic potential energy and is, of course, independent of the adiabatic state but depends on the distance x along the MEP. $E_{vib}(v,x)$ is the energy associated with motion or motions orthogonal to the reaction path. If, for the moment, we consider only a collinear A + BC reaction, $E_{vib}(v,x)$ is the energy of the single vibrational motion orthogonal to x and correlates in the two asymptotic limits to the energies of BC(v) and AB($v' = v$). In general, the frequency of this motion and hence $E_{vib}(v,x)$ decreases but $V_{el}(x)$ increases as the saddlepoint is approached. Moreover, as v increases the importance of $E_{vib}(v,x)$ relative to $V_{el}(x)$ increases and as a result the maximum in $V_{ad}(v,x)$ moves further away from the potential energy barrier to an

[28] E. Pollak, in 'Theory of Chemical Reaction Dynamics', ed. M. Baer, CRC Press, Boca Raton, 1985, Vol. 3, Chap. 2.

[29] D. G. Truhlar, A. D. Isaacson, and B. C. Garrett, in 'Theory of Chemical Reaction Dynamics', ed. M. Baer, CRC Press, Boca Raton, 1985, Vol. 4, Chap. 2.

earlier position on the MEP, and becomes lower relative to the reagent asymptotic limit.

The vibrationally adiabatic analysis can then be used to estimate state-selected reaction rate constants using TST at any of several levels of sophistication. In the simplest version of TST the transition-state is placed at the adiabatic maximum or saddlepoint and the rate constant is evaluated via the customary TST expression:

$$k_v(T) = (k_B T/h)\left(Q_v^{\ddagger}/\prod_{\text{reag}} Q_v\right)\exp[-V_{\text{ad}}^{\max}(v)/k_B T] \tag{2}$$

where the Q are 'per volume' partition functions for the state-selected transition state and reagents, and are usually evaluated within the simple harmonic oscillator–rigid rotor approximation, and $V_{\text{ad}}^{\max}(v)$ is the energy of the vibrationally adiabatic maximum relative to its reagent asymptotic value for the same adiabatic channel. Even at this simplest level, it is apparent that both the activation energy and the Arrhenius A factor can be altered by vibrational excitation: $V_{\text{ad}}^{\max}(v)$ is clearly a function of v and so is $Q_v^{\ddagger}$, and the configuration at the adiabatic maximum is different for different v.

Three improvements can be made to the basic theory. First, $Q_v^{\ddagger}$ can be evaluated more carefully taking due account of anharmonicities, including quartic terms in the bending potential, and mode–mode and rotation–vibration couplings.[30] Second, one can adopt the variational approach. This treatment is based on the notion that generalized transition states exist at all points along the MEP. Each yields an estimate of the rate constant via the usual TST expression and the lowest value is the best estimate: it will correspond to placing the transition state where the free energy along the MEP reaches a maximum.[29,31] To apply the variational method, it is necessary to perform a mode analysis at points along the MEP and use the results to estimate a rate constant for a particular state-selected channel, according to equation (2). Neglecting the tunnelling correction, variational transition-state theory should provide an upper bound to the true rate constant. It is an upper bound because there is always the possibility that trajectories that have passed through the critical dividing surface will be reflected in the strong interaction zone and pass back through the transition-state region, hence yielding reagents. This phenomenon of 'recrossing' becomes generally more likely for high energies and therefore in the reactions of species that are vibrationally excited.

The third improvement that can be made to simplistic TST estimates of $k_v(T)$ is to allow for non-quantum effects along the reaction

[30] (*a*) Ref. 29, pp. 87—92; (*b*) B. C. Garrett and D. G. Truhlar, *J. Phys. Chem.*, 1979, **83,** 1915; (*c*) B. C. Garrett and D. G. Truhlar, *J. Chem. Phys.*, 1984, **81,** 309.

[31] (*a*) D. G. Truhlar and B. C. Garrett, *Acc. Chem. Res.*, 1980, **13,** 440; (*b*) D. G. Truhlar and B. C. Garrett, *Annu. Rev. Phys. Chem.*, 1984, **35,** 159.

co-ordinate: what is usually referred to as quantum mechanical tunnelling. Garrett and Truhlar and their co-workers[32] have been responsible for developing and testing several methods of incorporating tunnelling corrections into transition-state theory, and some of these will be mentioned in the next section.

Two related limitations of VA-TST should be mentioned at this stage. First, as it is not a truly dynamical theory it provides no information at the state-to-state level, only at the state-selected level. Second, there is the question of whether the assumption of vibrational adiabaticity is valid. Evidence for this will be examined in the next section. In general terms, the theory will be most soundly based when the adiabatic maximum is displaced well into the entrance valley on the potential energy surface and hence before there is appreciable curvature in the MEP. On the other hand, in these circumstances one might have to be especially careful in evaluating $Q_v^{\ddagger}$, since the force constant for bending will be small and there will be particularly strong coupling between bending and rotation. In some cases, there will be more than one vibrationally adiabatic maximum. To estimate the state-selected rate one should select the maximum (or its variational equivalent) closest to the reagent side of the potential energy surface. The region between adiabatic maxima is likely to be that of particularly strong coupling between the internal modes of motion and that along the reaction co-ordinate.

2 Aspects of Theory

Microscopic Reversibility[3,18] *and the Information Theoretic Approach*[4,18,33]

Because of the principle of microscopic reversibility, there exist relationships between the rate constants and cross-sections which describe elementary reactions occurring in opposite directions. For example, for the elementary reaction:

$$\mathrm{A} + \mathrm{BC}(v,j) \rightarrow \mathrm{AB}(v',j') + \mathrm{C} \tag{3}$$

the rate constants describing reaction at the rovibrational state-to-rovibrational state level are related by the equation:

$$\frac{k(v',j'|v,j)}{k(v,j|v',j')} = \frac{(2j'+1)}{(2j+1)}\left(\frac{\mu'}{\mu}\right)^{3/2}\exp(-\Delta E_{\mathrm{int}}/k_{\mathrm{B}}T) \tag{4}$$

[32] (*a*) Ref. 29, pp. 92—110; (*b*) D. G. Truhlar, A. D. Isaacson, R. T. Skodje and B. C. Garrett, *J. Phys. Chem.*, 1982, **86,** 2252; (*c*) B. C. Garrett, N. Abusalbi, D. J. Kouri, and D. G. Truhlar, *J. Chem. Phys.*, 1985, **83,** 2252.

[33] R. B. Bernstein and R. D. Levine, in 'Modern Theoretical Chemistry. Dynamics of Molecular Collisions, Part B', ed. W. H. Miller, Plenum Press, New York, 1975, Vol. 2, Chap. 7.

where ΔE_{int} is the energy difference between the v', j' and v, j states, and μ and μ' are the collisional reduced masses for A + BC and AB + C. If one considers the rotational distributions of reagents to be thermalized and those of products to be undetermined, then one can take a weighted sum of j and a full sum of j' and derive:

$$\frac{k(v'|v)}{k(v|v')} = \frac{Q_{v',\text{rot}}}{Q_{v,\text{rot}}}\left(\frac{\mu'}{\mu}\right)^{3/2} \exp(-\Delta E_{\text{int}}/k_{\text{B}}T) \tag{5}$$

where ΔE_{int} is now the energy difference between the v', $j' = 0$ and v, $j = 0$ states, and $Q_{v',\text{rot}}$ and $Q_{v,\text{rot}}$ are rotational partition functions for AB(v') and BC(v), respectively.

Equations (4) and (5) allow one to determine state-to-state rate constants for reaction in one direction if the corresponding rate constant for reaction in the reverse direction is known. This procedure has chiefly been applied[10,18] to the results of infrared chemiluminescence experiments and quasi-classical trajectory calculations on energy disposal in exothermic reactions to derive information about the rate constants, and how they vary with vibrational state, for reaction in the reverse endothermic direction. Since the data in the exothermic direction are usually obtained from thermalized reagents, with the great majority of BC reagents in $v = 0$, they yield absolute or relative values of $k(v'|v = 0)$ and application of equation (5) yields $k(v = 0|v')$. These rate constants for endothermic reactions are usually much too small to measure directly.

In general, the values of $k(v = 0|v')$ rise steeply with v' but, to see what dynamic information is revealed by this increase, it is useful to consider the ratio of rate constants for reaction from two neighbouring levels $(v' + 1)$ and v':

$$\begin{aligned}\frac{k(v = 0|v' + 1)}{k(v = 0|v')} &= \frac{k(v' + 1|v = 0)}{k(v'|v = 0)}\frac{Q_{v'+1,\text{rot}}}{Q_{v',\text{rot}}}\exp(\Delta E_{v'+1,v'}/k_{\text{B}}T) \\ &\simeq \frac{k(v' + 1|v = 0)}{k(v'|v = 0)}\exp(\Delta E_{v'+1,v'}/k_{\text{B}}T)\end{aligned} \tag{6}$$

where $\Delta E_{v'+1,v'}$ is the energy difference between levels $(v' + 1)$ and v'. This expression demonstrates that the sharp increase in rate with vibrational quantum number in the endothermic regime is largely a reflection of the essentially uninteresting increase in $E_{v'}$. Any specific effects are reflected in the ratio of rate constants for reaction in the reverse, exothermic direction.

A related but different viewpoint on vibrational excitation effects in bimolecular reactions is provided by the information theoretic or surprisal analysis introduced by Bernstein and Levine.[4,33] In this treatment the state-specific rate constants $k(v'|v = 0)$, or those for reaction from selected states $k(v = 0|v')$, are compared with values that might be expected on prior, 'least-biased', grounds. Usually the distribution of prior rate constants, $k°(v'|v = 0)$ and $k°(v = 0|v')$, is calculated on the 'democratic' basis that all energetically accessible states are equally likely: angular momentum constraints are normally not included.

For many exothermic reactions of the A + BC type – that between $O(^3P)$ and CS,[34]

$$O(^3P) + CS \rightarrow CO(v') + S \quad (\Delta H_0^\circ = -85\,\text{kcal mol}^{-1}) \tag{7}$$

is an oft-quoted example[18,33,35] – the surprisal defined by

$$I(v'|v = 0) = -\ln[k(v'|v = 0)/k°(v'|v = 0)] \tag{8}$$

varies linearly with $f_{v'}$, the energy of the v' level expressed as a fraction of the average total energy available. The whole vibrational distribution is then characterized by a single, temperature-like, parameter $\lambda_{v'}$:

$$k(v'|v = 0) = k°(v'|v = 0)\exp(-\lambda_{v'}f_{v'})/\exp(\lambda°_{v'}) \tag{9}$$

Moreover, because the prior rates as well as the actual rates must conform to equation (5), the surprisal is independent of the direction of reaction with the result that one can also derive[18] the equation:

$$k'(v = 0|v') = k'°(v = 0|v')\exp(-\lambda_{v'}^\circ - \Lambda_{v'}E_{v'}/k_\text{B}T) \tag{10}$$

where $\Lambda_{v'} = \lambda_{v'}(k_\text{B}T/E_\text{total})$. Then, if equation (5) is used to relate $k'°(v = 0|v')$ to $k°(v'|v = 0)$, equation (10) becomes:

$$k'(v = 0|v') = \left(\frac{\mu}{\mu'}\right)^{3/2}\left(\frac{Q_{v,\text{rot}}}{Q_{v',\text{rot}}}\right)k°(v'|v = 0)\exp(-\lambda_{v'}^\circ - E_0/k_\text{B}T)$$
$$\times \exp[E_{v'}(1 - \Lambda_{v'})/k_\text{B}T] \tag{11}$$

Most, but not all, of the variation of $k'(v = 0|v')$ with v' evolves from the last term on the right-hand side of this equation. $\Lambda_{v'}$ is usually negative and much smaller than 1, since the average energy available, E_total, is normally much greater than $k_\text{B}T$. Consequently, in the strongly endothermic region a plot of $\ln k'(v = 0|v')$ against $E_{v'}/k_\text{B}T$ is linear, with a slope only slightly different from unity. This conclusion re-emphasizes the fact that it is easier to determine information about state-selected kinetics – and indeed about state-to-state kinetics – by

[34] (*a*) G. Hancock, C. Morley, and I. W. M. Smith, *Chem. Phys. Lett.*, 1971, **12,** 193; (*b*) G. Hancock, B. A. Ridley, and I. W. M. Smith, *J. Chem. Soc., Faraday Trans.* 2, 1972, **68,** 2117.

[35] A. Ben-Shaul, *Chem. Phys.*, 1973, **1,** 244.

observing the state distributions arising from exoergic reactions.

The foregoing discussion may suggest that experiments on the kinetics of state-selected species are of little value. This is not so, because the methods outlined above have their limitations, arising from the fact that experiments on energy disposal are usually performed with thermally equilibrated reagents – which essentially means that reagents are in their ground vibrational state. Absolute or relative values of $k(v'|v = 0)$, and hence of $k'(v = 0|v')$ can only be determined for product levels which are accessible energetically; that is, for which $E_{v'} < E_{v=0}$. To discover the valuable information about the kinetics of species with higher internal energies requires experiments involving direct excitation of the reagents.

Vibrationally Adiabatic Transition-state Theory and Other Treatments of Collinear Reactions

Even if truly accurate potential energy surfaces were available, accurate quantum scattering calculations remain impossible for any chemically reactive system comprising more than three atoms. Indeed, to date the only reports of converged quantal calculations of reaction cross-sections are those for $H + H_2(v = 0)$[36] and, very recently, $F + H_2(v = 0)$.[37] Quantal calculations on reactions involving excited reagents need to be larger because of the greater number of accessible states. It is these difficulties which lead to the wide interest in, and wide use of, approximate methods.

In contrast to the situation for fully dimensional systems, quantal calculations on collinear A + BC collisions have become fairly routine. The results provide an excellent testing ground for approximate theories. If the one-dimensional rate constants from an approximate theory do not satisfactorily match the quantal results, then the results of three-dimensional calculations using the same model must be viewed with some suspicion. On the other hand, agreement in the collinear world does not necessarily imply agreement in the real three-dimensional world. Many of the tests that have been carried out on VA-TST, chiefly by Garrett and Truhlar and their colleagues,[29] have been performed on reactions of species in their vibrational ground states. The tests on reactions from excited vibrational states across a vibrationally adiabatic potential are, in a sense, more stringent, as any non-adiabaticity could lead to significant reaction below the adiabatic threshold and hence to a much enhanced rate of reaction. In tests of models made on collinear systems the accuracy of the potential energy surface is not, of course, of great importance.

[36] (*a*) G. C. Schatz and A. Kuppermann, *J. Chem. Phys.*, 1976, **65**, 4642 and 4668; (*b*) M. C. Colton and G. C. Schatz, *Int. J. Chem. Kinet.*, 1986, **18,** 961: (c) J. Z. H. Zhang and W. H. Miller, *Chem. Phys. Lett.*, 1988, **153**, 465.

[37] J. Z. H. Zhang and W. H. Miller, *J. Chem. Phys.*, 1988, in press.

As indicated earlier, for a collinear reaction $E_{vib}(v,x)$ in equation (1) is the energy in the motion orthogonal to the MEP on the properly mass-scaled potential energy surface for reaction. The simple VA-TST expression for the one-dimensional, state-selected rate constant (usually in units of cm^3 $molecule^{-1}$ s^{-1}) is:

$$k_v(T) = (k_B T/h)(1/Q_{tr}) \exp[-V_{ad}^{max}(v)/k_B T] \quad (12)$$

where Q_{tr} is a partition function 'per unit length' associated with relative motion of the reagents in one dimension. Note that partition functions for the vibrations of BC and the transition state are not included since a state-selected, adiabatic process is under consideration. However, to evaluate $E_{vib}(v,x)$, and hence $V_{ad}(v,x)$ and $V_{ad}^{max}(v)$, it is necessary to analyse the motion orthogonal to x allowing for anharmonicity. Garrett and Truhlar have employed a Morse expression in many calculations of this kind,[30a] but recently they have found a significant improvement using the WKB approximation.[30c]

The second improvement to be made on the simplest TST approach is to adopt the generalized transition-state or variational approach.[31] A series of TST calculations are made for various points along the MEP to find the minimum calculated value of $k_v(T)$, or its equivalent the location of the maximum in free energy. In some systems, this procedure may lower the level of agreement with the quantal result compared with that found by adopting the simpler method. However, this is no reason for rejecting the undoubtedly correct variational technique. In any case, the agreement is often restored when a quantum tunnelling correction is applied.

Garrett and Truhlar have developed and tested several methods of allowing for quantum mechanical tunnelling or quantum effects associated with reaction co-ordinate motion.[29] Their preferred method appears to be one based on the least action principle.[30c] Sequences of parameterized paths leading from the entrance to the exit valley are tested to see which gives the least exponential damping in the tunnelling region. This path can deviate appreciably from the MEP, especially when the skew angle on the potential energy surface is small. Then the 'corner cutting' is severe. In general, there are vibrational as well as reaction co-ordinate contributions to the tunnelling motion.

Garrett and Truhlar have tested their TST calculations incorporating the corrections enumerated above against accurate quantal results for a wide variety of systems reacting in the vibrational ground state. For 13 systems tested,[30c] the average ratio of the accurate TST rate constant to the quantal rate constant ($\pm$ one standard deviation) was 1.03 ± 0.26 at 300 K and 1.04 ± 0.13 at 600 K. Calculations on collinear H,D,T + $F_2(v = 1)$[38] give excellent agreement between TST and quantal estimates of the rate constants.

[38] R. Steckler, D. G. Truhlar, and B. C. Garrett, *J. Chem. Phys.*, 1986, **84**, 6712.

The comparison of collinear quantal calculations and VA-TST in its variational form with tunnelling corrections suggests that the assumption of vibrational adiabaticity up to the transition state is valid, at least in many collinear systems. Of course, the transition-state theory estimate of the rate constant remains in principle an upper bound, as no allowance is made for 'recrossing', *i.e.* for trajectories which enter the strong interaction zone but turn back and yield reagents. Garrett *et al.*[39] have described an approximate method of allowing for recrossing using the unified statistical model of Miller.[40]

A different method of accounting for recrossing effects and of testing the fundamental assumption of VA-TST has been provided by Frost and Smith.[41,42] They have developed a method which combines TST and quasi-classical trajectories (QCT). Instead of sampling the reagent-phase space quasi-classically to provide the starting conditions for individual trajectories, they do this at the transition state, thereby allowing quantization to be imposed at the most crucial stage of reactive trajectories. QCTs are then run from each selected phase space point in the transition state both 'forwards in time' (*i.e.* towards products) and 'backwards in time' (*i.e.* towards reagents).

Frost and Smith[41] found that the best method to define the transition state for a collinear reaction between A and BC(v) was to find the first pods (periodic orbiting dividing surface) along the MEP for which the action over one cycle of the pods is equal to $(v + \frac{1}{2})h$. They tested their method on collinear systems colliding on three different potential energy surfaces and with a wide variety of relative masses. In only a few cases were the results obtained from the combined TST–QCT method significantly different from those obtained by standard QCT calculations. These cases were where the pods crossed the MEP after substantial curvature in the MEP. Except for these cases, the vibrational energies in BC(v) obtained by running trajectories back from the transition state clustered very closely indeed about the quantum energy of isolated BC(v), confirming that motion up to the transition state was strongly vibrationally adiabatic. In cases where the transition state was beyond the region of strong curvature, a suitable procedure for estimating the rate constant would be to find it from the rate constant for the reverse process, using the principle of detailed balance.

So far it has been implied that there is little difficulty over the choice of transition state, at least for collinear reactions. In practice this is not so because the transition state is not simply a point on the MEP but is a

[39] B. C. Garrett, D. G. Truhlar, R. S. Grev, A. W. Magnuson, and J. N. L. Connor, *J. Chem. Phys.*, 1980, **73,** 1721.

[40] (*a*) W. H. Miller, *J. Chem. Phys.*, 1976, **65**, 2216; (*b*) E. Pollak and P. Pechukas, *ibid.*, 1979, **70,** 325.

[41] R. J. Frost and I. W. M. Smith, *Chem. Phys.*, 1987, **111**, 389.

[42] (*a*) R. J. Frost and I. W. M. Smith, *Chem. Phys.*, 1987, **117,** 421; (*b*) R. J. Frost and I. W. M. Smith, *Chem. Phys. Lett.*, 1987, **140,**, 499.

line across the potential energy surface representing collinear interaction. As both Steckler *et al.*[38] and Frost and Smith[41] have pointed out, and as Figure 2 shows, if this line is straight and at right angles to the MEP on the mass-scaled surface, the turning point for motion along the line can frequently lie in the exit valley, even at quite modest energies. Pollak[28] has shown that a preferable choice is the trajectory of a pods whose action is $(v + \frac{1}{2})h$, and Frost and Smith[41] demonstrated that this was a better choice in their combined TST–QCT calculations. In some cases the pods will closely correspond to the variational transition state defined by a line orthogonal to the MEP, but this will not always be the case.

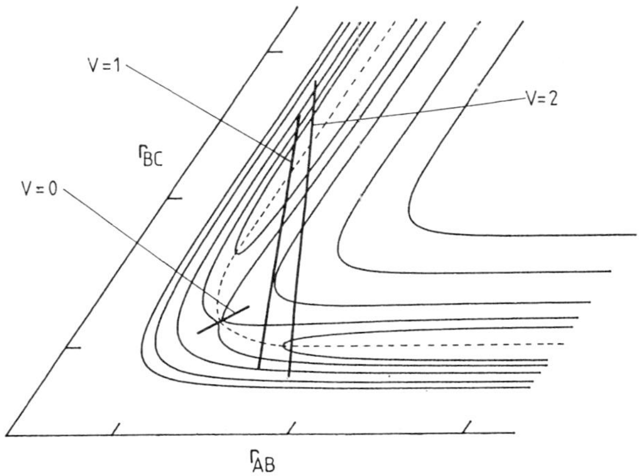

Figure 2 *Diagram illustrating the difficulty of defining vibrationally adiabatic transition states in certain systems. The mass-scaled LEPS surface represents collinear L + L L* (i.e. *each atom has a mass of one atomic mass unit*) *with a symmetrically placed potential barrier of* 10 kcal mol^{-1}. *The lines are orthogonal to the MEP and intersect the MEP at the location of the vibrationally adiabatic maxima. The limits of the lines are where* $E_{\text{vib}}(v, x) = V_{\text{el}}$. *Clearly, the lines cannot be used to define vibrationally adiabatic transition states for* $v = 1$ *and* $v = 2$
(Reproduced with permission from *Chem. Phys.*, 1987, **111,** 389)

The relationship between the assumption of vibrational adiabaticity and reaction path or MEP curvature has been examined by Truhlar and co-workers.[38,43] They have tested the result of assuming adiabaticity only as far as the first maximum in the plot of MEP curvature against reaction co-ordinate; it being assumed that all flux at this point is transferred to the lowest adiabatic route, so that reaction becomes

[43] (*a*) B. C. Garrett, D. G. Truhlar, J. M. Bowman, and A. F. Wagner, *J. Phys. Chem.*, 1986, **90,** 435; (*b*) B. C. Garrett and D. G. Truhlar, *Int. J. Quantum Chem.*, 1986, **29,** 1463.

possible. Although this model gives apparently better agreement with experiment in some cases, it is not clear that the assumption is justified. In other systems, the adiabatic maximum precedes the point of maximum curvature, so the two models give identical results.

Vibrationally Adiabatic Transition-State Theory and Other Treatments of Reactions in Three Dimensions

The requirement for good, if approximate, theoretical models of bimolecular reactions is especially urgent as one proceeds into the real world of three spatial dimensions, and also when one considers reactions involving more than three atoms. Any method, like those based on TST, which requires information about only limited regions of the potential energy surface is especially attractive. This aspect of TST meshes well with *ab initio* methods which incorporate 'gradient methods'. In these calculations, not just energies but also first and second derivatives of those energies are calculated for points on the potential energy surface. As a result, one can track the MEP across the surface and concentrate the calculations around the crucial regions along the MEP.

In addition to the time saved and accuracy gained in *ab initio* calculations if one can focus on a small part of the complete potential energy surface, VA-TST calculations are much less time-consuming than scattering calculations, and this gain increases with the complexity of the system. The difficulty of 3D quantal calculations has already been mentioned. There is no fundamental reason why QCT calculations cannot be extended to bimolecular reactions involving reagents of arbitrary complexity. However, the efficiency of these calculations is likely to fall. Not only will each individual trajectory take longer to compute but also a lower fraction of them will lead to an 'interesting' result; that is, to reaction.

Recognizing the limited efficiency of the standard QCT technique, Frost and Smith[42,44] have extended their combined TST-QCT calculations to reactions in three dimensions. Although they have only treated the three-atom reactions, $H + H_2(v \leqslant 9)$, $N + N_2(v \leqslant 15)$, and $F + H_2(v = 0)$, there seems no reason, in principle, why the general method should not be extended to much more complex reactions, although the description of a suitable transition state and a proper statistical sampling over it to find starting points for trajectories will be quite difficult. The calculations of Frost and Smith yield much dynamical information, *e.g.* about the partitioning of energy among product degrees of freedom, but they do not immediately yield absolute rate constants for the state-to-state processes. For this they needed to adopt

[44] (*a*) R. J. Frost and I. W. M. Smith, *J. Chem. Soc., Faraday Trans.* 2, 1988, **84**, 1825. (*b*) R. J. Frost and I. W. M. Smith, *ibid.*, 1988, in press.

a VA-TST approach, and the method chosen was strongly related to that used to define the 3D transition states.

To explain the difficulties encountered, it is useful to return to equation (1) and write it in a slightly different form explicitly allowing for contributions from i modes orthogonal to the MEP:

$$V_{\mathrm{ad}}(v_1, v_2, \ldots, x) = V_{\mathrm{el}}(x) + \sum E_{\mathrm{vib}}(v_i, x) \tag{13}$$

where $E_{\mathrm{vib}}(v_i,x)$ is the energy associated with the ith vibration in state v_i and at position x on the MEP. Thus, for a three-atom reaction proceeding along a collinear MEP, the last term in this equation would include a contribution from the doubly degenerate bending motion, as well as the bond-stretch mode which correlates with BC(v). VA-TST calculations based on this formulation of the MEP have been successfully carried out for Cl + $H_2(v)$,[38] O + $H_2(v)$,[43] and $OH(v_1)$ + $H_2(v_2)$.[45] These systems are all quite tightly constrained to a particular pathway, collinear in the case of the three-atom reactions.

The problems of defining three-dimensional vibrationally adiabatic transition states have been discussed in considerable detail by Pollak[28] and by Wyatt and Pollak.[46a] The problem arises because of the strong coupling between the bending and bond-stretching motions in typical transition states. Consequently, a suitable critical dividing surface in the full configuration space of the system is not provided by taking the r_{AB}, r_{BC} (or $r_{\mathrm{A,BC}}$, r_{BC}) distances most suitable for the collinear case and assuming that they remain the best choice for all angles.

Pollak and Wyatt[46a] have introduced an adiabatic periodic reduction sudden (ADRS) version of TST and this has been used by Frost and Smith[42,44] in their combined TST–QCT calculations, both to define the optimum transition states and also to make estimates of *absolute* rate constants. The ADRS–TST theory makes two main assumptions. The first concerns a separation of time-scales: namely that the characteristic translational time is much shorter than the period of the bending motion but much longer than the period of bond-stretching. This allows the overall reaction probability to be decomposed into γ-dependent elements, where γ is the angle between r_{BC} and the line joining A to the centre-of-mass of BC. The second assumption is that the reaction probability is unity if the system has sufficient energy of motion along the MEP to overcome the adiabatic barrier on the γ-dependent potential. Within this framework, the value of $V_{\mathrm{ad}}^{\mathrm{max}}(v,x)$ could be found for surfaces with a different γ by seeking the pods for which the action over one complete cycle equals $(v + \frac{1}{2})h$. A difficulty in applying the ADRS

[45] (*a*) A. D. Isaacson and D. G. Truhlar, *J. Chem. Phys.*, 1982, **76,** 1380; (*b*) D. G. Truhlar and A. D. Isaacson, *ibid.*, 1982, **77,** 3516.

[46] (*a*) E. Pollak and R. E. Wyatt, *J. Chem. Phys.*, 1983, **78,** 4464; (*b*) R. B. Walker and E. Pollak, *ibid.*, 1985, **83,** 2851.

method is to decide whether the estimate of the rate is for reagents in their lowest rotational state or is averaged over *j*. These alternative points of view are discussed elsewhere.[42,46b]

Pollak and Wyatt[46] applied their ADRS method, in a semi-classical adiabatic (SCAD) version, to estimate rate constants for the reactions of $H + H_2(v = 1)$ and $D + H_2(v = 1)$. The results of these calculations are compared with other theoretical predictions and with experimental results in the next section. For $H + H_2(v \leqslant 9)$ collisions in three dimensions, as well as for $N + N_2(v \leqslant 15)$ and $F + H_2(v = 0)$, Frost and Smith[42,44] have successfully used the ADRS scheme to choose transition states and estimate rate constants. Their ADRS–TST rate constants can, of course, be corrected for recrossing by counting the fraction of trajectories which, starting from the transition state towards products, are reflected and separate as reactants. These trajectories are of two types. The first class are reflected having reached the strong interaction zone. The second group are trajectories which are reflected because of the tightening of the transition state as the ABC system moves away from collinearity. As a result, trajectories initiated in near-collinear geometry can be reflected quite soon, and before the strong interaction zone is reached, as the system bends. These facts reflect the difficulty of finding true, optimized transition states for systems reacting in three dimensions.

3 Case Histories

Isotopic Variants of the Reaction $H + H_2(v) \rightarrow H_2(v') + H$

Because of its electronic simplicity, the reaction:

$$H + H_2(v) \rightarrow H_2(v') + H \tag{14}$$

and its isotopic variants have long been recognized as, at least in one sense, the simplest bimolecular reactions between neutral species and, as such, they occupy a special place in the hearts and minds of theoretically inclined kineticists.[47] The pioneering work on these reactions by Farkas and Farkas was celebrated in a special issue of the *International Journal of Chemical Kinetics*,[48] which demonstrated that theoretical interest in the reaction is still intense. Recent developments in the study of this prototypical reaction are reviewed in Chapter 1.

There have now been several very accurate *ab initio* calculations on the H_3 potential energy surface. Of these the one that has been used most in recent theoretical estimates of rate constants and cross-sections is the Truhlar–Horowitz[49] parameterization of the calculations by Liu

[47] D. G. Truhlar and R. E. Wyatt, *Annu. Rev. Phys. Chem.*, 1976, **27,** 1.
[48] Issue No. 9, *Int. J. Chem. Kinet.*, 1986, **18,** 919—1112.
[49] D. G. Truhlar and C. J. Horowitz, *J. Chem. Phys.*, 1978, **68,** 2468; 1979, **71,** 1514.

and Siegbahn:[50] the LSTH surface. It has a barrier height of 9.80 kcal mol^{-1}, a value which, according to later extensive computations[51] around the saddle point, is 0.15 kcal mol^{-1} too high. More recently, Varandas *et al.*[52] have reported calculations using a double many-body expansion (DMBE), which was chiefly designed to be accurate at high energies but also has the 'improved' barrier height of 9.65 kcal mol^{-1}.

Although the three-atom reactions involving only H and/or D atoms are a natural testing ground for theories of bimolecular reactions,[47] some of their characteristics make comparison with experiment difficult. First, there is the fact that quantum effects in the scattering dynamics[53] are especially large because of the light nuclear masses. Consequently, one has to be particularly cautious about the results of QCT calculations – although this has not prevented a large number of such studies[54] being carried out! On the other hand, the wide spacing of the internal energy levels reduces the number of internal states that need to be included in coupled-channel[36a] or coupled-states[36b] calculations on the $H + H_2$ reaction, allowing accurate quantum scattering calculations to be performed, although no such calculations have been reported for reactions involving vibrationally excited reagents.

The major difficulty in comparing detailed results of experiment and theory on vibrational effects in these reactions is, however, with the experiments. Thermal rate constants for four of the reactions were measured over a wide temperature range in the 1960s. The results of those experiments remain unchallenged and agree well with theory, including the results of 'full scale' TST calculations.[55] However, state-to-state or even state-selected rate constants have proved difficult to obtain. The lack of infrared transitions and of electronic absorptions, except in the vacuum ultraviolet, has made the use of spectroscopic techniques for state-specific excitation and/or detection of H_2 and its isotopomers very difficult. CARS (coherent anti-Stokes Raman scattering) spectroscopy has been used[56] to observe the products from (*a*) and $H + D_2$ reaction at enhanced collision energies, and (*b*) the reaction of

[50] (*a*) B. Liu, *J. Chem. Phys.*, 1973, **58,** 1925; (*b*) P. Siegbahn and B. Liu, *ibid.*, 1978, **68,** 2457.

[51] (*a*) B. Liu, *J. Chem. Phys.*, 1984, **80,** 581; (*b*) D. M. Ceperley and B. J. Alder, *ibid.*, 1984, **81,** 5833; (*c*) R. N. Bartlett, P. J. Reynolds, and W. A. Lester, jun., *ibid.*, 1985, **82,** 2700.

[52] A. J. Varandas, F. B. Brown, C. A. Mead, D. G. Truhlar, and N. C. Blais, *J. Chem. Phys.*, 1987, **86,** 6258.

[53] G. C. Schatz, *Annu. Rev. Phys. Chem.*, 1988, **39**, 317.

[54] For example: (*a*) M. Karplus, R. N. Porter, and R. D. Sharma, *J. Chem. Phys.*, 1965, **43,** 3259; (*b*) H. R. Mayne and J. P. Toennies, *ibid.*, 1981, **75,** 1794; (*c*) E. Garcia and A. Lagana, *Chem. Phys. Lett.*, 1986, **123,** 356; (*d*) A. Lagana, *Int. J. Chem. Kinet.*, 1986, **18,** 1009.

[55] B. C. Garrett, D. G. Truhlar, A. J. C. Varandas, and N. C. Blais, *Int. J. Chem. Kinet.*, 1986, **18,** 1065.

[56] (*a*) J. J. Valentini and D. P. Gerrity, *Int. J. Chem. Kinet.*, 1986, **18,** 937; (*b*) J.-C. Nieh and J. J. Valentini, *Phys. Rev. Lett.*, 1988, **60**, 519, and references therein.

D atoms with unexcited and vibrationally excited H_2.[57] Moreover, it has been demonstrated that high fractions of $H_2(v = 1)$ and $D_2(v = 1)$ can be generated using stimulated Raman pumping,[58,59] but this technique has yet to be applied in studies of the reactions of $H_2(v = 1)$ or $D_2(v = 1)$.

The product rovibrational distributions from

$$H + D_2 \rightarrow HD(v',j') + D \tag{15}$$

at high collision energies have been measured by CARS[56] and REMPI[60] (resonance-enhanced multiphoton ionization) spectroscopies. Translationally hot H atoms were produced by laser photolysis of HI, the spectra of HD being observed using a 'probe' laser pulse delayed by a few ns from the photolysis laser pulse. At 1.3 eV, Valentini and Gerrity[56] measured rovibrational state distributions which corresponded to average yields of energy into product translation, rotation and vibration of $\langle f_{tr'} \rangle = 0.70$, $\langle f_{rot'} \rangle = 0.21$, and $\langle f_{vib'} \rangle = 0.09$, with the fraction of HD into specific product vibrational states of $f_{v'=0} = 0.18$, and $f_{v'=2} = 0.02$. In their words: 'the dynamical bias of the reaction is against product rotational and vibrational excitation'.

The results of Valentini and Gerrity[56] are in fair agreement with those of Zare's group,[60] and in quite remarkable agreement with the results of QCT calculations.[61] These findings also testify that the dynamics of this reaction are vibrationally adiabatic, *i.e.* $v = 0 \rightarrow v' = 0$ dominates, to a surprising degree, given the curvature of the MEP and the high initial velocity along that path.

The agreement between experimental and theoretical results on the reaction of D atoms with vibrationally excited H_2 has, at least until recently, been unsatisfactory. Before considering these data, it is worth pointing out that the comparison is not quite one of like with like, since theoretical treatments calculate constants for the defined process:

$$D + H_2(v = 1) \rightarrow HD + H \tag{16}$$

whereas the experiments have been performed under conditions where they measure a rate constant for reaction at an enhanced vibrational temperature. In other words, they determine

57 (*a*) T. Dreier and J. Wolfrum, *Int. J. Chem. Kinet.*, 1986, **18,** 919; (*b*) J. Wolfrum, *Faraday Discuss. Chem. Soc.*, 1987, **84**, 191.
58 W. Meier, G. Ahlers, and H. Zacharias, *J. Chem. Phys.*, 1986, **85,** 2599.
59 (*a*) E. A. Rohlfing, D. W. Chandler, and D. H. Parker, *J. Chem. Phys.*, 1987, **87** 5229; (*b*) D. W. Chandler and R. L. Farrow, *ibid.*, 1986, **85,** 810.
60 E. E. Marinero, C. T. Rettner, and R. N. Zare, *J. Chem. Phys.*, 1984, **80,** 4142.
61 (*a*) N. C. Blais and D. G. Truhlar, *Chem. Phys. Lett.*, 1983, **102,** 120; (*b*) N. C. Blais, D. G. Truhlar, and B. C. Garrett, *J. Chem. Phys.*, 1985, **82,** 2300; (*c*) N. C. Blais and D. G. Truhlar, *ibid.*, 1985, **83,** 2201.

$$k(T_{\mathrm{vib}}) = \sum_{v} f_v(T_{\mathrm{vib}})k(|v) \tag{17}$$

where $f_v(T_{\mathrm{vib}})$ is the fraction of H_2 molecules in level v at the vibrational temperature T_{vib} and $k(|v)$ is the rate constant for reaction out of the specific level v. However, the results of QCT calculations[54] suggest that, at the low fractional excitations of H_2 in the experiments, the difference between $k(|v = 1)$ and $\sum_v f_v(T_{\mathrm{vib}})k(|v)$ will be quite small, because the acceleration in the reaction rate for $v > 1$ is much less than the increase in going from $v = 0$ to $v = 1$.

Experimental results for reaction (16) are summarized in Table 1 and compared with the results of some theoretical treatments in Figure 3. A more extensive compilation of theoretical results has been given recently by Schatz.[53] There have been some measurements on vibrational effects in other isotopic variants of this system but they are not discussed here. All the experiments to date,[57,62–64] have used thermal or discharge excitation to create $H_2(v)$, but it will be surprising if more direct measurements using Raman pumping are not reported in the near future. Dreier and Wolfrum's[57] experiments enabled them to follow the concentrations of both the molecular reagents and products, but it is nevertheless not clear what causes their result to differ so radically from earlier experimental data. Comparison of the rates of $H_2(v = 1)$ loss and $HD(v' = 0,1)$ formation gave rate constants of $(1.9 \pm 0.2) \times 10^{-13}$ and $(1.0 \pm 0.4) \times 10^{-13}$ cm^3 molecule^{-1} s^{-1} at 310 K, respectively, indicating comparable rates for reactive and inelastic loss of $H_2(v = 1)$

Table 1 *Measured rate constants for* $D + H_2(v = 1) \rightarrow HD + H$ *at room temperature*

Methods[a] *A*	*B*	$k/10^{-13}$ cm^3 molecule^{-1} s^{-1}	*Reference*
Discharge	e.p.r. on H,D	9.8 ± 3.0	62
Thermal	e.p.r. on H,D; thermometric for $H_2(v'' = 1)$	16 ± 5	63
Discharge	e.p.r. on H,D; thermometric for $H_2(v'' = 1)$	12 ± 5	64
Discharge	CARS on $H_2(v = 1)$, $HD(v = 1)$	1.0 ± 0.4	57

[a] *A*: production of $H_2(v = 1)$; *B*: detection of reagents and products

[62] G. P. Glass and B. K. Chaturvedi, *J. Chem. Phys.*, 1982, **77,** 3478.
[63] (*a*) V.B Rozenshtein, Yu. M. Gershenzon, A. V. Ivanov, and S. I. Kucheryavii, *Chem. Phys. Lett.*, 1984, **105,** 423; (*b*) V.B. Rozenshtein, Yu. M. Gershenzon, A. V. Ivanov, S. D. Il'in, S. I. Kucheryavii, and S. Ya. Umanskii, *ibid.*, 1985, **121,** 89.
[64] U. Wellenhausen and J. Wolfrum, *Ber. Bunsenges, Phys. Chem.*, 1985, **89,** 314.

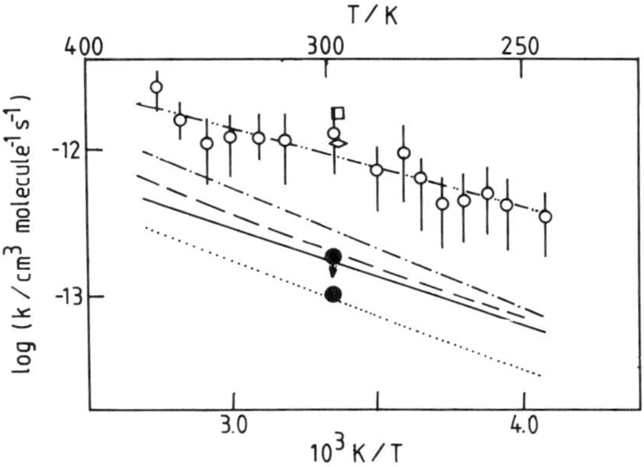

Figure 3 *Comparison of experimental and theoretical rate constants for the reaction* $D + H_2(v = 1) \rightarrow HD + H$. *Experiment: Glass and Chaturvedi*[62] (◇); *Rozenshtein* et al.[63] (□); *Wellenhausen and Wolfrum*[64] (○); *Dreier and Wolfrum*[57] (●, ⬇), *the upper point being a measure of the rate constant for reaction plus relaxation. Theory: Pollak* et al.[69] (··—··—); *Walker and Hayes*[67] (·—·—·); *Garrett* et al.[55] (– – –); *Mayne and Toennies*[54(b)] (——); *Bowman* et al.[66] (········)

with D atoms. Comparison of the CARS signals from $HD(v' = 0)$ and $HD(v' = 1)$ showed that the adiabatic, $H_2(v = 1) \rightarrow HD(v' = 1)$, and non-adiabatic, $H_2(v = 1) \rightarrow HD(v' = 0)$, reaction rates are also of comparable magnitude, and therefore that reaction involving $H_2(v = 1)$ + D atoms exhibits considerably less vibrational adiabaticity than that between translationally hot H atoms and $D_2(v = 0)$.

Theoretical attempts to calculate the rate of reaction between D atoms and $H_2(v = 1)$ include:

(*a*) QCT calculations on the LSTH surface;[54]
(*b*) calculations using variational VA-TST with thorough treatments of anharmonicities and tunnelling,[56,65] and the ADRS version of VA-TST;[46]
(*c*) methods based on interpolating the results of one-dimensional exact quantal calculations to three dimensions;[66,67] and
(*d*) approximate quantal methods.[68]

Although there are differences between the results from various theoretical models, they are all (with one exception[69]) much lower than all the experimental rate constants except the latest value from Wolfrum's laboratory.[57] The highest theoretical value was obtained from calculations[69] based on the infinite order sudden approximation and

[65] B. C. Garrett and D. G. Truhlar, *J. Phys. Chem.*, 1985, **89,** 2204.
[66] J. M. Bowman, K. T. Lee, and R. B. Walker, *J. Chem. Phys.*, 1983, **79**, 3742.
[67] R. B. Walker and E. F. Hayes, *J. Phys. Chem.*, 1983, **87,** 1255.
[68] N. Abu Salbi, D. J. Kouri, and M. Baer, *Chem. Phys. Lett.*, 1984, **105,** 472.
[69] E. Pollak, N. Abu Salbi, and D. J. Kouri, *Chem. Phys. Lett.* 1985, **113,** 585.

assuming the results to be for rotationally averaged reagents, not just the $j = 0$ level. This difference is closely connected to the question of whether or not the bending motion at the $DH_2(v = 1)$ transition state is properly quantized. The presumption that it may not be was justified on the grounds that the period of this motion is much longer than the collision interaction time required to reach the transition state. However, an interesting paper by Schatz[70] points out that quantum dynamics appears more vibrationally adiabatic than classical mechanics because the real justification for vibrationally adiabatic treatments, and particularly for including a bending zero-point energy in the vibrationally adiabatic potential for three dimensions, is the uncertainty principle, not the adiabatic principle.

The weight of theoretical opinion is undoubtedly on the side of quantizing the bend motion, and it appears that VA-TST applied with full rigour can provide a good quantum description of the effects of vibrational excitation on the $D + H_2(v = 1)$ reaction. Figure 4 shows vibrationally adiabatic potentials for $DH_2(v = 0)$ and $DH_2(v = 1)$. As in all the isotopic variants of the H_3 system, the VA barrier is lowered and moved to an earlier position as the vibrational excitation is increased.

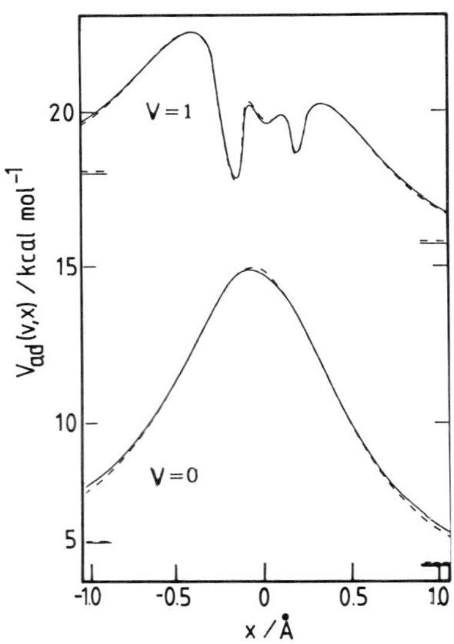

Figure 4 *Zero angular momentum, ground-bend-state, vibrationally adiabatic potential energy curves for* $D + H_2(v) \rightarrow DH + H$. *The solid curves are for the DMBE potential energy surface and the dashed lines for the LSTH surface. The long horizontal lines show the asymptotes of the adiabatic curves*

(Adapted with permission from *Int. J. Chem. Kinet.*, 1986, **18,** 1065)

[70] G. C. Schatz, *J. Chem. Phys.*, 1983, **79,** 5386.

Garrett *et al.*[55] have addressed the question of whether vibrational adiabaticity up to the transition state is valid, and conclude that it is.

Analysis of the H_3 system leads, of course, to vibrationally adiabatic curves that are symmetric about the position of the saddlepoint, with two equal maxima for all states but $v = 0$. Frost and Smith[42] have applied their combined TST–QCT method to the calculation of rate constants for reactive and inelastic collisions between H atoms and H_2 with up to nine vibrational quanta. The results are in good agreement with those obtained[54b,c] using standard quasi-classical trajectory techniques. They show that the effect of reagent excitation falls monotonically with each additional vibrational quantum that is added and they lend support to the concept of vibrational adiabaticity up to the transition state (*i.e.* the 'first' vibrationally adiabatic maximum) since trajectories run 'backwards in time' to reagents yield H_2 molecules with vibrational energies closely clustered around the corresponding quantum value in isolated H_2.

Reactions of F, O(^3P), *and* O(^1D) *Atoms with* H_2

F + H_2. The reaction of F atoms with H_2 to yield HF,

$$F + H_2(v) \rightarrow HF(v') + H \quad (\Delta H_0^\circ = -32.1 \text{ kcal mol}^{-1}) \tag{18}$$

has become very important as a testing ground for theories of chemical reaction dynamics.[71–73] A major objective has been to reproduce the distributions of product energy states and scattering angles that have been measured in chemical laser, infrared chemiluminescence and molecular beam experiments.[9] A large number of *ab initio* and semi-empirical potential energy surfaces have been calculated and proposed.[73,74] They have been employed in various scattering calculations, especially QCT[72] and modified TST–QCT calculations.[44] A preliminary report of an accurate 3D quantal scattering calculation has just been published.[37]

Curiously, it has proved easier to find potential energy surfaces which satisfactorily reproduce the experimental data on the reaction dynamics than the thermal rate constants and their variation with temperature.[75] Despite the relative electronic simplicity of the FH_2 system, it has

[71] J. B. Anderson, *Adv. Chem. Phys.*, 1980, **41**, 299.

[72] J. T. Muckerman, in 'Theoretical Chemistry: Advances and Perspectives, Vol. 6, Part A, Theory of Scattering', ed. D. Henderson, Academic Press, New York, 1981, Chap. 1.

[73] H. F. Schaefer III, *J. Phys. Chem.*, 1985, **89**, 5336.

[74] C. W. Bauschlicher, jun., S. P. Walch, S. R. Langhoff, P. R. Taylor, and R. L. Jaffe, *J. Chem. Phys.*, 1988, **88**, 1743.

[75] (*a*) E. Wurzberg and P. L. Houston, *J. Chem. Phys.*, 1980, **72**, 4811; (*b*) R. F. Heidner, J. F. Bott, C. E. Gardner, and J.E. Melzer, *ibid.*, 1980, **72**, 4815.

proved very difficult[73] to compute an *ab initio* surface that has saddlepoint and MEP properties which then give satisfactory agreement with experimental rate constants when used in a VA-TST analysis. These problems may have been solved in very recent and very extensive calculations by Bauschlicher *et al.*,[74] who have obtained a classical barrier height of 1.86 kcal mol^{-1} and conclude that the true value is probably close to 1.4 kcal mol^{-1}. There is little doubt that the saddlepoint is at an early position along the MEP. The extensions of the F—H and H—H internuclear distances at the saddlepoint compared with their values in the isolated molecules are: $\Delta r_{\mathrm{HH}}^{\mathrm{sp}} = 0.023$ Å and $\Delta r_{\mathrm{FH}}^{\mathrm{sp}} = 0.62$ Å.

Many of the dynamical calculations on reaction (18) have been performed on a semi-empirical surface proposed some time ago by Muckerman and generally referred to as 'Muckerman 5'.[72] The results of QCT calculations on this surface adequately reproduce the average extent of energy release into the HF product vibration.[72] The relative yields of HF(v) measured experimentally from the thermal reaction are $f_{v'=0}$=0.03, $f_{v'=1} = 0.15$, $f_{v'=2} = 0.53$, and $f_{v'=3} = 0.29$.[9] QCT calculations do not easily match this distribution because the 'binning' procedure used to derive a quantum-state distribution from the classical energy distribution is not entirely satisfactory. Formation of HF($v' = 3$) from F + H_2($v = 0$) is actually slightly endothermic. However, any product molecule from a QCT which has a vibrational energy corresponding to a 'quantum number' between 2.5 and 3.5 is assigned to the $v' = 3$ state.

There do not appear to have been any experimental studies of the effects of H_2 vibrational excitation on the rate of the F + H_2 reaction or on the distribution of energy over product degrees of freedom when H_2 is excited. Pollak[76] has reported a vibrationally adiabatic analysis of the potential surface for collinear reaction, using the pods procedure for determining the transition states. For H_2($v = 0,1,2$) there is a single adiabatic maximum (transition state) located well into the entrance channel on the Muckerman 5 surface. These maxima are slight, *e.g.* for $v = 0$ the energy is only 0.58 kcal mol^{-1}. They move to slightly earlier positions on the MEP as v increases and all are located sufficiently early that motion up to the transition state is vibrationally adiabatic. Figure 5 compares the position of transition states for F + H_2($v = 0$) and F + H_2($v = 1$) given by the semiclassical pods analysis with curves calculated via a quantal analysis of the collinear MEP. Based solely on the lowering of the adiabatic barrier (*i.e.* making no allowance for its shift in position), the rate constant for F + H_2($v = 1$) should be ~1.7 times that for F + H_2($v = 0$). This very low enhancement is the result of both the low barrier to reaction and its early position along the MEP – the frequency of bond-stretch motion orthogonal to the MEP is only slightly modified as the system approaches the transition state. Such a

[76] E. Pollak, *J. Chem. Phys.*, 1981, **74**, 5586.

small acceleration in the rate will be very difficult to measure, given the experimental difficulty of creating large fractional concentrations of vibrationally excited H_2.

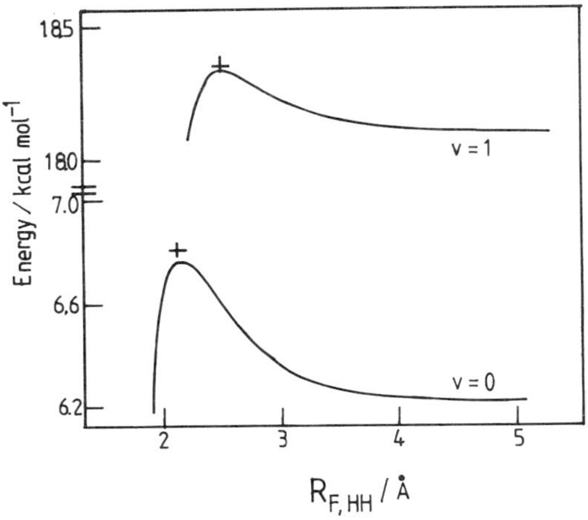

Figure 5 *Comparison of transition states for collinear* F + $H_2(v = 0{,}1)$ *from a semiclassical pods analysis and from an approximate quantal analysis of the bound motion orthogonal to the MEP*
(Adapted with permission from *J. Chem. Phys.*, 1981, **75**, 4435)

For $v = 3$ in the collinear FH_2 system, Pollak's vibrationally adiabatic analysis[76] reveals that, after an entrance channel barrier, there is a potential minimum and then an exit channel maximum. This late adiabatic maximum is ~1.8 kcal mol^{-1} above both the HF($v' = 3$) + H asymptotic energy and the height of the early $v = 0$ adiabatic maximum. The presence of the $v' = 3$ barrier is connected to a number of interesting dynamical effects. For example, it has been suggested that in both collinear and three-dimensional calculations,[77,78] the state-to-state rate constant $k(v' = 3|v = 0)$ should be estimated by applying detailed balance to the results of calculations on H + HF($v' = 3$) collisions. Although this does seem a preferred procedure, it should be noted that Frost and Smith's ADRS analysis of transition states in the FH_2 system shows that the late adiabatic $v = 3$ maximum is only defined for a limited range of geometries close to collinear.[44] At moderate and high bend angles, the only pods with action equal to $3.5h$ is in the entrance valley of the surface before the strong interaction region.

The existence of the exit channel maximum for $v' = 3$ is also related to the existence in the FH_2 system of quantum mechanical

[77] G. C. Schatz, J. M. Bowman, and A. Kuppermann, *J. Chem. Phys.*, 1975, **63**, 674 and 685.

[78] C. J. Ashton, J. T. Muckerman, and F. E. Schubert, *J. Chem. Phys.*, 1984, **81**, 5786.

resonances.[4,53] These correspond to long-lived states of the system whose existence is correlated with wide angle scattering of $HF(v' = 3) + H$ compared with the sharply peaked backward scattering of HF molecules in lower vibrational states.

$O(^3P) + H_2$. The reaction between ground-state oxygen atoms and H_2 is important in combustion chemistry, and its thermal rate constant has been measured in a large number of studies over a wide range of temperature.[79] The reaction is very nearly thermoneutral:

$$O(^3P) + H_2 \rightarrow OH + H \quad (\Delta H_0^\circ = 1.8\ \text{kcal mol}^{-1}) \qquad (19)$$

As such it might be expected to have a barrier to reaction roughly half-way along the MEP, as on the H_3 surface, but now of course the kinematics and the scaled and skewed surface for collinear reaction are quite different. The rate constant for reaction is strongly temperature dependent, indicating a high barrier to reaction.

The dynamics of the $O(^3P) + H_2$ reaction have been investigated quite extensively in theoretical calculations based on quasi-classical trajectories,[80] variational transition-state theory,[43] and reduced dimensionality quantum scattering[81] methods. Various potential energy surfaces have been employed and some emphasis has been placed on investigating systematic trends in the reaction dynamics as the saddle-point moves to a progressively later location on the potential energy surface. The results from all these studies are in keeping with Polanyi's general rule: the later the position of the barrier the greater is the enhancement of the reaction rate as the H_2 reagent is raised from $v = 0$ to $v = 1$.

The experimental evidence concerning the effect of reagent vibrational excitation is restricted to the one measurement by Light,[82] who found that the reaction rate at 298 K is enhanced by a factor of $\sim 2.6 \times 10^3$ for $H_2(v = 1)$. The OH radicals produced were observed by laser-induced fluorescence and remarkably 95% of them were found in $v' = 1$ despite the fact that the $v = 1 \rightarrow v' = 1$ process is only just thermoneutral, indicating a high degree of vibrational adiabaticity in this system. This finding is in at least qualitative agreement with results from quasi-classical trajectories[80] and collinear quantum calculations.[81]

It appears that the so-called ModPolCI surface[81] is currently the most accurate for $O(^3P) + H_2$. It has a classical barrier height of 12.6 kcal mol^{-1} and $\Delta r_{HH}^{sp} = 0.18$ Å, $\Delta r_{OH}^{sp} = 0.24$ Å. The vibrationally

[79] K. Westberg and N. Cohen, *J. Phys. Chem. Ref. Data*, 1983, **12,** 531.
[80] M. Broida and A. Persky, *J. Chem. Phys.*, 1984, **80,** 3587.
[81] (*a*) K. T. Lee, J. M. Bowman, A. F. Wagner, and G. C. Schatz, *J. Chem. Phys.*, 1982, **76,** 3563 and 3583; (*b*) J. M. Bowman, A. F. Wagner, S. P. Walch, and T. H. Dunning, jun., *ibid.*, 1984, **81,** 1739.
[82] G. C. Light, *J. Chem. Phys.*, 1978, **68,** 2831.

adiabatic barrier heights (neglecting bending) decrease from 11.0 kcal mol^{-1} for $H_2(v = 0)$, to 5.9 kcal mol^{-1} for $v = 1$, to 3.9 kcal mol^{-1} for $v = 2$. As with H_3, the position of these barriers migrates from the saddlepoint for $v = 0$ to the entrance valley for higher vibrational states.

The observed effect of H_2 vibrational excitation on the O + H_2 reaction rate[82] is reproduced quite well both (*a*) by calculations using generalized variational transition-state theory assuming that the excited degree of freedom (*i.e.* the bond-stretching vibration) is adiabatic from the beginning of the collision up to the bottleneck defined by the free energy maximum,[43] and (*b*) by transition-state theory calculations in which the tunnelling correction is estimated by collinear exact quantum calculations in which an adiabatic bend energy is added to the potential (the TST–CEQB method).[81] Garrett *et al.*[43] have also examined a modified version of TST for vibrationally excited reactions in which it is assumed that all flux is diverted to the vibrational ground-state channel at the point on the reaction path where the path curvature reaches its first maximum. Comparisons were made with the results of collinear exact quantum calculations. The result of the comparison depends on the surface and in particular whether the point of maximum curvature came before or after the vibrationally adiabatic maximum. Whether this method is a valid way to treat the effect of vibrational non-adiabaticity in state-selected reactions would still appear to require confirmation.

$O(^1D)$ + H_2. The reaction of $O(^1D)$ atoms with H_2 is quite different in virtually every respect from that between ground-state $O(^3P)$ and H_2.[83] It proceeds extremely rapidly over a barrierless surface which apparently corresponds to the electronic ground state of H_2O:

$$O(^1D) + H_2 \rightarrow (H_2O)^\dagger \rightarrow OH + H \quad (\Delta H_0^\circ = -43.5 \text{ kcal mol}^{-1}) \quad (20)$$

although other surfaces correlating with reagents and products are available. Despite the presence of the deep potential minimum in this surface, it has not yet been determined unequivocally whether the energy partitioning is statistically or dynamically controlled.[83] The reaction is considered here to draw attention to two studies of the OH product states which utilize significant advances in experimental technique. (A more comprehensive review of the dynamics of the reaction is included in a recent article by Sloan.[83])

The first of these methods is time-resolved infrared chemiluminescence using a Fourier transform infrared (FT-i.r.) spectrometer to record emission from the rovibrationally excited reaction products.[83,84] This

[83] J. J. Sloan, *J. Phys. Chem.*, 1988, **92,** 18.

[84] (*a*) J. E. Butler, R. G. Macdonald, D. J. Donaldson, and J. J. Sloan, *J. Chem. Phys.*, 1986, **84,** 5365; (*b*) P. M. Aker and J. J. Sloan, *ibid.*, 1986, **85,** 1412.

development means that i.r. chemiluminescence techniques can be applied at high resolution to the products of reactions involving species [such as $O(^1D)$] that can only be created by pulsed photolysis. The measured distribution in levels $v' = 1$—4 (the population in $v' = 0$ cannot be measured and is ignored) is $f_{v'=1} = 0.29$, $f_{v'=2} = 0.32$, $f_{v'=3} = 0.25$, and $f_{v'=4} = 0.13$. Another recent measurement based on chemical laser techniques[85] (from which $f_{v'=0}$ can be derived) gives $f_{v'=0} = 0.51$, $f_{v'=1} = 0.30$, $f_{v'=2} = 0.32$, $f_{v'=3} = 0.26$, and $f_{v'=4} < 0.18$, if the distribution is normalized to give the same number for $f_{v'=2}$ as that derived from emission spectroscopy. These results indicate non-statistical energy partitioning, but recent laser-induced fluorescence (LIF) measurements do suggest that the overall vibrational excitation may have been overestimated by a failure to observe extremely high rotational states.

The LIF experiments monitor initial OH populations in vibrational, rotational, spin–orbit, and Λ-doublet levels. A difficulty in applying LIF techniques to OH is that many of the rovibrational levels of the excited $A\,^2\Sigma^+$ state are strongly predissociated and that the transition moments for off-diagonal ($\Delta v \neq 0$) bands in the A—X system are very small. Wiesenfeld and his co-workers,[86] however, have demonstrated that good quality LIF spectra can be obtained under single collision conditions in pulse-and-probe experiments using off-diagonal transitions with Δv at least as high as 2. Their results indicate high yields of rotational excitation and that $f_{v'=3}/f_{v'=2} = 0.39$, rather than ~ 0.8 as derived from the i.r. emission[84] and chemical laser[85] experiments.

Trajectories run on a representation of the lowest H_2O potential energy surface indicate that, despite the depth of the well, the intermediates formed from $O(^1D) + H_2$ usually dissociate after only one or two vibrations. This is insufficient for complete randomization of energy: hence the non-statistical distribution. The most recent calculations[87] support the view that it is the insertion/elimination mechanism which predominates, with little contribution from an abstraction route via linear or nearly collinear attack by the atom on H_2. The factors which control energy partitioning in systems where reaction can proceed across an initially attractive surface with a deep potential well are not fully understood, and definitive experiments which examine the dynamics of such systems are badly needed. This topic is raised again below under a brief discussion of radical–radical reactions.

[85] Y. Huang, Y. Gu, C. Liu, X. Yang, and Y. Tao, *Chem. Phys. Lett.*, 1986, **127**, 432.

[86] (*a*) G. M. Jurisch and J. R. Wiesenfeld, *Chem. Phys. Lett.*, 1985, **119**, 511; (*b*) J. E. Butler, G. M. Jurisch, I. A. Watson, and J. R. Wiesenfeld, *J. Chem. Phys.*, 1986, **84**, 5365; (*c*) C. B. Cleveland, G. M. Jurisch, M. Trolier, and J. R. Wiesenfeld, *ibid.*, 1987, **86**, 3253.

[87] M. S. Fitzcharles and G. C. Schatz, *J. Phys. Chem.*, 1986, **90**, 3634.

Reactions of O(^3P) *with* HCl *and* HBr

The reactions of ground-state oxygen atoms with HCl:

$$O(^3P) + HCl \rightarrow OH + Cl \quad (\Delta H_0^\circ = +0.9\ \text{kcal mol}^{-1}) \tag{21}$$

and with HBr:

$$O(^3P) + HBr \rightarrow OH + Br \quad (\Delta H_0^\circ = -14.7\ \text{kcal mol}^{-1}) \tag{22}$$

are examples of the many important bimolecular reactions in which a hydrogen atom is transferred between two heavier moieties. They belong to what are often called,[5,22] the heavy–light–heavy (*HLH*) class of reactions which exhibit a number of unusual dynamic and kinematic features. The first of these is that the scaled and skewed potential energy surfaces for collinear reaction are very heavily skewed, *e.g.* the angles between the skewed axes for O + HCl and O + HBr are 16.9 and 15.4°, respectively. As a result, trajectories, and the equivalent quantum mechanical flux, may not pass between the reagent and product valleys via the saddlepoint region but cut the corner and cross the ridge separating the two valleys at some distance from the saddlepoint. Gertitschke *et al.*[88] have coined the phrase 'dynamical white spot' to describe the region near the saddlepoint and in the corner of the skewed surface.

A second feature of *HLH* systems is that trajectories may make several traversals of the potential energy ridge before escaping down one or other (product or reagent) valleys. Classically, this corresponds to the hydrogen atom bouncing between its two ponderous potential partners before the collision is completed. Accurate quantum scattering calculations[89] indicate that this type of behaviour, at least in collinear systems, leads to a high propensity for vibrationally adiabatic (*i.e.* $v \rightarrow v' = v$) reaction. Moreover the potentials, at least for symmetric, collinear *HLH* systems, can support several bound states despite the fact that the classical potential energy is higher than at the asymptotic *HL* + *H* limit.[90]

A third feature of the unusual kinematics in the *HLH* systems is that the vibrationally adiabatic curves can differ markedly in shape from one another and from the potential energy profile $V_{el}(x)$. This is because the vibration orthogonal to the MEP is associated with a very light reduced mass at either end of the MEP but in between corresponds to motion of the two heavy atoms with the hydrogen atom motionless or nearly so.

There do not appear to be any accurate *ab initio* potential energy surfaces available for reactions (21) and (22), and most dynamical

88 P. L. Gertitschke, P. Kiprof, and J. Manz, *J. Chem. Phys.*, 1987, **87,** 941.
89 D. K. Bondi, J. N. L. Connor, J. Manz, and J. Romelt, *Mol. Phys.*, 1983, **50,** 467.
90 D. C. Clary and J. N. L. Connor, *J. Phys. Chem.*, 1984, **88,** 2758.

calculations have been performed on LEPS or modified LEPS surfaces. Studies of the thermal rate constants yield activation energies of 6.7 and 3.1 kcal mol^{-1} for the two reactions. These energies contrast with the excitation energies of HCl($v = 1$), 8.25 kcal mol^{-1}, and of HBr($v = 1$), 7.3 kcal mol^{-1}.

Studies of the kinetics of HCl($v = 1$) and HBr($v = 1$) with O(^3P) atoms were among the first of their kind. The experimental method combined the creation of known concentrations of oxygen atoms in a discharge–flow system with the measurement of i.r. fluorescence decay lifetimes of vibrationally excited molecules generated by a pulse of i.r. laser radiation. The results of such measurements at 298 K are compared in Table 2 with those for the ground-state reaction at the same temperature. However, it has to be appreciated that, in these relatively primitive experiments, the measured rate constant is one for the sum of two processes, reaction and non-reactive relaxation. On the grounds that DCl($v = 1$) was removed by O atoms at a comparable rate to HCl($v = 1$), Brown *et al.*[92] suggested that relaxation made the dominant contribution to the overall rate in both systems, and this was confirmed by later experiments in which the products of chemical reaction were detected by LIF, molecular beam mass spectroscopy, and resonance absorption.[93,94]

Table 2 *Rate constants for reaction of* O(^3P) + HCl($v = 0$) *and for relaxation plus reaction of* HCl($v > 0$) *by* O *atoms at* 298 K

v	Process	k/cm^3 molecule^{-1} s^{-1}	*Reference*
Experiment			
0	reaction	1.4×10^{-16}	91
1	reaction + relaxation	8.9×10^{-13}	92
1	reaction + relaxation	$10. \times 10^{-13}$	
2	reaction	1.5×10^{-12}	93
2	relaxation	3.7×10^{-12}	
1	reaction	1.0×10^{-13}	
1	relaxation	9.0×10^{-13}	94
2	reaction + relaxation	6.3×10^{-12}	
1	reaction	2.0×10^{-14}	
1	reaction + relaxation	8.9×10^{-13}	95
2	reaction	$< 2.8 \times 10^{-12}$	
2	relaxation	4.2×10^{-12}	
QCT calculations			
	$k_{v=2}/k_{v=1}/k_{v=0}$ (reaction)		
	3080 : 230 : 1		97
	1840 : 150 : 1		98
	2120 : 390 : 1		98

[91] D. L. Baulch, R. A Cox, R. F. Hampson, jun., J. A. Kerr, J. Troe, and R. T. Watson, *J. Phys. Chem. Ref. Data*, 1984, **13,** 1259.

[92] R. D. H. Brown, G. P. Glass, and I. W. M. Smith, *Chem. Phys. Lett.*, 1975, **32,** 517.

[93] R. G. Macdonald and C. B. Moore, *J. Chem. Phys.*, 1978, **68,** 513.

[94] M. Kneba and J. Wolfrum, 'Proceedings of the 17th International Symposium on Combustion', Combustion Institute, Pittsburgh, 1978, p. 497.

There have also been studies on $O(^3P)$ + HCl, the most direct being that in which HCl molecules are raised directly to $v = 2$ by the tuned output from an optical parametric oscillator (OPO).[93] The agreement of the results as regards *total* removal of HCl from level $v = 2$ with those from other less direct studies[94,95] is excellent. However, as Table 2 shows, there is less agreement over the rate constants associated just with the reactive channel. Whereas Butler *et al.*[95] measured roughly equal enhancements of reaction rate for excitations from $v = 0$ to $v = 1$ and from $v = 1$ to $v = 2$, the results of Kneba and Wolfrum[94] indicate a major enhancement for addition of the first quantum, $\sim 10^3$ as against 6 for $v = 1$ to $v = 2$.

In a very recent paper, Rakestraw *et al.*[96] report on some very elegant experiments which promise to give quite exceptional insight into the $O(^3P) + HCl(v)$ reaction. They simultaneously produced O atoms by laser photolysis of NO_2 and HCl molecules in a selected rotational level of $v = 2$ using radiation from an OPO. They then monitored the formation of OH in selected rovibrational states. They found comparable amounts of OH in $v' = 0$ and $v' = 1$, a high degree of rotational excitation in the OH products, and a significant change in the distribution over rotational states when the rotational state of the HCl reagent is changed. These experiments appear to open the way to truly state-to-state $[(v,j)\text{–}(v',j')]$ rate constants, at least in some favourable reaction systems.

Quasi-classical trajectory calculations have been carried out on $O(^3P) + HCl(v)$ collisions using various LEPS potential energy surfaces.[97,98] On all of these surfaces, the barrier was centrally located, as would be expected for a reaction which is very nearly thermoneutral, and its height was chosen so that the calculated thermal rate constant matched the experimental value at 298 K. Despite differences in the two surfaces used by Persky and Broida[98] and that used by Brown and Smith,[97] the calculated results of the effects of reagent vibration are quite similar. The ratios of the rate constants for *reaction* from $v = 2$, $v = 1$, and $v = 0$ are given in Table 2. The surface with the latest barrier (the second listed of Persky and Broida) gives the largest enhancement for $v = 1$ but surprisingly this is not so for $v = 2$. The observation that the rate enhancement brought about by addition of the second quantum is much less than the effect of the first quantum is in keeping with expectation based on the analysis of several systems,[18] with the experimental results of Macdonald and Moore[93] and of Kneba and Wolfrum,[94] but not with those of Butler *et al.*[95]

Although there has not apparently been any analysis of the

[95] J. E. Butler, J. W. Hudgens, M. C. Lin, and G. K. Smith, *Chem. Phys. Lett.*, 1978, **58,** 216.

[96] D. J. Rakestraw, K. G. McKendrick, and R. N. Zare, *J. Chem. Phys.*, 1987, **87,** 7341.

[97] R. D. H. Brown and I. W. M. Smith, *Int. J. Chem. Kinet.*, 1978, **10**, 1.

[98] A. Persky and M. Broida, *J. Chem. Phys.*, 1984, **81,** 4352.

O + HCl(v) system by VA-TST, Persky and Broida[98] have reported excitation functions for O + HCl(v = 0,1,2,4) on both the potential energy surfaces that they used. The effect of increasing v is both to lower the threshold energy and to increase the steepness of the rise in the excitation function immediately post-threshold. This is qualitatively what one would anticipate from a VA-TST analysis, as the vibrationally adiabatic maxima are lowered and move to an earlier position as v increases. Finally, it should be noted that the trajectories show a rather high degree of vibrational adiabaticity (*i.e.* a propensity for $v \rightarrow v' = v$) and rather small yields of rotational energy in the OH product. Neither of these findings appears to agree with the experimental results of Rakestraw *et al.*,[96] but the observation of high vibrational adiabaticity is consistent with what has been found[99] for the reverse reaction between Cl atoms and highly excited OH($v \leqslant 9$).

In contrast to the reaction between O atoms and HCl, that between O(^3P) and HBr is quite strongly exothermic. Nevertheless, reaction at thermal energies can only populate OH(v' = 0,1); the excitation energy of OH(v' = 2) is 5.2 kcal mol^{-1} higher than the magnitude of ΔH°_0. A discharge–flow/e.p.r. study of the reaction several years ago by Spencer and Glass[100] led to the conclusion that 97% of the OH product was created in v' = 1. This astonishing result remained unconfirmed until very recent laser experiments in Zare's laboratory[101] confirmed that the reaction does indeed have a remarkable propensity for forming molecules in a single vibrational state.

McKendrick *et al.*[101] report two kinds of experiment. In one, the O(^3P) atoms are formed by pulsed laser photolysis of NO_2; in the other, they are formed in an effusive jet following dissociation of O_2 in a microwave discharge. In both sets of experiments, the OH product states are sampled by LIF. The photolysis source produces translationally hot O atoms and the energy of their first collisions with HBr is ~ 5.9 kcal mol^{-1}. This exceeds the activation energy for the reaction and also increases the accessibility of OH(v' = 2).

In the laser photolysis experiments, the relative populations ($N_{v'}$) of OH in v' = 1 and v' = 2 were determined to be 9.4 ± 3.1 and the population in v' = 0 was conservatively estimated to be < 10% of $N_{v'=1}$. In the crossed-beam experiments, only populations in v' = 1 were detected. These results confirm the essential correctness of the finding of Spencer and Glass.[100] In addition, however, McKendrick *et al.*[101] were able to measure detailed rotational and fine-structure distributions. A high degree of rotational excitation was found with an average yield $\langle f_{rot'} \rangle$ of 0.24.

[99] B. A. Blackwell, J. C. Polanyi, and J. J Sloan, *Chem. Phys.*, 1977, **24**, 25.

[100] J. E. Spencer and G. P. Glass, *Int. J. Chem. Kinet.*, 1977, **11**, 97.

[101] (*a*) K. G. McKendrick, D. J. Rakestraw, and R. N. Zare, *Faraday Discuss. Chem. Soc.*, 1987, **84**, 39; (*b*) K. G. McKendrick, D. J. Rakestraw, and R. N. Zare, *J. Phys. Chem.*, 1988, **92**, 5530.

McKendrick *et al.*[101] have compared their experimental results with those from two sets of quasi-classical trajectory calculations on a LEPS surface derived by Broida *et al.*[102] This surface was chosen to fit the available kinetic data after extensive trials and has a barrier height of 3.16 kcal mol^{-1} in an early position ($\Delta r^{sp}_{OH} = 0.40$ Å; $\Delta r^{sp}_{HBr} = 0.06$ Å) on the MEP. Quasi-classical trajectories on this surface reproduce the experimentally observed state distributions quite well, although the division of the classical vibrational energy distribution over widely spaced OH product states is clearly problematical. The substantial degree of OH rotational excitation, compared with those found in studies of O(^3P) atoms with alkanes, is attributed to a quite different angular dependence of the potential energy surfaces. In the case of O + HBr, reaction can occur through appreciably bent OHBr geometries as the potential varies quite slowly with change in the bend angle away from collinearity.

There appear to have been no experiments on the kinetics of O(^3P) + HBr($v > 1$) or any analyses of the effects of reagent vibrational excitation on this reaction. Broida *et al.*[102] did calculate excitation functions for reaction of HBr($v = 1$) and HBr($v = 0$). These show that the reaction threshold is lowered by 1.1 kcal mol^{-1} (≡ 385 cm^{-1}) compared with the additional 7.3 kcal mol^{-1} vibrational energy in HBr($v = 1$), and that the cross-sections rise more steeply from threshold when the HBr is vibrationally excited. This is entirely consistent with what a VA-TST analysis might be expected to predict: namely, a modest lowering of the adiabatic barrier and its shift to an earlier position along the MEP as the extent of reagent vibrational excitation in increased. For this exothermic reaction, the trajectory calculations predict an increase in the rate constant of only a factor of 11.4 on raising the HBr level from $v = 0$ to $v = 1$.

Reactions Involving Four Atoms

Bimolecular reactions involving four atoms can be of three types:

$$\mathrm{AB + CD \rightarrow ABC + D} \tag{23}$$

$$\mathrm{A + BCD \rightarrow AB + CD} \tag{24}$$

or the four-centre reaction:

$$\mathrm{AB + CD \rightarrow AC + BD} \tag{25}$$

In the last case, if neither AB nor CD are radicals, the reactions are characterized by very high barriers. Indeed in the simplest case:

$$\mathrm{H_2 + D_2 \rightarrow 2HD} \tag{26}$$

[102] M. Broida, M. Tamir, and A. Persky, *Chem. Phys.*, 1986, **110**, 83.

it is doubtful if the four-centre mechanism plays any role in the process of isotopic exchange. The available evidence about vibrational effects in this reaction does not appear to have increased significantly since an earlier review[18] and it will not be considered further here.

Metathetical reactions of type (23) with AB a diatomic radical have been studied in the case of OH and CN. The earliest work on reactions of vibrationally excited OH radicals was by Glass and co-workers[103] using a flow–discharge apparatus with detection of transient concentrations by e.p.r. spectroscopy. Among other data they found[103c] that reaction with H_2:

$$OH(v_1) + H_2(v_2) \rightarrow H_2O + H \quad (\Delta H^\circ_0 = -14.7\,\text{kcal mol}^{-1}) \quad (27)$$

was accelerated by a factor of $< 50\%$ by vibrational excitation of the OH. This contrasts with a rate enhancement of ~120—160 for excitation of the H_2 reagent to $v_2 = 1$.[104,105]

The potential energy surface for reaction (27) has been examined in a large-scale *ab initio* calculation by Walch and Dunning.[106] The saddle-point geometry is found to be co-planar with an early barrier, *i.e.* the HO—H separation is 36% greater than in the H_2O product, whereas the H—H separation is only stretched by 10% from its value in isolated H_2. The early position of the barrier is consistent with the relatively modest effect on the reaction rate of promoting H_2 to $v = 1$.

Walch and Dunning's calculations yield a barrier height for reaction (27) of 6.2 kcal mol^{-1} compared with an experimental activation energy of ~4.6 kcal mol^{-1}. Schatz has empirically fitted the *ab initio* points calculated by Walch and Dunning and then performed quasi-classical trajectory computations on the complete surface.[107] Isaacson and Truhlar[45] have used the same surface in a variational TST theory analysis of this reaction, including the effects of vibrational excitation. They tracked the MEP over the multidimensional potential energy surface. At closely spaced points along the path they performed a mode analysis on which they based the value of the partition function for a generalized transition state at that location. With tunnelling corrections applied using a small curvature approximation, the calculated results were in reasonable agreement with the absolute magnitude of the experimental rate constants and their temperature dependence, and with the measured kinetic isotope effects.

[103] (*a*) J. E. Spencer and G. P. Glass, *Int. J. Chem. Kinet.*, 1977, **9**, 111; (*b*) J. E. Spencer, H. Endo, and G. P. Glass, 'Proceedings of the 16th International Symposium on Combustion', Combustion Institute, Pittsburgh, 1977, p. 289; (*c*) G. P. Glass, H. Endo, and B. K. Chaturvedi, *J. Chem. Phys.*, 1982, **77**, 5450; (*d*) G. P. Glass and B. K. Chaturvedi, *ibid.*, 1981, **75**, 2749.

[104] R. Zellner, *J. Phys. Chem.*, 1979, **83**, 18.

[105] G. C. Light and J. H. Matsumoto, *Chem. Phys. Lett.*, 1978, **58**, 578.

[106] S. P. Walch and T. H. Dunning, jun., *J. Chem. Phys.*, 1980, **72**, 1303.

[107] (*a*) G. C. Schatz and H. Elgersma, *Chem. Phys. Lett.*, 1980, **73**, 21; (*b*) G. C. Schatz, *J. Chem. Phys.*, 1981, **73**, 1133.

The results of Isaacson and Truhlar also provide a rationale for understanding the effects of vibrational excitation in this reaction, though some care must be taken in such polyatomic systems to correlate the vibrational modes correctly between the reagent and product limits.[45b] The surface calculations show that the OH separation in the OH moiety remains virtually unchanged as the reaction proceeds. The Isaacson–Truhlar analysis indicates that the vibrationally adiabatic curves for $OH(v = 1) + H_2$ and $OH(v = 0) + H_2$ are virtually parallel, with the frequency correlating with the OH stretch being lowered by only $\sim 90\ cm^{-1}$ by the time the transition state is reached. Vibrationally adiabatic curves for $OH(v_1 = 0) + H_2(v_2 = 0)$, $OH(v_1 = 1) + H_2(v_2 = 0)$, and $OH(v_1 = 0) + H_2(v_2 = 1)$ are shown in Figure 6. Because the (0,0) and (1,0) curves are virtually parallel, their maxima are at essentially identical positions along the MEP and even a simplistic TST estimate of the rate constant ratio:

$$k_{OH(v=1)}/k_{OH(v=0)} = \exp(+90hc/k_BT)$$

$$= 1.6 \text{ at } 298\ K \tag{28}$$

is in agreement with experiment. Truhlar and Isaacson's full analysis,[45] including tunnelling, lowers this crude estimate of the ratio of rate constants to 1.42—1.58, the exact value depending on which approximate method was used to estimate the quantal correction to reaction path motion.

For excitation of the H_2 reagent the lowering of the adiabatic maximum is much greater ($\sim 1100\ cm^{-1}$), the shape of the adiabatic potential curve is altered, and the maximum is shifted to an earlier

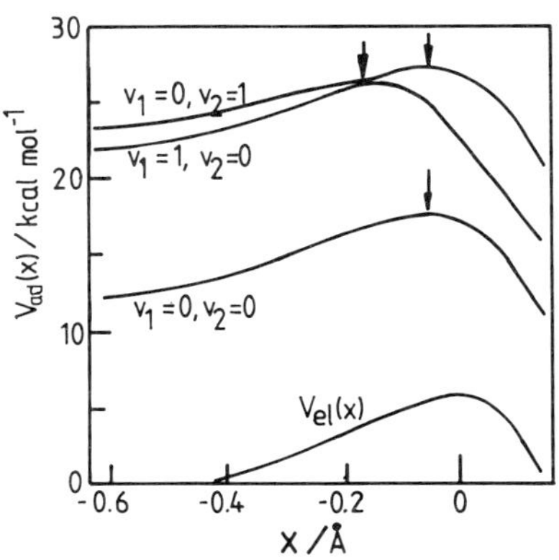

Figure 6 *Vibrationally adiabatic potential energy curves for* $OH(v_1) + H_2(v_2) \rightarrow H_2O + H$. *The positions of the variational vibrationally adiabatic transition states are indicated by the arrows*
(Adapted with permission from *J. Chem. Phys.*, 1982, **77**, 3516)

position. Consequently, the changes in the TST rate expression are more complex. The estimated increase in rate constant at 298 K varies from 108 (with no tunnelling correction applied) to 75.5 and 26.9 with different tunnelling corrections.[45] These results are in fair agreement with experiment.

The quasi-classical trajectory calculations of Schatz[107] lead to similar conclusions regarding the effects of reagent excitation on reaction (27) as the TST estimates. Schatz found enhancement factors of 1.28 for OH excitation and 393 for H_2 excitation, both results corresponding to thermal reaction at 300 K. A major advantage of trajectory calculations is, of course, that they yield product vibrational state or energy distributions. Converting the internal energies of a polyatomic product molecule into a distribution over rovibrational states is not straightforward. Schatz proceeded via the 'vibrational good actions'. At 6.9 kcal mol^{-1} collision energy and reaction between vibrationally unexcited reagents, the yield of vibrational energy in H_2O corresponds to $\langle f_{v'} \rangle = 0.56$ when one counts the vibrational energy from the classical potential energy associated with the products. The reaction demonstrates relatively little specificity regarding the individual modes, the energy being divided in the ratios 0.19 : 0.51 : 0.31 between v_1, v_2, and v_3.[107a] However, any additional energy coming from reagent vibrational excitation was found to end up as product vibration, with all of it going to the H_2O stretching modes when OH is initially excited and a more random distribution when $H_2(v = 1)$ reacts. This result is further evidence of how the OH moiety acts virtually as a spectator as reaction takes place.

The proposition that, in radical–molecule reactions, the radical is scarcely coupled to the crucial motions carrying the system from reagent to product configuration space is given strong support by the results of a number of recent studies of reactions of the OH[108] and CN[109] radicals. These experiments use pulsed photolysis (of HNO_3 or NCNO) in processes that yield a small fraction of radicals in $v = 1$, as well as the great majority in the lowest vibrational state. The subsequent kinetic decays are followed using time-resolved, state-specific LIF. The decays of vibrationally excited radicals do, of course, correspond to those for the sum of two processes, one reactive, *e.g.*

$$OH(v = 1) + HX \rightarrow H_2O + X \tag{29a}$$

and one the energy transfer process

[108] (*a*) B. D. Cannon, J. S. Robertshaw, I. W. M. Smith, and M. D. Williams, *Chem. Phys. Lett.*, 1984, **105,** 380; (*b*) I. W. M. Smith and M. D. Williams, *J. Chem. Soc., Faraday Trans.* 2, 1986, **82,** 1043.

[109] (*a*) J. de Juan, I. W. M. Smith, and B. Veyret, *Chem. Phys. Lett.*, 1986, **132,** 108; (*b*) I. R. Sims and I. W. M. Smith, *ibid.*, 1988, **149**, 565; (*c*) I. R. Sims and I. W. M. Smith, ibid., 1988, **151**, 481.

$$OH(v = 1) + HX \rightarrow OH(v = 0) + HX \qquad (29b)$$

Consequently, the ratio of observed rate constants, $k_{v=1}/k_{v=0}$, should be looked upon as an upper limit to the ratio of rate constants for *reaction* of $OH(v = 1)$ and $OH(v = 0)$.

Table 3 *Rate constants at 298 K for reaction of* $OH(v = 0)$ *and* $CN(v = 0)$ *with various molecules, and the ratios of the rate constants for total*[a] *removal of the radicals in* $v = 1$ *to those for* $v = 0$

Reactants	k/cm^3 molecule^{-1} s^{-1}	$k_{v=1}/k_{v=0}$	*Reference*
$OH + H_2$	6.7×10^{-15}	<1.5	103
$OH + HBr$	1.1×10^{-11}	0.72	108*a*
$OH + HCl$	6.8×10^{-13}	1.4	108*b*
$OH + DCl$	1.1×10^{-13}	~1.1	108*b*
$OD + DCl$	1.3×10^{-13}	2.3[b]	108*b*
$CN + H_2$	2.5×10^{-14}	1.16	109*b*
$CN + D_2$	8.5×10^{-15}	0.70	109*b*
$CN + HCl$	6.0×10^{-15}	4.5[b]	109*c*
$CN + HBr$	9.2×10^{-13}	0.96	109*c*
$CN + HI$	7.0×10^{-11}	1.15	109*c*

[a]Relaxation *plus* reaction. [b] There is some evidence for significant, non-reactive relaxation in these systems

Table 3 summarizes the results of the studies on OH and CN with saturated molecular species where it is reasonable to assume that reaction occurs by a direct, atom-transfer, mechanism. It is clear that the additional vibrational energy in the radical, amounting to 10.2 kcal mol^{-1} in the case of OH and 5.8 kcal mol^{-1} for CN, brings about at most a very small change in reaction rate. The result of Cannon *et al.*[108a] for the $OH + HBr$ system is in sharp conflict with Spencer and Glass's conclusion[103a] that the rate of this reaction is increased by a factor of ~9 when the state of the radical is increased from $v = 0$ to $v = 1$. However, this earlier finding is quite incompatible with the ideas of vibrational adiabaticity, especially if one bears in mind that thermal reaction between OH and HBr proceeds rapidly and that the rate constant displays virtually no activation energy.

The reaction between $CN(v = 0,1)$ and H_2 has been studied by Li *et al.*[110] as well as by Smith and co-workers.[109] They agree that the reaction rate is scarcely affected by excitation in the CN and these direct studies suggest that an earlier deduction of a large effect[111] must have been mistaken. An *ab initio* calculation of the potential energy surface

[110] X. Li, N. Sayah, and W. M. Jackson, *J. Chem. Phys.*, 1984, **81**, 833.

[111] H. Schacke, H. Gg. Wagner, and J. Wolfrum, *Ber. Bunsenges. Phys. Chem.*, 1977, **81**, 670.

for this system has been carried out by Bair and Dunning.[112] The barrier is fairly early, the H—H distance having increased by only 0.07 Å at the saddlepoint whereas the H—CN distance is 0.38 Å longer than its eventual value in HCN. However, the H—H stretch frequency is already lowered by $\sim 1510\ \mathrm{cm}^{-1}$ at the saddlepoint. No full mode analysis along the MEP is yet available so the energy of the $CN + H_2(v = 1)$ adiabatic maximum is not known. Nevertheless, it appears likely that this reaction will be substantially accelerated if the H_2 is vibrationally excited.

The barrier on the Bair–Dunning potential energy surface has been scaled down to bring TST estimates of the rate of reaction at room temperature into agreement with experiment, and then the results of further TST calculations[113,109c] have been compared with the temperature dependence of the rates, the kinetic isotope effect, and the effect of reagent vibration. The CN group vibrational frequency is predicted by the *ab initio* calculations to change very little along the reaction path, and Wagner and Bair estimate[113] that excitation of CN to $v = 1$ will increase the reaction rate by 50% – in reasonable accord with experiment.

The observation that the kinetics of reactions like (23) are scarcely affected by vibrational excitation of AB, *i.e.* that the radical reagent remains a virtual spectator to the reaction, is in keeping not only with the general notions of vibrational adiabaticity and of weak coupling between the AB stretch and the reaction co-ordinate motion, but it also accords with the limited information available regarding energy partitioning in exothermic reactions like (24).[18] For such reactions, it has been realised for some time – from both theoretical and experimental evidence – that the new (AB) bond will generally be vibrationally excited but that little energy will be released into the vibration of the 'old' (CD) bond. However, it has to be appreciated that experimental evidence on this point is extremely sparse. Most of the studies of energy partitioning in reactions between atomic radicals and triatomic species are for systems where the triatomic is itself a free radical (*e.g.* NO_2, ClO_2) or contains multiple bonds (*e.g.* O_3, NOCl), so that a mechanism involving initial formation of a collision complex may influence the final partitioning of reaction energy.

It is also true that there have been no direct, 'clean', studies on the effects of reagent vibrational excitation in reactions like (24). In cases where the normal modes of BCD are virtually local modes (as in HCN), it is to be expected on the grounds of microscopic reversibility that reaction would be promoted most effectively by excitation of the vibration stretching the bond which is broken in the reaction. Excitation of the other bond stretch is likely to have much less effect. The result of exciting the bending mode is likely to reflect the angular dependence of

[112] R. A. Bair and T. H. Dunning, *J. Chem. Phys.*, 1985, **82,** 2280.
[113] A. F. Wagner and R. A. Bair, *Int. J. Chem. Kinet.*, 1986, **18,** 473.

the potential energy surface for reaction.

The results of studies on four-atom reactions of the AB + CD → ABC + D type strongly affirm that direct bimolecular reactions are generally strongly mode-specific. Studies of the dependence of the rates of A + BCD → AB + CD reactions on the extent and mode of BCD excitation would considerably further our knowledge about mode selectivity and, via VA-TST analyses, about the potential energy hypersurfaces for particular reactions of this kind.

Four-Centre Reactions

Under this heading will be discussed vibrational excitation effects in reactions which involve, in one direction, the unimolecular elimination of a hydrogen halide from a halogenated hydrocarbon and, in the reverse direction, the addition of a hydrogen halide across a double or triple bond. The energy partitioning in the products of four-centre unimolecular reactions has been studied in a limited number of cases using chemical or photochemical activation to prepare molecules with very high internal energy and i.r. chemiluminescence to obtain partial information about the energy disposal.

Examples of particular experiments are the measurement by Clough *et al.*[114] of the HF vibrational distribution from the reaction:

$$CH_3 + CF_3 \rightarrow (H_3C{-}CF_3)^{\dagger} \rightarrow CH_2{=}CF_2 + HF \qquad (\Delta H = -72\ \text{kcal mol}^{-1}) \qquad (30)$$

and the chemical laser study of Berry[115] and the time-resolved i.r. chemiluminescence experiments of Donaldson and Leone[116] on photoelimination of HCl from chlorinated ethenes, *e.g.*

$$CH_2{=}CHCl + h\nu \rightarrow HCl + HC{\equiv}CH \qquad (31)$$

In these cases, the average yield of vibrational energy to the vibration of the hydrogen halide was ~10%; in reaction (30), $\langle f_{v'} \rangle = 0.12$ and in (31), $\langle f_{v'} \rangle = 0.11$. These relatively low yields, although they exceed statistical expectation, indicate that there is no strong propensity for excitation of that particular product mode.

A qualitative explanation for the energy distributions found for reactions of this kind distinguishes between statistical and direct regimes along the reaction path. Excess energy over the minimum required to surmount the exit channel barrier is assumed to be distributed randomly among the modes of the system. Having passed through this region, the

[114] P. N. Clough, J. C Polanyi, and R. T. Tauguchi, *Can. J. Chem.*, 1970, **48,** 2919.
[115] M. J. Berry, *J. Chem. Phys.*, 1974, **61,** 3114.
[116] D. J. Donaldson and S. R. Leone, *Chem. Phys. Lett.*, 1986, **132,** 240.

system proceeds along the multidimensional exit valley in a direct manner. If the framework of either product is markedly different in the transition state from that in the isolated molecule, then energy will be released 'attractively' and excite internal vibrational modes. It is this late, but attractive, energy release following passage through a transition state in which the H—X separation is greater than its final equilibrium value $r_{e,HX}$, which is probably responsible for the majority of the HX excitation observed in reactions (30) and (31). However, the total energy released as the products separate can be estimated from the magnitude of the activation energy for the reverse process. For reaction (31), this is ~56 kcal mol^{-1}, so that the HCl vibrational excitation corresponds to ~25% of this directly released energy, indicating substantial energy release into other product degrees of freedom. The translational energy release has been measured in photofragmentation experiments by Umemoto *et al.*[117] They find a broad distribution of product translational energies with an average release corresponding to 12% of the total energy available and ~27% of that which is released directly as the system recedes from the exit channel barrier.

Theoretical treatments of four-centred elimination reactions have been reported by Kato and Morokuma.[118] They have carried out *ab initio* calculations of the potential energy surfaces for HF elimination from CH_3CH_2F and $CH_2{=}CHF$ and, using the energy gradient method, they have obtained the stationary points on the surfaces, and the changes in energy and geometrical parameters along the MEP (which they call the 'intrinsic reaction co-ordinate'). Some of their results for the CH_3CH_2F system are given in Figure 7. These demonstrate that the adjustments which the system makes as the products separate from the region of the exit channel barrier are by no means simple. In particular, it is to be noted that the intramolecular geometric changes are virtually complete at $s = 2$, where 16 kcal mol^{-1} of energy is still to be released, and indeed even at $s = 1$ (where the energy on the intrinsic reaction co-ordinate is 30 kcal mol^{-1} above its asymptotic value) most of the geometric parameters are quite close to their final values. Kato and Morokuma did not perform a mode analysis along the MEP so it is not possible to construct vibrationally adiabatic curves. However, the analyses of simpler systems (see above) suggest that both the change in frequencies orthogonal to the reaction path and the changes in bond lengths (and bond angles) provide a measure of the strength of couplings between motion along the reaction path and the other motions orthogonal to it. On this basis, it seems fair to assume that motion with respect to the HF stretching vibration will be adiabatic for $s > 1$, and that virtually all the energy released beyond $s = 2$ will go into relative

[117] M. Umemoto, K. Seki, H. Shinohara, U. Nagashima, N. Nishi, M. Kinoshita, and R. Shimada, *J. Chem. Phys.*, 1985, **83**, 1657.

[118] S. Kato and K. Morokuma, *J. Chem. Phys.*, 1980, **73**, 3900; (*b*) S. Kato and K. Morokuma, *ibid.*, 1981, **74**, 6285.

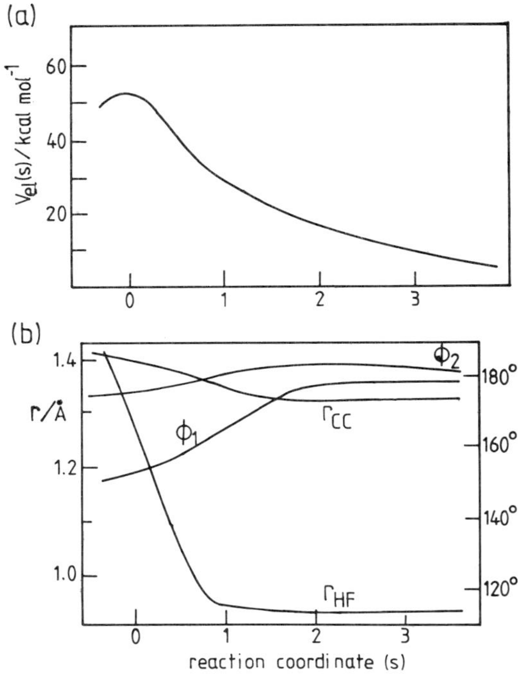

Figure 7 *Variation of* (a) *electronic potential energy and* (b) *some of the intermolecular co-ordinates as a function of distance along the MEP for the reaction*: $CH_3CH_2F \rightarrow CH_2{=}CH_2 + HF$. ϕ_1 *and* ϕ_2 *are the angles between the planes of the* CH_2 *groups and the* CC *bond*; *s is the reaction co-ordinate in mass-weighted units*
(Adapted with permission from *J. Chem. Phys.*, 1980, **73**, 3900)

translational motion of the $CH_2{=}CH_2 + HF$ products. Internal excitation of the HF is likely to be concentrated in the region between $0 > s > 1$, during which a total of only 23 kcal mol^{-1} is available for distribution. These considerations help one to understand why the vibrational energy yield in reactions (30) and (31) is so low.

Microscopic reversibility applied to the results on energy disposal in four-centre elimination reactions suggests that vibrational excitation of the diatomic species alone (or indeed of any other single mode in the reagents) will not very effectively promote the reverse reaction of addition of HX across a double or triple bond. Experiments exploring this possibility have been performed by Herman and Marling[119] on

$$C_2H_2 + HCl,HBr \rightarrow CH_2{=}CHCl,CH_2{=}CHBr \qquad (32)$$

[119] I. P. Herman and J. B. Marling, *J. Chem. Phys.*, 1979, **71**, 643.

and by Klenerman and Zare[120] on

$$C_4H_6 + HCl \rightarrow CH_2{=}CHCHClCH_3 \quad (33a)$$

$$C_4H_6 + HCl \rightarrow CH_3CH{=}CHCH_2Cl \quad (33b)$$

The experimental technique involves excitation of high overtones ($\Delta v = 5,6$) of either the C—H stretch modes or the HX vibration using the tuned radiation of a high-power, continuous-wave dye laser. After irradiation for periods of the order of one hour, the gas is analysed by gas chromatography. In none of these studies was there any evidence for reaction as a result of enhancement of the rate of the selective vibrational excitation. Klenerman and Zare[120] both re-analysed the earlier experiments on reactions (32), in the light of new information about the competing processes of energy transfer, and also deduced from their own experiments an upper limit of 2.1×10^{-14} cm^3 molecule^{-1} s^{-1} for reaction between HCl and butadiene raised to the $v = 5$ level of a C—H stretch mode and therefore containing energy $\sim$16 kcal mol^{-1} in excess of the thermal activation energies for reactions (33a) and (33b).

With some caution, the results of Herman and Marling[119] and of Klenerman and Zare[120] can be understood in terms of the theoretical analysis of similar reactions by Kato and Morokuma.[118] Returning to Figure 7 and considering the addition of HF to $CH_2{=}CH_2$, it would seem that the vibrational motions of HF and the C—C stretch are likely to remain adiabatic up to a position of $s \sim 1$, and that the vibrationally adiabatic curves are likely to be roughly parallel to one another with no great lowering of the adiabatic barriers as the levels of excitation increase. Only once this region of the potential energy surface is reached can energy in the HF vibration be effectively channelled into motion along the reaction co-ordinate. Furthermore, it seems that localized C—H excitations are likely to be even less useful in enabling any reaction system of this type to surmount the potential energy barriers. This is confirmed by the fact that the calculations yield C—H frequencies (in cm^{-1}) in the transition state of 3480 (A''), 3393 (A''), 3377 (A'), and 3307 (A'), compared with calculated values of 3404 (A''), 3374 (A''), 3327 (A'), and 3302 (A') in the isolated $CH_2{=}CH_2$ molecule. This comparison confirms that the C—H stretching modes are only very weakly coupled to motion along the reaction co-ordinate in the $CH_2{=}CH_2$ + HF addition reaction, and probably also in similar reactions. On a vibrationally adiabatic basis it is unlikely that C—H excitations bring about other than a slight lowering of the very high activation energy for these four-centre addition reactions.

Of course, a further aspect of reactions involving highly excited large molecules is that the energy originally deposited in a particular mode

[120] D. Klenerman and R. N. Zare, *Chem. Phys. Lett.*, 1987, **130**, 190.

may be randomized among several or all modes of the molecule on a time-scale which is short compared with the time characteristic of relaxation and reaction for the excited molecules. To estimate a rate constant for addition in the limit of this behaviour, one could estimate the probability $P_j(E)$ of a particular state j with energy E being occupied (using standard techniques borrowed from unimolecular reaction rate theory), then use transition-state theory to calculate a rate constant $k_j(E)$ for each adiabatic channel. The overall, energy-specific, rate constant would be:

$$k(E) = \sum_j k_j(E)P_j(E) \tag{34}$$

Such calculations would be quite straightforward to perform if the necessary information was available about how the mode frequencies change along the reaction path, as it is in the case of HF elimination from CH_2FOH.[121] In the absence of these data for any four-centre elimination like those studied by Herman and Marling[119] and Klenerman and Zare,[120] one can only fall back again on qualitative arguments based on the calculations of HF elimination from CH_3CH_2F. As has already been stated, it seems that for $s > 2$, all the internal co-ordinates are close to their asymptotic, $CH_2{=}CH_2$ + HF, values. Therefore, it is likely that all the vibrationally adiabatic curves will be nearly parallel in this region and it is very unlikely that the vibrationally adiabatic maximum of any of them will be substantially lower than the value of 16 kcal mol^{-1}, the energy on the MEP at $s = 2$. Although the maxima on some of the adiabatic potential curves may be significantly lowered from that on the ground-state curve, this decrease will still leave a substantial barrier to reaction, with the result that the rate of reaction is quite unable to compete with that of the collisional relaxation.

Reactions over Attractive Potentials: Radical–Radical Reactions

So far emphasis has been placed on direct bimolecular reactions with no deep potential minima – or 'wells'. The only exception has been reaction (20) between $O(^1D)$, and H_2, which apparently takes place over a surface corresponding to the electronic ground state of H_2O.[83] The deep potential energy well on this surface is accessed via sideways approach of the $O(^1D)$ atom to H_2 and its transitory insertion between the two H atoms. Sloan and his co-workers[122] have investigated the product vibrational state distributions from several other reactions of $O(^1D)$

[121] K. Yamashita, T. Yamabe, and K. Fukui, *Theoret. Chim. Acta* (*Berlin*), 1982, **60,** 523.

[122] (*a*)P. M. Aker, B. I. Niefer, and J. J. Sloan, *J. Chem. Phys*., 1987, **87,** 203; (*b*) E. J. Kruus, B. I. Niefer, and J. J. Sloan, *ibid*., 1988, **88,** 985.

which proceed by initial insertion of the atom into a single bond in the other reagent.

An interaction via a deep potential well could, as long as there is no exit channel barrier, give rise to a statistical or phase space distribution over product energy states, but this will only happen if the complex survives for at least several vibrational periods. How factors, such as well depth, overall reaction exothermicity, and complexity (*i.e.* number of atoms) determine the transition between direct and statistical dynamics is not yet fully understood.

Reactions between free radicals[123] often proceed across a potential energy surface with a deep well. In general, more than one potential energy surface will correlate with the separated radicals – since they must possess at least spin degeneracy – and the lowest of the surfaces will usually be attractive as electrons from each radical pair-up to form a chemical bond. Frequently, the reaction dynamics and chemical kinetics will be dominated by collisions that occur on this lowest surface. Broadly speaking, the result of such complex-forming collisions will depend on whether there is an exit channel from this surface which is lower in energy than the entrance channel. Two examples where this is the case are reactions whose dynamics will be discussed in this section:

$$H + NO_2 \rightarrow OH + NO \quad (\Delta H_0^\circ = -29.5\ \mathrm{kcal\,mol^{-1}}) \tag{35}$$

$$CN + O_2 \rightarrow NCO + O \quad (\Delta H_0^\circ = -6.5\ \mathrm{kcal\,mol^{-1}}) \tag{36}$$

when there is no lower exit from the potential surface than the channel by which the complex was formed, then the only possibility for chemical reaction is radical–radical association or recombination. An example that has recently been studied[124,125,109c] is

$$CN + NO(+ M) \rightarrow NCNO\ (+ M) \tag{37}$$

Its rate is dependent on total pressure, and to some extent on the nature of the third-body gas M, because collisions are required to stabilize the energized adduct or collision complex formed in binary collisions between CN and NO.

Of course, other types of radical–radical reaction do occur. An example is that between F atoms and HCO to yield HF and CO,[83] but this proceeds via addition and subsequent elimination of HF, a second stage which requires progress over a high and well-defined exit channel barrier. The dynamics are like those discussed earlier in respect of four-centre reactions and different from those in reactions like (36),

[123] M. J. Howard and I. W. M. Smith, *Prog. React. Kinet.*, 1983, **12,** 55.
[124] I. R. Sims and I. W. M. Smith, *J. Chem. Soc., Faraday Trans. 2*, 1988, **84,** 527.
[125] X. Li, N. Sayah, and W. M. Jackson, *J. Chem. Phys.*, 1985, **83,** 616.

which proceed through one step involving formation of one bond and a second step in which one other bond breaks.

The relationship between the kinetics of radical–radical association and of vibrational relaxation in collisions between the same pair of radicals has been discussed at length by Howard and Smith,[123] and only a brief summary of the reasons for this connection will be given here. The basic notion is that the rate constants for association in the limit of high pressure (k^{∞}) and for vibrational relaxation (k_{relax}) should be virtually the same, as they essentially measure the formation of the energized complex. In the case of high-pressure association, all the complexes undergo collisional stabilization; in the case of relaxation, the assumption is that energy, including initial vibrational excitation in one of the radicals, is completely randomized during the lifetime of the complex. Redissociation will overwhelmingly yield radicals in their vibrational ground states, since energy in a single vibrational mode or state occupies much less phase space than the same energy distributed over rotation and translation.

The supposition that $k^{\infty} \approx k_{\text{relax}}$ rests on one further assumption of particular importance in the present context: namely, that the rates of complex formation are approximately independent of the initial vibrational states of the radical reagents. For reaction over a monotonically attractive potential, vibrationally adiabatic transition-state theory becomes the statistical adiabatic channel model (SACM).[126] The essential difference between the two is that the SACM treatment takes explicit account of angular momentum effects by including in the expression for the adiabatic energy a term for the energy associated with the conserved angular momentum (J). According to the SACM theory, a specific rate constant for reaction in the direction of dissociation is defined by the expression from microcanonical TST:

$$k(E,J) \leqslant N^{*}(E,J)/h\rho(E,J) \tag{38}$$

where $N^{*}(E,J)$ is the number of 'open' channels, *i.e.* those for which E exceeds the adiabatic maximum, and $\rho(E,J)$ is the density of states in the molecule. A thermal rate constant is obtained by taking a properly weighted sum over J and integrating over E. For our purposes, it is sufficient to note that the adiabatic maxima are likely to lie at large separations between the radicals so that the coupling between the internal motions and relative translation will be very weak. Consequently, channel potentials differing only in the vibrational excitation of one of the radicals will be very nearly parallel. Indeed, this situation corresponds to the limiting behaviour of adiabatic potentials as the

[126] (*a*) M. Quack and J. Troe, *Ber. Bunsenges. Phys. Chem.*, 1974, **78,** 240; 1975, **79,** 170; (*b*) M. Quack and J. Troe, in 'Gas Kinetics and Energy Transfer', eds. P. G. Ashmore and R. J. Donovan (Specialist Periodical Reports), The Chemical Society, London, 1977, Vol. 2, Chap. 5.

electronic barrier height diminishes and moves to an earlier position along the MEP.

The picture that has just been described appears to be borne out by recent measurements[124] on the kinetics of CN(v = 0) and CN(v = 1) with O_2 and with NO. With O_2, CN(v = 0) is removed by reaction (36) with a rate constant $k = 2.5 \times 10^{-11}$ $(T/298)^{-0.45}$ cm^3 molecule^{-1} s^{-1} (*i.e.* which decreases slightly with temperature). The rate constants for CN(v = 1) over the same range of temperatures ($295 \leqslant T/\mathrm{K} \leqslant 761$) are ~25% larger. These results are consistent with a radical–radical reaction occurring over an attractive potential and via a short-lived intermediate. The very slight increase for CN(v = 1) suggests some weak coupling of the CN vibration to the reaction co-ordinate in the transition-state region.

For CN + NO, reaction can occur by association, or possibly to CO + N_2, though Sims and Smith[124] find no evidence to support this latter pathway. The removal of CN(v = 1) by NO is extremely rapid, $k = 7.6 \times 10^{-11}$ $(T/298)^{-0.3}$ cm^3 molecule^{-1} s^{-1}, which is entirely and only consistent with a relaxation mechanism involving formation of a collision complex in which energy randomization is facile.[123] At the total pressures of the experiments (10—30 Torr argon), removal of CN(v = 0) by NO is much slower, the rate is pressure-dependent and decreases sharply with temperature, findings which are all consistent with a radical association reaction in or close to its limiting low-pressure regime.

There have been relatively few studies of the product-state distributions from radical–radical reactions. Reaction (35) between H atoms and NO_2 has, however, been studied several times by various techniques. (It is, of course, no accident that the three radical–radical reactions discussed in this section each involve a stable free radical – O_2, NO, or NO_2 – as one of the reagents.) Klenerman and Smith[127] provide a summary of the various measurements of the OH vibrational distributions. The large rate constant for the reaction and its independence of temperature suggests that reaction occurs predominantly on the singlet surface corresponding to the ground state of HONO. HONO has dissociation energies of 77.9 and 48.4 kcal mol^{-1} with respect to H + NO_2 and OH + NO, respectively. Molecular beam experiments[128] yield a nearly isotropic angular distribution and reveal no evidence for reaction across the triplet surface which additionally correlates with reagents and products. The product energy distributions that are found do not correspond entirely to those expected if full randomization occurs in a long-lived collision complex, but the energy released in the reaction is distributed quite 'evenly' among the products' degrees of freedom. As

[127] D. Klenerman and I. W. M. Smith, *J. Chem. Soc., Faraday Trans. 2*, 1987, **83**, 229, and references therein.

[128] H. Haberland, W. V. Lucadou, and P. Rohwer, *Ber. Bunsenges. Phys. Chem.*, 1980, **84**, 507.

a whole, the observations suggest that reaction takes place via a complex which usually survives for a few, but not many, vibrational periods before dissociating to OH + NO.

Reaction Between OH *and* CO

The final case history to be considered is the reaction between OH radicals and CO. It plays an important role in atmospheric chemistry and in combustion, being the final step in the oxidation of carbon-based fuels to CO_2. Here it is considered as an example of a reaction whose first stage is the addition of a free radical to a multiple bond. These reactions usually have low but significant activation energies ($< 1\ \text{kcal mol}^{-1}$) and appear to involve tighter transition states than those in radical–radical reactions.

In order to explain the unusual temperature dependence of the reaction rate, it was suggested[129] over a decade ago that reaction between OH and CO proceeds via the formation of a transitory (HOCO)† complex which can decompose at comparable rates back to OH + CO or to H + CO_2. More recently, and in keeping with that proposition, it has been established that the overall rate of reaction depends on total pressure, so that the complete mechanism for reaction is:

$$\begin{array}{ccccc} \text{OH} + \text{CO} & \underset{k_{-c}}{\overset{k_c}{\rightleftarrows}} & (\text{HOCO})^\dagger & \overset{k_p}{\rightarrow} & \text{H} + \text{CO}_2 \\ & & k_m \downarrow +\text{M} & & \\ & & \text{HOCO} & & \end{array} \tag{39}$$

Although the HOCO radical has not yet been identified in gas-phase experiments, it has been observed in matrices[130] and its role in the reaction between H and CO_2 has been clearly established in experiments by Zewail and co-workers.[131] Photolysing HBr in a CO_2–HBr van der Waals complex, they measured a delay of several picoseconds for the appearance of the OH reaction product.

Ab initio calculations on the HOCO surface have been carried out,[132] with effort being concentrated on the maxima and minima of the potential energy. Results for the reaction pathway via HOCO are shown in Figure 8. It is clear that the energies of the classical barriers and of the adiabatic vibrational ground-state maxima, TSl separating OH + CO

[129] (*a*) I. W. M. Smith and R. Zellner, *J. Chem. Soc., Faraday Trans. 2*, 1973, **69,** 1617; (*b*) I. W. M. Smith, *Chem. Phys. Lett.*, 1977, **49,** 112.

[130] (*a*) D. E. Milligan and M. E. Jacox, *J. Chem. Phys.*, 1971, **54,** 927; (*b*) M. E. Jacox, *ibid.*, 1988, **88,** 4598.

[131] N. F. Scherer, L. R. Khundkar, R. B. Bernstein, and A. H. Zewail, *J. Chem. Phys.*, 1987, **87,** 1451.

[132] G. C. Schatz, M. S. Fitzcharles, and L. B. Harding, *Faraday Discuss. Chem. Soc.*, 1987, **84,** 359.

from HOCO and TS2 separating HOCO from H + CO_2, are very similar, as suggested earlier.[129]

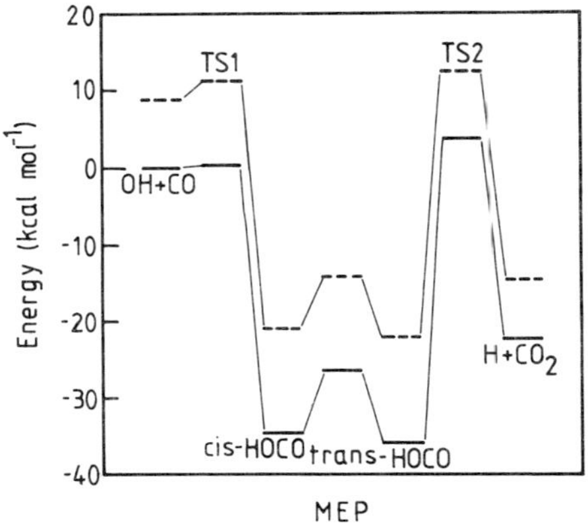

Figure 8 *Energy level diagram showing minima and saddlepoint energies on the MEP for* OH + CO → H + CO_2. *Full lines show potential energies from the best estimates of Schatz* et al.;[132] *dashed lines show levels including zero-point energies estimated from the calculated vibrational frequencies*

There have been extensive measurements on the thermal kinetics of OH + CO and OD + CO. Here only vibrational effects, and what they reveal about the reaction dynamics, are considered in any detail. An attempt to observe the CO_2 vibrational state distribution was made some time ago[133] but no i.r. chemiluminescence was detected and only upper limits could be set on the extent of vibrational excitation. In fact, the form of the potential energy surface suggests that rather little vibrational excitation of the CO_2 reaction product is to be expected. At TS2, the CO bond lengths are extended by only 0.03 Å from their equilibrium values in CO_2, so there will be very little 'attractive' energy release into the CO_2 bond-stretch modes. On the other hand, the OCO bond angle is calculated to be 157.4° at TS2 so that some excitation of the low-frequency bending vibration is to be anticipated.

Dreier and Wolfrum[134] have shown that the rate of reaction between OH and CO is not significantly increased if the CO vibrational temperature is raised to 1800 K. Recent measurements by Brunning *et al.*[135] have shown that the total removal rates of OH(v = 1) and OD(v = 1)

[133] D. W. Trainor and C. W. Von Rosenberg, *Chem. Phys. Lett.*, 1974, **29**, 35.

[134] T. Dreier and J. Wolfrum, 'Proceedings of the 18th International Symposium on Combustion', Combustion Institute, Pittsburgh, 1981, p. 801.

[135] J. Brunning, D. W. Derbyshire, I. W. M. Smith, and M. D. Williams, *J. Chem. Soc., Faraday Trans. 2*, 1988, **84**, 105.

are, respectively, factors of ~7 and 17 greater than the *reaction* rates for radicals in their vibrational ground states. These results are compatible if it is assumed that Brunning's results measure rate constants k_c for the sum of reaction and relaxation, both proceeding via the formation of strongly bound (HOCO)† complexes, whereas Dreier and Wolfrum measure $k_c[k_p/(k_{-c} + k_p)]$ [ignoring the possibility of collisional stabilization of (HOCO)†] and how this purely reactive rate constant depends on the vibrational temperature of CO.

The results obtained for OH(v = 1),OD(v = 1) + CO illustrate how measurements on vibrationally excited species can provide important insights into the detailed mechanism of some reactions. Based on Schatz's[132] potential energy surface calculations and their own measurements of k_c for OH(v = 1) and OD(v = 1), Brunning *et al.*[135] performed TST calculations which satisfactorily reproduce all the available kinetic data for the OH,OD + CO reactions. The only adjustable factors were the level of the OH(v = 1) + CO adiabatic maximum in TS1 and the zero-point energy level of TS2. Their calculations suggest that the rate of reaction between OH and CO(v = 1) is actually less than that between OH and CO(v = 0) because a higher fraction of the complexes formed in the first case decompose back to OH + CO. However, at a CO vibrational temperature of 1800 K, the overall rate of reaction would be decreased by ~10% compared with the thermal value at 300 K, in good agreement with the measurements.[134]

4 Summary and Conclusions

This article has attempted to demonstrate that vibrationally adiabatic transition-state theory provides a theoretical framework for understanding many of the effects of vibrational excitation in bimolecular reactions. It appears that for most exothermic or thermoneutral reactions the assumption that motion up to the transition state is vibrationally adiabatic is at least approximately valid, and that such a transition state is best defined by finding appropriate pods for which the action over one cycle of its motion is $(v + \frac{1}{2})h$.

Where the initial excitation is of the vibration along the bond broken in the reaction, the adiabatic maxima may be lowered appreciably as the vibrational quantum number associated with this mode is increased. Furthermore, the maxima move to progressively earlier positions along the MEP with the result that the A factor, as well as the activation energy for reaction, depends on v. The extent of any rate enhancement will depend on details of the particular potential energy surface. However, the lower and earlier the saddlepoint, the more parallel the vibrational adiabatic potential energy curves, and the less the rate of reaction depends on reagent vibrational excitation. This is the situation for the F + H_2 reaction. Finally, where the potential at long-range is monotonically attractive, as in the case of reaction between two radicals proceeding via a collision complex, the effect of vibrational excitation of the reagents is likely to be very slight.

Vibrationally adiabatic transition-state theory, and the notions which lie behind it, both predict and explain strong mode-selectivity in direct, bimolecular reactions. This was illustrated earlier by considering the $OH + H_2$ reaction. The OH vibration is only very weakly coupled to motion along the reaction co-ordinate, *i.e.* its frequency remains almost unchanged throughout the reaction. As a result, energy added to the OH as initial vibrational excitation is not made available to help the system surmount the potential energy barrier to reaction. By contrast, excitation of the H_2 vibration can couple to the reaction co-ordinate motion quite efficiently. Although the system probably stays in the same H_2 bond-stretch state up to the transition state, the frequency of this mode decreases, allowing energy to flow into the motion along the reaction co-ordinate.

Mode-specific effects certainly persist for more complex reactions like the four-centre reactions considered in Section 3. The problem here is that although the rate constant may be very much greater than that for reaction between unexcited species it remains very small, and consequently reaction cannot compete favourably with collisional relaxation.

Finally, it is re-emphasized that measurements on the kinetics of vibrationally excited species in potentially reactive collisions may provide insight into the dynamics of those collisions even if the dominant process for loss is non-reactive vibrational relaxation.

CHAPTER 3

Association Reactions

J. W. DAVIES and M. J. PILLING

1 Introduction

Association reactions involving atoms and radicals present an interesting challenge to both experimentalists and theoreticians. In this chapter five reactions will be discussed which are chosen to illustrate a range of attributes shown by reactions that produce or proceed via a stable association complex:

$$H + C_2H_4 \xrightarrow{k_1} C_2H_5 \qquad [1]$$

$$CH_3 + CH_3 \xrightarrow{k_2} C_2H_6 \qquad [2]$$

$$CH_3 + H \xrightarrow{k_3} CH_4 \qquad [3]$$

$$O + OH \xrightarrow{k_4} H + O_2 \qquad [4]$$

$$HO_2 + HO_2 \xrightarrow{k_5} H_2O_2 + O_2 \qquad [5]$$

The reactions occur on three types of potential energy surface (PES), illustrated schematically in Figure 1. Reaction [1] occurs on a Type I PES, in which there is an energy barrier located at a quite small C_2H_4—H bond distance, producing a tight transition state that is 'prescribed', *i.e.* its geometry is independent of the energy of the reaction system at all energies which are significant in a thermal reaction.

Reactions [2] and [3] proceed on a Type II PES, in which no electronic energy barrier faces the incoming reactants. The location of the transition state on such a surface is dependent on the internal energy and moves to small bond distances as this energy increases: variational or adiabatic channel techniques are required to model the reaction.

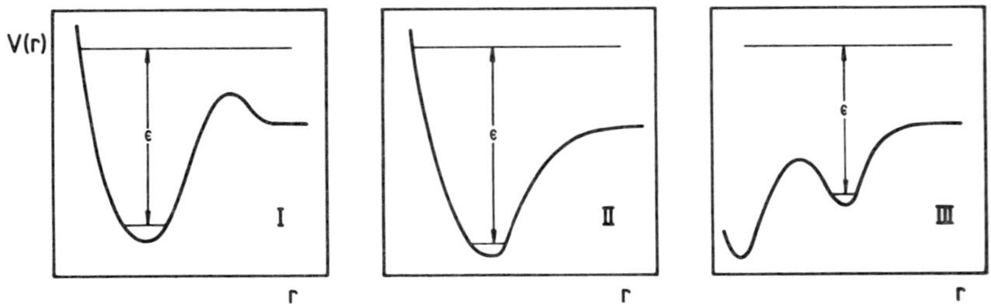

Figure 1 *A schematic representation of the potential energy surfaces for the three reaction types considered*: I – *association with an energy barrier* (e.g. $H + C_2H_4$); II – *association with no energy barrier* (e.g. $CH_3 + CH_3$, $CH_3 + H$); III – *association with a second, lower-energy dissociation channel* (e.g. $O + OH$, $HO_2 + HO_2$)

Reactions [4] and [5] proceed on a Type III PES which involves the formation of a stable, or comparatively stable, intermediate (HO_2 or H_2O_4), followed by dissociation via a lower energy channel. Reactions [4] and [5] show very different dependences on pressure, determined by the relative magnitudes of the rate coefficients for reverse and forward dissociation of the intermediate. At high temperatures, a second channel also opens up for the $CH_3 + CH_3$ system:

$$CH_3 + CH_3 \rightarrow C_2H_5 + H \qquad [6]$$

Thus this reaction also proceeds on a Type III surface, the difference from reactions [4] and [5] being that the second channel corresponding to reaction [6] is endothermic, with neither channel showing an energy maximum.

Experimental Investigations

In an ideal kinetic experiment, the removal of a reactant A is examined in the presence of an excess of reactant B, so that the reaction shows pseudo-first-order kinetics in isolation from any other potential reactions. Alternatively, the second-order reaction, A + A, is studied in isolation. These ideals are rarely met and in the study of the five chosen reactions a range of experimental difficulties are encountered.

Reaction [1] is a classic system that has been investigated by both pulsed (flash photolysis, pulse radiolysis) and discharge–flow techniques. It is ideally studied at low atom and high molecule densities, otherwise subsequent atom–radical reactions contribute to the H atom decay. Such conditions have rarely been fully established because of restrictions of detection sensitivity or accessible timescale, and allowances need to be made for the contributions from these secondary reactions.

Reactions [2] and [5] are second-order in radical concentration and present two difficulties: first, the establishment of the *absolute* radical density and second, the provision of conditions under which the radical is generated in isolation from other reactive species. For reaction [2], the latter difficulty was easily overcome because of the availability of clean photolytic radical sources. HO_2, on the other hand, is much more difficult to generate and this problem has, for several years, hampered the investigation of reaction [5] over a wide range of temperatures.

In many ways, reactions [3] and [4] present the most exacting experimental problems, since they require the generation of two different radicals. The development of numerical methods of integration of coupled differential rate equations led to a series of investigations in which reaction [3] was one of several contributors to a spectrum of stable products or to the decay of a transient species. Such investigations, although of great value and sometimes the only approach available, often suffer from mechanistic assumptions and fall well short of our idealized experiment. In recent years, however, techniques have been developed which allow such reactions to be studied in isolation, and three such techniques that have been applied to reactions [3] and [4] are described below.

Data Fitting and Extrapolation

Reactions [1]—[3] are classic termolecular reactions and have, in consequence, pressure-dependent rate coefficients:

$$A + B \underset{k_d}{\overset{k_a}{\rightleftharpoons}} AB^*$$

$$AB^* + M \xrightarrow{k_s[M]} AB + M$$

As the pressure increases, a larger fraction of the nascent adduct AB^* is stabilized and the rate coefficient, k, increases towards a limiting value, k^∞ (Figure 2). A rise in temperature leads to increases in the energies of (A + B) and AB^* and, consequently, to an increase in k_d. Higher third-body pressures are then needed to stabilize AB^*. At a given pressure, the reaction moves further into the fall-off region, and $k(P,T)$ further from the high-pressure limit, as the temperature increases. Figure 2b shows this effect for a reaction occurring on a Type II surface, demonstrating the negative temperature dependence of the rate coefficient at fixed [M].

Reactions [1]—[3] are of significance in combustion, and rate coefficient data are required for combustion modelling at elevated temperatures, often (indeed usually) at temperatures well in excess of those accessible in the laboratory. It is necessary, therefore, to develop a concise but accurate representation of $k(P,T)$ that provides a reliable means of extrapolation from the experimental to the combustion (P,T)

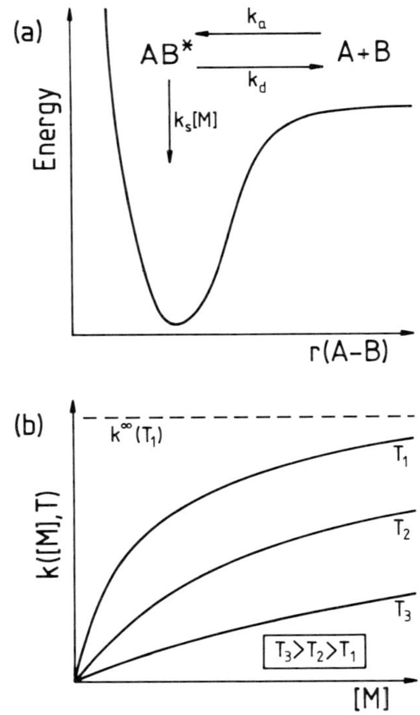

Figure 2 (a) *Schematic representation of the rate processes involved in an association reaction.* (b) *Schematic representation of the pressure and temperature dependence of the rate coefficient for an association reaction occurring on a Type II surface*

regime. It has become customary to employ a representation developed by Troe and co-workers[1–3] which is discussed in Section 2, while in Section 3 a technique based on this representation is described for fitting experimental $k(P, T)$ data.

Theoretical Modelling

Strictly speaking, the Troe representation generates empirical parameters whose validity outside the experimental regime is questionable. Extrapolation is likely to be much more reliable if the experimental measurements are supplemented with a detailed theoretical model. The development of such models is not only a desirable, or even necessary, prerequisite for extrapolation, but also a challenging theoretical exercise in its own right. Each of the reactions [1]—[5] has been subject to detailed theoretical investigation, and together these studies illustrate

[1] J. Troe, *J. Phys. Chem.*, 1979, **83,** 114.
[2] J. Troe, *Ber. Bunsenges. Phys. Chem.*, 1983, **87,** 161.
[3] R. G. Gilbert, K. Luther, and J. Troe, *Ber. Bunsenges. Phys. Chem.*, 1983, **87,** 169.

the wide range of techniques available. Rather than review these techniques separately, we have chosen to discuss them in the specific sections below. We summarize here the approaches adopted and the connections between them.

Since reaction [1] occurs on a Type I PES, conventional RRKM theory can be applied to calculate the microcanonical rate coefficients $k_{\mathrm{d}}(\varepsilon)$ or $k_{\mathrm{d}}(\varepsilon,J)$ for dissociation of the adduct AB*, where ε is the total internal energy and J the overall rotational quantum number of AB*. The limiting high-pressure dissociation rate coefficient, $k_{\mathrm{diss}}^{\infty}$, can then be calculated by integration over a Boltzmann distribution, and the limiting high-pressure rate coefficient for association, $k_{\mathrm{ass}}^{\infty}$, can be obtained from $k_{\mathrm{diss}}^{\infty}$ and the equilibrium constant, K_c. The problem reduces, therefore, to one of determining the geometry and vibrational frequencies of the complex, and *ab initio* molecular-orbital calculations have begun to play a significant role in this respect.

The positions of the transition states in reactions [2] and [3] are not prescribed since Type II surfaces are involved, and a variational approach is necessary to determine $k_{\mathrm{d}}(\varepsilon,J)$. The techniques can be illustrated by reference to one of the early microcanonical calculations on reaction [2],[4] which serves to demonstrate the difficulties involved, and to a recent model, the flexible transition-state model,[5,6] in which many of these difficulties have been overcome. An alternative approach is that of the statistical adiabatic channel model (SACM) adopted by Quack and Troe,[7] which relies on tracing individual quantum-state-resolved channels through from reactants to products and determining the number of such channels which are 'open' at a given energy. Another approach, which provides an easier calculation of $k_{\mathrm{ass}}^{\infty}$, is to use a variational canonical technique, and such calculations will be discussed for reaction [3].

Thus far, we have concentrated on the calculation of the microcanonical dissociation rate coefficients and not mentioned the other important aspect of modelling of association reactions, that of energy transfer. One approach is to use the Troe procedure to obtain $k_{\mathrm{ass}}^{\infty}$ by extrapolation of the experimental rate data and to compare these values with those predicted by theory. An alternative and superior (although more time-consuming) approach is to incorporate model $k_{\mathrm{d}}(\varepsilon,J)$ values into a master equation which describes the energy-dependent population distribution of AB*. Such an approach is described briefly by reference to reaction [3], where disagreement between theory and extrapolated experimental data demanded a more detailed investigation which, it must be admitted, did not resolve the anomaly.[8]

[4] W. L. Hase, *J. Chem. Phys.*, 1976, **64,** 2442.
[5] D. M. Wardlaw and R. A. Marcus, *J. Phys. Chem.*, 1986, **90,** 5383.
[6] D. M. Wardlaw and R. A. Marcus, *J. Chem. Phys.*, 1985, **83,** 3462.
[7] M. Quack and J. Troe, *Ber. Bunsenges. Phys. Chem.*, 1974, **78,** 240.
[8] M. Brouard and M. J. Pilling, *Chem. Phys. Lett.*, 1986, **129,** 439.

Finally, we turn to theoretical models of reactions occurring on a Type III surface. The range of approaches possible is well illustrated by reference to reaction [4]. Variational transition-state theory,[9] concentrating on the flux bottleneck for formation of HO_2^*, provides an upper limit for k_4 since it does not allow for recrossing of the transition state region. Such an approach can also lead to a poor representation of the temperature dependence of the rate coefficient, since the recrossing probability is usually energy dependent. Two ways of examining this probability are trajectory calculations[10] and two-channel SACM calculations,[11] both of which have been applied to reaction [4]. This reaction also serves to illustrate recent developments in the statistical adiabatic channel model[11] and the detailed quantum transition-state theory employed by Clary,[12] together with its simplified variant, adiabatic capture theory, which has been applied with great success to ion–molecule reactions.[13]

The dissociation rate coefficients for HO_2^* are very fast, and pressure effects are not evident under conditions investigated experimentally. Reaction [5], by contrast, shows a strong pressure dependence, and models of the reaction must, in consequence, also incorporate collisional energy transfer. It has also been suggested that this reaction has a direct channel which becomes important at high temperatures.[14] Recent experimental data support this contention.[15]

In the following sections reactions [1]—[5] are examined in turn, with descriptions of the experimental investigations of the temperature and, if appropriate, the pressure dependences of their rate coefficients. Theoretical investigations of the reactions are then examined and comparisons made of theoretical and experimental rate coefficients. The reverse dissociation rate coefficients have not been considered in detail, except in the case of reaction [1].

2 $H + C_2H_4 \rightarrow C_2H_5$

Introduction

The addition of hydrogen atoms to ethene to form ethyl radicals:

$$H + C_2H_4 \xrightarrow{k_1} C_2H_5 \qquad [1]$$

[9] S. M. Rai and D.G. Truhlar, *J. Chem. Phys.*, 1983, **79**, 6046.
[10] J. A. Miller, *J. Chem. Phys.*, 1986, **84**, 6170.
[11] J. Troe, *J. Phys. Chem.*, 1986, **90**, 3485.
[12] D. C. Clary, *Mol. Phys.*, 1984, **53**, 3.
[13] D. C. Clary, *Mol. Phys.*, 1985, **54**, 605.
[14] R. R. Baldwin, C. E. Dean, M. R. Honeyman, and R. W. Walker, *J. Chem. Soc., Faraday Trans*. 1, 1986, **82**, 89.
[15] P. D. Lightfoot, B. Veyret and R. Lesclaux, *Chem. Phys. Lett.*, 1988, **150**, 120.

takes place on a Type I potential energy surface, with a prescribed configuration for the activated complex located at the potential maximum (Figure 1). The reaction has been extensively studied experimentally, especially at lower temperatures, and was the first reaction to which the technique of flash photolysis/resonance fluorescence was applied, by Braun and Lenzi.[16a] Reaction [1] and its reverse are central reactions in the industrially important conversion of ethane to ethene,[17] in which high temperatures (1000—1200 K) are employed at pressures of ~2—3 atmospheres. Under these conditions, the reactions are well into the fall-off regime, so that parameterization of $k_1(P,T)$ and $k_{-1}(P,T)$ is necessary. The value of k_{-1} is also important in low-temperature combustion, since it determines the (P,T) regime over which decomposition competes with oxidation of the ethyl radical. Finally, largely through the efforts of Hase and co-workers, reaction [1] has been the subject of numerous theoretical studies.

Experimental Investigations

Experimental studies of reaction [1] which have employed discharge–flow, flash photolysis, and pulse radiolysis are summarized in Table 1. The discharge–flow experiments have used both Lyman–α resonance absorption and time-of-flight (TOF) mass spectroscopy as detection techniques.[18–20] The principal drawback of the discharge–flow method is the significant contribution of secondary reactions to the H atom decay:

$$H + C_2H_5 \rightarrow CH_3 + CH_3 \qquad [-6]$$

$$CH_3 + CH_3 \rightarrow C_2H_6 \qquad [2]$$

$$CH_3 + H \rightarrow CH_4 \qquad [3]$$

so that

$$-\frac{d[H]}{dt} = k_{app}[C_2H_4][H]$$

where $k_{app} = sk_1$ and s is a 'stoichiometric number' greater than unity. Much of the experimental effort of Michael and co-workers[19,20] was directed at the establishment of s by conducting experiments with H or

[16] (*a*) W. Braun and M. Lenzi, *Discuss. Faraday Soc.*, 1967, **44,** 252; (*b*) M. J. Kurylo, N. C. Peterson, and W. Braun, *J. Chem. Phys.*, 1970, **53,** 2776.

[17] M. J. Pilling, in 'Modern Gas Kinetics', eds. M. J. Pilling and I. W. M. Smith, Blackwell Scientific Publications, Oxford, 1987, Chap. C5, p. 316.

[18] J. R. Barker, D. G. Keil, J. V. Michael, and D. T. Osborne, *J. Chem. Phys.*, 1970, **52,** 2079.

[19] J. V. Michael, D. T. Osborne, and G. N. Suess, *J. Chem. Phys.*, 1973, **58,** 2800 and 2807.

[20] J. A. Cowfer and J. V. Michael, *J. Chem. Phys.*, 1975, **62**, 3504.

Table 1 *Experimental studies of* $H + C_2H_4$ *and its isotopic analogues*

Technique[a]	T/K	P/Torr	*Reference*
$\mathbf{H + C_2H_4}$			
FP/TRRF	298	5—75 (He,N_2)	16*a*
FP/TRRF	298	5—500 (He)	16*b*
DF/RA, DF/TOF, HgP/TRRA	298	0.4—25 (He)	18
HgP/TRRA, DF/TOF	298	0.3—600 (He,Ne,Ar,Kr, H_2,N_2,SF_6)	19
FP/TRRF	198—320	300—700 (Ar)	28
PR/TRRA	206—461	200—1200 (H_2)	29
LFP/TRRF	285—600	25—600 (He)	21
Isotopic studies			
HgP/TRRA, DF/TOF	298	1—5 (He)	20
PR/TRRA	206—461	200—1200 (D_2)	29

[a] Key: FP, flash photolysis; LFP, laser flash photolysis; HgP, mercury photosensitization; DF, discharge flow; PR, pulse radiolysis; TRRF, time-resolved resonance fluorescence; TRRA, time-resolved resonance absorption; RA, resonance absorption; TOF, time-of-flight mass spectrometry

with C_2H_4 in excess, and by careful measurements of the concentrations of H, CH_3, CH_4 and C_2H_6, coupled with computer simulations. This latter study generated one of the more reliable estimates of $k_{-6}(= 3 \times 10^{-11}\ \text{cm}^3\ \text{molecule}^{-1}\ \text{s}^{-1})$.[19] Once s has been established, and values of ~1.7 are typical for excess C_2H_4, room-temperature values for k_1 are obtained in agreement with pulsed experiments. The discharge–flow technique provides the more accurate means of obtaining rate data at low pressures.

Pulsed measurements have employed both resonance absorption and resonance fluorescence as detection techniques. These measurements can also be affected by the secondary reactions [2], [3], and [−6], as is illustrated by the recent work of Lightfoot and Pilling.[21] They used exciplex laser flash photolysis (193 nm) of N_2O–H_2 mixtures, which, at least at higher temperatures, is a clean source of H and does not

[21] P. D. Lightfoot and M. J. Pilling, *J. Phys. Chem.*, 1987, **91,** 3373.

produce any co-radicals:

$$N_2O \rightarrow N_2 + O(^1D)$$
$$O(^1D) + H_2 \rightarrow H + OH$$
$$OH + H_2 \rightarrow H + H_2O$$

and found that, although their plots of the first-order decay coefficient for [H] versus $[C_2H_4]$ were linear, they extrapolated ($[C_2H_4] \rightarrow 0$) to rate coefficients significantly greater than that expected for diffusive loss of H from the monitoring zone. Simulations demonstrated that reactions [2], [3] and [−6] played a perceptible role in the H atom decay and accounted for the enhanced apparent decay on extrapolation to $[C_2H_4] = 0$. The simulations also demonstrated that, provided experiments were conducted at constant $[H]_0$ (*i.e.* at constant $[N_2O]$ and constant laser energy), the slopes of plots of the apparent first-order decay coefficient, k_{app}, versus $[C_2H_4]$ were very close approximations (within a few per cent) to k_1. This observation enabled an accurate technique to be devised for determining k_1, despite the presence of a non-zero contribution from reactions [2], [3] and [−6]. Figure 3 shows fall-off plots for k_1, in a helium diluent, over the temperature range 285—600 K, demonstrating the enhanced fall-off behaviour (over this pressure range) at the higher temperatures.

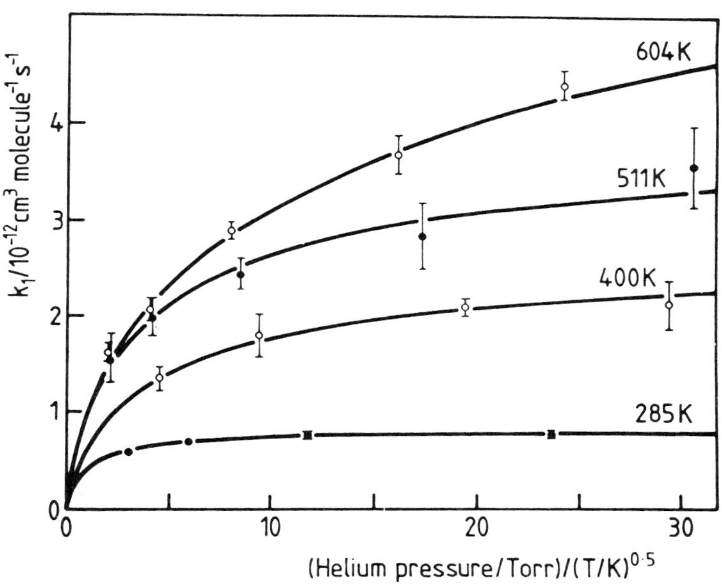

Figure 3 *Dependence of k_1 on temperature and pressure in a helium diluent.*[21] *(Error bars ≡ 95.5% confidence limits)*

Data Fitting and Extrapolation to Higher Temperatures

If we are to estimate rate data for use in hydrocarbon cracking models at temperatures of 1000—1200 K, then we must fit the experimental data, in some way, to provide a realistic parameterization of both the pressure dependence and temperature dependence of k_1. There are several techniques, of varying degrees of sophistication, available for effecting such a fit and most of them will be discussed in this chapter. The technique which has been used for $H + C_2H_4$ is based on a widely employed procedure developed by Troe and co-workers.[1-3]

The Troe Formalism. Troe's technique is based on a modification of the original theory of unimolecular reactions developed by Lindemann[22] and Hinshelwood[23]:

$$\begin{aligned} AB + M &\rightarrow AB^* + M \\ AB^* + M &\rightarrow AB + M \\ AB^* &\rightarrow A + B \end{aligned}$$

A reduced form for the pressure-dependent overall rate coefficient, k, for decomposition of AB may be expressed in the form:

$$\frac{k}{k^\infty} = \frac{k^0[M]/k^\infty}{1 + k^0[M]/k^\infty} = F_{LH} \tag{1}$$

where k^0 is the second-order limiting low-pressure rate coefficient, k^∞ is the first-order limiting high-pressure rate coefficient, and [M] is the third-body concentration.

Equation (1) predicts a sharper fall-off in k than is found experimentally. Two effects lead to a broadening of the curve:

(*a*) The Lindemann mechanism requires that all energized molecules, AB^*, decompose with the same rate coefficient, regardless of energy. In a real molecular system, by contrast, the microcanonical rate coefficient increases with energy, as described by RRKM theory.[24]

(*b*) Lindemann–Hinshelwood and conventional RRKM theories both apply the strong collision approximation, whereby the energies of AB^* before and after collision are uncorrelated and deactivation may be considered to occur in a single collision. A more realistic description invokes weak collisions where energy is removed in a sequence of collision-induced energy 'jumps'.

[22] F. A. Lindemann, *Trans. Faraday Soc.*, 1922, **17,** 598.

[23] C. N. Hinshelwood, 'The Kinetics of Chemical Change in Gaseous Systems', 3rd Ed., Clarendon Press, Oxford, 1933.

[24] R. A. Marcus, *J. Chem. Phys.*, 1952, **20**, 359.

Troe and co-workers[1-3] developed a modification to equation (1) which incorporates corrections for these two effects and is simple to implement. The reduced fall-off curves are expressed in terms of the product of F_{LH} and two correction factors, F^{SC} and F^{WC}:

$$k/k^{\infty} = F_{LH}F^{SC}F^{WC} \tag{2}$$

The strong collision broadening factor, F^{SC}, accounts for the broadening of fall-off curves as a result of the energy dependence of the microcanonical dissociation rate coefficient for AB*. F^{SC} is a function of the ratio $k^0[M]/k^{\infty}$ and the modified Kassel parameters S_T and B_T.[2]

S_T is related to the internal energy of the activated complex, $U^{\dagger}_{vib}$:

$$S_T = 1 + \frac{U^{\dagger}_{vib}}{k_B T} \tag{3}$$

and B_T can be approximated to:

$$B_T \approx \left[\frac{S_T - 1}{s - 1}\right]^{1.6}\left[\frac{\varepsilon_0 + a(\varepsilon_0)\varepsilon_z}{k_B T}\right] \tag{4}$$

where s is the number of oscillators in AB, ε_0 is the threshold energy, ε_z is the zero-point vibrational energy of AB, and $a(\varepsilon_0)$ is the Whitten–Rabinovitch correction factor.[25]

The broadening factor, F^{SC} is a nearly symmetrical function relative to $k^0[M]/k^{\infty} = 1$, the 'centre of the fall-off curve', and can thus be characterized by its value at the centre, $F^{SC}_{cent} \equiv F^{SC}$ ($[M] = k^{\infty}/k^0$), and a width parameter, N^{SC}:

$$\log F^{SC} \approx \frac{\log F^{SC}_{cent}}{1 + \left[\dfrac{\log(k^0[M]/k^{\infty})}{N^{SC}}\right]^2} \tag{5}$$

Troe[1,2] analysed twenty molecular systems using RRKM theory and found that F^{SC}_{cent} was best represented by the empirical expression

$$-\log F^{SC}_{cent} = \frac{(1.06 \log S_T)^{2.2}}{1 + C_1 S_T^{C_2}} \tag{6}$$

with

$$C_1 = 0.10 \exp[2.5B_T^{-1} - 0.22B_T - (6 \times 10^{-10}B_T^6)] \tag{7}$$

$$C_2 = 1.9 + (4.6 \times 10^{-5}B_T^{2.8}) \tag{8}$$

[25] G. Z. Whitten and B. S. Rabinovitch, *J. Chem. Phys.*, 1963, **38**, 2466.

The width parameter, N^{SC}, results from the dependence of F^{SC} on the reduced pressure $k^0[M]/k^\infty$, and is related to F_{cent}^{SC} by

$$N^{SC} \approx 0.75 - (1.27 \log F_{cent}^{SC}) \tag{9}$$

More complex expressions, allowing for slight asymmetries in F^{SC}, may be found in reference 2.

Troe[1,2] defined a parameter β_c which is related to the collision efficiency of the third body, M, ($\beta_c = 1$ for strong collisions). The broadening associated with weak collisions is accounted for by the factor F^{WC}, whose form was determined empirically by comparison with full master equation calculations for a wide variety of unimolecular reactions. As for F^{SC}, F^{WC} is expressed in terms of its value at the centre of the fall-off curve, F_{cent}^{WC}:

$$\log F^{WC} \approx \frac{\log F_{cent}^{WC}}{1 + \left\{ \dfrac{\log(k^0[M]/k^\infty) + c}{N^{WC} - d[\log(k^0[M]/k^\infty) + c]} \right\}^2} \tag{10}$$

By definition, $\log F_{cent}^{WC} = 0$ for strong collisions, *i.e.* when $\beta_c = 1$, and for decreasing β_c, $\log F_{cent}^{WC}$ decreases:

$$\log F_{cent}^{WC} \approx [0.16 \log \beta_c + 0.04(\log \beta_c)^2][1 + (\log F_E)^{2.2}] \tag{11}$$

where F_E corrects for the energy dependence of the density of states above the threshold, and is approximated by:

$$F_E \approx \sum_{i=0}^{s-1} \frac{(s-1)!}{(s-1-i)!} \left[\frac{k_B T}{\varepsilon_0 + a(\varepsilon_0)\varepsilon_z} \right]^i \tag{12}$$

The width, shift, and asymmetry parameters N^{WC}, c, and d are calculated from the relationships:

$$N^{WC} \approx 1 - (0.086 S_T + 1.7) \log F_{cent}^{SC} - 0.3 \log \beta_c \tag{13}$$

$$c \approx 1.5 (\log F_{cent}^{SC})^2 (1 - 0.5 \log \beta_c) \tag{14}$$

$$d \approx - 0.03 S_T + 0.45[1 - \exp(- 0.3 S_T)] \tag{15}$$

In summary, the Troe semi-empirical representation (2) broadens the Lindemann–Hinshelwood fall-off curve via the two parameters F^{SC}, which depends primarily on S_T, and therefore on the structure of the activated complex, and F^{WC}, which depends on S_T and β_c. The above discussion relates to dissociation reactions, but is easily applied to the reverse association reactions via the equilibrium constant:

$$K_c = k_{ass}/k_{diss}$$

Analysis of Experimental Data for k_1. Equation (2) may be expresed in the form:

$$k/k^{\infty} = F_{\mathrm{LH}}(k^0/k^{\infty})F^{\mathrm{SC}}(k^0/k^{\infty},S_{\mathrm{T}},B_{\mathrm{T}})F^{\mathrm{WC}}(k^0/k^{\infty},S_{\mathrm{T}},\beta_c) \quad (16)$$

thereby emphasizing the dependence of k on four principal parameters: k^0, k^{∞}, S_{T}, and β_c. B_{T} depends primarily on S_{T} and on known molecular parameters. There is insufficient information in a fall-off curve to determine all of these parameters and some assumptions must, therefore, be made.

As will be discussed below, Hase and Schlegel[26] have developed a model of the transition state which enables S_{T} to be calculated via equation (3). Lightfoot and Pilling[21] fitted their data at each temperature with k_1^0 and k_1^{∞} as variable parameters, using a non-linear least squares fitting routine.[27] They found that their fits were comparatively insensitive to β_c, although they quoted their limiting parameters, k_1^0 and k_1^{∞}, with $\langle\Delta\varepsilon\rangle_{\mathrm{down}}$ fixed at a temperature-independent value of 285 cm^{-1}. $\langle\Delta\varepsilon\rangle_{\mathrm{down}}$ is the average energy transferred per collision in a downwards direction and is related to β_c via:[3]

$$\beta_c = \left[\frac{\langle\Delta\varepsilon\rangle_{\mathrm{down}}}{\langle\Delta\varepsilon\rangle_{\mathrm{down}} + F_{\mathrm{E}}k_{\mathrm{B}}T}\right]^2$$

There are some deficiencies in Hase and Schlegel's transition state (see later discussion), but Lightfoot and Pilling demonstrated that the consequent uncertainties in S_{T} had a minimal effect on the optimal values of k_1^0 and k_1^{∞}. They combined their own $k_1^{\infty}(T)$ values with those of Lee *et al.*[28] and Sugawara *et al.*[29] to produce excellent agreement on a linear Arrhenius plot (Figure 4) which gave:

$$k_1^{\infty} = (4.4 \pm 0.6) \times 10^{-11} \exp[-(1090 \pm 40)\ \mathrm{K}/T]\ \mathrm{cm^3\ molecule^{-1}\ s^{-1}}$$

while their own low-pressure data gave:

$$k_1^0 = (1.4 \pm 0.8) \times 10^{-29} \exp[-(570 \pm 220)\ \mathrm{K}/T]\ \mathrm{cm^6\ molecule^{-2}\ s^{-1}}$$

(error limits $\equiv$ 95.5% confidence limits).

These Arrhenius expressions could be used together with $\langle\Delta\varepsilon\rangle_{\mathrm{down}} = 285\ \mathrm{cm}^{-1}$, the vibrational frequencies of the activated complex,[26] and Troe's equations (1)—(15) to calculate $k_1([\mathrm{M}],T)$ directly. Such a procedure is time-consuming, however, and a more explicit

[26] W. L. Hase and H. B. Schlegel, *J. Phys. Chem.*, 1982, **86**, 3901.

[27] P. R. Bevington, 'Data Reduction and Error Analysis for the Physical Sciences', McGraw-Hill, New York, 1969.

[28] J. H. Lee, J. V. Michael, W. A. Payne, and L. J. Stief, *J. Chem. Phys.*, 1978, **68**, 1817.

[29] K. Sugawara, K. Okazaki, and S. Sato, *Bull. Chem. Soc. Jpn.*, 1981, **54**, 2872.

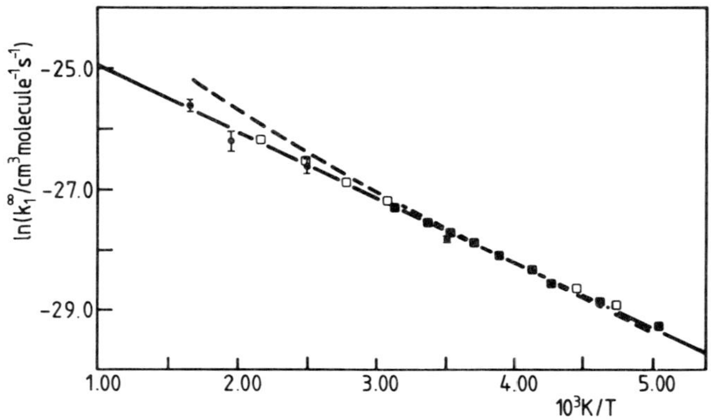

Figure 4 *Arrhenius plot for k_1^∞. Experimental data:* ●, *Lightfoot and Pilling;*[21] ■, *Lee* et al.;[28] □, *Sugawara* et al.[29] ——, *Linear least squares fit;* – – –, *fit using the transition-state model of Hase and Schlegel,*[26] *including a tunnelling correction*

formulation is required if the data are to be realistically, yet efficiently, incorporated into models of complex kinetic processes, such as hydrocarbon cracking. Gilbert *et al.*[3] provided such a formulation, suggesting that F^{SC} and F^{WC} be combined in the form:

$$\log(F^{SC}F^{WC}) \approx \frac{\log F_{cent}}{1 + \left\{\dfrac{\log(k^0[M]/k^\infty) + c}{N - d(\log(k^0[M]/k^\infty) + c]}\right\}^2} \tag{17}$$

with

$$N \approx 0.75 - 1.27 \log F_{cent} \tag{18}$$

$$c = -0.4 - 0.67 \log F_{cent} \tag{19}$$

$$d = 0.14 \tag{20}$$

and F_{cent} expressed in the parameterized form:

$$F_{cent} \approx (1 - a)\exp(-T/T_1) + a\exp(-T/T_2) + \exp(-T_3/T) \tag{21}$$

where a, T_1, T_2, and T_3 are fitting parameters.

Lightfoot and Pilling[21] fitted F_{cent} from fall-off curves calculated using the Arrhenius expressions for $k_1^\infty(T)$ and $k_1^0(T)$ and the molecular parameters for C_2H_5, C_2H_4, and $C_2H_5^\dagger$. They examined N_2 and C_2H_6 as third bodies, making the necessary corrections for changes in $\langle\Delta\varepsilon\rangle_{down}$, and obtained the parameters given in Table 2. These parameters, together with the Arrhenius expressions for the limiting rate

Table 2 *Fitting parameters for calculating rate data for reaction*[1][a,b]

Diluent	$\langle\Delta\varepsilon\rangle_{down}/cm^{-1}$	$k_1^0/cm^6\ molecule^{-2}\ s^{-1}$	T_1/K	T_3/K	*a*
Nitrogen	285	$1.1 \times 10^{-29} \exp(-564\ K/T)$	786	3250	0
Ethane	1670	$5.8 \times 10^{-29} \exp(-880\ K/T)$	908	3177	0

[a] From ref 21. [b] $k_1^{\infty} = 4.4 \times 10^{-11} \exp(-1090\ K/T)\ cm^3\ molecule^{-1}\ s^{-1}$

coefficients, provide an accurate (within 5%) and concise representation of the fall-off curves calculated using the full expressions [equations (16)—(21)] over the temperature range 600—1200 K and the pressure range 1—4000 Torr. They should, therefore, be of value in modelling studies. It must be emphasized, however, that they represent extrapolations outside the experimental range and so must be subject to uncertainty on that account.

Theoretical Studies

Hase and co-workers have carried out a number of theoretical studies on reaction [1] and on the generic reaction X + C=C. They first calculated *ab initio* potential energy surfaces[30,31] at various levels of optimization, which enabled them to determine the vibrational frequencies of C_2H_4, C_2H_5, and, having located it on the surface, the transition state. The surface also provides the geometry of the transition state. The frequencies of C_2H_4 are, of course, well known, while most of those of C_2H_5 have been measured by Pacansky ad Dupuis,[32] except for the symmetric and asymmetric CH_2 bends and the CH_3 rocking mode. These frequencies can be estimated from the surface, along with those of the transition state, having first applied an empirical scaling factor of 0.89. Table 3 lists the frequencies and moments of inertia of the activated complex.

Hase *et al.* developed an analytic surface from the numerical *ab initio* surfaces[30] which they used in a series of Monte Carlo quasiclassical trajectory studies.[33–36] For our present purpose, the most relevant of these is that by Swamy and Hase[36] in which the rate coefficient was calculated and compared with that estimated from transition-state theory. The agreement between the two types of calculation was very good, although agreement with experiment was not perfect, largely because the threshold energy on the *ab initio* surface is too high.

[30] W. L. Hase, G. Mrowka, R. J. Brudzynski, and C. S. Sloane, *J. Chem. Phys.*, 1978, **69**, 3548.
[31] H. B. Schlegel, *J. Phys. Chem.*, 1982, **86**, 4878.
[32] J. Pacansky and M. Dupuis, *J. Am. Chem. Soc.*, 1982, **104**, 415.
[33] W. L. Hase, R. J. Wolf, and C. S. Sloane, *J. Chem. Phys.*, 1979, **71**, 2911.
[34] W. L. Hase, D. M. Ludlow, R. J. Wolf, and T. Schlick, *J. Phys. Chem.*, 1981, **85**, 958.
[35] W. L. Hase, D. G. Buckowski, and K. N. Swamy, *J. Phys. Chem.*, 1983, **87**, 2754.
[36] K. N. Swamy and W. L. Hase, *J. Phys. Chem.*, 1983, **87**, 4715.

Table 3 *Parameters for the activated complex in reaction* [1]

Hase and Schlegel[26]	*Nagase* et al.[38]	*Michael* et al.[19]
Vibrational frequencies/cm^{-1}		
3036	3416[a]	3047(4)
3009	3387	
2955	3322	
2944	3309	
1504	1697	1466(3)
1446	1626	
1218	1326	
1185	1341	1023(2)
944	1059	
938	1042	933(3)
869	980	
821	915	
399	432	100(2)
369	404	
546i	587i	[b]
Moments of inertia, $I_A I_B I_C$/$u^3 Å^6$		
3021	3419	3675
Threshold energy (0 K), ε_0/cm^{-1}		
876	730	1165[c]

[a] The frequencies of Nagase *et al*. do not appear to have been scaled downwards in the manner employed by Hase and Schlegel. [b]No estimate given: Hase and Schlegel's value used in Figure 5. [c]Michael *et al.*[19] estimated their threshold energy on the basis of C_2H_5 decomposition data. The present estimate is based on optimizing the agreement between experimental data and theoretical estimates of k_1 at 300 K

Hase and Schlegel[26] then took the molecular geometries and vibrational frequencies (Table 3) and calculated transition-state theory rate coefficients over the experimental range covered in references 28 and 29. They included a Wigner tunnelling correction:[37]

$$\kappa = 1 + \frac{|h\nu/k_B T|^2}{24}$$

where ν is the imaginary frequency associated with the reaction coordinate (Table 3). They then adjusted the threshold energy, ε_0, and obtained optimal agreement with the experimental data with $\varepsilon_0 = 10$ kJ mol^{-1}.

Figure 5 compares the model calculations with experimental data, plotting $k_1^\infty(\text{th})/k_1^\infty(\text{expt})$ over the now extended experimental range 200—600 K. In addition, the theoretical plot is shown in Figure 4. The

[37] H. S. Johnston, 'Gas Phase Reaction Rate Theory', Ronald Press, New York, 1966.

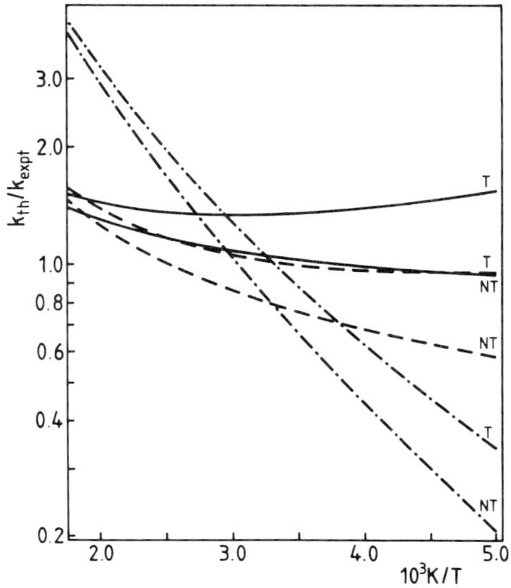

Figure 5 *A comparison of theoretical and experimental data for* k_1^∞. *The plot shows* k_1^∞(th) / k_1^∞(expt) *for the models of Michael* et al., ·–·–·,[19] *Hase and Schlegel*, – – –,[26] *and Nagase* et al., ——.[38] *Plots are shown with* (T) *and without* (NT) *a tunnelling correction*

model produces a stronger temperature dependence and significantly more curvature than do the experimental data.

Ab initio calculations have also been carried out by Nagase *et al.*[38]: the transition-state parameters are shown in Table 3 and the rate coefficients in Figure 5. Over the temperature range 198—320 K, they give

$$k_1^\infty = 5 \times 10^{-11} \exp(-1160 \text{ K}/T) \text{ cm}^3 \text{ molecule}^{-1} \text{ s}^{-1}$$

in good agreement with experiment, but, once again, the temperature dependence is steeper and this effect is accentuated at higher temperatures. Figure 5 also shows the $k_1^\infty(\text{th})/k_1^\infty(\text{expt})$ ratios calculated using the empirical data of Michael *et al.*[19]

C_2H_5 *Decomposition*

The reverse of the atom addition reaction:

$$C_2H_5 \rightarrow C_2H_4 + H \qquad [-1]$$

is arguably the more important reaction in cracking and is undoubtedly

[38] S. Nagase, T. Fueno, and K. Morokuma, *J. Am. Chem. Soc.*, 1979, **101**, 5849.

the more important in combustion. It can, in principle, be calculated from k_1 using tabulated thermodynamic parameters:

$$k_1/k_{-1} = K_{p,1}(T)RT = RT\exp\left[\frac{\Delta S^\circ(T)}{R} - \frac{\Delta H^\circ(T)}{RT}\right] \tag{22}$$

or from statistical mechanics:

$$k_1/k_{-1} = K_{c,1}(T) = \frac{(q^0_{C_2H_5}/V)}{(q^0_H q^0_{C_2H_4}/V^2)}\exp\left[-\frac{\Delta\varepsilon_0}{k_B T}\right] \tag{23}$$

where q^0_X is the molecular partition function of X, referred to its zero point level, and $\Delta\varepsilon_0$ is the difference in zero-point energies between reactants and products; $\Delta\varepsilon_0 \equiv \Delta H^\circ_0/N_A$. The molar thermodynamic parameters may be found in polynomial form in the NASA tables[39] for most species of interest. The parameters are generally derived from statistical mechanics using rotational constants and vibrational frequencies determined spectroscopically or computationally. In addition, the enthalpy of reaction is required and must be determined experimentally, by bond additivity principles,[40] or by *ab initio* potential energy surface calculations.[41] Thus equations (22) and (23) are essentially equivalent, with $\Delta S^\circ(T)$ and the temperature dependence of $\Delta H^\circ(T)$ being well defined through spectroscopic parameters; the major uncertainty lies in the determination of $\Delta H^\circ(T)$ at any fixed temperature, *e.g.* ΔH^0_{298} or ΔH°_0.

There has been some controversy about $\Delta H^\circ_{f,298}$ for C_2H_5, with estimates ranging from 110[42] to 119[43] kJ mol^{-1}. Most data have been determined kinetically using the radical buffer technique or from separate measurements of decomposition and radical combination reactions.[42]

The experimental techniques described earlier in this section for measuring k_1 may be extended to higher temperatures to provide direct estimates of $K_{c,1}$, as is illustrated by the work of Brouard *et al.*[44] using laser flash photolysis/resonance fluorescence. Around 800 K, the ethyl radical begins to decompose on the experimental timescale and, instead of being removed completely by reaction [1], the hydrogen atoms react to generate an equilibrium mixture of H and C_2H_5, with both present in measurable concentrations. Such a reaction system corresponds to a

[39] W. C. Gardiner, jun., ed., 'Combustion Chemistry', Springer-Verlag, New York, 1984, Appendix C, p. 485.

[40] S. W. Benson, 'Thermochemical Kinetics', 2nd Ed., John Wiley and Sons, New York, 1976.

[41] C. F. Melius and J. S. Binkley, 'Proceedings of the 20th International Symposium on Combustion', Combustion Institute, Pittsburgh, 1984, p. 575.

[42] D. F. McMillen and D. M. Golden, *Annu. Rev. Phys. Chem.*, 1982, **33,** 493.

[43] W. Tsang, *Int. J. Chem. Kinet.*, 1978, **10,** 821.

[44] M. Brouard, P. D. Lightfoot, and M. J. Pilling, *J. Phys. Chem.*, 1986, **90,** 445.

classical relaxation process, illustrated in Figure 6, with ([H] − $[H]_\infty$) decreasing exponentially with a time constant τ_1, where $\tau_1^{-1} = k_1[C_2H_4] + k_{-1}$, and with the limiting concentration ratio, $[C_2H_5]_\infty : [H]_\infty$, equal to $K_{c,1}[C_2H_4]$. Thus k_1, k_{-1}, and $K_{c,1}$ can all be determined from equilibration measurements of this type.

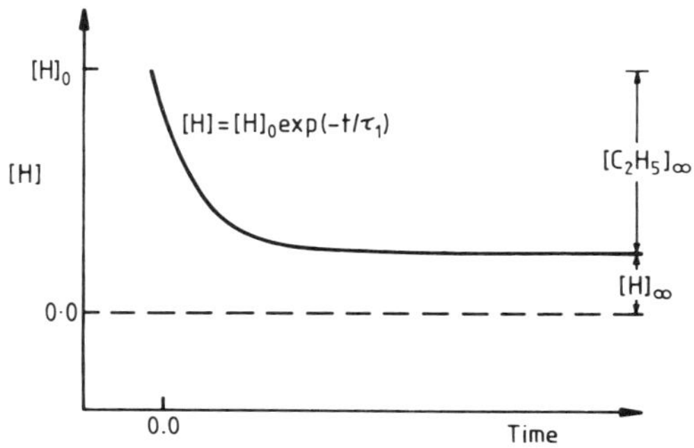

Figure 6 *Schematic diagram of the approach to equilibrium in a relaxation experiment on reaction* [1]. *In this idealized experiment, the relaxation is exponential, with a time constant* $\tau_1^{-1} = k_1[C_2H_4] + k_{-1}$

The real experimental system is not quite so simple because the hydrogen atoms are lost from the monitoring zone by diffusion:

$$\mathrm{H} \xrightarrow{k_7} \text{diffusive loss} \qquad [7]$$

so that their decay is biexponential:

$$[\mathrm{H}] = [\mathrm{H}]_0 \{A \exp(\lambda_+ t) + B \exp(\lambda_- t)\} \qquad (24)$$
$$A = (k_1[C_2H_4] + k_7 + \lambda_-)/(\lambda_- - \lambda_+)$$
$$B = (k_1[C_2H_4] + k_7 + \lambda_+)/(\lambda_+ - \lambda_-)$$
$$\lambda_\pm = \{-b \pm (b^2 - 4c)^{1/2}\}/2$$
$$b = k_1[C_2H_4] + k_{-1} + k_7$$
$$c = k_{-1}k_7$$

The initial decay is determined principally by the relaxation to equilibrium and the later, slower decay, by diffusion. Brouard *et al.*[44] applied equations (24) to the H atom decays using non-linear least squares fitting techniques to generate k_1, k_{-1}, and $K_{c,1}$ (see Table 4). The equilibrium constants were then converted, with the aid of spectroscopic data, into ΔH°_{298} and then, using established heats of formation for H and C_2H_4, into $\Delta H^\circ_{f,298}$ (C_2H_5). Table 4 shows that the data strongly support the higher literature value.

Finally, estimates of k_1 are shown in Table 4 which are based on the Troe fitting function [equations (16)—(21)] and the parameters determined by the experiments described earlier in this section, which covered the range $200 \leq T/\mathrm{K} \leq 600$. They compare favourably with those measured directly in the relaxation experiments. At least over this limited extrapolation range, the performance of the parameterized expression is very satisfactory.

Table 4 *Rate coefficients obtained from a study of* $H + C_2H_4 \rightleftharpoons C_2H_5$ *in the equilibrium regime at a helium pressure of* 200 Torr[a]

T/K	$10^{12}k_1$/ cm^3 molecule^{-1} s^{-1}	k_{-1}/s^{-1}	$\Delta H^\circ_{f,298}(C_2H_5)$/ kJ mol^{-1}	$10^{12}k_1$ [b]/ cm^3 molecule^{-1} s^{-1}
775	3.23	95.7	119.2	2.69
800	2.97	162.0	118.4	2.76
825	2.79	362.0	119.2	2.79

[a]Data from refs. 21 and 44. [b]Calculated from the lower temperature data (285—600 K) using the Troe fitting procedure described in Section 2

Conclusions

Experimental data on reaction [1] are well defined over the temperature range 200—600 K, and the Troe parameterization appears to work well, at least at temperatures up to 800 K. There has been some comparison with steady-state measurements of the reverse rate coefficient, k_{-1},[21] showing satisfactory agreement, but a more complete analysis is needed. Such an analysis, and also direct measurements of C_2H_5 decomposition at high temperatures (by monitoring, for example, the rate of H formation), would be of value in extending the range of experimentally validated fall-off parameters. Comparisons with theoretical models are qualitatively satisfactory, although the high temperature curvature predicted theoretically does not appear to be confirmed by experiment.

3 $CH_3 + CH_3 \rightarrow C_2H_6$

Introduction

Unlike the $H + C_2H_4$ reaction dealt with in Section 2, the recombination reaction of methyl radicals:

$$CH_3 + CH_3 \xrightarrow{k_2} C_2H_6 \qquad [2]$$

occurs on a Type II potential energy surface (Figure 1). The potential energy falls monotonically as r decreases, so the location of the transition state is not prescribed and variational methods are required to model the reaction. These techniques are discussed below.

Reaction [2] plays a major role in the pyrolysis and oxidation of hydrocarbons.[45–47] It competes with the oxidation reactions of CH_3, and as such is an important reaction for flame propagation and ignition. In the combustion of methane, it is the main source of C_2 hydrocarbons. The recombination reaction of methyl radicals is a standard reference reaction for relative kinetic measurements of radical—molecule reactions,[48,49] and is often used to test and improve theories of unimolecular reactions.[4–7,50–53]

Measurements of the rate coefficient for reaction [2], k_2, were reviewed by Baulch and Duxbury[54] in 1980. Most of the results reported were at or near the high-pressure limit, and showed that a detailed study of the reaction in the fall-off region was required. Some more recent experimental determinations of k_2 are described below, along with a suitable method of representing k_2 over a wide range of temperature and pressure.

This section ends with a discussion of the multichannel system comprising reaction [2] and the activated bimolecular process

$$CH_3 + CH_3 \rightarrow C_2H_5 + H \qquad [6]$$

Experimental Investigations

Determination of k_2(P,T). Macpherson *et al.*[55,56] studied the recombination of methyl radicals in the fall-off region over the temperature range 296—577 K and at pressures between 5 and 500 Torr in an argon diluent, using laser flash photolysis/u.v. absorption spectroscopy. Photolysis of azomethane by 193 nm radiation from an argon fluoride exciplex

45 J. Warnatz 'Proceedings of the 18th International Symposium on Combustion'. Combustion Institute, Pittsburgh, 1981, p. 369.
46 J. Warnatz, in 'Combustion Chemistry', ed. W. C. Gardiner, Springer-Verlag, New York, 1984, Chap. 5, p. 197.
47 C. K. Westbrook and F. L Dryer, *Prog. Energy Combust. Sci.*, 1984, **10,** 1.
48 J. A. Kerr, in 'Free Radicals', ed. J. K. Kochi, Wiley-Interscience, New York, 1973, Vol. 1, Chap. 1.
49 J. A. Kerr and S. J. Moss, 'CRC Handbook of Bimolecular and Termolecular Gas Reactions', CRC Press, Boca Raton, 1981.
50 E. V. Waage and B. S Rabinovitch, *Int. J. Chem. Kinet.*, 1971, **3,** 105.
51 K. Glänzer, M. Quack, and J. Troe, *Chem. Phys. Lett.*, 1976, **39,** 304.
52 D. M. Wardlaw and R. A. Marcus, *Adv. Chem. Phys.*, 1988, **70,** 231.
53 A. F. Wagner and D.M Wardlaw, *J. Phys. Chem.*, 1988, **92,** 2462.
54 D. L. Baulch and J. Duxbury, *Combust. Flame*, 1980, **37,** 313.
55 M. T. Macpherson, M. J. Pilling, and M. J. C. Smith, *Chem. Phys. Lett.*, 1983, **94,** 430.
56 M. T. Macpherson, M. J. Pilling, and M. J. C. Smith, *J. Phys. Chem.*, 1985, **89,** 2268.

laser was used as the methyl radical source. The time dependence of the methyl radical concentration was monitored by its u.v. absorption at 216.36 nm. Light from a 450 W xenon lamp passed through the spectrosil cell and a monochromator to a photomultiplier, and the resultant signals were digitized and averaged. A non-linear least squares fit to the absorption decay traces gave k_2/σ, where σ is the absorption cross-section of CH_3. In the same apparatus, they measured σ at temperatures between 296 and 537 K using both end-product analysis and a repetitive photolysis method. The experimental temperature dependence[56] agreed well with that calculated by Quack[57] and by Ashfold,[58] and the calculated dependences were used to extend the temperature range of σ. The absolute value of σ, when combined with the measured values of k_2/σ, led to a general upward revision of k_2 from that recommended by Baulch and Duxbury.[54]

Both the temperature and pressure ranges of the work of Macpherson *et al.*[55,56] have now been extended in a collaborative paper by Slagle and Gutman at the Illinois Institute of Technology and Davies and Pilling at Oxford.[59] Using the same apparatus as that used by Macpherson *et al.*, but with a reaction cell capable of sustaining higher temperatures (Figure 7), they studied reaction [2] at temperatures up to 906 K and pressures between 25 and 600 Torr using 193 nm photolysis of acetone to generate the methyl radicals. Lightfoot *et al.*[60] have shown that >96% of the photolysis occurs via the process:

$$CH_3COCH_3 + h\nu(193\text{ nm}) \rightarrow CH_3 + CH_3 + CO$$

These results were combined with those from a laser flash photolysis/photoionization mass spectrometry (PIMS) technique (Figure 8). Here again, acetone was photolysed at 193 nm to produce the methyl radicals. The reacting gas mixture was sampled through a hole in the reactor wall and formed into a beam by a conical skimmer. The beam then entered a vacuum chamber containing a photoionization mass spectrometer; a portion of it was ionized and then mass-selected. The time dependence of the methyl radical concentration was determined from the averaged temporal ion signal profiles. Absolute methyl radical concentrations were determined by monitoring the depletion of the acetone signal. The PIMS technique is particularly suitable for low-pressure measurements and experiments were performed over the 1—5 Torr range at temperatures between 296 and 906 K.

Combining the work of Macpherson *et al.*[55,56] and Slagle *et al.*[59] gives fall-off curves at nine different temperatures, three of which are shown in Figure 9.

[57] M. Quack, private communication.

[58] M. N. R. Ashfold, private communication.

[59] I. R. Slagle, D. Gutman, J. W. Davies, and M. J. Pilling, *J. Phys. Chem.*, 1988, **92,** 2455.

[60] P. D. Lightfoot, S. P. Kirwan, and M. J. Pilling, *J. Phys. Chem.*, 1988, **92**, 4938.

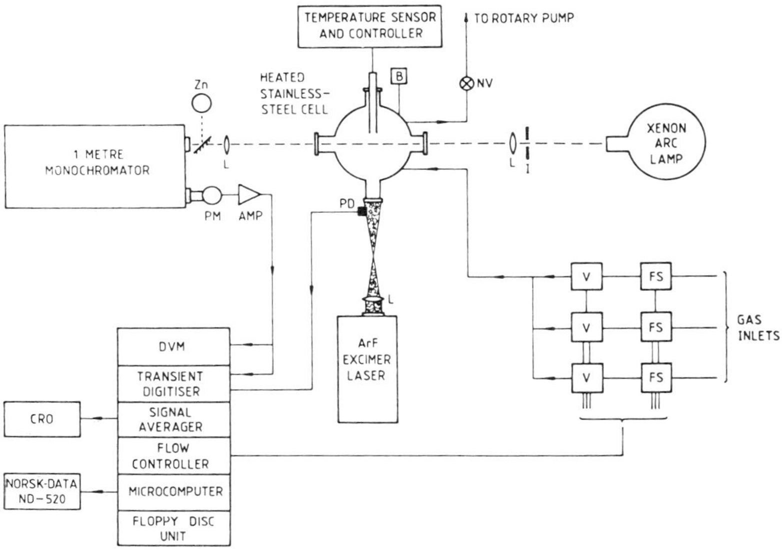

Figure 7 *Schematic diagram of the laser flash photolysis*/u.v. *absorption spectroscopy apparatus employed by Slagle* et al.[59] *to study reaction* [2]. B, *Baratron*; FS, *flow sensor*; I, *Iris*; L, *lens*; NV, *needle valve*; PD, *photodiode*; PM, *photomultiplier*; V, *solenoid valve*; Zn, *zinc lamp for calibration*

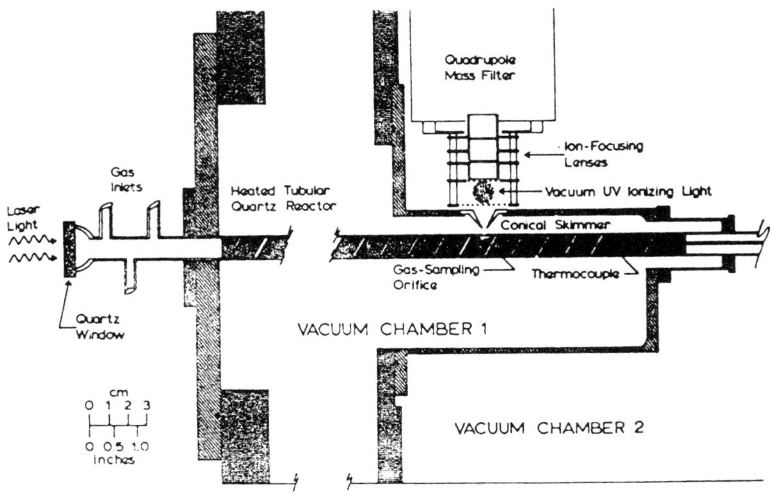

Figure 8 *Schematic diagram of the laser flash photolysis/photoionization mass spectrometry apparatus employed by Slagle* et al.[59] *to study reaction* [2]
(Reproduced with permission from *J. Am. Chem. Soc.*, 1985, **107,** 5342)

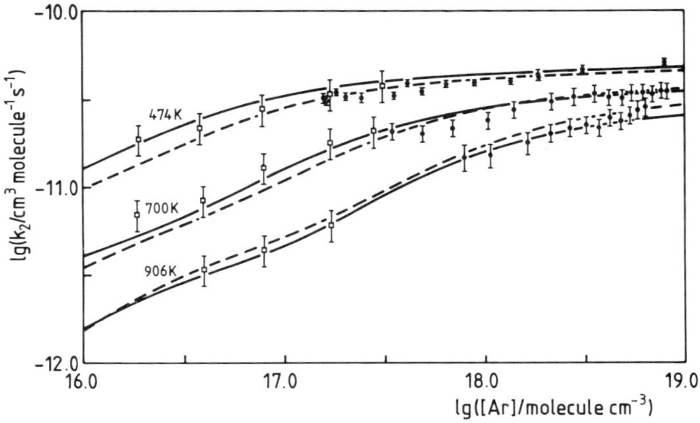

Figure 9 *Dependence of* k_2 *on pressure at three temperatures*: 474, 700, *and* 906 K. □, *Laser flash photolysis/PIMS experiments*;[59] ●, *laser flash photolysis/u.v. absorption spectroscopy experiments*;[55,56,59] ——, *flexible transition-state theory fit to the experimental data*;[53] ---, *global fit*[65] *to the experimental data*.[59,66] (*Error bars* ≡ 95.5% *confidence limits*)

At room temperature, the limiting high-pressure rate coefficient, k_2^∞, has been measured directly in a study by Hippler *et al.*[61] They also used photolysis of azomethane and u.v. absorption at 216.2 nm. Argon was used as the third body and a pressure range of 1—200 bar was covered. Agreement with the room-temperature measurements of Macpherson *et al.*[55,56] was good. k_2 was shown to have no significant pressure dependence between 1 and 10 bar, and analysis of these results gave a limiting high-pressure rate coefficient of

$$k_2^\infty = 5.8 \times 10^{-11}\ \text{cm}^3\,\text{molecule}^{-1}\,\text{s}^{-1}$$

At pressures greater than 30 bar, k_2 decreases slightly due to the onset of diffusion-controlled kinetics. Figure 10 shows these results together with the room-temperature determinations of Macpherson *et al.*[55,56] and Slagle *et al.*[59]

An additional experimental study of the fall-off region is that of Glänzer *et al.*,[62] who measured k_2 at 1350 K and argon pressures of 0.2—25 bar. Their shock-tube experiments used the fast thermal decomposition of azomethane at temperatures above 1300 K as the methyl radical source. The methyl concentration was monitored by u.v. absorption at 216 nm, with σ being measured directly in the shock tube at 1400 K. The results show that reaction [2] is still in the fall-off region at 1350 K, even at 20 bar.

[61] H. Hippler, K. Luther, A. R. Ravishankara, and J. Troe, *Z. Phys. Chem.* (*Neue Folge*), 1984, **142,** 1.

[62] K. Glänzer, M. Quack, and J. Troe, 'Proceedings of the 16th International Symposium on Combustion', Combustion Institute, Pittsburgh, 1976, p. 949.

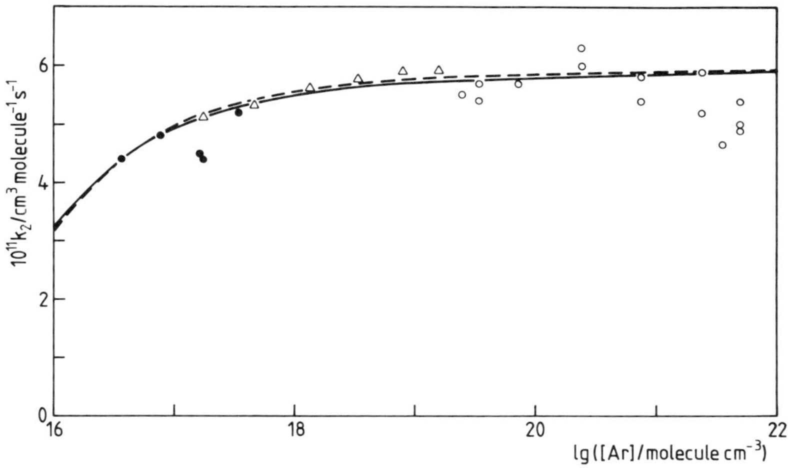

Figure 10 *Dependence of k_2 on pressure at room temperature.* ●, *Slagle* et al.;[59] △, *Macpherson* et al.;[55,56] ○, *Hippler* et al.;[61] ——, *flexible transition-state theory fit*[53] *to the experimental data*; ---, *global fit*[65] *to the experimental data of Slagle* et al.[59] *and Macpherson* et al.[55,56]

Extrapolation of Experimental Data. The limiting low- and high-pressure rate coefficients for the recombination of methyl radicals, k_2^0 and k_2^∞, respectively, are not usually experimentally obtainable, especially at high temperatures. It is then necessary to extrapolate data determined in the fall-off region. The Troe factorization method[1–3] described in Section 2 is now widely used for extrapolation of both unimolecular dissociation and termolecular recombination rate data. Reduced fall-off curves can be described by equation (16). In general, calculated fall-off curves depend on k^0, k^∞, S_T, and β_c. Values of S_T and β_c are therefore required to determine k^0 and k^∞ from experimental data. β_c can often be estimated from other determinations of the collision efficiency of the third body, or can be treated as a variable parameter. In most cases, however, the shape of the fall-off curve is not very sensitive to β_c and the main problem in fitting experimental data reduces to that of estimating S_T.

S_T depends on the vibrational frequencies of the activated complex [equation (3)], and so if the structure of the complex is known, S_T can be calculated. If such information is not available, as is the case for reaction [2], then the vibrational frequencies of the activated complex must be estimated, or S_T treated as a variable parameter. In the latter case, there are then three variable parameters: k^0, k^∞, and S_T. These parameters can only be obtained simultaneously when there is significant curvature in the experimental data, as was demonstrated by Pilling and

Smith[63] who combined their own experimental data at 298 K for the reaction

$$CH_3 + O_2 + M \rightarrow CH_3O_2 + M \qquad [8]$$

with the lower pressure data of Selzer and Bayes.[64]

However, in less favourable cases, it is impossible to determine k^0, k^∞, and S_T simultaneously from single fall-off curves. Recently, Keiffer *et al.*[65] developed a global fitting procedure which enables fall-off curves determined at several temperatures to be analysed simultaneously. They make use of the fact that S_T is related to the temperature dependence of k^∞, which requires that the fitted values of $k^\infty(T)$ and S_T must be compatible.

If the temperature dependence of k^∞ for the general recombination reaction

$$A + B + M \rightarrow AB + M$$

is expressed either in the form

$$k^\infty = A^\infty\,(T/300\text{ K})^{n^\infty} \qquad (25)$$

or

$$k^\infty = A^\infty\,(T/300\text{ K})^{n^\infty}\exp(-\,\varepsilon_r/k_\mathrm{B}T) \qquad (26)$$

where ε_r is the activation barrier to association, then Keiffer *et al.*[65] showed that

$$S_T = n^\infty + C + \frac{U_{\mathrm{vib}}^{\mathrm{A}}}{k_\mathrm{B}T} + \frac{U_{\mathrm{vib}}^{\mathrm{B}}}{k_\mathrm{B}T} \qquad (27)$$

Thus, S_T is determined from the vibrational energies of A and B – $U_{\mathrm{vib}}^{\mathrm{A}}$ and $U_{\mathrm{vib}}^{\mathrm{B}}$, respectively – the experimental parameter n^∞, and a constant C. The value of C depends on the geometries of A, B and the activated complex $AB^\dagger$, and the various forms of equation (27) are given in Table 5. Equation (27) enables S_T to be calculated without the need to make any assumptions about the activated complex.

If equation (25) is employed and k^0 is expressed in the form $k^0 = A^0(T/300\text{ K})^{n^0}$, then several fall-off curves can be analysed in a single non-linear least squares fitting procedure, with variable parameters A^0, n^0, A^∞, and n^∞. There is more information in the several fall-off curves taken together than in a single fall-off curve, and thus all

[63] M. J. Pilling and M. J. C. Smith, *J. Phys. Chem.*, 1985, **89,** 4713.
[64] E. A. Selzer and K. D. Bayes, *J. Phys. Chem.*, 1983, **87,** 392.
[65] M. Keiffer, M. J. Pilling, and M. J. C. Smith, *J. Phys. Chem.*, 1987, **91,** 6028.

Table 5 *Expressions for* S_T [*equation* (27)] *for different types of association reaction* A + B → AB† → AB[a]

A	B	AB†	*Expression*
N	N	N	$S_T = n^\infty + 3.0 + U^A_{vib}/k_BT + U^B_{vib}/k_BT$
N	L	N	$S_T = n^\infty + 2.5 + U^A_{vib}/k_BT + U^B_{vib}/k_BT$
N	A	N	$S_T = n^\infty + 1.5 + U^A_{vib}/k_BT$
L	L	N	$S_T = n^\infty + 2.0 + U^A_{vib}/k_BT + U^B_{vib}/k_BT$
L	L	L	$S_T = n^\infty + 2.5 + U^A_{vib}/k_BT + U^B_{vib}/k_BT$
L	A	N	$S_T = n^\infty + 1.0 + U^A_{vib}/k_BT$
L	A	L	$S_T = n^\infty + 1.5 + U^A_{vib}/k_BT$
A	A	L	$S_T = n^\infty + 0.5$

Key: N = non-linear molecule; L = linear molecule; A = atom
[a] M. Keiffer, D. Phil. Thesis, Oxford University, 1987

four parameters can be determined. Those temperatures at which the data show the greatest curvature aid the fitting of data at the other temperatures, so greater constraints are placed on the fitting parameters.

If equation (26) is used for the temperature dependence of k^∞ instead of equation (25), *i.e.* if it is suspected that there is an activation barrier to association, ε_r, then this introduces a fifth parameter. In this instance, with $k^0 = A^0\ (T/300\ \text{K})^{n^0}\exp(-\varepsilon_r/k_BT)$, Keiffer *et al.*[65] adopted a grid search in which ε_r was fixed in each fit (with A^0, n^0, A^∞, and n^∞ as the variable parameters) and then ε_r was changed and the analysis repeated. The resulting value of ε_r was that which produced the minimum χ^2. If β_c is also unknown, then this can also be determined by a grid search procedure, although Keiffer *et al.* found that the fits were comparatively insensitive to β_c.

Slagle and co-workers[59,66] applied this global fitting procedure to their data for reaction [2], along with those of Macpherson *et al.*[55,56] The variable parameters were A^0, n^0, A^∞, and n^∞, and a grid search was used to determine $\langle\Delta\varepsilon\rangle_{down}$. The resulting fitted parameters were:

$$k_2^0 = (2.7 \pm 0.9) \times 10^{-26}\ (T/300\ \text{K})^{-(4.2 \pm 0.2)}\ \text{cm}^6\,\text{molecule}^{-2}\,\text{s}^{-1}$$

$$k_2^\infty = (6.0 \pm 0.2) \times 10^{-11}\ (T/300\ \text{K})^{-(0.39 \pm 0.03)}\ \text{cm}^6\,\text{molecule}^{-1}\,\text{s}^{-1}$$

$$F_{cent} = 0.601\exp(-T/1460\ \text{K}) + 0.399\exp(-T/113\ \text{K})$$

where the error limits are two standard deviations. The fits were found to be very insensitive to $\langle\Delta\varepsilon\rangle_{down}$; the values of k_2^0 and k_2^∞ lying within the above error limits for values of $\langle\Delta\varepsilon\rangle_{down}$ over the range 100—600 cm^{-1}. Figures 9 and 10 show the calculated fall-off curves at four temperatures.

[66] J. W. Davies, D. Phil. Thesis, Oxford University, 1989.

Pilling[67] has pointed out deficiencies in this treatment arising from an implicitly oversimplified treatment of the heat capacity of the complex. It is also not clear that it can be applied to a reaction occurring on a Type II PES whose activated complex geometry varies with energy, as is found for reaction [2], because the temperature dependence of k^{∞} is not then related in a simple way to S_T.

Tests of the global procedure, incorporating a correct treatment of the heat capacity of the activated complex, have been carried out by Collingwood *et al.*[68] for H + C_2H_4 (reaction [1]), where the transition state is prescribed and the above criticism does not apply. They generated (P, T)-dependent rate coefficients via a full RRKM/master equation treatment and then, employing these coefficients as 'experimental' data, analysed them using the global fitting routine. They demonstrated satisfactory recovery of the limiting RRKM k_1^0 and k_1^{∞} values.

Theoretical Studies

The global procedure for fitting experimental data, based on the Troe factorization model, suggests that k_2^{∞} decreases with temperature. A similar conclusion was reached by Macpherson *et al.*,[55,56] using fits to individual fall-off curves over a range of temperatures. It is of interest to examine the temperature dependences predicted by theoretical models. In this section, several of the different approaches that have been applied to the methyl radical recombination reaction are discussed.

Variational RRKM Theory. The variational approach was first introduced to unimolecular rate theory by Bunker and Pattengill.[69] A discussion of the development of theories of this type may be found in the article by Truhlar *et al.*[70]

The Type II potential energy surface, $V(r)$, on which reaction [2] occurs is illustrated in Figure 1. The microscopic unimolecular rate coefficient, $k(\varepsilon)$, is given by:

$$k(\varepsilon) = \frac{N^{\dagger}(\varepsilon)}{h\rho(\varepsilon)} \tag{28}$$

where $\rho(\varepsilon)$ is the density of internal states of the energized molecule with energy ε, and $N^{\dagger}(\varepsilon)$ is the number of ways of arranging the energy $(\varepsilon - V)$ amongst the internal states of the activated complex. As the carbon–carbon bond length, r, increases, $(\varepsilon - V)$ decreases, while the frequencies of certain so-called *transitional* modes decrease.

Figure 11 shows the correlation between the vibrational modes in

[67] M. J. Pilling, *Int. J. Chem. Kinet.*, to be published.
[68] E. Collingwood, A. L. Davis, and M. J. Pilling, unpublished results.
[69] D. L. Bunker and M. Pattengill, *J. Chem. Phys.*, 1968, **48**, 772.
[70] D. G. Truhlar, W. L. Hase, and J. T. Hynes, *J. Phys. Chem.*, 1983, **87**, 2664.

C_2H_6 and the vibrational and rotational modes in CH_3. Most of the frequencies change little; the exceptions are the transitional modes, which are transformed from rocking and torsional vibrations in C_2H_6 to rotations in CH_3. The only 'non-transitional' mode which shows a significant variation is that correlating with the out-of-plane bending vibration in CH_3.

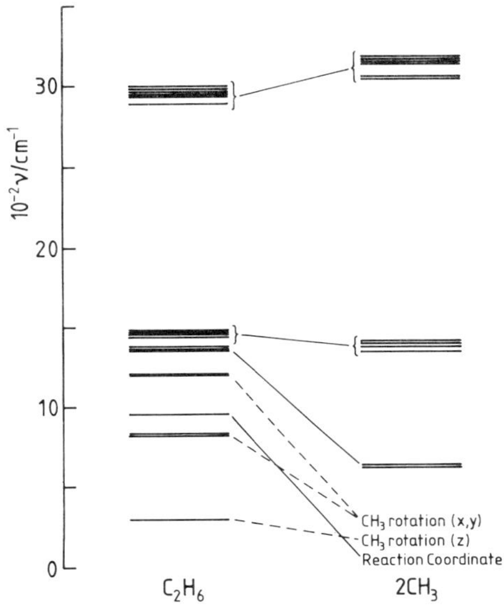

Figure 11 *Correlation diagram for* $C_2H_6 \rightarrow CH_3 + CH_3$, *showing the correlation in vibrational frequencies between the molecule and the radical products. The correlations of the transitional modes are shown as dotted lines*

The decrease in $(\varepsilon - V)$ with increasing r leads to a decrease in $N^{\dagger}(\varepsilon)$, while the decrease in the vibrational frequencies of the transitional vibrations leads to an increase in $N^{\dagger}(\varepsilon)$. The former dominates at small r, the latter at large r, so that a minimum occurs in $N^{\dagger}(\varepsilon)$ at $r^{\dagger}$, and the activated complex is located at this 'bottleneck'. Having determined $r^{\dagger}$ (which depends on ε, tending to decrease as ε increases) and $N^{\dagger}(\varepsilon)$, then $k(\varepsilon)$ can be evaluated via equation (28).

The limiting high-pressure rate coefficient for recombination, k_2^{∞}, is then readily calculated from that for dissociation, k_{-2}^{∞}, using:

$$k_{-2}^{\infty} = \int_{\varepsilon_0}^{\infty} \frac{k(\varepsilon)\rho(\varepsilon)\exp(-\varepsilon/k_B T)}{q^0} \, d\varepsilon \tag{29}$$

$$k_2^{\infty} = k_{-2}^{\infty}.K_c \tag{30}$$

where q^0 is the internal molecular partition function for ethane and K_c

is the equilibrium constant for reactions [2,−2].

Theories of reactions occurring on Type II potentials therefore need to consider the PES, the dependence of the transitional frequencies on r, and methods of calculating $\rho(\varepsilon)$ and $N^{\dagger}(\varepsilon)$.

Bunker and Pattengill[69] initially suggested that the activated complex be located at the position of minimum *density*, rather than minimum *sum* of states. The approach adopted is illustrated by Hase's calculations on ethane dissociation;[4] a discussion of these calculations also demonstrates the simplifying assumptions that have commonly been made to describe $V(r)$ and the r-dependence of the transitional mode frequencies. The classical potential energy along the reaction co-ordinate, the C—C bond extension, r, was represented by a Morse function:

$$V(r) = D\{1 - \exp[-\beta(r - r_{\mathrm{eq}})]\}^2 + \varepsilon_{\mathrm{z}}(r) - \varepsilon_{\mathrm{z}}^{\mathrm{r}}$$

$$D = \Delta H_0 + \varepsilon_{\mathrm{z}}^{\mathrm{r}} - \varepsilon_{\mathrm{z}}^{\mathrm{p}}$$

$$\beta = (f/2D)^{1/2}$$

ΔH_0 is the bond dissociation energy of ethane at 0 K, $\varepsilon_{\mathrm{z}}(r)$ is the r-dependent zero-point energy of all the vibrational modes in the distorted ethane molecule other than the reaction co-ordinate, and $\varepsilon_{\mathrm{z}}^{\mathrm{r}}$ and $\varepsilon_{\mathrm{z}}^{\mathrm{p}}$ are the zero-point energies of ethane and the two methyl radicals, respectively. r_{eq} is the C—C equilibrium bond length, and f is the C—C stretching force constant. Hase[4] also assumed that the carbon atoms remained sp^3 hybridized throughout the dissociation, while the internal rotor was assumed to become a free rotor for $r > 2.9$ Å, which covers the range of $r^{\dagger}$ values. Thus only the two degenerate rocking modes are transitional and Hase assumed r-dependences for the frequencies, ν, of the form:

$$\nu(r) = \nu(r_{\mathrm{eq}})g(r) \tag{31}$$

where

$$g(r) = \exp[-\alpha(r - r_{\mathrm{eq}})] \tag{32}$$

and α is an adjustable parameter. He used the Whitten–Rabinovitch semiclassical approximation[71] to calculate $N^{\dagger}(\varepsilon)$ and $\rho(\varepsilon)$, determining $N^{\dagger}(\varepsilon)$ as a function of r and locating the transition state at the minimum. The procedure was repeated, with α as a variable parameter, until the calculated microcanonical dissociation rate coefficient, for $\varepsilon = 481$ kJ mol^{-1}, agreed with that obtained experimentally using chemical activation [$CH_2(\tilde{a}\,^1A_1) + CH_4$],[72] giving $\alpha = 0.82$ Å^{-1}. Hase then

[71] P. J. Robinson and K. A. Holbrook, 'Unimolecular Reactions', John Wiley and Sons, New York, 1972.

determined $k(\varepsilon)$ over a range of energies, using $\alpha = 0.82$ Å^{-1}, showing, as expected, that $r^{\dagger}$ decreases as ε increases. He determined k_2^{∞} via equations (29) and (30), obtaining the positive temperature dependence shown in Figure 12.

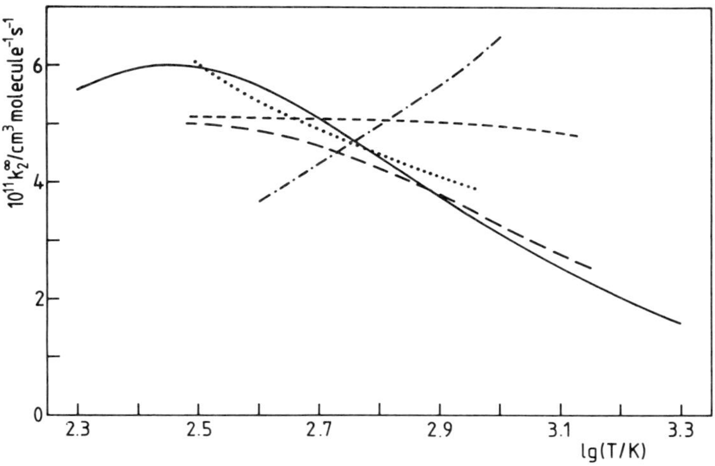

Figure 12 *Plot of* $k_2^{\infty}(T)$ *against* log *T*: ·–·–·, *variational RRKM model* (*Hase*[4]); ------, *statistical adiabatic channel model* (*Quack and Troe*[7]); ——, *flexible transition-state theory* (*Wagner and Wardlaw*[53]); ········, *global fit to experimental data* (*Slagle* et al.,[59] *Davies*[66]); – –, *modified Gorin model* (*Stewart* et al.[82])

There are several criticisms that can be made of this early variational model:

(*a*) The bottleneck is located at the position of minimum flux and is therefore more correctly determined using a minimum *sum*, rather than *density*, of states criterion.[70]

(*b*) Angular momentum conservation was included in a rather *ad hoc* manner, by considering the change in average rotational energy between molecule and complex. A more complete description would consider the rovibrational sums of states, with angular momentum conservation incorporated, or evaluate $k(\varepsilon,J)$ explicitly (see below).

(*c*) The geometry of the CH_3 moieties changes from sp^3 to sp^2 hybridization as the molecule dissociates, and these changes affect the moments of inertia. A simple exponential correlation could be employed [*cf.* equation (32)] or the results of *ab initio* potential energy calculations could be used. Section 4 deals with the CH_3–H system, where the change in CH_3 geometry has been calculated and incorporated into kinetic models.

[72] F. B. Growcock, W. L. Hase, and J. W. Simons, *Int. J. Chem. Kinet.*, 1973, **5**, 77.

(*d*) The Whitten–Rabinovitch approximation[71] provides a very useful analytic expression for sums and densities of states, but one which only becomes accurate at comparatively high energies. It does not, therefore, provide an acceptable representation of $N^{\dagger}(\varepsilon)$. It is nowadays common to use direct counting techniques, usually based on the Beyer–Swinehart algorithm[73] or its derivative developed by Stein and Rabinovitch,[74] to determine both $N^{\dagger}(\varepsilon)$ and $\rho(\varepsilon)$.

(*e*) The rocking modes correlate with radical rotations so that the eigenvalues associated with these modes do not tend to zero as $r \rightarrow \infty$, in contrast to the representation adopted in equations (31) and (32). Thus Hase's model overestimates $N^{\dagger}(\varepsilon)$ at large r and is likely to introduce errors if the transition state is very loose.

All of these limitations can be overcome and an efficient, variational microcanonical RRKM model, incorporating the necessary modifications, has been developed by Legg *et al.*[75] and applied by Brouard *et al.*[76] to the CH_3 + H reaction [3] (see Section 4).

The Statistical Adiabatic Channel Model. Quack and Troe[7] developed a microcanonical model of unimolecular reactions in which the idea of a localized activated complex is replaced by open reaction channels that adiabatically correlate reactant and product states. These adiabatic potential energy curves are open only if their energy never exceeds the total internal energy of the system, ε. The sum of the open channels, $W(\varepsilon,J)$, replaces $N^{\dagger}(\varepsilon)$ in equation (28), giving the analogous, but rotationally resolved, equation

$$k(\varepsilon,J) = \frac{W(\varepsilon,J)}{h\rho(\varepsilon,J)}$$

In calculating the adiabatic potential curves, Quack and Troe[7] used a Morse potential and an exponential interpolation method, *cf.* equation (32), to interpolate between the reactant and product states, α being treated, once more, as a variable parameter. As only the transitional modes vary appreciably as the reaction co-ordinate, r, increases, only the number of open channels of these modes, $W_t(\varepsilon,J)$, need to be calculated by this method. $W_t(\varepsilon,J)$ is then convoluted with the density of states, $\rho_r(\varepsilon' - X,J)$, of the other modes to give $W(\varepsilon,J)$:

$$W(\varepsilon,J) = \int_0^{\varepsilon'} \rho_r(\varepsilon' - X,J)W_t(X,J)\,\mathrm{d}X \tag{33}$$

where ε', the available energy, is equal to ε minus the effective potential energy at the given value of the reaction co-ordinate.

[73] T. Beyer and D. F. Swineheart, *Commun. Assoc. Comp. Machin.*, 1973, **16,** 379.
[74] S. E. Stein and B. S. Rabinovitch, *J. Chem. Phys.*, 1973, **58,** 2438.
[75] A. F. Legg, M. J. Pilling, and C. J. Whitham, to be published.
[76] M. Brouard, M. T. Macpherson, and M. J. Pilling, *J. Phys. Chem.* to be published.

The densities of states, $\rho(\varepsilon,J)$ and $\rho_r(\varepsilon' - X,J)$, were calculated for a symmetric top molecule using:

$$\rho(\varepsilon,J) = \sum_{K=0}^{J} g_K \rho_{vib}[\varepsilon - BJ(J + 1) - (A - B)K^2 - \varepsilon_z^r]$$

where A and B are the rotational constants and $g_K = 1$ for $K = 0$ and 2 for $K > 0$. ρ_{vib} was calculated from the formulae for anharmonic vibrational densities of states given by Haarhoff[77,78] and Forst.[79] The resulting $\rho(\varepsilon,J)$ were then divided by the appropriate symmetry number.

The limiting high-pressure rate coefficient for ethane dissociation, k_{-2}^{∞}, was determined by thermal averaging over the calculated $k(\varepsilon,J)$, equation (29). Equation (30) then gives the recombination rate coefficient, k_2^{∞}. With $\alpha = 1.0$ Å^{-1}, Quack and Troe[7] obtained:

$$k_{-2}^{\infty} = 1.58 \times 10^{17} \exp(-45\,089 \text{ K}/T) \text{ s}^{-1}$$

giving values of k_2^{∞} of 5.1×10^{-11} cm^3 molecule^{-1} s^{-1} at 300 K, 5.0×10^{-11} cm^3 molecule^{-1} s^{-1} at 800—900 K, and 4.8×10^{-11} cm^3 moleçule^{-1} s^{-1} at 1300—1400 K. Their results are shown in Figure 12.

SACM has been widely applied to a range of reactions and has recently been extended to incorporate a more complete description of the angular potential. The model is considered in more detail in Section 5, where its application to the system $O + OH \rightarrow HO_2 \rightarrow H + O_2$ is discussed.

Flexible Transition-State Theory. Wardlaw and Marcus[5,6] have recently developed a flexible transition-state theory (FTST) employing a parameterized potential energy surface which takes angular contributions into account and accommodates the interaction between the transitional modes. The potential, V, was treated as the sum of that due to the conserved vibrational modes, V_c, and that due to the transitional modes (carbon–carbon stretch, torsion, and two methyl rocks), V_t, *i.e.*,

$$V = V_c + V_t$$

V_c was assumed to be separable and quadratic, and the frequencies of the complex, $\nu_i(r^{\dagger})$, were obtained by interpolating between the reactant and product frequencies, ν_i^r and ν_i^p, respectively:

$$\nu_i(r^{\dagger}) = \nu_i^r + (\nu_i^p - \nu_i^r)g(r^{\dagger})$$

where $g(r^{\dagger})$ is defined by equation (32).

[77] P. C. Haarhoff, *Mol. Phys.*, 1963, **6**, 337.
[78] P. C. Haarhoff, *Mol. Phys.*, 1963, **7**, 101.
[79] W. Forst, *Chem. Rev.*, 1971, **71**, 339.

The potential, V_t arises from non-bonded, V_{NB}, and bonded, V_B, interactions:

$$V_t = V_{NB} + V_B$$

V_{NB} is the sum of the Lennard–Jones potentials, V_{LJ}, for all the non-bonded H—H and C—H interactions:

$$V_{NB} = \sum_{m,n=1}^{4}{}' V_{LJ}(r_{mn})$$

m and *n* label atoms on different methyl fragments and the prime indicates that the C—C interaction is not included. The bonding C—C potential, V_B, was described by a modified Morse function, which included an orientational factor accounting for the alignment of each methyl radical C_3 axis with the C—C axis:

$$V_B = V_M^{ef} r \cos^2\theta_1 \cos^2\theta_2$$

$$V_M^{ef} = D^{ef}\{1 - \exp[-\beta^{ef}(r - r_{eq}^{ef})]\}^2 - D^{ef}$$

where θ_i is the angle between the C—C axis and the C_3 axis of the *i*th ($i = 1,2$) methyl fragment.

The change in shape of the methyl radical as *r* increases, from that in ethane to the planar structure of the isolated methyl radical, was obtained by interpolation:

$$r_{CH}(r^\dagger) = r_{CH}^r + [r_{CH}^p - r_{CH}^r]g(r^\dagger)$$

$$\alpha_{HCH}(r^\dagger) = \alpha_{HCH}^r + [\alpha_{HCH}^p - \alpha_{HCH}^r]g(r^\dagger)$$

where r_{CH} is the carbon–hydrogen bond length and α_{HCH} is the H—C—H bond angle. The superscripts r and p denote the equilibrium values in the reactants ($CH_3 + CH_3$) and product (C_2H_6) respectively.

The minimum-sum-of-states criterion was applied to locate the transition state. A similar convolution to that given in equation (33) was used to evaluate $N(\varepsilon,J)$. $\rho_r(\varepsilon' - X,J)$ was calculated by a direct count, but $N_t(X,J)$ was determined by Monte Carlo methods. With $\alpha = 1.0$ Å^{-1}, Wardlaw and Marcus obtained a negative temperature dependence for k_2^∞, in good agreement with the experimental determinations of Macpherson *et al.*[56]

This theory was recently applied by Wagner and Wardlaw[53] to the pressure dependence of the methyl radical recombination reaction. They performed a microcanonical variational RRKM calculation based on the FTST model with two parameters, α and $\langle\Delta\varepsilon\rangle_{tot}$, the average total energy transferred per collision for $C_2H_6^*$. These two parameters were essentially uncoupled by determining α from the value of the limiting high-pressure rate coefficient at 298 K,[61] and then $\langle \Delta\varepsilon \rangle_{tot}$ was varied to give the best fit of the calculated values of k_2 to the experimental data of Macpherson *et al.*[55,56] and of Slagle *et al.*[59] The final parameter values for argon as the third body were $\alpha = 0.70 \pm 0.13$ Å^{-1} and $\langle \Delta\varepsilon \rangle_{tot} = -205 \pm 65$ cm^{-1}, giving the following expression for the temperature dependence of k_2^∞ (see Figure 12):

$$k_2^\infty = 1.50 \times 10^{-7}\, T^{-1.18} \exp(-329\ \mathrm{K}/T)\ \mathrm{cm^3\,molecule^{-1}\,s^{-1}}$$

A fall-off curve calculated at 1350 K using the model with these parameters was in good agreement with the experimental data of Glänzer *et al.*[62]

The theoretical rate coefficients were then fitted to the simplified Troe factorization expressions [equations (16)—(21)] over the temperature range 200—2000 K, giving:

$$k_2^0 = 8.76 \times 10^{-7} T^{-7.03} \exp(-\ 1390\ \mathrm{K}/T)\ [\mathrm{Ar}]\ \mathrm{cm^6\,molecule^{-2}\,s^{-1}}$$

$$F_{cent} = 0.381 \exp(-\ T/73.2\ \mathrm{K}) + 0.619 \exp(-\ T/1180\ \mathrm{K})$$

These expressions, together with that for k_2^∞, provide a concise description of $k_2(P,T)$. Figures 9 and 10 show the behaviour of this parameterized form over the experimental range, where it performs similarly to that determined from the global fit. At higher temperatures, it predicts lower values for k_2 whose accuracies depend on the validity of the FTST model.

The Modified Gorin Model. Gorin's loose transition-state model[80] assumes that the transition state occurs at the maximum of the thermally averaged centrifugal barrier. It was originally developed to describe the methyl radical recombination process, reaction [2], but then generalized for all radical–radical recombinations by Benson[40], who showed that the distance between the two radicals at the transition state, $r^\dagger$, is given by:

$$r^\dagger / r_{eq} = (6D_e/RT)^{1/6} \tag{34}$$

where D_e is the dissociation energy and r_{eq} the equilibrium bond length in the molecular product. From equation (34), it can be seen that the transition state becomes tighter, *i.e.* $r^\dagger$ decreases, as the temperature increases. As a result, the rotation of the radical fragments becomes more restricted, and Smith and Golden[81] introduced a hindrance parameter, η, to describe this effect. The internal modes of the transition state are then simply the vibrations and rotations of the independent fragments, but with the internal rotations reduced by the hindrance

[80] E. Gorin, *Acta Physicochim.*, 1938, **9**, 681.
[81] G. P. Smith and D. M. Golden, *Int. J. Chem. Kinet.*, 1978, **10**, 489.

parameter. The effective moment of inertia of a fragment, I', is given by:

$$I' = (1 - \eta)^{1/2} I$$

where I is the moment of inertia of the isolated fragment.

Using this model for the transition state, Stewart *et al.*[82] have performed RRKM calculations for the ethane dissociation reaction [−2]. The value of the hindrance parameter, η, was fixed by comparing the calculated rate coefficients with the limiting high-pressure rate coefficients of Baulch and Duxbury,[54] Quack and Troe,[83] and Gilbert *et al.*[3] (via the equilibrium constant). These results covered the temperature range 300—1400 K, and the hindrance parameters could be expressed by:

$$\eta = 152.4 - 224.3\ T^{-1/6}$$

The pressure-dependence of reaction [−2], determined by multiplying the experimental data of Slagle *et al.*[59] for reaction [2] by the equilibrium constant, was also modelled by Stewart *et al.*[82] A value of the average total energy transferred per collision, $\langle \Delta\varepsilon \rangle_{\text{tot}}$, of $-112 \pm 35\ \text{cm}^{-1}$ was required to reproduce the data. They obtained expressions for the limiting high-pressure rate coefficients for reaction [−2] and the equilibrium constant for reactions [2,−2], from which the limiting high-pressure rate coefficient for methyl radical recombination is given by:

$$k_2^{\infty} = 3.53 \times 10^{-8}\ T^{-0.967} \exp(-312\ \text{K}/T)\ \text{cm}^3\,\text{molecule}^{-1}\,\text{s}^{-1}$$

This temperature dependence of k_2^{∞} is illustrated in Figure 12, and the agreement with Slagle and co-workers[59,66] and Wagner and Wardlaw[53] is good considering the relative simplicity of the model.

The Two-Channel System

At high temperatures ($T > 1500$ K), reaction [2] is well into the fall-off region and at atmospheric pressure the rate coefficient begins to decrease significantly. Eventually, as the temperature is increased, the endothermic bimolecular process

$$CH_3 + CH_3 \xrightarrow{k_6} H + C_2H_5 \qquad [6]$$

[82] P. H. Stewart, C. W. Larson, and D. M. Golden, submitted to *Combust. Flame*, 1989, **75**, 25.

[83] M. Quack and J. Troe, in 'Gas Kinetics and Energy Transfer', eds. P. G. Ashmore and R. J. Donovan (Specialist Periodical Reports), The Chemical Society, London, 1977, Vol. 2, Chap. 5, p. 175.

begins to compete and eventually becomes the dominant channel.

Reaction [6] has recently been investigated by Frank and Braun-Unkhoff[84] over the temperature range 1320—2300 K. They used atomic resonance absorption spectroscopy to probe the time-dependent H atom profiles behind reflected shock waves, generating the CH_3 radicals thermally from various precursors. They simulated the H atom profiles using numerical integration of a 34 rection mechanism, although rate coefficients for only a small number of reactions were significant. They obtained

$$k_6 = (4.6 \pm 1.0) \times 10^{-11} \exp[-(6800 \pm 300)\ \mathrm{K}/T)\ \mathrm{cm^3\ molecule^{-1}\ s^{-1}}$$

by optimization of the simulated profiles relative to those obtained by experiment.

The reverse reaction:

$$H + C_2H_5 \rightarrow CH_3 + CH_3 \qquad [-6]$$

has already been discussed in Section 2. Most estimates of the rate coefficient have been obtained indirectly from computer simulation of atom/radical contributions to the removal of H in an H/C_2H_4 system. There is reasonably good agreement between the estimates, and a temperature-independent rate coefficient of $\sim 5 \times 10^{-11}$ cm^3 molecule^{-1} s^{-1} is indicated. Combination of this value with estimates of K_p gives an Arrhenius expression for k_6 of $8 \times 10^{-12} \exp(-6530\ \mathrm{K}/T)$ cm^3 molecule^{-1} s^{-1} over Frank and Braun-Unkhoff's experimental temperature range. The agreement is reasonable given the uncertainty in the rate coefficients and also in the thermodynamic parameters.

Multichannel reactions can, in principle, show quite complex pressure dependences, especially if the threshold energies are close. The crucial factors are the relative magnitudes of $k_n(\varepsilon)$, $k_p(\varepsilon)$, and of the pseudo-first-order stabilization rate coefficient $k_s[M]$ at energies close to the higher threshold.

$$CH_3 + CH_3 \underset{k_n(\varepsilon)}{\overset{k_m(\varepsilon)}{\rightleftharpoons}} C_2H_6^*$$

$$C_2H_6^* \xrightarrow{k_p(\varepsilon)} H + C_2H_5$$

$$C_2H_6^* \xrightarrow{k_s[M]} C_2H_6$$

In the present system, there is a large discrepancy in the threshold energies so that $k_n(\varepsilon) \gg k_p(\varepsilon)$ except at very high energies; in addition,

[84] P. Frank and M. Braun-Unkhoff, 'Proceedings of the 16th Symposium on Shock Waves' Aachen, 1987, in press.

$k_n(\varepsilon) \gg k_s[M]$, except at very high pressures. In consequence, neither the overall forward reaction [6] nor the backward reaction [−6] shows a significant pressure dependence. In more complex systems, where the threshold energies and first-order rate coefficients are more comparable, the reactions are best treated using a master equation formulation (see Section 4), and a general discussion has been presented by Just and Troe.[85] Larson *et al.*[86] have developed an approximate analytic procedure which has been applied by Stewart *et al.*[82] to the $C_2H_6/CH_3 + CH_3/H + C_2H_5$ system. Figure 13 shows their calculated values for k_2 and k_6 for a range of P and T.

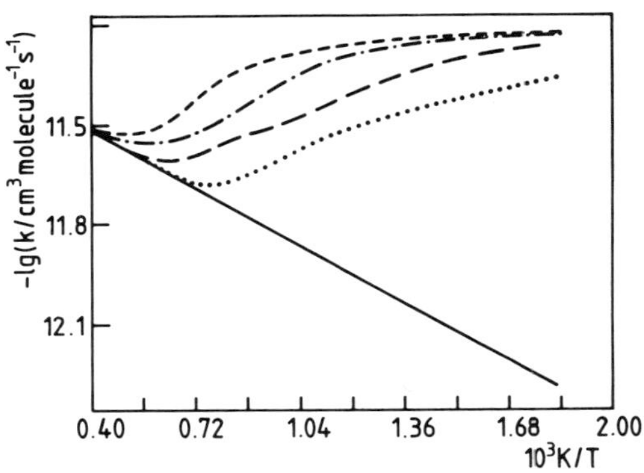

Figure 13 *Arrhenius plot for the two-channel reaction* $CH_3 + CH_3 \rightarrow C_2H_6$ (*reaction* [2]) *and* $CH_3 + CH_3 \rightarrow H + C_2H_5$. (*reaction* [6]). *Reaction* [2]: ········, 1 Torr; – – –, 10 Torr; ·–·–·, 100 Torr; ------, 760 Torr, *where the pressures refer to the argon diluent. Reaction* [6]: ——. (Reproduced with permission from *Combust. Flame*, 1989, **75**, 25.)

Conclusions

Reaction [2] is now well characterized experimentally over the temperature range 300—900 K, with some data also available at 1350 K. The room-temperature limiting high-pressure rate coefficient, k_2^∞, is well established at a value of 6×10^{-11} cm^3 molecule^{-1} s^{-1}. At high temperatures the reaction is well into the fall-off region and extrapolation to k_2^∞ is more uncertain, but there does appear to be a small but significant decrease in k_2^∞ with temperature. Such a decrease is predicted by the flexible transition-state theory of Marcus and Wardlaw, although the detailed SACM calculations of Quack and Troe suggest that k_2^∞ should be approximately independent of temperature. Further high-temperature experiments, preferably at high pressures, and low-temperature experiments would be of value.

[85] Th. Just and J. Troe, *J. Phys. Chem.*, 1980, **84,** 3068.
[86] C. W. Larson, P. H. Stewart, and D. M. Golden, *Int. J. Chem. Kinet.*, 1988, **20,** 27.

4 $CH_3 + H \rightarrow CH_4$

Introduction

The CH_3 + H reaction is the simplest association reaction involving an alkyl radical:

$$CH_3 + H \xrightarrow{k_3} CH_4 \qquad [3]$$

Like reaction [2], it has received extensive theoretical investigation, using canonical RRKM and SACM calculations; these studies have primarily concentrated on the determination of k_3^∞.[87–89] Thanks to the comparative simplicity of the molecular system, detailed *ab initio* calculations of the potential energy surface have been made which include studies of the important region close to the dissociation limit. The reaction is more difficult to study experimentally than methyl radical recombination, and in most investigations it has been only one of several contributing reactions whose combined behaviour was monitored by either end-product analysis or time-resolved absorption spectroscopy. Recently, Brouard *et al.*[76,90] have studied the reaction under essentially isolated conditions using laser flash photolysis coupled with time-resolved resonance fluorescence and absorption spectroscopy.

Experimental Investigations

Many of the initial estimates of k_3 were made at low pressures using discharge–flow systems to study the H + C_2H_4 system. The technique may be illustrated by the work of Pratt and Veltman,[91] who studied the reaction at a single pressure of 8 Torr (helium) over the temperature range 321—521 K, using mass spectrometry to detect the products CH_4, C_2H_6, and C_3H_8 resulting from the reaction sequence:

$$H + C_2H_4 \rightarrow C_2H_5 \qquad [1]$$
$$H + C_2H_5 \rightarrow CH_3 + CH_3 \qquad [-6]$$
$$CH_3 + H \rightarrow CH_4 \qquad [3]$$
$$CH_3 + CH_3 \rightarrow C_2H_6 \qquad [2]$$
$$CH_3 + C_2H_5 \rightarrow C_3H_8 \qquad [9]$$
$$C_2H_5 + C_2H_5 \rightarrow C_4H_{10} \qquad [10]$$

k_1 was found to be independent of temperature and was fixed at a value

87 W. L. Hase, S. L. Mondro, R. J. Duchovic, and D. M. Hirst, *J. Am. Chem. Soc.*, 1987, **109,** 2916.

88 M. Quack, *J. Phys. Chem.*, 1979, **83,** 150.

89 C. J. Cobos, *J. Chem. Phys.*, 1986, **85,** 5644.

90 M. Brouard, M. T. Macpherson, M. J. Pilling, J. M. Tulloch, and A. P. Williamson, *Chem. Phys. Lett.*, 1985, **113,** 413.

91 G. Pratt and I. Veltman, *J. Chem. Soc., Faraday Trans.* 1, 1976, **72,** 1733.

of 2×10^{-13} cm^3 molecule^{-1} s^{-1}. Literature values were assumed for k_2, k_9, and k_{10} and the end-product data fitted with k_3 and k_{-6} as variable parameters. k_3 was assumed to be at the low-pressure limit and the following temperature-dependent expressions obtained:

$$k_3^0 = 2 \times 10^{-28}(T/\mathrm{K})^{-0.33}\ \mathrm{cm}^6\ \mathrm{molecule}^{-2}\ \mathrm{s}^{-1}$$

$$k_{-6} = 1 \times 10^{-10} \exp(-\ 110\ \mathrm{K}/T)\ \mathrm{cm}^3\ \mathrm{molecule}^{-1}\ \mathrm{s}^{-1}$$

Figure 14 shows k_3 at 300 K.

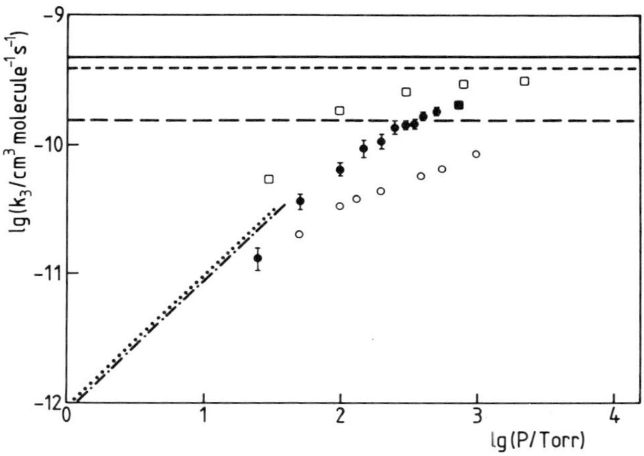

Figure 14 *Comparison of* CH_3 + H *rate coefficient data at* 300 K: □, *Cheng* et al.[92] (*ethane diluent*); ○, *Patrick* et al.[94] (*argon diluent*); ■, *Sworski* et al.[93] (*nitrogen and hydrogen diluents*); ●, *Brouard* et al.[76,90] (*helium diluent*). *Limiting rate coefficients*: k_3^∞, ------, *Cheng* et al.;[92] – –, *Patrick* et al.;[94] ——, *Brouard* et al.;[76,90] k_3^0[M], ········, *Pratt and Veltman*[91] (*helium*); —·—·, *Brouard* et al.[76,90] (*helium*)

Cheng, Lee, and Yeh[92] studied the reaction at higher pressures using the steady-state, high-intensity, mercury-photosensitized decomposition of ethane, where, in place of reaction [1], the reaction sequence discussed above is now initiated by:

$$H + C_2H_6 \rightarrow C_2H_5 + H_2 \qquad [11]$$

Cheng *et al.* used end-product analysis at 318 K, over the pressure range 30—2300 Torr, and determined k_3 from the methane, ethane, propane, and butane yields, based on a steady-state analysis. The determination was most sensitive to the assumed values of k_2 and k_{11}. Figure 14 shows

[92] J. -T. Cheng, Y. -S. Lee, and C. -T. Yeh, *J. Phys. Chem.*, 1977, **81**, 687.

the pressure dependence of k_3 and the proposed high-pressure limit of 3.3×10^{-10} cm^3 molecule^{-1} s^{-1}.

Sworski *et al.*[93] developed a time-resolved technique for determining k_3, albeit based on a rather complex system. They generated H and OH, at concentrations of $\sim 5 \times 10^{14}$ cm^{-3}, by the flash photolysis of H_2O, and CH_3 by the reaction of OH with CH_4:

$$OH + CH_4 \rightarrow CH_3 + H_2O \qquad [12]$$

In addition to reactions [2] and [3], reactions [13]—[17] occur:

$$\begin{aligned} H + CH_4 &\rightarrow CH_3 + H_2 &\qquad [13]\\ H + H &\rightarrow H_2 &\qquad [14]\\ OH + CH_3 &\rightarrow \text{products} &\qquad [15]\\ OH + OH &\rightarrow H_2O + O &\qquad [16]\\ OH + OH &\rightarrow H_2O_2 &\qquad [17] \end{aligned}$$

and, in the prescence of H_2, reaction [18] becomes important:

$$OH + H_2 \rightarrow H_2O + H \qquad [18]$$

Sworski *et al.* probed the reaction via the CH_3 absorption at 216 nm and fitted the decay curves numerically, using k_3, k_{12}, and k_{15} as variable parameters. To define the rate coefficients more precisely, they operated either in excess N_2, where the decays were particularly sensitive to k_{15}, or in excess H_2, where, by changing the rate of reaction [18], the $[CH_3]_0 : [H]_0$ ratio and hence the relative rates of reactions [2] and [3] could be varied. They performed a full sensitivity analysis on the system, demonstrating that k_3 was most dependent on the assumed values of k_2 and σ_{CH_3}, the CH_3 absorption cross-section. Their data are shown in Figure 14.

Figure 14 also shows the fall-off data obtained by Patrick *et al.*,[94] in an argon diluent, using the flash photolysis of ethylene–azomethane mixtures, coupled with end-product analysis by gas chromatography. Their estimates were, once again, determined by numerical integration, comparing simulated product yields with experimental values, and relied primarily on an assumed value for k_2. Their values for k_3 are almost certainly too small, partly because they used a value for k_2 of 4.2×10^{-11} cm^3 molecule^{-1} s^{-1}, well below the currently accepted value (Section 3), and also because they neglected the surprisingly rapid reaction of H with azomethane.

These results demonstrate a common problem of extracting rate data from complex kinetic systems. Such determinations rely heavily on

[93] T. J. Sworski, C. T. Hochanadel, and P. J. Ogren, *J. Phys. Chem.*, 1980, **84,** 129.
[94] R. Patrick, M. J. Pilling, and G. J. Rogers, *Chem. Phys.*, 1980, **53,** 279.

literature rate coefficients, whose uncertainties make a significant contribution to the overall uncertainty in the estimate of k_3. In addition, unrecognized complications, such as the fast rate of the H + azomethane reaction, can lead to significant systematic errors.

Brouard *et al.*[76,90] developed a pulsed technique for measuring k_3 directly, with minimal contributions from competing reactions, and at pressures over the range $25 \leq P/\text{Torr} \leq 600$. The basis of the technique is to generate CH_3 and H simultaneously by laser flash photolysis with $[CH_3]_0 \gg [H]_0$. The 193 nm photolysis of acetone provides an almost ideal source, with the quantum yields of the two radical-producing channels:

$$CH_3COCH_3 \rightarrow 2\,CH_3 + CO$$
$$CH_3COCH_3 \rightarrow H + CH_2COCH_3$$

in the ratio $> 30:1$. Under these conditions, methyl radicals decay primarily by recombination (reaction [2]), and H by reaction [3] and by diffusion (reaction [7]), so that the hydrogen atom decay takes the form:

$$[H] = [H]_0 \exp\left[-\int_0^t k_3[CH_3]\,dt - k_7 t\right]$$

or

$$[H] = [H]_0[1 + 2k_2[CH_3]_0 t]^{-k_3/2k_2} \exp(-k_7 t)$$

It is necessary, therefore, to determine experimentally the time-dependence of the hydrogen atom concentration (in relative units), the absolute value of the zero-time methyl radical concentration, and to establish that the methyl decay is genuinely second-order, with reaction [2] predominating.

Figure 15 shows the apparatus employed by Brouard *et al.*[8,76,90] in which the gases [radical precursor(s) plus the helium diluent] flowed slowly through a stainless steel, multiport cell capable of reaching temperatures of ~900 K. The atom and radical were generated by exciplex laser photolysis and their time dependences monitored, respectively, by Lyman-α resonance fluorescence and by absorption at 216.36 nm. The requirements of the two detection techniques are somewhat contradictory and substantial compromises were required, the most notable of which was the use of very small absorptions ($\lesssim 0.3\%$). Analysis of the absorption decay profiles returned good second-order decays with rate coefficients in agreement with the k_2 values determined previously (Section 3), albeit with greater uncertainty. The most satisfactory way to generate $[CH_3]_0$ was found to be to fit the decays to a second-order rate law, with k_2/σ_{CH_3} and σ_{CH_3} as known parameters, using the values discussed in Section 3, and $[CH_3]_0$ as the only unknown

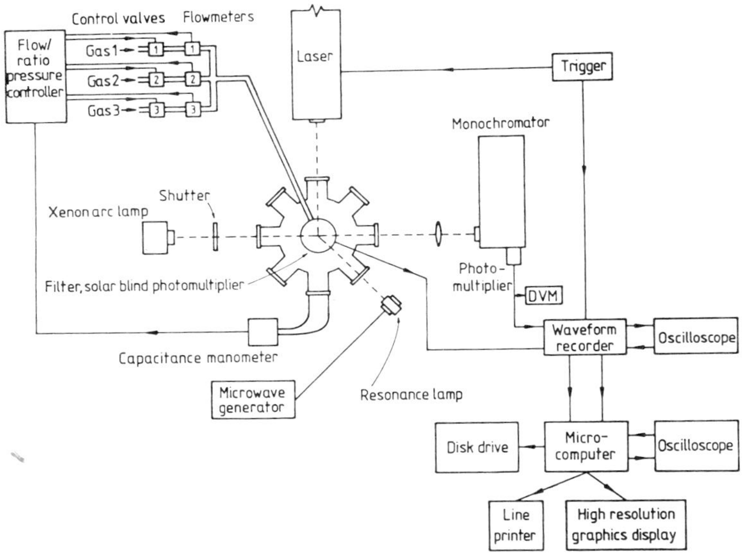

Figure 15 *Experimental system for the study of reaction* [3], $CH_3 + H \rightarrow CH_4$ (*Brouard* et al.[76,90])

parameter. $[CH_3]_0$ was then used as an input parameter in the analysis of the simultaneously determined hydrogen atom decay profiles, with $[H]_0$ (relative) and k_3 as variable parameters.

Figure 16 shows a fall-off plot at 300 K. Data were also generated over the range 400—600 K.

The rate coefficients all lie well into the fall-off region and extrapolation to k_3^∞ is subject to significant uncertainties. Brouard and Pilling[8] argued that this difficulty could be overcome by measuring the rate coefficient for CH_3 + D:

$$CH_3 + D \underset{k_r}{\overset{k_q}{\rightleftharpoons}} \begin{array}{c} CH_3D^* \\ \downarrow k_s[M] \\ CH_3D \end{array} \overset{k_t}{\rightarrow} CH_2D + H$$

Because of zero-point energy differences, $k_t \gg k_r$ at accessible energies, so that $(k_t + k_s[M])$ should always greatly exceed k_r, and $k_3(D) = k_q = k_3^\infty(D)$ at all pressures. Thus if the rate coefficient for the disappearance of D via the reaction:

$$CH_3 + D \rightarrow \text{products}$$

were measured, it should be found to be independent of pressure and,

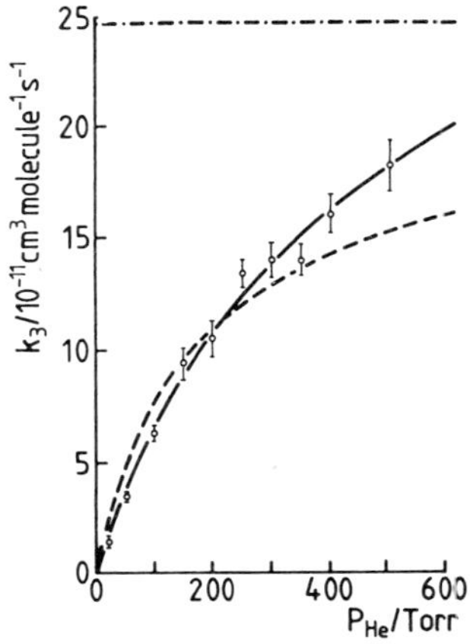

Figure 16 *Comparison of fall-off data for k_3(H) with models based on k_3(D).* ○, *Experimental data for k_3 (error bars ≡ 95.5% confidence limits).* ——, *Best fit with k_3^∞ and $\langle \Delta\varepsilon \rangle_{down}$ varied:* $k_3^\infty = 4.68 \times 10^{-10}$ cm^3 molecule^{-1} s^{-1}, $\langle \Delta\varepsilon \rangle_{down} = 208$ cm^{-1}. – – –, *Best fit with k_3^∞ fixed at the value calculated from k_3(D) and $\langle \Delta\varepsilon \rangle_{down}$ varied:* $\langle \Delta\varepsilon \rangle_{down} =$ 394 cm^{-1}. ·—·—·, *k_3^∞ calculated from k_3(D). The pressure dependences were calculated using an energy-grained master equation, with the microcanonical dissociation constants calculated using a variational RRKM procedure. The only variable in the latter calculation was the parameter α, which describes the variation of the transitional vibrational frequencies with bond distance. The value of α required for* $k_3^\infty = 4.68 \times 10^{-10}$ cm^3 molecule^{-1} s^{-1} *was* 0.91 Å^{-1}

by applying transition-state theory, it should then be possible to estimate k_3^∞(H) from the measured value of k_3^∞(D).

Brouard and Pilling[8] generated D from the 193 nm photolysis of N_2O in the presence of D_2:

$$
\begin{aligned}
N_2O + h\nu &\rightarrow O(^1D) + N_2 \\
O(^1D) + D_2 &\rightarrow OD + D \\
OD + D_2 &\rightarrow D_2O + D
\end{aligned}
$$

while CH_3 was produced from the co-photolysis of acetone. D was detected by resonance fluorescence, H atom Lyman-α fluorescence being eliminated by means of a hydrogen atom filter. Measurements were conducted over the pressure range 50—600 Torr and at temperatures between 289 and 401 K. The system was not entirely clean at the lower temperatures, primarily because the OD + D_2 reaction occurred on the experimental timescale. Brouard and Pilling included this reaction in their analysis by using numerical integration. Consistent results were obtained over a 7-fold range in [D_2], demonstrating the acceptable accommodation of OD + D_2. k_3(D) was found, as expected, to be independent of pressure; it was also shown to be independent of temperature over the experimental range, with a mean value of 1.75×10^{-10} cm^3 molecule^{-1} s^{-1}.

If we assume identical geometries for the transition states for the H + CH_3 and D + CH_3 reactions, then it is readily shown that:

$$\frac{k_3^\infty(\mathrm{H})}{k_3^\infty(\mathrm{D})} = \sqrt{2}\left[\frac{q_\mathrm{v}^0(\mathrm{CH}_4^\dagger)}{q_\mathrm{v}^0(\mathrm{CH_3D}^\dagger)}\right]\exp\left[-\frac{(\varepsilon_\mathrm{H}^0 - \varepsilon_\mathrm{D}^0)}{k_B T}\right]$$

where $q_\mathrm{v}^0(\mathrm{X}^\dagger)$ is the vibrational partition function in the complex and ε_X^0 is the activation barrier, arising from zero-point energy effects, for CH_3 + X. More detailed considerations demonstrate that $(\varepsilon_\mathrm{H}^0 - \varepsilon_\mathrm{D}^0)$ is negligible, while a normal mode analysis of $\mathrm{X}^\dagger$ shows that its vibrational frequencies, exclusive of that associated with the reaction co-ordinate, are unaffected by isotopic substitution. Thus $k_3^\infty(\mathrm{H}) = \sqrt{2}k_3^\infty(\mathrm{D}) = 2.5 \times 10^{-10}$ cm^3 molecule^{-1} s^{-1}. A more detailed variational analysis (see below) confirms this conclusion.

Figure 16 shows the 300 K experimental fall-off data for CH_3 + H determined by Brouard and Pilling[8] and the calculated high-pressure limit. In the next section we shall discuss the incompatibility of this limit with the experimental data.

Modelling and Theoretical Studies

Fits to Experimental Data. Most of the theoretical studies of reaction [3] have concentrated on the high-pressure limit. It is, therefore, necessary to extrapolate the experimental data to obtain k_3^∞ and this has produced substantial difficulties. The most thorough analysis is that of Brouard *et al.*,[76] which will be briefly reviewed here.

Brouard *et al.* first attempted fits to the experimental rate coefficients for k_3(H) using the Troe factorization model (Section 2), calculating S_T from a variational RRKM model, assuming a temperature-independent value of 300 cm^{-1} for $\langle\Delta\varepsilon\rangle_\mathrm{down}$ and floating k_3^0 and k_3^∞. The fitted parameters were insensitive to $\langle\Delta\varepsilon\rangle_\mathrm{down}$, but sensitive to S_T, so an iterative procedure was adopted, refining the RRKM model until the values of S_T and k_3^∞ were compatible at each temperature. This procedure gave $k_3^\infty(\mathrm{H}) = (7.6 \pm 1.6) \times 10^{-10}$ cm^3 molecule^{-1} s^{-1} at 300 K, a factor of 3 greater than that predicted from the measurements of k_3^∞(D).

This disagreement could, in principle, derive from the comparatively crude extrapolation procedure. The extrapolation is a long one (the highest experimental value at 300 K is 1.8×10^{-10} cm^3 molecule^{-1} s^{-1}) and the Troe extrapolation procedure has not been tested quantitatively for specific systems – it is a general, semi-empirical method which has been optimized against a whole series of reactions. Nevertheless, it is worth noting that if $\langle\Delta\varepsilon\rangle_\mathrm{down}$ is taken as 300 cm^{-1}, then a value of $F_\mathrm{cent}^\mathrm{SC}$ *greater than unity* is required if the data are to extrapolate to

$k_3^\infty(\mathrm{H}) = 2.5 \times 10^{-10}\ \mathrm{cm^3\ molecule^{-1}\ s^{-1}}$, *i.e.* negative broadening is needed!

Accordingly, Brouard *et al.*[76] undertook detailed microcanonical variational RRKM calculations, coupled with master equation calculations of the fall-off behaviour. Their variational model is described in detail by Legg *et al.*[75] Suffice it here to say that they used detailed correlations to generate energy level expressions for calculating sums and densities of states, and that they demonstrated that their model produced microcanonical rate coefficients in good agreement with those obtained from the statistical adiabatic channel model. They used a single variable parameter, α, [equation (32)] which was optimized to reproduce the $k_3^\infty(\mathrm{D})$, giving $\alpha = 0.80\ \text{Å}^{-1}$.

The most realistic models of energy transfer and reaction for an association mechanism adopt a master equation description. The reactants, CH_3 and H in the present case, are usually assumed to be in a Boltzmann distribution and the quasi-continuum of states in the energized adduct are usually discretized into a set of energy grains (Figure 17). The master equation then becomes the set of coupled differential equations:

$$\frac{\mathrm{d}n_i}{\mathrm{d}t} = g_i + \omega \sum_j P_{ij} n_j - \omega n_i - k_i n_i \qquad (35)$$

where n_i is the number density of molecules in the ith grain, g_i the rate of formation of such molecules from the reactants, ω the collision frequency, k_i the microcanonical dissociation rate coefficient, and P_{ij} the probability of collisional energy transfer from grain j to grain i. An exponential down model is generally assumed for P_{ij} ($j > i$), with P_{ji} calculated using microscopic reversibility. g_i is readily evaluated from k_3^∞ coupled with the microcanonical dissociation rate coefficients, provided the reactants are in a Boltzmann distribution.[71] Several studies on CH_3 relaxation demonstrate that this will be the case for reaction [3]

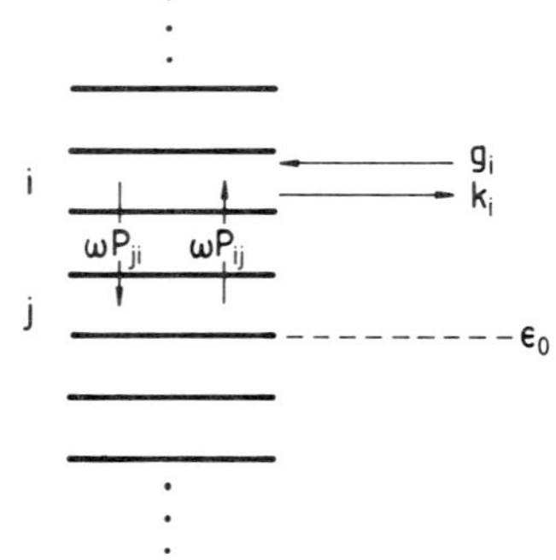

Figure 17 *Schematic diagram of the microscopic reaction scheme employed in an energy-grained master equation*

under the experimental conditions. Gaynor *et al.*[95] and Clifford *et al.*[96] have discussed the proper discretization of the master equation.

Solution of the master equation at times of interest corresponds to invoking the steady-state approximation ($dn_i/dt = 0$) or, equivalently, to determining the smallest non-zero eigenvalue of the matrix representation of equation (35). Brouard *et al.*[76] determined the k_i values using their microcanonical variational RRKM model, with α set equal to 0.80 Å^{-1}, giving $k_3^\infty(\text{H}) = 2.3 \times 10^{-10}$ $\text{cm}^3\,\text{molecule}^{-1}\,\text{s}^{-1}$, a value slightly less than that predicted from the simple transition-state theory analysis presented above. Figure 16 shows the best fit to the experimental data with this constrained set of k_i, with $\langle\Delta\varepsilon\rangle_{\text{down}}$, the parameter defining the efficiency of energy transfer:

$$P_{ij} = A \exp\left[\frac{-(\varepsilon_j - \varepsilon_i)}{\langle\Delta\varepsilon\rangle_{\text{down}}}\right], \quad j > i$$

as a variable parameter. Figure 16 demonstrates the poorness of this optimal fit, confirming the incompatibility of the $k_3(\text{H})$ data and the limit, $k_3^\infty(\text{H})$, calculated on the basis of the $k_3^\infty(\text{D})$ value. Figure 16 also shows a fit in which both α (and therefore k_i) and $\langle\Delta\varepsilon\rangle_{\text{down}}$ were varied. The fit is much improved, but requires $\alpha = 0.91$ Å^{-1} and $k_3^\infty(\text{H}) = 4.7 \times 10^{-10}$ $\text{cm}^3\,\text{molecule}^{-1}\,\text{s}^{-1}$.

The origin of this inconsistency is not evident. It might derive from errors in the experimental measurements. It will be recalled that the $CH_3 + D$ measurements at 300 K were made under the conditions where there was a significant contribution from $OD + D_2$. This reaction was explicitly recognized, however, and consistent fits to the data obtained over a wide range of conditions. Measurements on $CH_3 + H$, by contrast, were conducted under essentially 'clean' conditions. Nevertheless, further experimental measurements using an alternative technique would be of value.

More interesting explanations have also been invoked. Brouard and Pilling[8] explored the possibility of tunnelling through rotational barriers, but concluded that the contribution was small. They also suggested that the inconsistency might be due to differences in vibrational coupling in the two activated complexes. Hase *et al.*[87] also suggested that incomplete intramolecular vibrational relaxation might be responsible, or recrossing of the dividing surface separating reactants and products. It is clear that further theoretical work must also be considered.

Finally, it is notable that, where comparisons can be made, the experimental data for reaction [3] are compatible. Figure 14 shows the agreement between the k_3^∞ and k_3^0 values determined by Brouard *et al.*[76] at 300 K and those of Cheng *et al.*[92] and Pratt and Veltman,[91]

[95] B. J. Gaynor, R. G. Gilbert, and K. D. King, *Chem. Phys. Lett.*, 1978, **55**, 40.

[96] P. Clifford, N. J. B. Green, M. Bezant, and M. J. Pilling, *Chem. Phys. Lett.*, 1987, **135**, 477.

respectively. There is also good agreement with the measurement of Sworski *et al.*[93] There is a clear inconsistency with the data of Patrick *et al.*,[94] but problems associated with the latter work have already been discussed.

Theoretical Models of k_3^∞(H). The methane molecule contains a smaller number of electrons than either ethane or the ethyl radical and so is more amenable to *ab initio* molecular orbital calculations. There have, however, been very few microcanonical calculations of the kinetics of recombination on such surfaces and most modelling studies have adopted a canonical approach. Quack and Troe,[7] in their initial SACM paper, considered the CH_4 system, but in the absence, at that time, of data on recombination, they concentrated on high-temperature dissociation, where extrapolation to k^∞ is even more difficult. The microcanonical calculations of Brouard *et al.*,[76] designed both to calculate k^∞ and to be incorporated in a master equation for fitting to experimental fall-off data, have been discussed above.

The canonical variational approach to the calculation of k^∞ has been discussed in detail by Quack and Troe[97] and summarized concisely by Quack.[88] In their implementation, an exponential interpolation is made between the rovibrational partition functions of the reactants and adduct, in a similar manner to the interpolation of eigenvalues employed in the SACM model. Comparative calculations established that the canonical interpolation parameter, γ, was related to the microcanonical parameter α via $\gamma = 0.75\alpha$. Quack and Troe employed a Morse potential, adding on an interpolated zero-point energy, and incorporated a quasi-diatomic rotational partition function. The procedure then simply reduces to one of minimizing the total partition function q, and thus locating a canonical transition state, which, as expected, moves to smaller separations as the temperature increases. The calculations are very easy to perform, but, of course, do not yield $k(\varepsilon)$ values, only $k^\infty(T)$. Taking $\alpha = 1$ Å^{-1}, Quack calculated $k_3^\infty(\mathrm{H}) = 4.7 \times 10^{-10}$ cm^3 molecule^{-1} s^{-1} at 300 K, seven years before Brouard and Pilling[8] published their experimental value which was discussed above. As experimentalists, we would like to feel that this agreement is, to some extent, fortuitous!

More detailed canonical calculations have been performed by Hase *et al.*,[87] based on the *ab initio* surfaces of Hirst,[98] Brown and Truhlar,[99] and Schlegel.[100] They developed an analytic representation of the surfaces, using the approach previously adopted by Duchovic *et al.*[101] The reaction path was then found by determining the path of steepest descent in mass-weighted Cartesian co-ordinates between CH_3 + H and

[97] M. Quack and J. Troe, *Ber. Bunsenges. Phys. Chem.*, 1977, **81**, 329.
[98] D. M. Hirst, *Chem. Phys. Lett.*, 1985, **122**, 225.
[99] F. B. Brown and D. G. Truhlar, *Chem. Phys. Lett.*, 1985, **113**, 441.
[100] H. B. Schlegel, *J. Chem. Phys.*, 1986, **84**, 4530.
[101] R. J. Duchovic, W. L. Hase, and H. B. Schlegel, *J. Phys. Chem.*, 1984, **88**, 1339.

CH_4. The moments of inertia of the distorted CH_3–H species were calculated along the reaction path together with the harmonic frequencies of the normal modes orthogonal to the reaction co-ordinate. The partition function of the molecular system, q^0, could then be calculated and the position of the activated complex located from the minimum of $q^0(r)$. Two models were employed for the transitional rocking modes, corresponding to harmonic oscillators and to hindered rotors. The calculated values for k_3^∞ were little affected by the choice of model. Table 6 shows k_3^∞(H) and k_3^∞(D) versus temperature for the Hirst potential. Similar results, which bracketed the Hirst values, were found with the Brown–Truhlar and Schlegel surfaces.

Table 6 *Canonical variational rate coefficients for* CH_3 + H,D[a] *from the Hirst potential energy surface,*[b] *with a hindered rotor model for the transitional rocking modes*

T/K	$r^\ddagger$(H)/Å	$10^{10}k_3^\infty$(H)/cm^3 molecule^{-1} s^{-1}	$r^\ddagger$(D)/Å	$10^{10}k_3^\infty$(D)/cm^3 molecule^{-1} s^{-1}
200	3.67	2.27	3.66	1.64
300	3.55	2.64	3.54	1.89
400	3.45	2.87	3.44	2.06
500	3.37	3.05	3.35	2.19
600	3.29	3.19	3.27	2.27
800	3.17	3.40	3.16	2.46
1000	3.07	3.52	3.06	2.49

[a] Ref. 87. [b] Ref. 98

Table 6 demonstrates the tightening of the transition state with increasing temperature and, given that there were no adjustable parameters, shows remarkably good agreement with experiment. There is one major area of disagreement, since the calculations confirm the $\sqrt{2}$ ratio for k_3^∞(H) : k_3^∞(D) deduced by Brouard and Pilling,[8] and Hase *et al.*[87] comment that it is unlikely that transition-state theory can produce a k_3^∞(H) : k_3^∞(D) ratio as large as the experimental value of 2.5, which must arise from dynamical effects. The calculations show a slight increase in k_3^∞ with temperature – a common finding with canonical calculations of this type. Brouard and Pilling found that k_3^∞(D) was invariant over the range $300 \leq T/\mathrm{K} \leq 400$, but the small (9%) variation predicted by the model would not have been detected.

All of the *ab initio* C—H stretching potentials were found to be 'stiffer' than the Morse potential, *i.e.* to rise more steeply towards the dissociation limit and to bend more sharply close to the limit. Hase *et al.*[87] showed that the frequencies of the transitional rocking modes decrease more strongly with bond length than the exponential form which has generally been assumed [*cf.* equation (32)]. However, over the range of interest (3.0—3.5 Å) an exponential approximation is adequate with $\alpha \sim 1.65$ Å^{-1}, much higher than that employed by Quack and Troe.[97] Interestingly though, this behaviour conforms with a general observation made by Quack and Troe of the correlation between α and the Morse parameter β; thus the high α value found by Hase *et al.* is compatible with the steep C–H potential. Lewerenz and Quack,[102] on the other hand, performed *ab initio* calculations on CHD_3, comparing computed eigenvalues with overtone spectra at energies up to 20 000 cm^{-1}, and showed that at these energies their data are compatible with more 'normal' values for α and β.

The other recent calculations on CH_3 + H were made by Cobos[89] who employed a simplified version of SACM (SSACM), by Viswanathan *et al.*[103] who used a variational Monte Carlo technique, and by LeBlanc and Pacey[104] who used canonical variational transition-state theory. In the Cobos model a Boltzmann-averaged centrifugal barrier position is determined, enabling the phase space limit of k_3^∞ to be calculated. This limiting value is then moderated by a factor f_{rigid} which accounts for contributions from the transitional modes. Cobos examined a range of analytic and *ab initio* C—H potentials and compared his results with more detailed transition-state calculations. Viswanathan *et al.* employed a classical Monte Carlo sampling procedure, similar to that of Wardlaw and Marcus,[5,6] to calculate microcanonical variational rate constants, $k(\varepsilon,J)$ on the surface of Duchovic *et al.*[101] They found that an increase in angular momentum reduces $k(\varepsilon,J)$ in excess of that anticipated on the basis of the increase in the centrifugal barrier. LeBlanc and Pacey calculated canonical variational rate constants, k_3^∞, once again on the surface of Duchovic *et al.* They also used various potential energy functions for $V[r(\mathrm{C—H})]$. They found reasonable agreement with experiment provided they treated the bending motions as hindered rotors rather than harmonic oscillators.

Conclusions

The rate data for CH_3 + H cover only the limited temperature range 300—600 K and, at the higher temperatures, derive from a single source. Throughout this range the reaction is well into the fall-off region and extrapolation to k_3^∞ is imprecise. Nevertheless, there appears to be a considerable anomaly in the $k_3^\infty(\mathrm{H}) : k_3^\infty(\mathrm{D})$ ratio. More experimental work is required, especially at higher temperatures, as the reaction is of considerable significance in combustion.

It is surprising that there have been no detailed microcanonical calculations other than those of Brouard *et al.*[76] The use of, for

[102] M. Lewerenz and M. Quack, *J. Chem. Phys.*, 1988, **88,** 5408.
[103] R. Viswanathan, L. M. Raff, and D. L. Thompson, *J. Chem. Phys.*, 1985, **82,** 3083.
[104] J. F. LeBlanc and P. D. Pacey, *J. Chem. Phys.*, 1985, **83,** 4511.

example, flexible transition-state theory using an *ab inito* surface would be of interest.

5 $O + OH \rightarrow HO_2 \rightarrow H + O_2$

Introduction

The reaction of oxygen atoms with hydroxyl radicals occurs on an associative potential energy surface, via the ground electronic state of the HO_2 radical:

$$O + OH \rightarrow HO_2 \rightarrow H + O_2 \qquad [4]$$

Reaction [4] is important in the upper stratosphere and the mesosphere where it is a major source of H atoms. In combination with reaction [19], it is important in determining the concentration of ozone in the atmosphere, the net reaction being reaction [20]:

$$\begin{array}{rcll} O + OH & \rightarrow & H + O_2 & [4] \\ H + O_3 & \rightarrow & OH + O_2 & [19] \\ O + O_3 & \rightarrow & O_2 + O_2 & [20] \end{array}$$

The reverse endothermic reaction [−4] is significant as a chain branching step in combustion processes.

$$H + O_2 \rightarrow O + OH \qquad [-4]$$

Recent direct experimental determinations of the rate coefficient for reaction [4], k_4, and some of the theoretical approaches that have been applied to this reaction are discussed below.

Experimental Investigations

The first direct measurements of the rate coefficient for reaction [4] were performed by Howard and Smith at room temperature,[105] and then over a range of temperatures (250—515 K),[106] using a discharge–flow system. A schematic diagram of their apparatus is given in Figure 18. The temperature of the flow tube was varied by heating the brass block surrounding it with cartridge heaters, or by cooling the block with circulating cold methanol. A dilute mixture of N_2 in argon flowed through a microwave discharge cavity to generate N atoms whose concentration was determined by titration with NO. A known, steady-state concentration of O atoms was produced by adding NO past the

[105] M. J. Howard and I. W. M. Smith, *Chem. Phys. Lett.*, 1980, **69,** 40.
[106] M. J. Howard and I. W. M. Smith, *J. Chem. Soc., Faraday Trans.* 2, 1981, **77,** 997.

the end point of the titration reaction:

$$N + NO \rightarrow N_2 + O \quad [21]$$

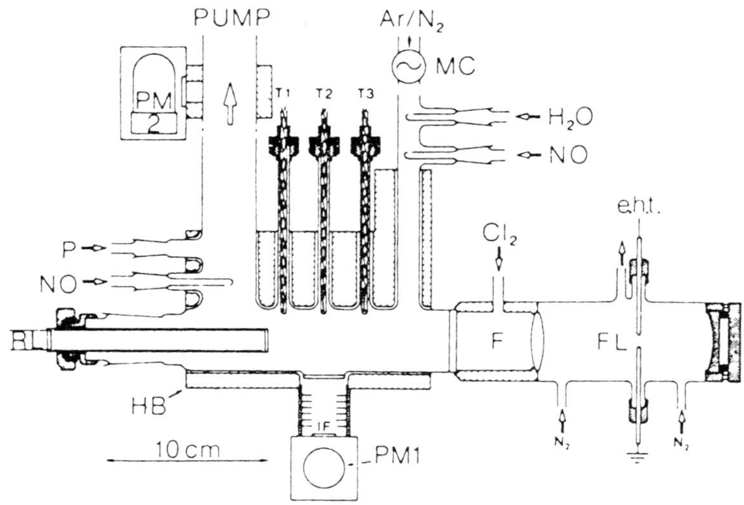

Figure 18 *Schematic diagram of the discharge-flow apparatus employed by Howard and Smith*[105,106] *to study reaction* [4]. F, Cl_2 *filter to remove* 308 nm *radiation*; FL, *flash lamp*; HB, *brass block*; RL, OH *resonance lamp*; IF, *interference filter*; PM1, *photomultiplier for detecting* OH *fluorescence*; PM2, *photomultiplier for detecting chemiluminescence for atom titrations*; MC, *microwave cavity*; P, *oil manometer*; T_1—T_3, *thermocouples*
(Reproduced with permission from *J. Chem. Soc., Faraday Trans.* 2, 1981, **77,** 997)

A much smaller concentration of OH radicals (1—15% of the O atom concentration) was generated by flash photolysis of H_2O injected into the gas flow. Pseudo-first-order conditions were therefore produced, and the rate coefficient $k'_4 = k_4[O]$ was determined by monitoring the depletion of the OH radical concentration using resonance fluorescence. Since [O] was known, k_4 could be evaluated.

Five temperatures between 250 and 515 K were studied and the temperature dependence of k_4 was found to be represented by:

$$k_4 = (6.65 \pm 0.23) \times 10^{-10} T^{-0.50 \pm 0.12} \text{ cm}^3 \text{ molecule}^{-1} \text{ s}^{-1}$$

where the error bars represent 95% confidence limits.

A similar slight negative temperature dependence was also found by Lewis and Watson:[107]

$$k_4 = (2.37^{+1.23}_{-0.81}) \times 10^{-10} T^{-0.362\pm0.072}\ \text{cm}^3\ \text{molecule}^{-1}\ \text{s}^{-1}$$

(error bars represent one standard deviation) over the temperature range 221—499 K, using a similar discharge–flow/resonance fluorescence technique. Both results are illustrated in Figure 19.

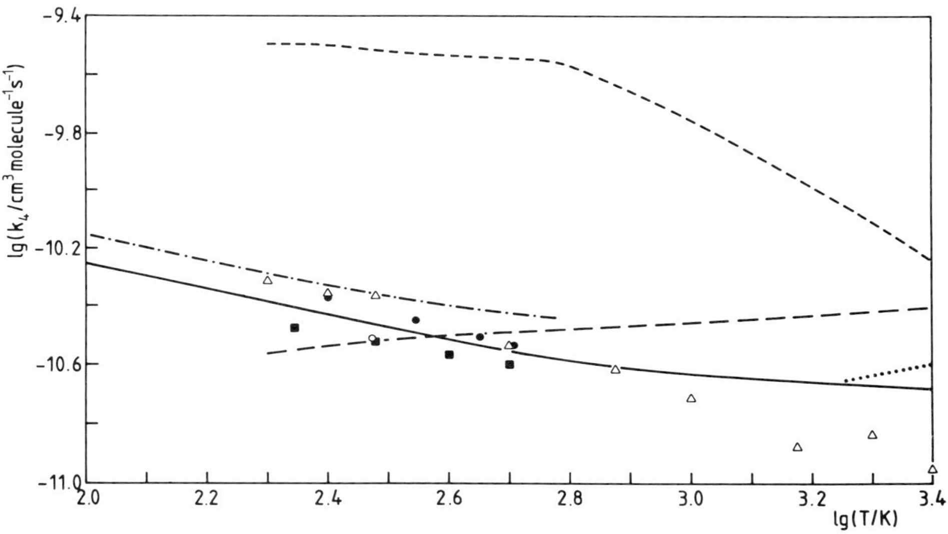

Figure 19 *Dependence of k_4 on temperature. Experimental data:* ●, *Howard and Smith;*[105,106] ■, *Lewis and Watson;*[107] ○, *Brune* et al.;[109] ········, *Frank and Just.*[110] *Theoretical studies:* ------, *canonical variational transition-state theory (MB surface), Rai and Truhlar;*[9] – –, *canonical variational transition-state theory (QT scheme), Rai and Truhlar;*[9] △, *quasi-classical trajectory calculations, Miller;*[10] ——, *statistical adiabatic channel model, Troe;*[11] —·—·, *adiabatic capture theory, Clary*[12,120]

Lewis and Watson[107] proposed that the negative temperature dependence could be explained by a mechanism in which the HO_2 collision complex could dissociate either into products or back into reactants:

$$O + OH \underset{k_v}{\overset{k_u}{\rightleftarrows}} (HO_2)^* \xrightarrow{k_w} H + O_2$$

[107] R. S. Lewis and R. T. Watson, *J. Phys. Chem.*, 1980, **84**, 3495.

They performed RRK calculations and showed that if k_w and k_v were of similar magnitude, but that k_v increased more rapidly with temperature than k_w, then the overall rate coefficient [$= k_u k_w/(k_v + k_w)$] would decrease with temperature.

However, Howard and Smith[106] applied the extended RRKM theory of Quack and Troe[97] and Troe[108] to this mechanism. Using the maximum free energy approach,[97] in which the transition state is located at the position which has the maximum local Helmholtz free energy, they estimated k_u, obtaining reasonable agreement with their observed value of k_4. The ratio $k_u : k_v$ was evaluated using the procedure developed by Troe[108] for calculating the rates of dissociation and association reactions in the limit of low pressure. Essentially, this ratio can be calculated from the equilibrium fraction of energized HO_2 molecules and the $O + OH/HO_2$ equilibrium constant. Troe has developed easily implemented approximations which do not require detailed state counting.[108] From the estimated value of $k_u : k_v$, k_v was determined, and k_w was calculated from equation (28), with $N^{\dagger}(\varepsilon)$ evaluated using the simple harmonic approximation. The frequencies of the transition state were estimated via the interpolation formulae of Quack and Troe[97] [equations (31) and (32)]. Their calculations gave $k_w \approx 15 \times k_v$, suggesting that redissociation of HO_2 to reactants is insignificant. Furthermore, even allowing $k_v = k_w$ requires an overall reaction cross-section, σ_4, of about 100 Å^2, which is significantly larger than the limit expected, even for a loose transition state.

An additional value for k_4 at room temperature has been determined by Brune *et al.*[109] in a flow system using laser magnetic resonance detection of OH, resonance fluorescence detection of OH and O, and resonance absorption detection of O. Figure 20 is a schematic diagram of their apparatus. The O atoms were produced by a microwave discharge of either an O_2–He mixture or a N_2–He–NO mixture. Fluorine atoms, generated by a microwave discharge of CF_4–He, were reacted with H_2O to produce OH radicals. Tests were carried out to ensure that the CF and CF_3 radicals produced did not interfere with reaction [4], which was studied, in an excess of O atoms, by monitoring the decay of OH and the production of H. A rate coefficient of $(3.1 \pm 0.5) \times 10^{-11}$ cm^3 molecule^{-1} s^{-1} was determined, independent of pressure over the range 1—3 Torr. The agreement of this result with those of Howard and Smith[106] and Lewis and Watson[107] is illustrated in Figure 19.

Frank and Just[110] have investigated the kinetics of the reverse reaction [−4] in shock-tube experiments at temperatures between 1700 and

[108] J. Troe, *J. Chem. Phys.*, 1977, **66,** 4758.
[109] W. H. Brune, J. J. Schwab, and J. G. Anderson, *J. Phys. Chem.*, 1983, **87**, 4503.
[110] P. Frank and Th. Just, *Ber. Bunsenges. Phys. Chem.*, 1985, **89,** 181.

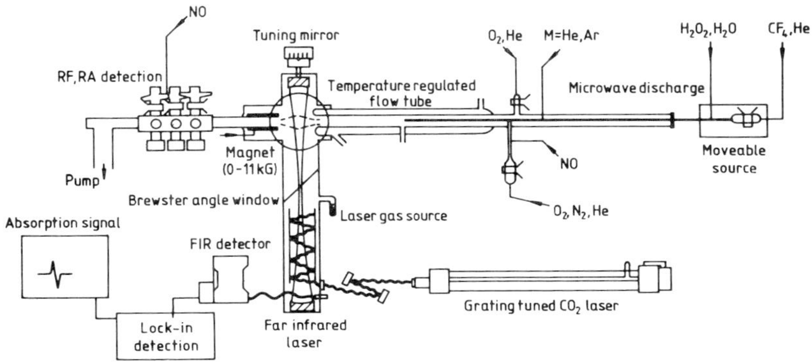

Figure 20 *Schematic diagram of the discharge–flow apparatus employed by Brune* et al.[109] *to study reaction* [4]
(Reproduced with permission from *J. Phys. Chem.*, 1983, **87,** 4503)

2500 K. Oxygen atoms were produced by the thermal dissociation of N_2O:

$$N_2O + Ar \rightarrow N_2 + O + Ar \qquad [22]$$

and then reacted with H_2 to produce hydrogen atoms

$$O + H_2 \rightarrow OH + H \qquad [23]$$

The concentrations of hydrogen and oxygen atoms were monitored by resonance absorption. A ten-step reaction scheme was used to analyse the absorption–time profiles. Sensitivity analysis revealed that the four most important reactions in this scheme were reactions [−4], [18], [22], and [23].

$$OH + H_2 \rightarrow H_2O + H \qquad [18]$$

Previous measurements of the rate coefficients for reactions [22][111,112] and [23][113–115] agree to within ±20%. With these rate coefficients fixed, the sensitivity analysis showed that the oxygen atom concentration was particularly sensitive to the value of k_{-4}, whereas reaction [18] influenced the concentrations of both the hydrogen atoms and the oxygen

[111] P. Roth and Th. Just, *Ber. Bunsenges. Phys. Chem.*, 1977, **81,** 572.
[112] H. A. Olschewski, J. Troe, and H. Gg. Wagner, *Ber. Bunsenges. Phys. Chem.*, 1966, **70,** 450.
[113] D. L. Baulch, D. D. Drysdale, D. G. Horne, and A. C. Lloyd, 'Evaluated Kinetic Data for High Temperature Reactions'. Butterworths, London, 1972, Vol 1.
[114] K. M. Pamidimukkala and G. B. Skinner, *J. Chem. Phys.*, 1982, **76,** 311.
[115] G. L. Schott, P. W. Getzinger, and W. Seitz, *Int. J. Chem. Kinet.*, 1974, **6,** 921.

atoms. Thus k_{-4} was deduced from the oxygen atom concentration profiles, and the following Arrhenius expression found for its temperature dependence over the range 1700—2500 K:

$$k_{-4} = (4.05 \pm 0.55) \times 10^{-10} \exp[-(8696 \pm 265)\ \text{K}/T]\ \text{cm}^3\ \text{molecule}^{-1}\ \text{s}^{-1}$$

The equilibrium constant for the equilibrium [4,−4] is given by Cohen and Westberg[116] as:

$$K_{eq} = k_4/k_{-4} = 2.7 \times 10^{-3}\ T^{0.4} \exp(8720\ \text{K}/T)$$

and hence the high-temperature values for k_4, shown in Figure 19, can be obtained.

Theoretical Studies

Variational Transition-State Theory. Rai and Truhlar[9] have calculated the rate coefficient for reaction [4] over the temperature range 200—2500 K using variational transition-state theory and the Melius–Blint *ab initio* potential energy surface.[117] This potential assumes that reaction [4] occurs via the $HO_2(\tilde{X}\,^2A'')$ ground state, the potential barrier to reaction on the quartet surface being too large. Both canonical and microcanonical variational theories were applied to the Melius–Blint (MB) surface, and canonical variational theory calculations were also performed using the adiabatic channel model of Quack and Troe (QT).[7,97]

The MB surface and the QT scheme gave qualitatively different potential energies in the region of the reaction bottleneck. The rate coefficients calculated using the MB surface were much larger than those given by the QT scheme owing to the more attractive nature of the MB potential energy surface. Figure 19 shows the results of the canonical variational transition-state theory calculations for both the MB surface and the QT scheme. The MB surface overestimates the long-range attractive forces between O and OH and the calculations show that this long-range potential is very important, especially at temperatures below 1000 K, where the reaction bottleneck occurs at O—OH distances where the MB surface is determined by extrapolation, not interpolation.

The QT calculations were in reasonable agreement with experiment at low temperatures, but at high temperatures the calculated values of k_4 were too high, probably because of inaccuracies in modelling the

[116] N. Cohen and K. R. Westberg, Aerospace Report No. ATR-82(7888)-3, El Segundo, California, 1982.
[117] C. F. Melius and R. J. Blint, *Chem. Phys. Lett.*, 1979, **64,** 183.

centrifugal effects or the minimum energy reaction path, or due to the neglect of recrossing of the transition state.

The negative temperature dependence of k_4 was predicted by the MB surface, but not by the QT scheme. In the MB calculations, the positive temperature dependence of the vibrational contribution to k_4 was overridden by the negative temperature dependence owing to the unbound degrees of freedom, *i.e.* the rotations of the complex and the reaction co-ordinate. However, in the QT scheme the inclusion of angular relative translational co-ordinates caused the temperature dependence of k_4 to be positive.

The MB variational transition-state calculations differed from early quasi-classical trajectory calculations performed by Miller[118] on the same surface for the reverse reaction [−4]. This was attributed to the neglect of recrossing of the dynamic bottleneck in the variational calculations, as the difference between the two calculations was greater at high tempertures and recrossing corrections become more important as the temperature rises.

Quasi-classical Trajectory Calculations. Miller[10] performed quasi-classical trajectory (QCT) calculations for reaction [4] in order to determine whether the discrepancies found by Rai and Truhlar[9] were due to errors in the Melius–Blint potential[117] or to the neglect of recrossing of the transition state, and also to see if his previous results[118] could be improved by computing trajectories in the direction of reaction [4].

Monte Carlo trajectory methods were used to calculate the thermal rate coefficients for both the forward and reverse reactions, k_4 and k_{-4}, respectively. The resulting rate coefficients satisfied detailed balance, *i.e.* $k_4/k_{-4} = K_c$, only for temperatures greater than 500 K. At low temperatures inaccuracies occurred in the calculation of k_{-4} due to the neglect of quantum effects near the threshold energy. $k_4(T)$ is shown in Figure 19 and is in good agreement with the experimental data. Agreement was also found with earlier indirect experimental determinations of k_4 at temperatures up to 2500 K.

These QCT calculations showed that only 40% of the HO_2 complexes form products at 500 K, and only 30% at 2500 K, implying that redissociation into reactants, O + OH, is significant. Miller[10] attributed the discrepancy between Rai and Truhlar's variational transition-state theory calculations[9] and experiment to this non-statistical recrossing of the transition-state dividing surfaces. The Melius–Blint potential[117] was seen to be adequate for describing the thermal rate coefficients. The deficiencies in the potential have little effect on k_4, especially at high temperatures, but they could be responsible for the recrossing of the transition state.

[118] J. A. Miller, *J. Chem. Phys.*, 1981, **74**, 5120.

The Statistical Adiabatic Channel Model. The statistical adiabatic channel model (SACM) discussed in Section 3 has been extended by Troe[11] to apply to reaction [4]. As illustrated by the calculations of Rai and Truhlar[9] described above, the pronounced long-range dipole–quadrupole interaction between OH and O plays an important role in determining the kinetics of this reaction.

In the original development of SACM,[7] the radical interaction potential was taken to be a Morse potential:

$$V(r) \approx D\{1 - \exp[-\beta(r - r_{eq})]\}^2$$

with the angular potential characterized by channel eigenvalues of the form:

$$\varepsilon_i(r) \approx \varepsilon_i(r_{eq})\exp[-\alpha(r - r_{eq})] + \varepsilon_i(\infty)\{1 - \exp[-\alpha(r - r_{eq})]\} \quad (36)$$

Although treated as a variable parameter, α can be shown to be related to the r dependence of the bending force constant, $f(r)$, by RKR inversion[119] of equation (36):

$$f(r) \approx f(r_{eq})\exp[-2\alpha(r - r_{eq})]$$

Troe[11] has modified the original SACM by including the long-range potential in $V(r)$. At small r, close to r_{eq}, the Morse potential was adequate, *i.e.* the short-range potential $V^{SR}(r)$ was given by:

$$V^{SR}(r) = D\{1 - \exp[-\beta(r - r_{eq})]\}^2$$

The long-range potential, $V^{LR}(r)$, was expressed as a power series in r^{-1}. The overall potential, $V(r)$, was then obtained by using a switching function, $F(r)$:

$$V(r) \approx V^{SR}(r)F(r) + V^{LR}(r)[1 - F(r)]$$

By comparing $V(r)$ with *ab initio* calculations, the best form for $F(r)$ was found to be $F(r) = g(r)$, equation (32), with $\alpha = 0.5\beta$.

The angular potentials were represented by expressing their force constants in a similar manner:

$$f(r) \approx f^{SR}(r)F(r) + f^{LR}(r)\,[1 - F(r)]$$

The method of calculation of the adiabatic channel eigenvalues was also modified. The separate channel eigenvalues for the short-range and long-range potentials, $\varepsilon_i^{SR}(r)$ and $\varepsilon_i^{LR}(r)$, respectively, were evaluated

[119] C. Schwartz and R. J. LeRoy, *J. Chem. Phys.*, 1984, **81**, 3996.

and the switching function, $F(r)$, used to give the overall eigenvalue:

$$\varepsilon_i(r) \approx \varepsilon_i^{SR}(r)F(r) + \varepsilon_i^{LR}(r)[1 - F(r)]$$

For transitional modes, $\varepsilon_i^{SR}(r)$ corresponds to harmonic oscillator eigenvalues and $\varepsilon_i^{LR}(r)$ to hindered rotor eigenvalues. Alternatively, $\varepsilon_i(r)$ could be calculated by using the complete short-range/long-range switching model for the angular potential. Both methods required the evaluation of the hindered rotor eigenvalues and this was done using approximate representations.

The complete channel potential, $V_i(r)$, was then determined from:

$$V_i(r) = V(r) + \varepsilon_i(r) + \varepsilon_{cent}(r)$$

where $V(r)$ is the radial potential, and $\varepsilon_{cent}(r)$ is the centrifugal energy given by:

$$\varepsilon_{cent}(r) \approx B_{eff}(r)P(P + 1)$$

P is the effective angular momentum quantum number:

$$P = J\exp[-\alpha(r - r_{eq})] + l\{1 - \exp[-\alpha(r - r_{eq})]\}$$

where J is the molecular angular momentum quantum number and l is the orbital quantum number for the relative motion of the two fragments. $B_{eff}(r)$ is the mean of the two smaller rotational constants.

The maxima, ε_{0i}, of the channel potentials $V_i(r)$ were calculated, giving a 'pseudo-partition function' for the activated complex:

$$q^{0\dagger}(T) = \sum_i \exp(-\varepsilon_{0i}/k_B T)$$

The limiting high-pressure rate coefficient, k_{24}^{∞}, for the recombination of O and OH:

$$O + OH \rightarrow HO_2 \qquad [24]$$

was given by:

$$k_{24}^{\infty} \approx \frac{k_B T}{h}\left[\frac{h^2}{2\pi\mu k_B T}\right]^{3/2} \frac{q_{el}^0(HO_2)}{q_{el}^0(OH)q_{el}^0(O)} \frac{q^{0\dagger\prime}}{q_{rot}^0(OH)} \qquad (37)$$

where μ is the reduced mass ($= m_{OH}m_O/m_{HO_2}$), and q_{el}^0 and q_{rot}^0 are the electronic and rotational partition functions, respectively. The vibrational contribution to $q^{0\dagger}$ from the conserved oscillator, *i.e.* the OH vibration in HO_2, approximately cancels with $q_{vib}^0(OH)$. Thus $q_{vib}^0(OH)$

is omitted from the denominator in equation (37) and $q^{0\dagger}$ replaced by $q^{0\dagger}$, which has this vibrational contribution removed.

The rate coefficients for the reaction to produce $H + O_2$ (reaction [4]) were calculated by multiplying k_{24} by dissociation factors that took into account both possible dissociation channels:

$$OH + O \underset{k_v}{\overset{k_u}{\rightleftharpoons}} HO_2^* \overset{k_w}{\rightarrow} H + O_2$$

Thus:

$$k_4 = k_{24}^{\infty} \frac{k(\varepsilon,J)_w}{k(\varepsilon,J)_v + k(\varepsilon,J)_w}$$

where the average energies, ε and the average angular momenta, J are those having the maximum contribution to k_{24}^{∞} at each temperature. This calculation predicted 4% redissociation at 100 K and 28% at 2500 K, contrary to the results of Miller.[10]

The calculated rate coefficient, k_4, is in good agreement with the experimental data and is illustrated in Figure 19.

Detailed Quantum Transition-State Theory and Adiabatic Capture Theory. The importance of the long-range interactions between O and OH in determining the value of the rate coefficient k_4 has also been illustrated by Clary.[12,120] He has performed three-dimensional quantum-mechanical calculations with a potential energy surface containing only long-range terms. The potential comprised a dipole–quadrupole interaction, a quadrupole–quadrupole interaction, and an isotropic factor depending on $\boldsymbol{R}^{-6}$, where $\boldsymbol{R}$ is the vector joining the O atom to the centre of mass of the OH radical. As in the Melius–Blint potential,[117] it was assumed that reaction [4] occurs on a single $^2A''$ surface.

Two techniques were used to calculate the reaction cross-sections. The first method combined the centrifugal sudden approximation[121] with a detailed quantum transition-state theory (DQTST) calculation. The scattering wavefunction for the entrance channel for reaction was expanded in terms of a close-coupled basis set of translational functions multiplied by rotational basis functions. The Schrödinger equation was then solved for values of $\boldsymbol{R}$ from a point well inside the centrifugal barrier to one in the asymptotic region of the entrance channel. The boundary conditions were applied by using a basis set of adiabatic rotational functions obtained by diagonalizing the atom–molecule Hamiltonian. The reaction cross-section, $\sigma(j)$, was then calculated from:

$$\sigma(j) = \frac{\pi}{k_j^2(2j+1)} \sum_J \sum_{\Omega=-J}^{J} (2J+1) \sum_n |S_{jn}^{J\Omega}|^2 \tag{38}$$

[120] D. C. Clary and H. -J. Werner, *Chem. Phys. Lett.*, 1984, **112,** 346.
[121] R. T. Pack, *J. Chem. Phys.*, 1974, **60,** 633.

k_j is the wavenumber for the initial rotational state, j, and Ω is the quantum number for the projection of both the total and rotational angular momenta along the body-fixed z axis (aligned along the $\boldsymbol{R}$ vector). $S_{jn}^{J\Omega}$ is a reactive S-matrix element for the transition $j \rightarrow n$ with total angular momentum J; n is the quantum number at the inner boundary. The probability of reaction with the given quantum numbers is $|S_{jn}^{J\Omega}|^2$. The third sum in equation (38) is only over open 'product' channels, n, which have positive translational kinetic energies. The centrifugal sudden approximation was assumed, *i.e.* coupling between different Ω levels was neglected.

The second technique, adiabatic capture (AC) theory, is a simpler method of calculating $\sigma(j)$. The effect on the reaction of energy transfer in the entrance channel was neglected, *i.e.* the initial (j,Ω) quantum numbers could be used to label the adiabatic potential energy curves $V_{j\Omega}(\boldsymbol{R})$. Tunnelling effects were ignored, since for high values of J these effects do not contribute significantly to $\sigma(j)$, and for low J values there is enough energy in the reactants to pass over the effective potential barrier. As a result, the reaction probabilities were approximated to:

$$\sum_n |S_{jn}^{J\Omega}|^2 = 0, \ \varepsilon < \bar{V}_j^{J\Omega}(\boldsymbol{R}) \tag{39a}$$

$$= 1, \ \varepsilon \geq \bar{V}_j^{J\Omega}(\boldsymbol{R}) \tag{39b}$$

ε is the total energy and $\bar{V}_j^{J\Omega}(\boldsymbol{R})$ is the effective potential given by:

$$\bar{V}_j^{J\Omega}(\boldsymbol{R}) = V_{j\Omega}(\boldsymbol{R}) + (\hbar^2/2MR^2)[J(J+1) + j(j+1) - 2\Omega^2]$$

where M is the collisional reduced mass and the $V_{j\Omega}(\boldsymbol{R})$ are the rotationally adiabatic potentials. Applying the conditions of equation (39) to equation (38) gave:

$$\sigma(j) = \frac{\pi}{k_j^2(2j+1)} \sum_{\Omega=-J}^{J} [J_{\max}(j,\Omega) + 1]^2$$

$J_{\max}(J,\Omega)$ is the maximum value for J for which condition (39b) holds, and it was assumed that $J_{\max}(j,\Omega)$ is very much greater than j.

In both methods, the state-selected rate coefficient, $k_j(T)$ was evaluated by Maxwell–Boltzmann averaging the cross-sections, $\sigma(j)$. The overall rate coefficient, $k(T)$, is then a Boltzmann average of $k_j(T)$ divided by a term which accounts for the electronic degeneracies.

The agreement between the DQTST and the AC calculations of the cross-sections was very good, and so the AC theory was used to calculate the $\sigma(j)$ over a range of energies in order that the rate coefficients could be obtained.

Figure 19 includes the calculated rate coefficients for reaction [4]. They are slightly higher than the experimental values because the AC theory neglects recrossing of the transition state and thus provides an upper bound to the exact rate coefficients. The calculated $k_4(T)$ has a negative temperature dependence, despite the fact that the state-selected rate coefficients, $k_j(T)$, increase with temperature. However, for a given temperature, $k_j(T)$ values decrease as j is increased, and so the Maxwell–Boltzmann average over the $k_j(T)$, together with the temperature dependence of the electronic partition functions, produces the negative temperature dependence in the overall rate coefficient.

Conclusions

There is excellent agreement between both theory and experiment for this reaction, especially at low temperatures. Troe's model[11] predicts that k_4 levels off quite sharply at temperatures above ~700 K and reproduces the reverse reaction data of Frank and Just[110] quite well. Miller's[10] theoretical rate coefficients, on the other hand, continue to decrease, although the agreement between the two models is still quite reasonable, even at 2500 K (within a factor of 2). Further high-temperature data would be of value, but it is clear that theoretical modelling, at least at the level of calculating thermal rate coefficients, is very satisfactory for a simple reaction of this type.

6 $HO_2 + HO_2 \rightarrow H_2O_2 + O_2$

The reactions of HO_2 with molecules, *e.g.* alkanes, have comparatively large activation barriers; its radical–radical reactions are, in consequence, of considerable significance, in both atmospheric chemistry and low-temperature combustion. Of these, reaction [5] has proved particularly intriguing.

$$HO_2 + HO_2 \xrightarrow{k_5} H_2O_2 + O_2 \qquad [5]$$

A major problem facing experimentalists is the development of a clean method of HO_2 generation. That most favoured in flash photolysis or molecular modulation[122,124–128] experiments is the photolysis of Cl_2 in the presence of CH_3OH and O_2:

122 R. A. Cox and J. P. Burrows, *J. Phys. Chem.*, 1979, **83,** 2560.
123 R. -R. Lii, R. A. Gorse, jun., M. C. Sauer, jun., and S. Gordon, *J. Phys. Chem.*, 1979, **83,** 1803.
124 B. A. Thrush and G. S. Tyndall, *J. Chem. Soc., Faraday Trans.* 2, 1982, **78,** 1469.
125 B. A. Thrush and G. S. Tyndall, *Chem. Phys. Lett.*, 1982, **92,** 232.
126 R. Patrick and M. J. Pilling, *Chem. Phys. Lett.*, 1982, **91,** 343.
127 C. C. Kircher and S. P. Sander, *J. Phys. Chem.*, 1984, **88,** 2082.
128 S. P. Sander, *J. Phys. Chem.*, 1984, **88,** 6018.

$$Cl_2 \xrightarrow{h\nu} Cl + Cl$$
$$Cl + CH_3OH \rightarrow CH_2OH + HCl$$
$$CH_2OH + O_2 \rightarrow HCHO + HO_2$$

A similar reaction sequence has been employed in discharge–flow experiments with $F–H_2O_2$ as an alternative source[129]:

$$F + H_2O_2 \rightarrow HO_2 + HF$$

The flash photolysis pulsed and modulation[122–129] experiments used u.v. absorption spectroscopy[122,123,126–128] and i.r. diode laser spectroscopy[124,125] to monitor HO_2, while in the discharge–flow studies, laser magnetic resonance (l.m.r.) was used.[129] Absolute concentrations were established by measuring the absorption cross-sections or by titration against NO, detecting the NO_2 product by l.m.r.:

$$HO_2 + NO \rightarrow HO + NO_2$$

The reaction shows an interesting and complex pressure and temperature dependence. The overall rate coefficient, k_5, increases linearly with [M], with a non-zero intercept at [M] = 0:[127]

$$k_5 = k'_5 + k''_5[M]$$

In addition, both k'_5 and k''_5 decrease with temperature. A series of experiments, using various techniques, shows good agreement for k'_5, typified by the Arrhenius expression obtained by Takacs and Howard[129] using discharge–flow/l.m.r. over the temperature range 253—390 K:

$$k'_5 = 2.0 \times 10^{-13} \exp(595\ K/T)\ cm^3\ molecule^{-1}\ s^{-1}$$

Kircher and Sander[127] obtained

$$k''_5 = 8.4 \times 10^{-34} \exp(1100\ K/T)\ cm^6\ molecule^{-2}\ s^{-1}$$

for an argon diluent.

The elucidation of the mechanism of reaction [5] has attracted a good deal of attention but not fully satisfactory mechanism has been developed. The most detailed is that of Patrick *et al.*,[130] who suggested that the reaction proceeds on a surface like that shown schematically in Figure 21, where H_2O_4 is a chain complex HOOOOH. They argued that the zero-pressure contribution arises from $H_2O_4{}^*$ species which isomerize directly to products, while the pressure-dependent component

[129] G. A. Takacs and C. J. Howard, *J. Phys. Chem.*, 1986, **90**, 687.
[130] R. Patrick, J. R. Barker, and D. M. Golden, *J. Phys. Chem.*, 1984, **88**, 128.

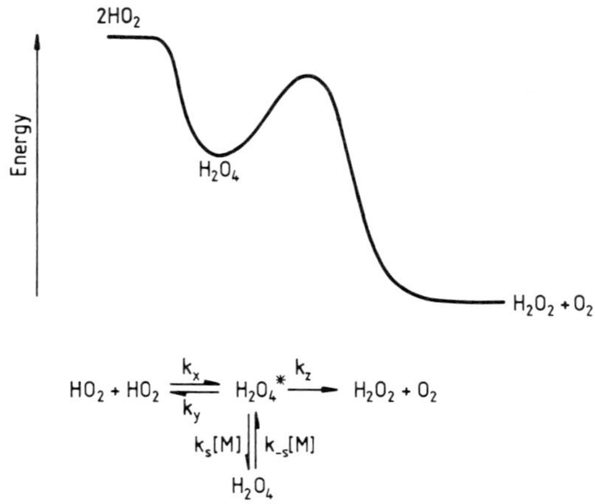

Figure 21 *Schematic potential energy diagram for* HO_2 *disproportionation* (Adapted from *J. Phys. Chem.*, 1984, **88,** 128)

corresponds to dissociation of thermalized H_2O_4 adducts. A steady-state analysis, using the phenomenological rate coefficients shown in Figure 21, gives:

$$k_5 = \frac{k_x k_z + k_z k_{-s}[M][H_2O_4]/[HO_2]^2}{k_s[M] + k_y + k_z}$$

Thus, with [M] = 0:

$$k_5 = k_5' = \frac{k_x k_z}{(k_y + k_z)}$$

and with $[M] \neq 0$ but $(k_y + k_z) \gg k_s[M]$:

$$k_5'' = \frac{k_z k_{-s}}{(k_y + k_z)} \frac{[H_2O_4]}{[HO_2]^2}$$

corresponding, qualitatively, to the observed behaviour.

Barker[131] examined the reaction using a Monte Carlo analysis, demonstrating the viability of a mechanism requiring a sharp division between short time, pressure-independent decomposition and subsequent activated decomposition of the thermalized H_2O_4 adduct. Patrick *et al.*[130] performed a detailed quasi-strong collision/RRKM analysis of the reaction scheme contained in Figure 21, demonstrating the feasibility of the mechanism, admittedly with a fair degree of free parameterization of the H_2O_4 molecular and thermodynamic constants. More research in

[131] J. R. Barker, *Chem. Phys.*, 1983, **77,** 301.

this area, coupled with multichannel master equation calculations, would be of value.

There was, until recently, only one experimental high-temperature determination of k_5, by Troe,[132] at 1100 K in a shock tube. His value, although carrying wide error limits (Figure 22), is not compatible with the lower-temperature data, and Baldwin *et al.*[14] suggested that at higher temperatures an activated, direct abstraction channel opens up. They estimated an A factor of 5.7×10^{-12} cm^3 molecule^{-1} s^{-1} for this channel and used Troe's single point to obtain an activation energy of 6.7 kJ mol^{-1} (Figure 22); they suggested that the highest temperature point determined by Patrick and Pilling[126] was in error in making this assignment, which was compatible with all the other low-temperature data.

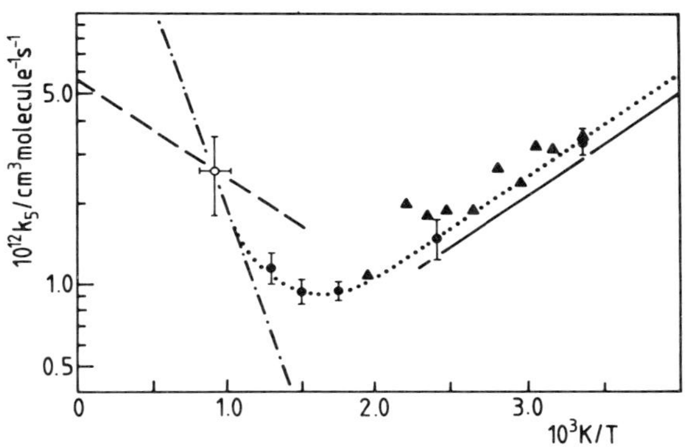

Figure 22 *Temperature dependence of k_5 at* 1 *atmosphere pressure.* ——, *Kircher and Sander;*[127] ▲, *Patrick and Pilling;*[126] ●, *Lesclaux* et al;[15] ○, *Troe.*[132] *Suggested temperature dependence of the high-temperature direct channel:* – – –, *Baldwin* et al.;[14] —·—·, *Lesclaux* et al.;[15] ·······, *composite (direct and associative) temperature dependence proposed by Lesclaux* et al.[15]

Recently, Lesclaux *et al.*[15] have performed experiments at temperatures up to 777 K. Their values for k_5 at 760 Torr agree well with the high-temperature data of Patrick and Pilling,[126] but at still higher temperatures show distinct curvature and confirm the onset of the direct channel. An analysis of all the data, including Troe's estimate, suggests a higher A factor [$(5.8 \pm 5.2) \times 10^{-11}$ cm^3 molecule^{-1} s^{-1}] than that proposed by Baldwin *et al.*[14]

After many years of extensive experimentation, reaction [5] is now well characterized at low temperatures, at least in simple diluents like

[132] J. Troe, *Ber. Bunsenges. Phys. Chem.*, 1969, **73,** 946.

argon. Further experiments are clearly required at higher temperatures, however, if even reasonably well-defined rate coefficients are to be provided for combustion modelling. In contrast to reactions [1]—[4], the modelling of k_5 has so far been only rudimentary, but the substantial experimental data set available at low temperatures for both $HO_2 + HO_2$ and $DO_2 + DO_2$ should make the system an attractive one for theoreticians.

CHAPTER 4

Transition-State Theory Rate Constants for Association Reactions Without Potential Energy Barriers

W. L. HASE and D. M. WARDLAW

1 Introduction

Association reactions which occur without potential energy barriers are of widespread importance. Included are processes as diverse as gas–surface adsorption, often an important step in heterogeneous catalysis,[1] and ion complexation. The latter process is of consequence in biochemistry and solution chemistry.[2] Examples of these two different types of association are NO + Pt(III)[3] and Cu(I) binding to cyclic polythiaethers.[4] Ion–molecule association, *e.g.*

$$H_3^+ + HCN \rightarrow H_3^+ \ldots HCN \rightarrow H_2 + H_2CN^+$$

is also of considerable significance. Such reactions are of general interest in physical and organic chemistry,[5] and are of more practical importance in atmospheric and interstellar kinetics.[6] Recent experiments have

[1] D. A. King and D. P. Woodruff, 'The Chemical Physics of Solid Surfaces and Heterogeneous Catalysis', Elsevier, New York, 1982.

[2] Y. Marcus, in 'Ion Solvation', John Wiley, New York, 1985; P. Portman, T. Maruizumi, M. Welti, M. Badertscher, A. Neszmelyi, W. Simon, and E. Pretsch, *J. Chem. Phys.*, 1987, **87**, 493.

[3] C. W. Muhlhausen, L. R. Williams, and J. C. Tully, *J. Chem. Phys.*, 1985, **83**, 2594.

[4] P. W. R. Corfield, C. Cellarelli, M. D. Glick, I. W.-Y. Moy, L. A. Ochrymowycz, and D. B. Rorabacher, *J. Am. Chem. Soc.*, 1985, **107**, 2399; J. A. R. Harmann and S. R. Cooper, *ibid.*, 1986, **108**, 1202.

[5] W. M. Olmstead and J. I. Brauman, *J. Am. Chem. Soc.*, 1977, **99**, 4219; W. R. Creasy and J. M. Farrar, *J. Chem. Phys.*, 1987, **87**, 5280.

[6] N. G. Adams, D. Smith, and D. C. Clary, *Astrophys. Lett.*, 1985, **296**, L31.

indicated that, at temperatures below 100 K, reactions of ions with dipolar molecules may play an important role in the synthesis of molecules in interstellar clouds.[7]

Radical–radical and radical–molecule association reactions, such as

$$CH_3 + H \rightarrow CH_4$$

and

$$H + O_2 \rightarrow HO_2$$

are of continuing importance in combustion and atmospheric kinetics.[8,9] Aggregation, which involves nucleation and cluster growth reactions such as

$$(H_2O)_n + H_2O \rightarrow (H_2O)_{n+1}$$

and

$$Al_n + Al \rightarrow Al_{n+1}$$

is another highly significant association process.[10] It is pertinent to solution kinetics and to understanding the transition from molecular to bulk properties.

Since there is such a wide variety of association reactions without potential energy barriers, it is not surprising that a number of different theoretical models have been proposed for calculating association rate constants. Table 1 suggests a classification of association rate constant calculations according to either a capture or a dynamical model. For each of these models the mechanism for formation of a stable association product (except for gas–surface adsorption) is

$$A + B \underset{k_b}{\overset{k_f}{\rightleftarrows}} AB^* \tag{1a}$$

$$AB^* + M \xrightarrow{k_s} AB + M \tag{1b}$$

Association involves the formation of a vibrationally/rotationally excited metastable AB^*. The association rate constant is k_f. A deactivating

[7] B. R. Rowe, J. B. Marquette, G. Dupeyrat, and E. E. Ferguson, *Chem. Phys. Lett.*, 1985, **113**, 403.

[8] D. L. Baulch and J. Duxbury, *Combust. Flame*, 1980, **37**, 313.

[9] M. T. Macpherson, M. J. Pilling, and M. J. C. Smith, *J. Phys. Chem.*, 1985, **89**, 2268.

[10] A. W. Castleman, jun. and R. G. Keese, *Chem. Rev.*, 1986, **86**, 589.

Table 1 *Classification of methods for calculating rate constants for association reactions without potential energy barriers*

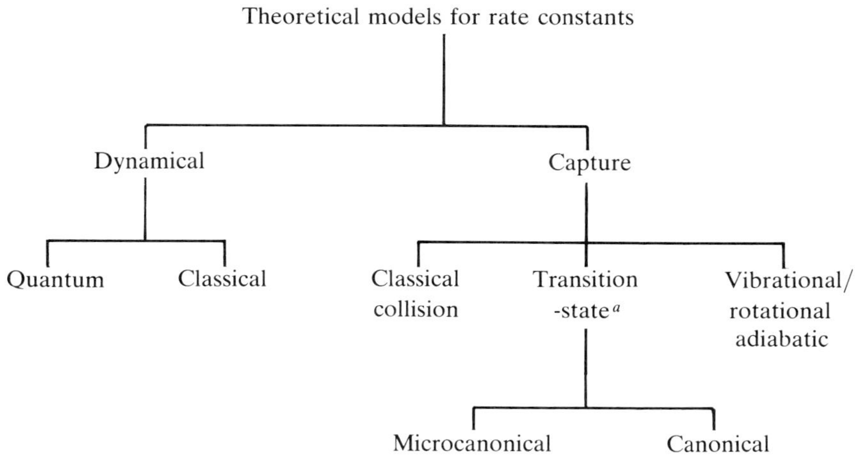

[a] Classical and quantum versions of transition-state theory for association reactions arise from either classical or quantum treatments of the low-frequency 'transitional modes' (see text)

collision, with rate constant k_s, is necessary to convert AB^* into the stable association product AB.

An assumption prevalent to each of the capture theoretical models is that inside a critical A + B separation all intramolecular degrees of freedom of the metastable AB^* are highly coupled and therefore readily exchange energy. Thus capture, *i.e.* attaining a critical A + B separation, is a criterion for formation of AB^*. The actual energy transfer pathways which give rise to the formation of the long-lived metastable AB^* are not considered.

This capture criterion is particularly evident in the Langevin,[11] average dipole orientation (ADO),[11] and perturbed rotational state[12] classical collision models which are widely used to calculate ion–molecule association rate constants. An orbiting impact parameter is found for collision on an effective potential energy curve. Association is assumed to occur for all smaller impact parameters. In transition-state theory,[13] the capture rate constant is found from the transition-state structure.

[11] T. Su and M. T. Bowers, in 'Gas Phase Ion Chemistry', ed. M. T. Bowers, Academic Press, New York, 1979, Vol. 1.

[12] K. Sakimoto, *Chem. Phys.*, 1981, **63**, 419; *Chem. Phys. Lett.*, 1985, **116**, 86.

[13] D. G. Truhlar, W. L. Hase, and J. T. Hynes, *J. Phys. Chem.*, 1983, **87**, 2664.

Adiabatic energy curves versus the A + B separation are used to determine the capture rate constant in vibrational/rotational adiabatic theory.[14–16] It becomes apparent that this is a capture theory when it is realized that association would not occur if the A + B collisions were truly adiabatic. Thus, extensive non-adiabatic coupling must be assumed at short A—B separations.

The dynamical theories for association attempt to treat the transfer of energy from A + B relative translation to the AB^* internal modes. The criterion for association is not capture, but the formation of a long-lived AB^* metastable. Presumably, if the dynamical calculations are quantum mechanical without extensive approximations, the correct value for k_f would be obtained. To date, the dynamical calculations have employed classical trajectories,[17–20] with varying degrees of coupling between the A + B relative translational motion and the AB^* intramolecular modes. In some cases, the A and B moieties have been assumed to be rigid with no internal degrees of freedom.[17] Another shortcoming in the classical calculations is that classical intramolecular energy transfer probabilities may differ from the correct quantum mechanical values.[21] For example, it is well known that classical trajectories may not treat the $A + B \rightarrow AB^*$ zero-point energy correctly during the association process.[22]

The above models make different assumptions concerning the dynamics of association reactions and thus complications arise in comparing their rate constants. In addition, some of the models are not applicable to all types of association reactions. A theory for association is most useful when it is generally applicable and when it has a well-defined theoretical basis. Transition-state theory, the theoretical model discussed here, fulfils both of these requirements.[13]

The variational version of transition-state theory[23–25] is required for reactions without potential energy barriers. Here, the transition state is

[14] R. A. Marcus, *J. Phys. Chem.*, 1979, **83**, 204.
[15] M. Quack and J. Troe, *Ber. Bunsenges. Phys. Chem.*, 1974, **78**, 240; 1975, **79**, 170 and 469; J. Troe, *J. Chem. Phys.*, 1987, **87**, 2773.
[16] D. C. Clary, *Mol. Phys.*, 1984, **53**, 3; 1985, **54**, 605.
[17] W. J. Chesnavich, T. Su, and M. T. Bowers, *J. Chem. Phys.*, 1980, **72**, 2641; T. Su and W. J. Chesnavich, *ibid.*, 1982, **76**, 5183; T. Su, *ibid.*, 1985, **82**, 2164; P. K. Bhowmik and T. Su, *ibid,*. 1986, **84**, 1432; S. Hu and T. Su, *ibid.*, 1986, **85**, 3127.
[18] K. N. Swamy and W. L. Hase, *J. Am. Chem. Soc.*, 1984, **106**, 4071; S. R. Vande Linde and W. L. Hase, *Comput. Phys. Commun.*, 1988, **51**, 17.
[19] L. M. Babcock and D. L. Thompson, *J. Chem. Phys.*, 1983, **78**, 2394; 1983, **79**, 4193.
[20] R. J. Duchovic and W. L. Hase, *J. Chem. Phys.*, 1985, **82**, 3599.
[21] D.-h. Lu and W. L. Hase, *Chem. Phys. Lett.*, 1987, **142**, 187; *J. Phys. Chem.*, 1988, **92**, 3217.
[22] R. A. Marcus, *Ber. Bunsenges. Phys. Chem.*, 1977, **81**, 190; W. L. Hase and D. G. Buckowski, *J. Comput. Chem.*, 1982, **3**, 335; W. L. Hase, *J. Phys. Chem.*, 1986, **90**, 365; G. C. Schatz, *J. Chem. Phys.*, 1983, **79**, 5386.
[23] D. G. Truhlar and B. C. Garrett, *Acc. Chem. Res.*, 1980, **13**, 440.
[24] W. L. Hase, in 'Dynamics of Molecular Collisions, Part B', ed. W. H. Miller, Plenum, New York, 1976, p. 121; W. L. Hase, *Acc. Chem. Res.*, 1983, **16**, 258.
[25] D. M. Wardlaw and R. A. Marcus, *Adv. Chem. Phys.* 1987, **70**, 231.

located at the point along the reaction path where the reaction flux is a minimum. Evaluation of the reaction path and its associated properties are discussed in the next section. This is followed by sections on: (*a*) a description of variational transition-state theory and minimization of the reaction path flux; (*b*) treating degrees of freedom in the transition state; (*c*) comparisons between different forms of variational transition-state theory; (*d*) general properties of variational transition-state theory structures and rate constants, and their dependence on potential energy surface features; (*e*) comparisons between variational transition-state theory and experimental rate constants, including a discussion on extrapolating experimental association rate constants to the high-pressure limit. In the final section, the assumptions of variational transition-state theory and their accuracy are reviewed.

2 The Reaction Path

In order to apply variational transition-state theory a knowledge of both the nature and properties associated with the reaction path is required. (The symbol R is used in this presentation to denote the reaction path. The position of the transition state along the reaction path is identified by $R^{\ddagger}$.) One approach is to postulate a reaction path Hamiltonian for which a particular co-ordinate is assumed to be the reaction co-ordinate.[26,27] A potential energy function is postulated for the reaction co-ordinate and properties are also postulated for the co-ordinates orthogonal to the reaction path.

Another approach involves finding the reaction path on the potential energy surface (PES) describing the association reaction. Here the reaction path, independent of the co-ordinate system, is defined by a trajectory that moves from reactants towards products in infinitesimal steps with its kinetic energy removed after each step.[28] For an association reaction without a saddlepoint the calculation of the reaction path is initialized with a large separation between the reactants. If the initial separation is sufficiently large the resulting path is independent of the starting separation.

In mass-weighted Cartesian co-ordinates, the reaction path trajectory traces out the steepest descent path and can be found by following the negative gradient vector of the $3N$ Cartesian co-ordinates from the initializing separation.[29-31] The negative gradient vector is given by:

[26] W. L. Hase, *J. Chem. Phys.*, 1976, **64**, 2442.

[27] D. M. Wardlaw and R. A. Marcus, *J. Phys. Chem.*, 1986, **90**, 5383.

[28] K. Fukui, S. Kato, and H. Fujimoto, *J. Am. Chem. Soc.*, 1975, **97**, 1; S. Kato, H. Kato, and K. Fukui, *ibid.*, 1977, **99**, 684; H. F. Schaefer, *Chem. Br.*, 1975, **11**, 227; K. Ishida, K. Morokuma, and A. Komornicki, *J. Chem. Phys.*, 1979, **66**, 2153.

[29] W. H. Miller, N. C. Handy, and J. E. Adams, *J. Chem. Phys.*, 1980, **72**, 99.

[30] S. Kato and K. Morokuma, *J. Chem. Phys.*, 1980, **73**, 3900; K. Morokuma and S. Kato, in 'Potential Energy Surfaces and Dynamics Calculations', ed. D. G. Truhlar, Plenum, New York, 1981, p. 243.

[31] A. D. Isaacson and D. G. Truhlar, *J. Chem. Phys.*, 1982, **76**, 1380.

$$\boldsymbol{v}_{i\gamma} = \frac{\mathrm{d}x_{i\gamma}}{\mathrm{d}R} = -c\,\frac{\partial V}{\partial x_{i\gamma}} \tag{2}$$

Here $i = 1,2,\ldots,N$ labels the atoms and $\gamma = x,y,z$ labels the Cartesian components, $x_{i\gamma} = \sqrt{m_i}\,q_{i\gamma}$ are the mass-weighted Cartesian co-ordinates, m_i is the mass of atom i, V is the potential energy function and c is a constant to normalize the $3N$-dimensional vector $\boldsymbol{v}_{i\gamma}$ to unity:

$$c = \left[\sum_{i=1}^{N} \sum_{\gamma=x,y,z} (\partial V/\partial x_{i\gamma})^2\right]^{-1/2} \tag{3}$$

For some association reactions the PES is represented by an analytic function which may be derived in part from *ab initio* calculations.[32] The partial derivatives $\partial V/\partial x_{i\gamma}$ are obtained analytically from the potential energy function. The $3N$ first-order differential equations given by equation (2), which define the reaction path, are solved numerically, *e.g.* by a Runge–Kutta or Gear integration algorithm. A normal reaction path step-size ΔR for the numerical integrations is 0.0005—0.001 amu$^{1/2}$ Å. Another approach for finding the reaction path is to evaluate it directly from *ab initio* calculations,[30,33] and thus circumvent the need for an analytic PES.

Properties found along the reaction path by solving equations (2) and (3) are the classical potential energy $V_{\mathrm{MEP}}(R)$ and Cartesian co-ordinates for each of the atoms. Internal co-ordinates such as bond lengths, valence angles, and torsion angles can be determined as a function of R from the atomic Cartesian co-ordinates. The Cartesian co-ordinates are also used to calculate the principal moments of inertia $I_\gamma(R)$, with $\gamma = x,y,z$ and $\gamma = x,y$ for non-linear and linear systems, respectively.

In addition to the above properties, generalized normal co-ordinates may be defined for the vibrational motions orthogonal to the reaction path.[29–31] Harmonic frequencies for these normal co-ordinates are found by diagonalizing the projected force constant matrix F^{P}:

$$F^{\mathrm{P}} = (1 - P)\cdot F\cdot(1 - P) \tag{4}$$

where F is a $3N \times 3N$ matrix of the mass-weighted Cartesian force constants $f_{i\gamma,i'\gamma'}$:

$$f_{i\gamma,i'\gamma'} = \frac{\partial^2 V}{\partial x_{i\gamma}\partial x_{i'\gamma'}} \tag{5}$$

[32] S. N. Rai and D. G. Truhlar, *J. Chem. Phys.*, 1983, **79**, 6046.

[33] T. Carrington, jun., L. M. Hubbard, H. F. Schaefer III, and W. H. Miller, *J. Chem. Phys.*, 1984, **80**, 4347; T. H. Dunning, jun., E. Kraka, and R. A. Eades, *Faraday Discuss. Chem. Soc.*, 1987, **84**, 427.

and P the projector. The role of P is to project out of F the direction along the reaction path and the directions corresponding to infinitesimal rotations and translations. As a result of this projection, F^P will have seven zero eigenvalues corresponding to infinitesimal rotations, translations, and motion along the reaction path. There will also be $3N - 7$ non-zero eigenvalues which give the harmonic frequencies $\nu_i(R)$ for the vibrations orthogonal to the reaction path.

In an actual calculation for an analytic PES, the $\nu_i(R)$ values are determined during the numerical integration of equation (2). The Cartesian force constants and the projector are evaluated at intervals along the reaction path. The Cartesian force constants are calculated numerically from the analytic first-derivatives $\partial V/\partial x_{i\gamma}$. Analytic expressions are used for determining components of the projector.[29] The F and P matrices are then inserted into equation (4) to form F^P, which is then diagonalized to obtain the $3N - 7$ harmonic frequencies. The vibrationally/rotationally adiabatic ground-state potential $V_a^G(R)$ is found from the harmonic frequencies and the classical potential, *i.e.*

$$V_a^G(R) = V_{MEP}(R) + \sum_{i=1}^{s-1} h\nu_i(R)/2 \tag{6}$$

where s equals $3N - 6$ if every degree of freedom is treated quantum mechanically. Zero-point energy $h\nu_i(R)/2$ is not included in equation (6) for modes treated classically (see the following sections).

In Figures 1—4 reaction path properties are illustrated for Na^+ complexation with the D_{3d} conformation of 18-crown-6,[34] and for $H + CH_3$ association.[35] Changes in molecular structure occur along the reaction path for each of these reactions. For the $H + CH_3$ reaction, the geometry of the CH_3 group changes from planar to pyramidal as the H atom associates. Structural changes that occur for the D_{3d} Na^+/18-crown-6 system are illustrated in Figures 1 and 2. For any position on the reaction path of this system, the carbon and oxygen atoms of the crown lie in four planes, each of which is perpendicular to the C_3 axis that passes through Na^+ and the crown centre of mass. These four planes are illustrated in Figure 1. In Figure 2 it is shown how the distances between these planes change along the Na^+—crown reaction path. The distance between the C,C planes changes very little. For the upper O,C planes, the major change is the compression that occurs at small centre-of-mass separations. The most significant effect is for the lower O,C planes, where the Na^+ cation pulls the O plane above the C plane.

[34] W. L. Hase, M. C. Richou, and S. L. Mondro, *J. Phys. Chem.*, 1989, **93**, 539.
[35] W. L. Hase, S. L. Mondro, R. J. Duchovic, and D. M. Hirst, *J. Am. Chem. Soc.*, 1987, **109**, 2916.

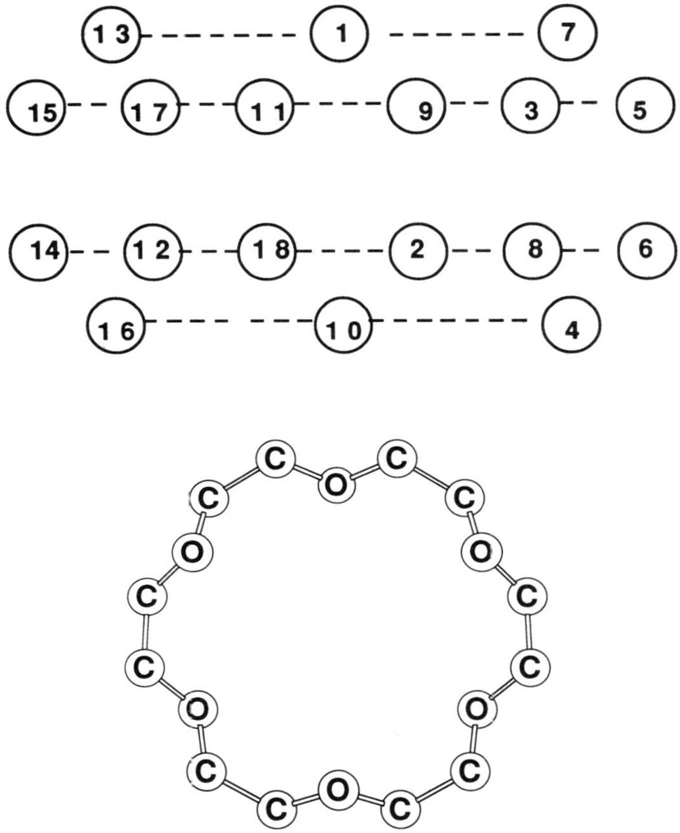

Figure 1 (a) *Depiction of the carbon and oxygen planes for* D_{3d} 18-*crown*-6. *The top and bottom rows refer to the* O *atoms*. (b) *Conformation of* D_{3d} 18-*crown*-6. *Atoms are numbered consecutively around the ring*

Changes in the molecular structures of the associating reactants along the reaction path can have a pronounced influence on the potential energy, $V_{\mathrm{MEP}}(R)$. This is illustrated in Figure 3, where the potential energy for the D_{3d} Na^+/18-crown-6 system is plotted against the separation between Na^- and the 18-crown-6 centre-of-mass. The lower-energy curve is for the reaction path, while the higher one is for 18-crown-6 constrained to its reactant conformation.

Vibrational frequencies orthogonal to the $H + CH_3 \rightarrow CH_4$ reaction path are shown in Figure 4. The two vibrational frequencies which change appreciably along the reaction path are those for the symmetric CH_3 out-of-plane bend and the degenerate H—CH_3 rocks. The frequency for these latter degenerate modes varies from zero at infinite separation where the motion is a free rotation, to $\sim$1300 cm^{-1} at the

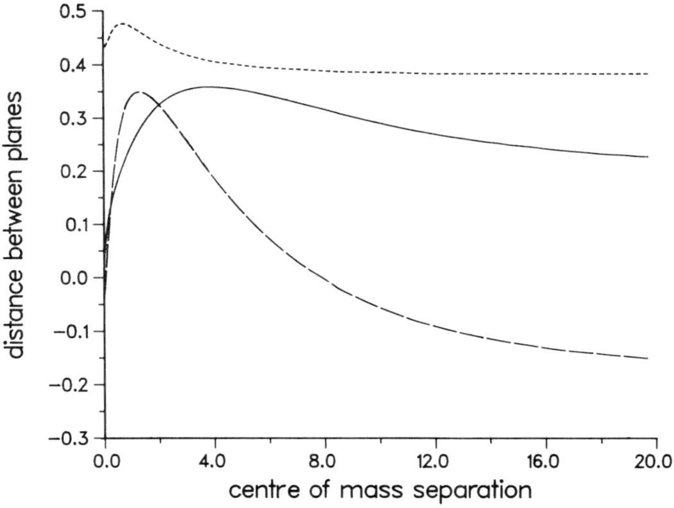

Figure 2 *Distance, (in Å) between the carbon and oxygen planes of D_{3d} 18-crown-6 as the* Na^+ *cation binds to the crown along the reaction path. The centre-of-mass separation, (in Å) is that between* Na^+ *and the crown.* (—), *Distance between the upper carbon and oxygen planes;* (– –), *distance between the lower carbon and oxygen planes;* (- - - -), *distance between the carbon planes. The planes are depicted in Figure 1a. The distance between the lower carbon and oxygen planes is negative when the oxygen plane is beneath the carbon plane*

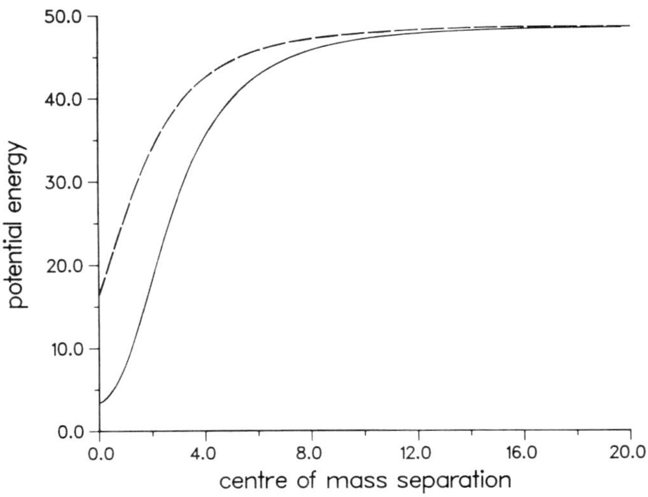

Figure 3 *Potential energy (kcal mol^{-1}) for* Na^+ *complexation with the D_{3d} conformation of* 18-*crown*-6. (—), *Potential energy along the reaction path;* (– –), *potential energy with* 18-*crown*-6 *fixed in its uncomplexed conformation. The centre of mass separation is in units of Å*

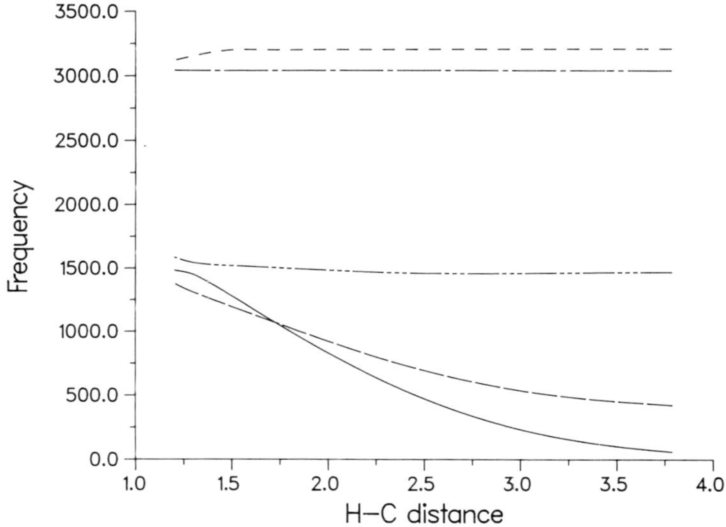

Figure 4 *Frequencies (in cm^{-1}) orthogonal to the reaction path for* $H + CH_3 \rightarrow CH_4$ *association. The potential energy function is the Hirst surface of ref.* 35. (—), *Doubly degenerate* H—CH_3 *bend;* (– –), *out-of-plane* CH_3 *bend;* (– - - -), *doubly degenerate* CH_3 *deformation;* (– - –), *symmetric* CH *stretch;* (- - -), *doubly degenerate* CH *stretch. The* H–C *distance is in* Å.

CH_4 potential minimum, where it is a CH_3 deformation. These degrees of freedom are called *transitional modes*[36,37] because of the pronounced change in the nature of their motion along the reaction path.

3 Variational Transition-State Theory

Outline of the Theory

The idea of varying the transition-state location to minimize the transition-state theory rate constant was first developed by Wigner[38] and Horiuti[39] in the late 1930s. Subsequently, Keck[40] established the classical dynamical foundations of variational transition-state theory (VTST), including the important result that (classical) VTST provides a rigorous upper bound to the true (classical) rate. One of the first applications to chemical reactions was by Bunker and Pattengill,[41] who used VTST to

[36] D. M. Wardlaw and R. A. Marcus, *Chem. Phys. Lett.*, 1984, **110**, 230.
[37] D. M. Wardlaw and R. A. Marcus, *J. Chem. Phys.*, 1985, **83**, 3462.
[38] E. Wigner, *J. Chem. Phys.*, 1937, **5**, 720.
[39] J. Horiuti, *Bull. Chem. Soc.*, 1938, **13**, 210.
[40] J. Keck, *Adv. Chem. Phys.*, 1967, **13**, 85.
[41] D. L. Bunker and M. Pattengill, *J. Chem. Phys.*, 1968, **48**, 772; D. L. Bunker, in 'Molecular Beams and Chemical Kinetics', ed. Ch. Schlier, Academic Press, New York, 1970, p. 315.

interpret classical trajectory calculations of unimolecular dissociation reactions. More recent investigations of the classical basis of transition-state theory have focused on the role of trajectories as dividing surfaces and as diagnostics for assessing the validity of the theory in collinear atom–diatom collisions.[42] An extension of these ideas to three-dimensional atom–diatom collisions has also been made.[43]

It is well known that the inclusion of certain quantum effects in VTST is essential to its accuracy, and hence utility, in predicting (quantum) dynamical rates. Difficulties in defining fully quantal versions of transition-state theory are noted in ref. 13, and attempts to establish the corresponding upper bound for quantal VTST have so far been unsuccessful.[44] The standard approximate approach to quantizing transition-state theory[13] is based on a hybrid model in which the reaction co-ordinate (RC) is treated classically and the transition-state dividing surface orthogonal to the RC is treated as a quantum system of dimension $N - 1$, where N is the number of degrees of freedom for the full system. It is this hybrid version of VTST to which we confine our discussion and applications, with the additional assumption that the reaction co-ordinate corresponds to the minimum energy path (MEP). Tunnelling along the RC, which is known to influence the rates near threshold in collision systems with potential barriers on the RC, is not expected to be important in reactions without potential barriers and is therefore ignored here. Transition-state recrossing effects which can be incorporated in unified statistical–dynamical reaction rate theories[45] are deemed to lie beyond the scope of this review (a brief discussion of these effects is given in the concluding section). Accordingly, and with the view that transition-state theory is a purely statistical model, transmission coefficients are not discussed, being implicitly set equal to unity throughout.

The VTST expression for the association rate constant k_f appearing in reaction (1a) can be written as a function of temperature and in the high-pressure limit:

$$k_f(T) = \frac{g_e Q^{\ddagger}(T)}{hQ_{AB}(T)} \tag{7}$$

where g_e is the ratio $g^{\ddagger}/g_A g_B$ of electronic partition functions for the transition state and infinitely separated reactants A and B; Q_{AB} is the partition function, in the centre-of-mass frame, for infinitely separated

[42] For excellent reviews of this work see P. Pechukas, *Annu. Rev. Phys. Chem.*, 1981, **32**, 159; *Ber. Bunsenges. Phys. Chem.*, 1982, **86**, 372.

[43] E. Pollak and R. E. Wyatt, *J. Chem. Phys.*, 1983, **78**, 446.

[44] P. Pechukas, in 'Dynamics of Molecular Collisions, Part B', ed. W. H. Miller, Plenum, New York, 1976, p. 239; F. J. McLafferty and P. Pechukas, *Chem. Phys. Lett.*, 1974, **27**, 511; W. H. Miller, *Acc. Chem. Res.*, 1976, **9**, 306; E. Pollak, *J. Chem. Phys.*, 1981, **74**, 6765; J. Costley and P. Pechukas, *Chem. Phys. Lett.*, 1981, **83**, 129.

A + B and is evaluated in standard fashion; $Q^{\ddagger}$ is a generalized or pseudo partition function for the $N - 1$ degree-of-freedom transition-state system. Implicit in $Q^{\ddagger}$ is the factor k_BT which appears explicitly in some formulations of VTST, *e.g.* equation (12). The crucial step in the estimation of k_f via equation (7) is the evaluation of $Q^{\ddagger}$, and it is at this point that various approximations may be invoked. The remainder of this section deals with approaches to the variational minimization of $Q^{\ddagger}$. Since a rigorous quantum treatment of the transition-state modes for three-dimensional polyatomic systems is not feasible, approximate treatments are required; these are considered in Section 4, although discussion is restricted throughout to association processes occurring on a single Born–Oppenheimer potential energy surface.

Microcanonical Versions (μVTST)

The most rigorous application of the variational criterion of VTST involves minimization of the reaction flux for each set of constants of the motion, followed by Boltzmann averaging of these fluxes to yield k_f. Since the total energy, E (in the centre-of-mass frame) and the total angular momentum, J are always conserved in reaction (1a), they will form our set of constants of the motion in the following discussion. (The projection J_z of J on a z-axis is also always a constant of the motion but the flux does not depend on it.) Additional constants of the motion, which may be exactly or approximately conserved, are of course possible and could be included in a straightforward fashion. The corresponding expression for $Q^{\ddagger}$, denoted $Q^{\ddagger}_{\mathscr{E}\mathscr{J}}$, is

$$Q^{\ddagger}_{\mathscr{E}\mathscr{J}}(T) = \sigma^{-1}\int_0^{\infty} \mathrm{d}E \sum_{J=0}^{\infty} N_{EJ}(R^{\ddagger}_{EJ}) \exp(-E/k_{\mathrm{B}}T) \tag{8}$$

where $N_{EJ}(R)$ is the sum of quantum states in all degrees of freedom except the reaction co-ordinate R for the given J and having energy less than or equal to E, $R^{\ddagger}_{EJ}$ is that value of R which minimizes N_{EJ} for a given E–J pair, and σ is a transition-state symmetry number to be discussed for particular association reactions in Section 7. It should be noted that the symbol G, rather than the symbol N used in this review, is widely used elsewhere to represent the sum of states.

A potentially useful alternative to $Q^{\ddagger}_{\mathscr{E}\mathscr{J}}$ is the related quantity

$$Q^{\ddagger}_{\mathscr{E}}(T) = \sigma^{-1} \int_0^{\infty} \mathrm{d}E\, N_E(R^{\ddagger}_E) \exp(-E/k_{\mathrm{B}}T) \tag{9}$$

where

$$N_E(R) = \sum_{J=0}^{\infty} N_{EJ}(R)$$

and $R^{\ddagger}_{E}$ is that value of R which minimizes N_E for a given E. $Q^{\ddagger}_{\mathscr{E}}$ will provide a good approximation to $Q^{\ddagger}_{\mathscr{E}\mathscr{J}}$ if $R^{\ddagger}_{E} \approx R^{\ddagger}_{EJ}$ for all J, *i.e.* if $R^{\ddagger}$ is practically independent of J. The advantage of using $Q^{\ddagger}_{\mathscr{E}}$ rather than $Q^{\ddagger}_{\mathscr{E}\mathscr{J}}$ in equation (7) is a significant reduction in the number of sum-of-state minimizations and an according reduction in computational effort. Unfortunately there is usually no *a priori* justification for using equation (9) instead of equation (8) and *a posteriori* numerical confirmation is required.

The integrals appearing in equations (8) and (9) must generally be solved by numerical quadrature. Various approaches to the evaluation of the integrands N_{EJ} and N_E for association reactions without potential barriers are presented in Section 4. The qualitative dependence of $R^{\ddagger}_{EJ}$ and $R^{\ddagger}_{E}$ on energy and angular momentum is discussed in Section 6.

Canonical Version (CVTST)

The most widely applied version of VTST is the canonical one, in which the dependence of the transition-state location on E and J is neglected. Instead a single transition state is found for each temperature. This is equivalent to minimizing the Boltzmann average of N_E with respect to R, *i.e.*

$$Q^{\ddagger}_{\mathscr{T}}(T) = \sigma^{-1} \int_0^{\infty} \mathrm{d}E\, N_E(R^{\ddagger}_{T}) \exp(-E/k_{\mathrm{B}}T) \tag{10}$$

$$\langle N_E(R) \rangle = \int_0^{\infty} \mathrm{d}E\, N_E(R) \exp(-E/k_{\mathrm{B}}T) \tag{11}$$

for a given T. The appeal of the canonical, relative to the microcanonical, approach is the reduction of the computational effort to a *single* minimization procedure at each temperature. The price to be paid for this saving is that $Q^{\ddagger}_{\mathscr{T}}$ generally exceeds $Q^{\ddagger}_{\mathscr{E}\mathscr{J}}$ (or $Q^{\ddagger}_{\mathscr{E}}$) and CVTST thus overestimates the μVTST prediction of k_{f}. Typical magnitudes of this overestimation are provided in Section 5; qualitative features of the temperature dependence of $R^{\ddagger}_{T}$ are discussed in Section 6.

In thermodynamic form, $k_{\mathrm{f}}(T)$ in equation (7) is expressed as:

$$k_{\mathrm{f}}(T) = \frac{k_{\mathrm{B}}T}{h} \exp[-\Delta G^{\ddagger}(R^{\ddagger}_{T})/k_{\mathrm{B}}T] \tag{12}$$

where $\Delta G^{\ddagger}$, the free energy of activation, is the free energy difference between the transition state and reactants. Thus, for CVTST a minimum in $k_{\mathrm{f}}(T)$ occurs when there is a maximum in the free energy $G(R)$ along the reaction path. This maximum gives the location $R^{\ddagger}_{T}$ of the transition state.

Minimization Procedure

The minimization of $N_{EJ}(R)$, $N_E(R)$, and $\langle N_E(R)\rangle$ in the evaluation of $Q^{\ddagger}_{\mathscr{E}\mathscr{J}}$, $Q^{\ddagger}_{\mathscr{E}}$, and $Q^{\ddagger}_{\mathscr{T}}$, respectively, must in general be done numerically. Only for highly simplified models, such as the one described in Section 6, can the minimization be done analytically. For the purposes of the present discussion, let us consider the evaluation of $Q^{\ddagger}_{\mathscr{E}\mathscr{J}}$. The standard approach is to calculate N_{EJ} on a grid of R values. If the grid is sufficiently fine, $N_{EJ}(R^{\ddagger}_{EJ})$ can be taken as the smallest value of N_{EJ} on the grid; practical experience indicates that a grid spacing of ~0.1 Å is usually small enough for this purpose. Alternatively, a coarser grid can be used to reduce the number of N_{EJ} evaluations and the N_{EJ} values fitted to a suitable functional form (*e.g.* a cubic spline) whose first derivative is set equal to zero, yielding $R^{\ddagger}_{EJ}$. Since $R^{\ddagger}$ depends on E and J (particularly on E), it is computationally expedient to adjust accordingly (usually in empirical fashion) the range of the grid on which N_{EJ} is calculated.

4 Treatment of the Transition-State Modes

A central feature in the application of VTST to association reactions without potential barriers is the recognition and classification of motion orthogonal to the reaction co-ordinate as 'conserved' or 'transitional' in nature. Conserved modes correspond to those motions whose form or character is not significantly altered as reactants approach along the reaction co-ordinate. They are generally identifiable as the internal vibrational degrees of freedom of the A and/or B moieties in reaction (1a) and are correlated (for example, by a reaction path analysis as outlined in Section 2) with vibrational modes of the AB* product. Transitional modes, on the other hand, correspond to motions whose form changes drastically in going from reactants to products; they include free rotations and orbital motion of the isolated A + B reactants which become vibrations, torsions, or hindered rotations of AB* product (see Figure 4). Since the transition-state location varies with energy and angular momentum in μVTST (or with temperature in CVTST), enumeration of the transitional modes is more difficult than in conventional transition-state theory where the transition-state location is fixed, *e.g.* at the maximum of a potential energy barrier. The situation is further compounded by the likelihood that, in the transition-state region, the transitional modes undergo large amplitude anharmonic motion and may be strongly coupled to each other.

The thrust of much recent theoretical work on barrierless association reactions has been to facilitate the VTST treatment of the transition-state modes, and in particular the transitional modes. An essential ingredient in any such treatment is a potential energy function valid over a wide range of reaction co-ordinate values and over the various transitional mode displacements. The derivation, accuracy, and availability of such potential surfaces are topics not covered in this review. Instead it is simply assumed that a suitable potential energy surface is, or could be made, available and the focus is on the treatment of the sum of states for the degrees of freedom orthogonal to the reaction co-ordinate. An essential first step in this direction is the MEP analysis described in Section 2. The remainder of this section reviews approaches to the sum-of-states calculation itself. Since we are unaware of any VTST applications to barrierless association reactions in which couplings between conserved and transitional modes are taken into account, the two groups of modes are treated as uncoupled throughout. The basis for such a separation of co-ordinates is conveniently discussed in terms of a reaction path Hamiltonian formulation[29] in which the Hamiltonian is separable at the quadratic level and couplings between degrees of freedom orthogonal to the reaction path are contained in perturbative terms. Small couplings are expected between pairs of modes having vastly different frequencies of motion, giving rise to what may be called an adiabatic separation of co-ordinates along the reaction path. Transitional mode-conserved mode couplings are therefore ignored on the grounds that conserved modes are typically of much higher frequency than the transitional modes.

Conserved Modes

An explicit quantum treatment of the conserved mode sum of states requires a direct count of all energetically allowed (vibrational) levels. In the simplest and most widely used approach, these levels are obtained by approximating the conserved modes as an ensemble of separable harmonic oscillators, *i.e.* normal modes. Harmonic frequencies as a function of the reaction co-ordinate are then the only data required and can be provided by MEP analysis of the potential energy surface or estimated by an interpolation between correlated reactant and product frequencies.[15] In the case of CVTST when the temperature is not too high, the direct count can be extended to infinite energy with the result that the Boltzmann-averaged sum of states can be expressed in customary form as a product of quantum harmonic oscillator partition functions.

The anharmonicity of the conserved modes can have a substantial effect on the VTST rate constant, both via a change in the transition-state zero-point energy and the non-uniform spacing of the anharmonic energy levels.[45] Under the assumption of separable conserved modes, various approximate anharmonicity corrections can be introduced,[45] the nature of which may depend on the type of vibrational motion. For

[45] D. G. Truhlar, A. D. Isaacson, and B. C. Garrett, in 'Theory of Chemical Reaction Dynamics', ed. M. Baer, CRC Press, Boca Raton, 1985, Vol. 4, p. 65.

example, stretching motions can be modelled as Morse oscillators and bending motions can be treated within a quadratic–quartic model for the bend potential. The effect of intermode coupling on the conserved mode sum of states is much more difficult to treat and has not been included in existing applications to barrierless association reactions. Several methods by which mode–mode couplings may be included in approximate fashion are discussed in ref. 45 and detailed analysis of this effect for a particular system is given in ref. 46.

Transitional Modes

The form of the transitional modes can be illustrated by considering the reactions $CH_3 + H \rightarrow CH_4$ and $CH_3 + CH_3 \rightarrow C_2H_6$, to which applications of VTST are discussed in Section 7. For $CH_3 + H$ there are two transitional modes, which become the CH_3—H degenerate rocking modes in CH_4 (Figure 4). The $CH_3 + CH_3$ reaction has five transitional modes. One is the H_3C—CH_3 torsion and the remaining four are two sets of degenerate CH_3 rocks. The number of transitional modes for various combinations of atomic and polyatomic products is listed in Table 2.

Table 2 *Number of transitional modes N in the association reaction* $A + B \rightarrow AB^*$

Nature of A	*Nature of* B	*Nature of* AB*	*N*
Atom	Linear polyatomic	Non-linear polyatomic	1
		Linear polyatomic	2
	Non-linear polyatomic	Non-linear polyatomic	2
		Linear polyatomic	3
Linear polyatomic	Linear polyatomic	Non-linear polyatomic	3
		Linear polyatomic	4
	Non-linear polyatomic	Non-linear polyatomic	4
		Linear polyatomic	5
Non-linear polyatomic	Non-linear polyatomic	Non-linear polyatomic	5
		Linear polyatomic	6

The very nature of the transitional modes renders treatment of their contribution to the overall sum of states at the transition state more difficult than the corresponding treatment of the conserved modes. Two fundamentally different approaches that have been employed in applications of VTST to barrierless association reactions are summarized below.

[46] L. B. Bhuiyan and W. L. Hase, *J. Chem. Phys.*, 1983, **78**, 5052.

In the first approach, which has been applied extensively by Hase and co-workers[35,47–50] and by others,[32,51] the standard form and calculational expediency of traditional transition-state theory have been retained by assuming separability of the transitional modes. The transitional modes are then modelled as oscillators, free rotors, or hindered rotors, depending on the mode. The physical constants (vibrational frequencies, moments of inertia, *etc.*) required to specify the associated energy levels have a reaction co-ordinate dependence which is most reliably provided by a MEP analysis of the potential energy surface. (In the absence of a suitable surface these quantities may be estimated by approximate interpolation formulae connecting corresponding reactant and product values.[15]) The ensuing sum-of-states calculation for the transitional modes is then straightforward. In μVTST, this sum of states must be convoluted with the associated conserved mode sum subject to the condition that the total energy of all transitional and conserved modes be less than or equal to a specified total energy. Furthermore, the total angular momentum J may also be held constant in the μVTST calculations. If all transitional modes are treated as vibrators, J is the external rotational angular momentum. Otherwise, constraints are placed on the angular momentum of those transitional modes treated as hindered or free rotors, so that the sum of the transitional mode angular momentum and external angular momentum equals J.

The simplifications resulting from assuming separability between the transitional modes and neglecting angular momentum coupling restrictions are quite evident in CVTST. The Boltzmann average of the transitional mode sum of states is most usefully expressed as a product of independent partition functions, so that the free energy along the reaction path is given by:

$$G(R) = G_{\text{vib}}(R) + G_{\text{ir}}(R) + G_{\text{rot}}(R) + V_{\text{MEP}}(R) \tag{13}$$

where the G_{vib}, G_{ir}, and G_{rot} terms are, respectively, the free energies for the vibrational, internal rotational (free and hindered), and external rotational degrees of freedom orthogonal to the reaction path, and $V_{\text{MEP}}(R)$ is the classical potential energy on the reaction path. Each G term in equation (13) is a sum of free energies for individual modes. The harmonic oscillator/rigid rotor approximation is usually made for conserved vibrational and external rotational modes. The most accurate treatment of separated transitional modes involves determining anharmonic (quantum) energy levels for each mode as a function of R and then evaluating partition functions and free energies from these levels.

[47] W. L. Hase and R. J. Duchovic, *J. Chem. Phys.*, 1985, **83**, 3448.
[48] S. L. Mondro, S. Vande Linde, and W. L. Hase, *J. Chem. Phys.*, 1986, **84**, 3783.
[49] S. R. Vande Linde, S. L. Mondro, and W. L. Hase, *J. Chem. Phys.*, 1987, **86**, 1348.
[50] W. L. Hase, *J. Chem. Phys.*, 1972, **57**, 730.
[51] J. F. LeBlanc and P. D. Pacey, *J. Chem. Phys.*, 1985, **83**, 4511.

Only limited applications[51,52] have been made using this approach.

In their applications of separable CVTST, Hase and co-workers[35,47–50] have made simplifying assumptions in treating the transitional modes. Torsions formed along the reaction path, such as the H_3C—CH_3 torsion, are treated as classical free rotors.[50] The transitional bending modes have been treated as harmonic oscillators (both classical and quantum) and as hindered rotors. Classical partition functions are evaluated numerically for the latter modes. Care must be exercised in using these simplifications, since it is unusual for a particular transitional mode to remain a vibration, free rotation, or hindered rotation over the entire range of reaction co-ordinate values. A hindered rotor treatment of a bending mode goes to the correct asymptotic limit of free rotation as the reaction co-ordinate is increased but does not go to the proper vibrator limit for small values of the reaction co-ordinate; the situation is reversed for a vibrator treatment of a transitional bending mode. The suitability of a particular model will depend on the tightness or looseness of the transition state, the latter feature also depending on E (and J) in μVTST, or T in CVTST. Fortunately, for the reactions which have been studied to date the vibrator and hindered rotor treatments of transitional bending modes have given very similar results. This is discussed in the next section.

A different approach, here denoted flexible transition-state theory (FTST), to the treatment of the transitional modes has been developed by Wardlaw and Marcus[25,27,36,37]. In FTST, the full anharmonicity and coupling of the transitional modes (as specified by the transitional mode potential energy function V_t) is retained; these modes are treated *classically*; modelling is confined to a rigid rotor description of the A and B moieties whose R-dependent moments of inertia are determined by their MEP configurations; conservation of angular momentum is rigorously included. The price to be paid for incorporation of any given V_t into the theory (and thus not being forced to separate the transitional modes) is an increase in computational effort and a classical, rather than the preferred quantum, treatment; each of these features is assessed in the next section. For the purpose of specific presentation we shall outline the FTST treatment of the sum of states N_{EJ}. A recent review[25] and original papers[27,36,37] provide further details as well as the FTST treatment of N_E and $\langle N_E \rangle$.

For each value of the reaction co-ordinate, N_{EJ} is written as the convolution

$$N_{EJ}(R) = \int_0^{E'} \mathrm{d}\varepsilon \, N_V(E' - \varepsilon)\rho_J(\varepsilon) \tag{14}$$

where $\rho_J(\varepsilon)$ is the density of states for the transitional modes in the

[52] P. D. Pacey, *J. Chem. Phys.*, 1982, **77**, 3540; P. D. Pacey and B. D. Wagner, *ibid.*, 1984, **80**, 1477.

energy interval $(\varepsilon,\varepsilon+d\varepsilon)$, and $N_V(E'-\varepsilon)$ is the number of conserved mode quantum states having energy less than or equal to $E'-\varepsilon$. The quantity E' in equation (14) is the available energy at a given value of R, *i.e.*

$$E' = E - V_{\text{MEP}}(R) - E_{\text{zp}}(R) \tag{15}$$

where E is the total energy available to infinitely separated reactants measured with respect to $V_{\text{MEP}}(\infty) = 0$, and E_{zp} is the zero-point energy of the conserved modes at the given value of R. The quantity $\rho_J(\varepsilon)$ is expressed as a $2(N+1)$ dimensional classical phase space integral, where N is the number of transitional degrees of freedom orthogonal to the reaction co-ordinate (see Table 2). The choice of variables of integration in the ρ_J expression is a matter of convenience. Most applications of FTST have employed action-angle variables,[37] but conventional Cartesian co-ordinates may also be used.[53] In either case, the multidimensional integral implicit in equation (14) is readily evaluated by a Monte Carlo method[37,53] and determination of the acceptance or rejection of each Monte Carlo selected phase space point involves a transformation from the integration variables to the internal co-ordinates in which the potential, V_t is customarily expressed.

An *ad hoc* semi-classical modification of N_{EJ}, designed to make it more closely approximate the energy-averaged behaviour of its quantum counterpart at small E' values,[54] is the following:

$$N_{EJ}^{\text{sc}}(R) = 1 + \int_{\varepsilon_{\text{zp}}}^{E'} d\varepsilon\, N_V(E'-\varepsilon)\, \rho_J(\varepsilon) \tag{16}$$

Here ε_{zp} is the R-dependent zero-point energy of the transitional modes (which can be determined, for example, by the MEP analysis discussed in Section 2), and the classical treatment of $\rho_J(\varepsilon)$ is retained. Comparison of microcanonical FTST rate constants obtained using equations (14) and (16) for N_{EJ} in equation (8) is provided in Section 7 for the $CH_3 + H \rightarrow CH_4$ reaction.

Each of the two approaches to the description of the transitional modes outlined in this section has distinct advantages over the other. FTST provides a more rigorous treatment since it makes no assumptions as to the separability of the transitional modes and always includes angular momentum conservation at each energy. The calculation of rate constants via FTST is, however, computationally intensive relative to the separable model approach. In FTST many Monte Carlo integrals, each typically requiring $\sim 10^3$—10^6 points for convergence (the actual number depends on R, on the association reaction, and on a microcanonical versus canonical implementation), and a commensurate number of

[53] S. J. Klippenstein and R. A. Marcus, *J. Phys. Chem.*, 1988, **92**, 3105.
[54] J. Troe, *J. Phys. Chem.*, 1979, **83**, 114.

co-ordinate transformations and potential function evaluations are involved in determining a rate constant. Selection of the separable model or FTST will depend on various factors such as the cost of calculation, ease of implementation, availability of a transitional mode potential function, and the required accuracy of the VTST rate constant predictions. Some comparisons of different versions of VTST, as applied to barrierless association reactions, are provided in Section 5.

5 Comparisons of Different Versions of VTST

Different approaches for applying VTST to association reactions without potential energy barriers have been outlined in the previous sections. In this section, results obtained with these different approaches will be compared. The comparisons include: (*a*) the microcanonical versus canonical versions of VTST; (*b*) FTST versus the separable version of VTST; (*c*) classical versus quantum mechanical treatments of the transitional modes; and (*d*) vibrator versus hindered rotor treatments of the transitional bending modes in the separable CVTST approach.

The first comparison of microcanonical and canonical versions of VTST was performed for the $2CH_3 \rightleftharpoons C_2H_6$ reactions.[26] In this work, the transitional bending modes were treated as separable vibrators, and microcanonical transition states were determined for individual values of E and J. At 860 K, the canonical rate constant was found to be 7% larger than the microcanonical rate constant. The separable microcanonical and canonical methods have also been compared for $Li^+ + (CH_3)_2O$ association,[49] for which the canonical rate constant is 37% and 6% larger at 200 and 1000 K, respectively.

Extensive comparisons of the microcanonical and canonical forms of FTST for the two reactions $2CH_3 \rightleftharpoons C_2H_6$[27] and $H + CH_3 \rightleftharpoons CH_4$[55] have been made. The calculated rate constants are listed in Tables 3 and 4. Microcanonical association rate constants were calculated by determining transition states for individual E–J pairs [equation (8)] and for fixed E with averaging over J [equation (9)], for which the rate constants are denoted by $k^{cl}_{\mu,E\text{–}J}$ and $k^{cl}_{\mu,E}$, respectively. The canonical rate constants are identified by k^{cl}_{C}. As shown in Table 3, very small differences are found between the two sets of microcanonical rate constants. Differences between the microcanonical and canonical rate constants are larger, and range from 15—20% for $2CH_3 \rightarrow C_2H_6$ (Table 3) and 5—10% for $H + CH_3 \rightarrow CH_4$ (Table 4).

In summarizing the above findings, one should note that there are differences between rate constants calculated by the microcanonical and canonical methods. However, the differences are not large and may in many cases be smaller than the uncertainty in an experimental rate constant.

[55] E. E. Aubanel and D. M. Wardlaw, *J. Phys. Chem.*, 1989, **93**, 000.

Table 3 *Comparison of μVTST and CVTST calculations of the* $2CH_3 \rightarrow C_2H_6$ *bimolecular rate constant*

Temperature/K	Rate constants[a]/10^{-11} cm^3 molecule^{-1} s^{-1} $k^{cl}_{\mu,E\text{-}J}$	$k^{cl}_{\mu,E}$	k^{cl}_{C}
300	4.33 ± 0.12	4.28 ± 0.10	5.08 ± 0.13
500	3.67 ± 0.08	3.58 ± 0.08	4.38 ± 0.06
1000	2.37 ± 0.06	2.29 ± 0.06	2.79 ± 0.06
2000	1.09 ± 0.04	1.08 ± 0.03	1.32 ± 0.03

[a] The three FTST rate constants are identified in the text. The uncertainties result from Monte Carlo averaging (see Section 4)

Table 4 *FTST rate constants for* $H + CH_3 \rightarrow CH_4$

Temperature/K	*Rate constants*[a]/10^{-10} cm^3 molecule^{-1} s^{-1} $k^{cl}_{\mu 0E\text{-}J}$	k^{cl}_{C}
300	2.36 ± 0.03	2.57 ± 0.08
400	2.64 ± 0.05	2.84 ± 0.10
500	2.87 ± 0.05	3.11 ± 0.08
600	3.02 ± 0.05	3.29 ± 0.10
800	3.20 ± 0.05	3.42 ± 0.10
1000	3.42 ± 0.05	3.60 ± 0.10
1250	3.52 ± 0.05	3.75 ± 0.10
1500	3.57 ± 0.07	3.82 ± 0.12
1750	3.59 ± 0.07	3.89 ± 0.12
2000	3.57 ± 0.07	3.89 ± 0.10
2500	3.65 ± 0.07	3.90 ± 0.12

[a] All rate constants are identified in the text. The uncertainties in the two sets of values result from Monte Carlo averaging

Only in very recent research[55] has a comparison of rate constants calculated with FTST and separable VTST become available. The comparison in Table 5 is at the canonical level and the reaction rate constant considered is that for bimolecular $H + CH_3$ combination. The two sets of separable VTST rate constants result from treating the transitional bending modes as either harmonic vibrations (vib) or hindered rotors (hr). Overall, the FTST and separable VTST rate constants and transition-state locations are in good agreement. At 200 K the separable vibrator model is in better agreement with FTST, while the separable hindered rotor model is in better agreement for $T = 300$–1000 K. Hopefully, more comparisons of FTST and the separable form of VTST will become available for calculations at both the microcanonical and canonical levels.

Table 5 *Comparison of canonical FTST and separable canonical VTST rate constants for* $H + CH_3 \rightarrow CH_4$ *association*

	Rate constants/ 10^{-10} cm^3 molecule^{-1} s^{-1}			*Transition-state locations* ($H—CH_3$)/Å		
Temperature/K	k_C^{cl}	k_{vib}	k_{hr}	$R^{\ddagger}_{T}$	$R^{\ddagger}_{vib}$	$R^{\ddagger}_{hr}$
200	2.11	2.06	2.27	3.7	3.54	3.67
300	2.51	2.37	2.64	3.7	3.42	3.55
400	2.84	2.59	2.87	3.5	3.33	3.45
500	3.11	2.72	3.06	3.5	3.25	3.37
600	3.29	2.82	3.19	3.3	3.19	3.29
800	3.42	2.96	3.40	3.2	3.09	3.17
1000	3.60	3.04	3.52	3.1	3.00	3.07

Several comparisons of VTST rate constants calculated with classical and quantum mechanical treatments of the transitional modes have been made. Comparisons made with the separable version of canonical VTST have considered the high-pressure rate constant for $H + CH_3 \rightarrow CH_4$. Quantum and classical anharmonic hindered rotor partition functions for the $H—CH_3$ transitional modes have been calculated[47,51] for a model potential energy function which gives 'loose' $H—CH_3$ transition states (*i.e.* large $R^{\ddagger}_{T}$). The resulting quantum and classical canonical VTST rate constants agree within 6% over the 200—400 K temperature range. For a model potential energy surface that gives significantly 'tighter' transition states, so that the transitional modes may be represented as vibrators, classical and quantum harmonic oscillator treatments of the transitional bending modes give canonical VTST rate constants which agree within 2% for temperatures ranging from 200—1000 K.[35]

The error introduced by the classical treatment of transitional modes in FTST has been assessed for the high-pressure rate constant of $2CH_3 \rightarrow C_2H_6$ recombination. Using a quantum Monte Carlo path integral method within the FTST framework, Klippenstein and Marcus[56] found the discrepancies between quantum and classical canonical rate constants to be <3% over the temperature range 300—2000K.

Applications of the separable version of canonical VTST have been made by treating the transitional bending modes as harmonic vibrators (classical or quantum) and classical hindered rotors. Rate constants and values of $R^{\ddagger}_{T}$ calculated with these two approaches are compared in Table 5 for $H + CH_3$ recombination[35] and in Table 6 for $Li^+ + H_2O$ and $Li^+ + (CH_3)_2O$ recombination.[48,49] For the results listed in these Tables the harmonic vibrators were treated quantum mechanically, although treating the harmonic vibrators classically gives nearly identical transition-state structures and rate constants which differ by less than

[56] S. J. Klippenstein and R. A. Marcus, *J. Chem. Phys.*, 1987, **87**, 3410.

Table 6 *Separable canonical VTST calculations for ion–molecule association*

	Rate constants 10^{-9} cm^3 molecule^{-1} s^{-1}		Transition-state locations/Å	
Temperature/K	k_{vib}	k_{hr}	$R^{\ddagger}_{vib}$	$R^{\ddagger}_{hr}$
$Li^+ + H_2O$				
200	13.5	13.6	14.6	15.4
300	11.2	11.0	11.9	12.6
400	9.85	9.66	10.4	10.9
500	8.90	8.68	9.3	9.8
750	7.39	7.12	7.6	8.0
1000	6.49	6.21	6.7	6.9
$Li^+ + (CH_3)_2O$				
200	57.0	61.8	17.5	20.6
300	32.7	37.9	13.6	15.5
400	22.9	27.6	11.4	12.8
500	17.6	22.1	10.0	11.1
750	11.5	15.1	7.9	8.7
1000	8.63	12.0	6.7	7.4

5%. Overall, there is good agreement between the harmonic oscillator and hindered rotor rate constants. The poorest agreement is for $Li^+ + (CH_3)_2O$ association at high temperatures. It should be noted that the type of agreement found here may not exist for association reactions with exceptionally tight transition states (for which the hindered rotor treatment is inappropriate) or with exceptionally loose transition states (for which the vibrator treatment is inappropriate).[47]

6 Properties of Transition States and Rate Constants Determined by VTST

Sensitivity to the Potential Energy Surface

For a reaction which occurs on a PES with a well-defined saddlepoint, VTST locates the transition state at the saddlepoint. Thus, the PES properties which affect the transition state and the rate constant are those localized about the saddlepoint. For the types of association reactions considered here, which do not have saddlepoints, the transition state is energy/temperature dependent and is not localized on the PES. Thus, global characteristics of the surface must be treated accurately to represent transition-state and rate-constant properties. Association reactions without potential energy barriers are principally influenced by the following two PES properties: (*a*) the attractiveness of the long-range potential along the reaction path; and (*b*) the attenuation of the anisotropy in the long-range potential.

Several studies have considered the effects of the above two properties on transition states and association rate constants.[35,47–49,57–59] Overall trends in the effects of these properties are summarized in Table 7. Increasing the attractiveness of the long-range reaction path potential, $V(R)$, has been studied for $H + CH_3$[35,47,59] and ion–molecule association.[48,49] The general trend is for the transition state to loosen and the association rate constant to increase. For $H + CH_3$ association, a more attractive $V(R)$ resulted in a rate constant with a smaller positive temperature dependence.[35] For the very attractive ion–molecule potentials, dominated by a R^{-2} term at long-range, the association rate constant has a substantial negative temperature dependence (Table 6).[48,49]

Table 7 *Effect of potential energy surface properties on bimolecular association transition states and rate constants*

Property	*Effect*		
	Transition state	*Rate constant*	*Temperature dependence of rate constant*
Make $V(R)$ more attractive, *e.g.* decrease n in $V = -c/R^n$	Becomes looser, $R^{\ddagger}$ lengthens	Increases	Becomes less positive, may be negative
Decrease attenuation of the long-range anisotropic potential	$R^{\ddagger}$ lengthens	Decreases	Becomes more positive[a]

[a] This effect may be sensitive to the form of the attenuation of the long-range anisotropic potential

Decreasing the attentuation of the anisotropic potential, and thus making it longer range, decreases the association rate constant and results in a transition state with a larger $R^{\ddagger}$.[57–59] The two switching functions, $S(R) = \exp[-a(R - R_0)]$ and $S'(R) = \exp[-a(R - R_0)^2]$ have been used to attenuate the anisotropic potential in model studies. Though both give association rate constants with a more positive temperature dependence as the attenuation is decreased, more work needs to be done before this is established as a general trend.

Energy, Angular Momentum, and Temperature Dependence of Transition-State Structures and Rate Constants

The normal trend observed in VTST calculations is for $R^{\ddagger}$ to decrease (*i.e.* the transition state to tighten) as either the energy, angular

[57] W. L. Hase, *Chem. Phys. Lett.*, 1987, **139**, 389.
[58] W. L. Hase and X. Hu, *Chem. Phys. Lett.*, in press.
[59] X. Hu and W. L. Hase, *J. Phys. Chem.*, submitted for publication.

momentum, or temperature is increased.[24,25] (For an exception see ref. 60.) The calculations for $H + CH_3 \rightarrow CH_4$ (Table 5) and ion–molecule association (Table 6) show that $R^{\ddagger}_T$ decreases with temperature. For $2CH_3 \rightarrow C_2H_6$, Wardlaw and Marcus[25,27] found $R^{\ddagger}$ to decrease with increasing energy but to be essentially independent of J, as illustrated by Figure 5. A related VTST calculation for $2CH_3 \rightarrow C_2H_6$ gives the same trend in $R^{\ddagger}$.[24,26]

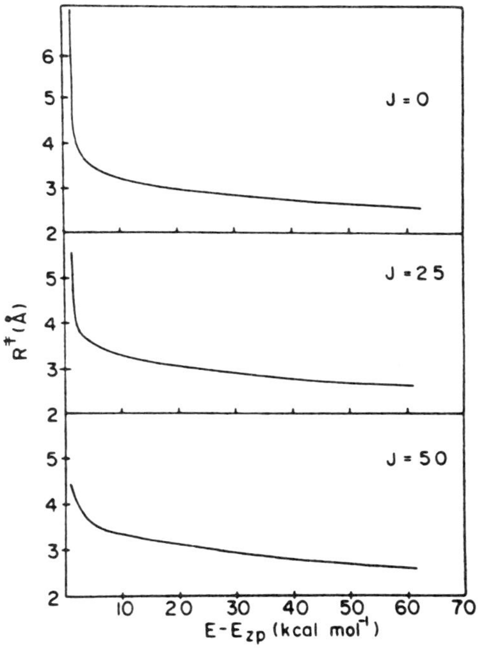

Figure 5 *Plot of the* $2CH_3 \rightarrow C_2H_6$ *transition-state location versus the energy in excess of the total zero-point energy of separated methyl radicals for several values of the total angular momentum* J
(Adapted with permission from *Adv. Chem. Phys.*, 1987, **70**, 231)

Tightening of the transition state with energy or temperature results from a balance between two PES properties (this is discussed in more detail in Section 5). One property is the decrease in the potential $V(R)$ as R is decreased so that more energy [*i.e.* E' in equation (15)] is available for the transitional modes. The other is an increase in the spacings between transitional mode energy levels as R is decreased (*i.e.* the forces for the transitional modes increase with decrease in R). The first property causes an increase in the number of states as R is

[60] R. Viswanathan, L. M. Raff, and D. L. Thompson, *J. Chem. Phys.*, 1984, **81**, 828.

decreased; the result of the second property is a decrease in the number of states with decrease in R. Normally, the balance between these two effects gives rise to transition states which tighten with increase in energy or temperature. Increasing the angular momentum decreases the amount of energy available for the transitional bending modes. This may cause a decrease in $R^{\ddagger}$.

As shown in Table 7, the temperature dependence of the bimolecular association rate constant depends on properties of the PES. Long-range radial potentials dominated by r^{-2} and/or r^{-4} terms, as in ion–molecule association, give rise to rate constants which decrease substantially with increasing temperature. This is illustrated in Table 6 for $Li^+ + H_2O$ and $Li^+ + (CH_3)_2O$ association. Covalent radial potentials for radical–radical association result in rate constants with a smaller temperature dependence, which can be positive ($H + CH_3 \rightarrow CH_4$ in Table 5) or negative ($2CH_3 \rightarrow C_2H_6$ in Table 3). Changing the attenuation of the association angular potential may also affect the temperature dependence of the rate constant.

Evidence for Multiple Transition States

In the above presentation it has been implied that there is only one microcanonical or canonical transition state for an association reaction at a particular energy or temperature. In actuality, multiple flux minima (*i.e.* transition states) may be present along the reaction path for an association reaction. However, for the $H + CH_3$, $2CH_3 \rightarrow C_2H_6$, $Li^+ + H_2O$, and $Li^+ + (CH_3)_2O$ association reactions considered above only one transition state is found at each energy and temperature.

Two transition states have been assumed for some ion–molecule association reactions to explain the observation that tight transition states are necessary to interpret experimental results at high energies/temperatures, while loose transition states are required at lower energies/temperatures.[61–63] It is not clear that this assumption is really necessary in light of the finding (Table 6) that ion–molecule transition-state structures are very dependent on energy/temperature. Multiple transition states were found in a VTST study of the $O + OH \rightarrow HO_2$ reaction.[32] However, they apparently arise from the pathological form of the O—O stretching potential in the analytic PES. A more convincing illustration of multiple transition states is the VTST calculation for the ion–molecule reaction $H + CH_3^+ \rightarrow CH_4^+$.[64] The possibility of multiple transition states in association reactions is intriguing, and will certainly receive consideration in future studies.

[61] W. J. Chesnavich, L. Bass, T. Su, and M. T. Bowers, *J. Chem. Phys.*, 1981, **74**, 2228.
[62] M. T. Bowers, M. F. Jarrold, W. Wagner-Redeker, P. R. Kemper, and L. M. Bass, *Faraday Discuss. Chem. Soc.*, 1983, **75**, 57.
[63] D. A. Webb and W. J. Chesnavich, *J. Phys. Chem.*, 1983, **87**, 3791.
[64] W. J. Chesnavich, *J. Chem. Phys.*, 1986, **84**, 2615.

Semi-empirical Model

Rather detailed numerical calculations, as described above, are required to determine variational transition states and rate constants. To analyse and interpret general trends in properties of variational transition states and rate constants, it is useful to have an analytic model. Such a model has been proposed for both microcanonical and canonical VTST.[57,58] Though the model is not quantitative, it is qualitatively correct and provides insight into the characteristics of VTST for association reactions.

The assumptions of the model are as follows.

(*a*) Vibrational frequencies for the conserved modes are assumed not to vary with R.

(*b*) Internal rotations formed as association occurs (*e.g.* the torsion for $H_3C—CH_3$) are free rotors. The validity of this assumption has been considered.[50]

(*c*) Following results from electronic structure theory calculations,[35,65] harmonic frequencies for the transitional bending modes are assumed to vary with R according to:

$$\nu_i(R) \propto \exp(-a_i R) \tag{17}$$

(*d*) The transitional bending modes are treated as classical harmonic oscillators. This assumption will be invalid for large R where the transitional bending mode has become a free rotor.

(*e*) The 'pseudo' diatomic approximation[41,66] is used to calculate the overall rotational energy along the $X + Y \rightarrow XY$ reaction path. Thus:

$$E_J = J^2/2\mu R^2 \tag{18}$$

where $\mu = m_X m_Y/(m_X + m_Y)$.

In the microcanonical model, the classical mechanical expression is used for the number of states along the reaction co-ordinate, *i.e.*

$$N_{EJ}(R) = \frac{[E - J^2/2\mu R^2 - V(R)]^s}{s!\prod_{i=1}^{s} h\nu_i(R)} \tag{19}$$

where s is the number of vibrations orthogonal to the reaction path. Inserting the above assumptions into equation (19) and solving $dN_{EJ}(R)/dR = 0$ to find the minimum flux gives:

[65] L. B. Harding, A. F. Wagner, and T. H. Dunning, jun., unpublished results; J. Troe, *J. Phys. Chem.*, 1986, **90**, 3485.

[66] R. A. Marcus, *J. Chem. Phys.*, 1965, **43**, 2658.

$$A = \sum_i a_i/s = \frac{[\mathrm{d}V(R)/\mathrm{d}R - J^2/\mu R^3]}{[E - J^2/2\mu R^2 - V(R)]} \tag{20}$$

where A is a constant for a particular reaction, i is summed over the transitional bending modes, and $V(R)$ is the classical potential energy on the reaction path $V_{\mathrm{MEP}}(R)$. Microcanonical variational transition states are found by solving this equation. Depending on the form of $V(R)$ and the values of E and J, there may be multiple solutions (*i.e.* transition states). It is obvious from equation (20) that a flat $V(R)$, *i.e.* $\mathrm{d}V(R)/\mathrm{d}R = 0$, will not have a transition state.

For normal long-range association potentials, $\mathrm{d}V(R)/\mathrm{d}R$ increases with decrease in R. It is instructive to consider transition-state properties for such potentials with $J = 0$. To find a transition state for a particular E, R is decreased until the right- and left-hand sides of equation (20) are equal. A transition state at a higher E will have a smaller R. Increasing the a_i parameters will also result in a tighter transition state. One also sees that the greater the change in the slope $\mathrm{d}V(R)/\mathrm{d}R$ with R, the less sensitive the transition-state structure is to E. Many association potentials are dominated by a long-range term of the form $-c/R^n$. If this term is inserted into equation (20), it may be shown[57] that there is only one transition state at each energy for $J = 0$. For $J \neq 0$ there may be multiple transition states.

The canonical model is derived by using the above assumptions to calculate the free energy along the reaction path $G(R)$, which is given by equation (13). Evaluating the minimum in the canonical flux, *i.e.* the maximum in $G(R)$, gives rise to the rather simple expression:[58]

$$R_g T = [\mathrm{d}V(R)/\mathrm{d}R]^{-1}[2/R + \sum_i a_i] \tag{21}$$

where R_g is the gas-law constant. The three contributing factors to the location of the transition state are the terms $\mathrm{d}V(R)/\mathrm{d}R$, $2/R$, and $\sum_i a_i$, which originate from the classical potential, external rotation modes, and transitional bending modes, respectively.

One of the significant aspects of equation (21) is that it explains the following trends observed in CVTST calculations for normal association potentials where $\mathrm{d}V(R)/\mathrm{d}R$ increases with decrease in R.

(*a*) The transition state 'tightens' (*i.e.* $R^{\ddagger}_T$ becomes smaller) as the temperature T is increased.
(*b*) Removing external rotational degrees of freedom results in a 'looser' transition state.
(*c*) The transition state loosens as the attractiveness of the potential is increased (*i.e.* increasing $\mathrm{d}V(R)/\mathrm{d}R$ at a particular R).
(*d*) Enhancing the attenuation of the transitional bending modes (*i.e.* increasing the a_i terms) results in a smaller $R^{\ddagger}_T$ at the transition state.

For a long-range association radial potential of the form $-c/R^n$, equation (21) may be written as:

$$1/R_{g}T = \frac{R^n}{cn}\left(R \sum_i a_i + 2\right) \tag{22}$$

Since the right-hand side of equation (22) monotonically increases with increasing R, the equation has only one root (*i.e.* transition state) for a particular temperature. For a more general $V(R)$, equation (21) shows that there may be multiple transition states.

The canonical model, equation (21), has been tested by comparing it with complete separable CVTST calculations for $H + CH_3 \rightarrow CH_4$ association.[58] Overall, the model gives very good agreement with the complete calculation. For temperatures in the range of 200—1000 K, the transition states have values for R which differ by a maximum of 0.05 Å. More comparisons between the semi-empirical canonical model and complete numerical calculations are expected.

7 Comparison with Experiment

One of the goals of transition-state theory is to provide rate-constant predictions that are reliably based on a conceptually useful and well-defined statistical model. In the ideal case, transition-state theory rate constants and experimentally derived ones would then be compared, with the extent of disagreement revealing the significance of, and possibly the nature of, dynamical effects in the actual reaction. Unfortunately, the ideal case is not realizable because of the approximate implementation of the theory (as discussed in Sections 3 and 4) as well as incomplete knowledge of potential energy surfaces. Comparisons are further complicated by the fact that experimental rates are often reliably measured only in the low- or medium-pressure region and not in the high-pressure limit to which the theoretical rate constants described in previous sections apply. A theoretical treatment of the pressure-dependent rate constants of reaction (1a) requires explicit consideration of the effect of reaction (1b), and this introduces further uncertainty into the comparison. Since one is often interested in the high-pressure limit of k_f, extrapolations of measured rate constants to this limit are required. Such extrapolations rely on a theoretical treatment of the pressure-dependent rate constants and can be sensitive to the particular theoretical description adopted.

To illustrate various aspects of comparisons between VTST and experimental rate constants, two barrierless association reactions, which have been studied in detail and which have come to be regarded as test cases, have been selected here. The first of these is the $CH_3 + H \rightarrow CH_4$ recombination, for which a PES based on *ab initio* calculations exists but whose pressure-dependent rate constants have received limited theoretical attention. The second is the $2CH_3 \rightarrow C_2H_6$ recombination, for which there is currently no *ab initio* PES but whose pressure-dependent rate constant has been extensively studied both experimentally and theoretically and whose high-pressure rate constant has received considerable theoretical attention over the years.

$CH_3 + H \rightarrow CH_4$

The availability of several realistic potential energy surfaces[35,67,68] for this reaction has inspired a great deal of theoretical effort aimed at elucidating the recombination kinetics. Here we consider only applications of VTST to the title reaction on the Hirst potential surface which was derived by Hase *et al.*[35] by refitting an earlier functional form[69] to Hirst's *ab initio* data.[70] Not discussed are LeBlanc and Pacey's separable CVTST treatment[51] on an earlier potential surface[69], Cobos and Troe's application[71] of a statistical adiabatic channel model, and Viswanathan, Raff, and Thompson's VTST treatment[60,72] of the reverse reaction $CH_4 \rightarrow CH_3 + H$. Also not discussed is the classical trajectory approach of Duchovic and Hase,[20,73] except to mention that it should be most useful in assessing the dynamical origins of the applicability or inapplicability of VTST.

An experimentally based value of the high-pressure rate constant k_{expt} for $CH_3 + H$ is given in Table 8; the rate constant is found not to depend on temperature over the indicated temperature range. The quoted value of k_{expt} was obtained by extrapolation of a fit of measured pressure-dependent rate constants to a variational RRKM model;[74] details of this theoretical treatment have not yet been published.[75] The remainder of Table 8 lists various $CH_3 + H$ rate constants as determined by FTST.[55] Details of the specification and computation of these quantities is not given here but can be found in ref. 55. The electronic partition function ratio g_e in equation (7) is 1/4, corresponding to the assumption that only reactants on the singlet electronic energy surface lead to recombination products. The symmetry factor σ implicit in $Q^{\ddagger}$ in equation (7) was taken to have its 'loose' value of 6 (*i.e.* the symmetry number of an isolated planar methyl radical) in all but one case, and the conserved modes were treated as separable harmonic oscillators, also in all but one case.

[67] F. B. Brown and D. G. Truhlar, *Chem. Phys. Lett.*, 1985, **113**, 441.
[68] H. B. Schlegel, *J. Chem. Phys.*, 1986, **84**, 4530.
[69] R. J. Duchovic, W. L. Hase, and H. B. Schlegel, *J. Phys. Chem.*, 1984, **88**, 1339.
[70] D. M. Hirst, *Chem. Phys. Lett.*, 1985, **122**, 225.
[71] C. J. Cobos and J. Troe, *Chem. Phys. Lett.*, 1985, **113**, 419; *J. Chem. Phys.*, 1985, **83**, 1010.
[72] R. Viswanathan, L. M. Raff, and D. L. Thompson, *J. Chem. Phys.*, 1985, **82**, 3083; 1984, **81**, 3118.
[73] R. J. Duchovic and W. L. Hase, *Chem. Phys. Lett.*, 1984, **110**, 474.
[74] M. Brouard and M. J. Pilling, *Chem. Phys. Lett.*, 1986, **129**, 439.
[75] A. F. Legg, M. J. Pilling, and C. J. Whitham, unpublished results.

Table 8 *Experimental and VTST high-pressure rate constants for* $CH_3 + H \rightarrow CH_4$

		Rate constants/10^{-10} cm^3 molecule^{-1} s^{-1}				
Temperature/K	k_{expt} [a]	k_{μ}^{cl}	k_{C}^{cl}	k_{μ}^{σ}	k_{μ}^{sc}	k_{μ}^{anh}
300	4.7	2.26	2.47	2.36	2.36	3.04
400		2.54	2.76	2.62	2.61	3.65
500		2.77	3.01	2.84	2.81	3.65
600		2.92	3.19	2.99	2.97	3.87

[a] From ref. 74

We begin the comparison between theory and experiment by considering the quantity k_{μ}^{cl} in Table 8. (In Table 3 the same quantity is denoted $k_{\mu,E\text{–}J}^{cl}$; here we drop the *E–J* subscript to simplify the notation). It is a microcanonical rate constant obtained by substituting $Q^{\ddagger}_{\mathscr{E}\mathscr{J}}$ [equation (8)] for $Q^{\ddagger}$ in equation (7) and adopting the purely classical treatment of the transitional modes specified in equation (14). Unlike k_{expt}, k_{μ}^{cl} has a positive temperature dependence, being approximately proportional to $T^{1/2}$ in the 300—600 K temperature range. More importantly, k_{μ}^{cl} is substantially less than k_{expt}, a point which will be further discussed below. The canonical rate constant corresponding to k_{μ}^{cl} is denoted k_{C}^{cl} in Table 8 and is seen to be ~9% larger than k_{μ}^{cl} in the 300—600 K range. The quantities k_{μ}^{sc}, k_{μ}^{σ}, and k_{μ}^{anh} are variations on k_{μ}^{cl} and are considered in order to investigate the effect of certain FTST assumptions on the disparity between k_{μ}^{cl} and k_{expt}. Both k_{μ}^{σ}, which incorporates an approximate symmetry correction to allow for deviations of σ from the assumed loose value of 6, and k_{μ}^{sc}, which incorporates the semi-classical correction to N_{EJ} specified by equation (16), are less than 5% larger than k_{μ}^{cl} and neither can therefore account for the disparity between theory and experiment. The rate constant k_{μ}^{anh} was obtained by treating one of the conserved modes as a separable anharmonic oscillator, specifically the CH_3 umbrella motion. Although k_{μ}^{anh} exceeds k_{μ}^{cl} by 13–19% in the 300—600 K range, indicating the significance of conserved mode anharmonicity for this system, it still underestimates k_{expt}.

Perhaps more interesting than the disparity between VTST rate constants and k_{expt} for $CH_3 + H \rightarrow CH_4$ is an apparent isotope anomaly in the rate constants for $CH_3 + H/D \rightarrow CH_4/CH_3D$. Brouard and Pilling[74] have measured rate constants for $CH_3 + D$ for a 289—401 K temperature range and a 50—600 Torr pressure range (see Chapter 3, Section 4). They conclude that the rate constant is essentially independent of pressure and effectively corresponds to the high-pressure limit under their experimental conditions. The reported high-pressure rate constant $k_{expt,D} = 1.75 \times 10^{-10}$ cm^3 molecule^{-1} s^{-1} for the 290—400 K temperature range is therefore free of the uncertainties associated with an extrapolation procedure. The ratio ρ_{expt} of their experimentally derived rate constant for $CH_3 + H$ in Table 8 to their directly measured rate constant for $CH_3 + D$ is 2.5. In the 300—400 K temperature range, the FTST rate constants $k_{\mu,D}^{cl}$ and $k_{\mu,D}^{anh}$ (which are the isotopic analogues of k_{μ}^{cl} and k_{μ}^{anh}) respectively agree with and slightly exceed $k_{expt,D}$. Comparison of the ratios ρ of rate constants for $CH_3 + H$ with those for $CH_3 + D$ in Table 9 reveals the anomaly that the ρ values for various VTST treatments (as specified in the footnotes to Table 9) are $\sim 1.4 \approx \sqrt{2} = \sqrt{m_D/m_H}$, as expected for an effectively loose transition state, whereas the experimental ρ value is 2.5. Various possibilities for the origin of the anomaly have been enumerated[35,55,74] but none has been substantiated. Further experimental and/or theoretical work will be required to resolve this situation.

Table 9 *Rate constant ratios for* $CH_3 + H \rightarrow CH_4$ *and* $CH_3 + D \rightarrow CH_3D$

Temperature/K	ρ_{expt} [a]	ρ_{μ}^{cl} [b]	ρ_{hr} [c]	ρ_{μ}^{anh} [d]
300	2.5	1.4	1.4	1.4
400		1.3	1.4	1.4

[a] From ref. 74. [b] Ratio of FTST rate constants k_{μ}^{cl}. [c] Ratio of separable CVTST rate constants calculated in ref. 35 by treating the transition modes as a classical hindered rotor. [d] Ratio of FTST rate constants k_{μ}^{anh}.

$2CH_3 \rightarrow C_2H_6$

The focus here is on recent experimental and complementary theoretical work on the pressure-dependent rate constant for the title reaction. It is informative, however, to begin by summarizing earlier predictions of the high-pressure rate constant, denoted here as k_1. Most of the experimentally based values for k_1 over the temperature range 250—2500 K and published between 1951 and 1985 are displayed in *Figure 2* of ref. 27 (a plot of $\log k_1$ versus $1/T$) and a comprehensive review of pre-1980 experimental work on methyl radical recombination has been published by Baulch and Duxbury.[8] The large scatter in the data in *Figure 2* of ref. 27 and the relatively large magnitudes of some of the associated uncertainties (not shown in that figure) are evidence of the difficulties in an experimental determination of k_1. Extrapolation of measured recombination rate constants at low pressures to the high-pressure limit is affected by the accuracy of the measured rates as well as being dependent on the particular theoretical treatment used to provide the extrapolation. A likely source of error in many earlier rate-constant determinations for $CH_3 + CH_3$ is the CH_3 absorption coefficient. Macpherson *et al.*[9] demonstrated the significance of this error by re-interpreting their earlier measurements,[76] thereby obtaining *different* values for k_1. Some estimates of k_1 are made by converting experimentally determined high-pressure C_2H_6 dissociation rate constants k_{-1} using the relationship $k_1/k_{-1} = K$, where K is the equilibrium constant for the title reaction. It is well known that such conversions are extremely sensitive to the C—C bond dissociation energy in C_2H_6, a quantity that is subject to some uncertainty.[77] Perhaps more striking than the scatter of experimental data in *Figure 2* of ref. 27 is the disparity in magnitude and temperature dependence of the various theoretical predictions for k_1. The collective theoretical treatments, which are described briefly in ref. 27 and include numerous versions of transition-state theory and one version of the statistical adiabatic channel model, illustrate the sensitivity of k_1 calculations to assumptions about the potential energy surface, about the nature of the transitional modes, and about the selection of the transition state. It is interesting to note that the first application of VTST to this system was performed in 1976 by Hase[26] and provided one of the earliest attempts at a realistic description of the evolution of the conserved and transitional modes along the reaction co-ordinate. On the whole, the pre-1986 situation for the $CH_3 + CH_3$ recombination rate constant was quite uncertain from both experimental and theoretical perspectives.

Available experimental data indicate that k_1 has a negative temperature dependence, although there is clearly uncertainty as to the degree of such a dependence. A microcanonical FTST treatment of k_1 by Wardlaw and Marcus[27] in 1986 qualitatively reproduced the magnitude and temperature dependence of the collective experimental data over the 300—2000 K temperature range, in contrast to all earlier transition-state theory treatments. A version of the statistical adiabatic channel model did, however, yield a slight negative temperature dependence within a reduced temperature range. Relevant aspects of this FTST treatment are provided below; otherwise the reader is referred to the original paper[27] and a summary in a recent review.[25] The availability of extensive new experimental measurements[78] of the fall-off behaviour of rate constants for the title reaction motivated the extension of the FTST treatment to the pressure-dependent regime,[79] providing a useful interplay between theory and experiment. The remainder of this section summarizes this extension and its ability to describe the experimental data.

Measured rate constants for the recombination $CH_3 + CH_3$, with argon as the bath or buffer gas M, have recently been reported by

[76] M. T. Macpherson, M. J. Pilling, and M. J. C. Smith, *Chem. Phys. Lett.*, 1984, **94**, 430.

[77] This is discussed in footnotes 25 and 26 of ref. 79.

[78] I. R. Slagle, D. Gutman, J. W. Davies, and M. J. Pilling, *J. Phys. Chem.*, 1988, **92**, 2455.

[79] A. F. Wagner and D. M. Wardlaw, *J. Phys. Chem.*, 1988, **92**, 2462.

Slagle *et al.*[78] at nine temperatures in the range 296—906 K and at Ar densities from 1.8×10^{16} to 1.4×10^{19} molecule cm^{-3}. At six of the nine temperatures, data were obtained by two independent experimental methods. Strong evidence for the accuracy of their data is provided by the observations that (*a*) rate constants measured by each method agree with each other under comparable conditions (*i.e.* same temperature and similar Ar gas densities), and (b) each set of rate constants is well described by the *same* theoretical model (which is defined below).

Theoretical rate constants for the overall recombination process

$$CH_3 + CH_3 + M \rightarrow C_2H_6 + M \tag{23}$$

were defined by the following expression:

$$k(T, [M]) = \frac{1}{4}\sigma^{-1}\frac{1}{hQ(T)^2}\int_0^\infty dE \int_0^\infty dJ\, N_{EJ}(R^{\ddagger}_{EJ})\, \frac{k_s[M]}{k_s[M] + k_{EJ}}\exp(-E/k_B T) \tag{24}$$

where $Q(T)$ is the partition function for an isolated methyl radical, $k_{EJ} = N_{EJ}/h\rho_{EJ}$ is the RRKM rate for the unimolecular dissociation of $C_2H_6^*$ at E and J, ρ_{EJ} is the density of states of $C_2H_6^*$, and k_s is an effective rate constant for stabilization of $C_2H_6^*$ by collisions with the buffer gas whose concentration (equivalent to pressure) is [M]. The factor 1/4 in equation (24) arises from the usual approximation that only systems initially on the singlet electronic energy surface lead to recombination products. The symmetry number σ in equation (24) is assumed to have its loose value of $2(6)^2 = 72$ (*i.e.* the total symmetry number for two identical, isolated methyl radicals); the deviation of σ from 72 is estimated in ref. 27 to be significant only for temperatures above 1000 K.

A conventional harmonic oscillator, rigid rotor model of C_2H_6 was used to evaluate ρ_{EJ}; the required frequencies, moments of inertia, and potential energy of C_2H_6 relative to $CH_3 + CH_3$ were all taken from experiment. The stabilization rate constant carries all information regarding the influence of the buffer gas on the rate constant. Buffer gas dynamics were treated in approximate fashion using a modified strong-coupling RRKM calculation followed by a weak-coupling correction, rather than in more rigorous fashion via a master equation model. This approximation applies if several collisions are required for stabilization and the rate constant is not near its low- or high-pressure limit, as is to be expected for the extensive temperature and pressure ranges in the measurements of Slagle *et al.*[78] The rate constant k_s in equation (24) is then given as a gas-kinetic rate constant multiplied by a correction factor β which depends on J and on the quantity $\langle \Delta E \rangle_{tot}$, the average total energy transferred by collisions with the buffer gas at temperature T. The weak-coupling correction was specified in terms of J-averaged values of β and J-averaged values of an effective number of degrees of freedom, the latter being determined from the FTST transition-state partition function. The only input parameter involved in obtaining k_s is $\langle \Delta E \rangle_{tot}$, which can be measured, calculated, or treated as adjustable. Since no measurements or calculations of $\langle \Delta E \rangle_{tot}$ are available for $C_2H_6^*$, it was adjusted to give the best fit to the measured rate constants.

For convenience, and to reduce the computational effort, values of the sum of states $N_{EJ}(R^{\ddagger}_{EJ})$ in equation (24) were determined only on a fixed grid of E and J values; $N_{EJ}(R^{\ddagger}_{EJ})$ at intermediate E and J values was then obtained by an interpolation scheme.[79] The grid values of $N_{EJ}(R^{\ddagger}_{EJ})$ were computed by variational minimization of equation (14) in which the transitional modes are treated in purely classical fashion. The potential energy surface controlling the recombination is a physically reasonable model but, unlike that for CH_3 + H, is not based on any *ab initio* electronic energy computations. The potential for the conserved modes is assumed to be separable and harmonic, in which case N_V in equation (14) is readily determined by a direct count of normal mode energy levels. The transitional mode potential consists of a modified bonded interaction between the two C atoms and atom-pair Lennard–Jones interactions between all non-bonded CH pairs. The evolution of conserved mode frequencies and of the equilibrium structure of each CH_3 group along the MEP is controlled by the function $\exp[-\alpha(R - R_e)]$, which interpolates between the separated reactants at $R = \infty$ and products at $R = R_e$. The quantity α is the only unspecified parameter in the model potential and it was adjusted to give the best fit to the experimental data. A distinguishing feature of this potential is that α influences the transitional degrees of freedom only weakly and indirectly by controlling small geometry changes in each CH_3 fragment (bond lengths and bond angles). This is to be compared with the statistical adiabatic channel model in which α directly controls the evolution of frequencies and moments of inertia describing the transitional modes. Full details of the PES and evaluation of equation (14) are given in ref. 27.

The final values of the parameters α and $\langle \Delta E \rangle_{tot}$ were selected by minimizing an average root mean square (r.m.s.) error between the measured and calculated rate constants and by the additional requirement that the high-pressure limit of equation (24) agree with the value directly measured in this limit at room temperature by Hippler *et al.*[80] This yields the physically plausible values of $\alpha = 0.70$ Å^{-1} and $\langle \Delta E \rangle_{tot} = -205$ cm^{-1} and a final average r.m.s. relative error of 9.9%.

[80] H. Hippler, K. Luther, A. R. Ravishankara, and J. Troe, *Z. Phys. Chem. Neue Folge*, 1984, **142**, 1.

High-pressure limiting rate constants calculated via equation (24) with this value of α are listed for a wide range of temperatures in Table 10. [In the high-pressure limit the value of k_s, and hence $\langle \Delta E \rangle_{tot}$, has no influence in equation (24)]. There is a marked negative temperature dependence with k_1 (296 K) being about a factor of four greater than k_1 (2000 K). Also listed (for comparison) in Table 10 are the only two available direct high-pressure measurements of k_1. A fit of the rate-constant calculations to a modification of the functional form of Gilbert *et al.*[81] over the 300—2000 K temperature range and the full pressure range is provided in ref. 79. This parameterization of the data produces an excellent analytic representation of the theoretical rate constants (average r.m.s. relative error of 6.2%) and is expected to be useful in modelling complex kinetic systems involving methyl radical recombination. The same parameterization accurately reproduces the experimental rate data of Slagle *et al.*[78] (average r.m.s. relative error of 7.1%).

The degree to which α and $\langle \Delta E \rangle_{tot}$ are constrained by the experimental results of ref. 78 is analysed in detail in ref. 79. The finding that the constraints are fairly mild, *i.e.* $\alpha = 0.70 \pm 0.13$ Å^{-1} and $\langle \Delta E \rangle_{tot} = -205 \pm 65$ cm^{-1}, suggests the need for further studies. To reduce the uncertainty in $\langle \Delta E \rangle_{tot}$ will require measurements over extended pressure ranges, especially at the lower temperatures. The range of α values lies below typical values of ~1.0 Å^{-1} attributed to many other recombination systems[71] whose semi-empirical potentials involve a loosely related α parameter. It is likely that the relatively low α range is an artifact of the $CH_3 + CH_3$ model potential constructed in ref. 27; plausible arguments to this effect are given in ref. 79. A potential energy surface based on *ab initio* electronic energy calculations is clearly desirable, and to this end preliminary comparisons of *ab initio* energies and model potential energies for selected CH_3–CH_3 configurations are underway.[82]

8 Future Directions

As discussed in the Introduction, VTST assumes each A + B trajectory that crosses the A—B transition state forms a long-lived vibrationally/rotationally excited metastable AB*. A topic for future study is the determination of those classes of reactions (*i.e.* potential energy surfaces) for which this assumption may be invalid. If there is weak-coupling past the transition state for an A + B trajectory so that some (or all) degrees of freedom are acting adiabatically, long-lived AB* metastables will not be formed and trajectories will immediately recross the transition state to form products. This type of behaviour has been observed in trajectory calculations of ion–molecule association.[18] As a

[81] R. G. Gilbert, K. Luther, and J. Troe, *Ber. Bunsenges. Phys. Chem,*, 1983, **87**, 169.
[82] D. M. Hirst, private communication.

Table 10 *High-pressure rate constants for* $CH_3 + CH_3 \rightarrow C_2H_6$

Temperature/K	*Rate constants*/10^{-11} cm^3 molecule^{-1} s^{-1} *Calculated*[a]	*Experimental*
296	5.78	5.8 ± 0.6[b]
407	5.35	
474	4.64	
513	5.06	
539	4.90	
577	4.78	
700	4.15	
810	3.76	
906	3.47	
1350	2.38	2.9 ± 1.5[c]
1600	1.95	
2000	1.47	

[a] $\alpha = 0.70$ Å^{-1}. [b] From ref. 80. [c] From ref. 83

result, the trajectory rate constant for forming the metastable AB* is significantly smaller than that of VTST.[18,48] For $H + CH_3 \rightarrow CH_4$, recrossing of the transition state is apparently not as important and the separable CVTST rate constant is only slightly larger than the quasi-classical trajectory value.[24,47] More comparisons between VTST and classical trajectories are expected in future studies of association reactions without potential energy barriers.

The VTST rate constant for the decomposition of the AB* metastable, equation (1a), is related to the association rate constant through the equilibrium constant:

$$K_{eq} = k_f/k_b \tag{25}$$

The microcanonical or canonical transition state used to calculate k_b at a particular energy or temperature is also used to calculate k_f at the same E or T. If recrossing of the transition state is important for association as a result of vibrational/rotational adiabaticity, a similar effect is expected for the decomposition reaction. Consider microcanonical transition-state theory where energy is randomized between degrees of freedom at the transition state. For the reactions considered here without potential energy barriers for association, the potential energy at the transition state A—B is less than that for the dissociation products A + B (for ion–molecule reactions the difference can be 4—5 kcal mol^{-1} for modest total energies[48,49]). Thus, there will be trajectories at the transition state without sufficient relative translational

[83] K. Glanzer, M. Quack, and J. Troe, *Chem. Phys. Lett.*, 1976, **39**, 304.

energy to overcome this potential energy difference unless there is substantial energy flow to relative translation from other degrees of freedom in proceeding from the transition state A—B to products A + B.[41] If this does not occur, trajectories will recross the transition state to reform the metastable AB* and VTST will overestimate the dissociation rate constant. Thus, substantial non-adiabaticity is necessary for VTST to be accurate for dissociation, as is the case for association. In future research, it seems worthwhile to consider in more detail the role of vibrational/rotational non-adiabaticity in validating VTST for association/dissociation equilibria where there is not a potential energy barrier for association.

Another aspect of the implementation of VTST for barrierless association/dissociation reactions which merits further study is the treatment of the conserved modes. Beginning in the 1970s, considerable effort has been devoted to improving the description of the transitional modes while the conserved modes have been routinely modelled as a collection of separable harmonic oscillators. In general, conserved mode anharmonicity alters both the transition state and reactant partition functions, making intuitive predictions as to the effect on the VTST rate constants difficult. Recent FTST calculations for CH_3 + H,[55] as discussed in Section 7, provide a striking example of the significance of anharmonicity within a separable vibrator model for the conserved modes. Systematic and thorough studies of different classes of association systems are needed to establish estimates of the effects of individual mode anharmonicity as well as intermode coupling. Very little is known about the influence of the latter on VTST rate constants because of the excessive theoretical and computational effort involved in obtaining approximate quantum energy levels for non-separable multidimensional vibrator systems. Further thought also needs to be given to the influence of coupling between a particular conserved mode(s) and the transitional modes. A good example is probably provided by the $CH_3 + H \rightarrow CH_4$ reaction for which the CH_3 umbrella motion is strongly influenced by the proximity of the approaching H atom (the frequency of the transition from the ground to first excited state of the umbrella mode more than doubles in going from reactants to product). The incorporation of transitional–conserved mode coupling in applications of VTST has not yet been attempted and appears to involve major reformulations of existing implementations of the theory. The typical influence of such coupling on the rate constant relative to the influence of conserved mode anharmonicity is unknown.

CHAPTER 5

Recent Direct Studies of Collisional Energy Transfer in Vibrationally Highly Excited Molecules in the Ground Electronic State

H. HIPPLER and J. TROE

1 Introduction

Intermolecular collisional energy transfer is of crucial importance in chemical reactions: reactants may be activated by collisions, reaction products may be deactivated. The reaction pathways of non-thermally activated reactants are determined by the competition between intra-molecular processes and collisional energy transfer.

The properties of collisional energy transfer in the sparse manifolds of rovibrational states at low degrees of excitation are relatively well understood both experimentally and theoretically. The recent excellent review by Krajnovich, Parmenter, and Catlett[1] summarizes the state of the art in this field. The situation appears much less well characterized in dense manifolds of rovibrational states at high degrees of excitation. Despite the particular importance of this range for large classes of chemical reactions, neither theory nor experiments provide a satisfactory picture of the phenomenon. For a long time, the available information was deduced from the results of reaction studies, either pressure-dependent reaction or fluorescence yields or pressure-dependent reaction rates. These 'indirect' studies have been extensively reviewed, *e.g.* by Tardy and Rabinovitch,[2] and by Quack and Troe.[3,4]

[1] D. J. Krajnovich, C. Parmenter, and D. L. Catlett, jun., *Chem. Rev.*, 1987, **87**, 237.
[2] D. C. Tardy and B. S. Rabinovitch, *Chem. Rev.*, 1977, **77**, 396.
[3] M. Quack and J. Troe, in 'Gas Kinetics and Energy Transfer', eds. P. G. Ashmore and R. J. Donovan (Specialist Periodical Reports), The Chemical Society, London, 1977, Vol. 2, p. 175.

A new chapter in the studies of collisional energy transfer of highly excited polyatomic molecules was opened in the last decade by the use of laser-flash excitation and time-resolved monitoring of excited-state populations. These experiments are 'direct' in the sense that overall rates for energy transfer in sequences of collisions are measured. The new experiments are not resolving state-to-state transitions in single collisions. The enormous density of states in highly excited polyatomic molecules would in any case prevent a study of a sufficient number of the contributing transitions. However, the average properties of energy transfer, such as are required for the implementation in chemical reaction models, are revealed in a new light by the 'direct' experiments. The present article reviews these results. In addition, a comparison of 'directly' and 'indirectly' obtained information can now be made. Examples of such comparisons will also be given in the present review.

Since the information on energy transfer of highly excited molecules to date always corresponds to averaged quantities, one has to ask which averages of single collision properties have been detected. This question becomes particularly important if the transition from energy transfer in the sparse level manifold at low energies to the dense manifold at high energies is considered. Likewise, this question is relevant in a comparison of theoretical results for single collisions with experimental results for collision sequences. The present review will therefore start with examples of theoretical results for collisional energy transfer of polyatomic molecules. A discussion of the relation between state-to-state cross-sections and the observable averages of experimental systems will lead on to a review of recent 'direct' experiments. 'Indirect' yield studies are only reviewed by comparison and insofar as they can be put on an absolute rate basis by direct measurements of the rates of the competing collision-free reference process.

This review will deal in particular with the following aspects of average energy transfer properties.

(*a*) The increase of the average energy transfer efficiency from the sparse low-energy manifold to the dense vibrational quasi-continuum at high energies.
(*b*) The dependence on the molecular complexity of the excited molecule and the generally unexcited bath-gas molecules.
(*c*) The dependence on the temperature of the bath-gas.
(*d*) The distribution of energies removed per collision from the excited molecule and accepted per collision in the bath-gas molecules.

2 From State-to-State Cross-sections to Energy Transfer Averages: Classical Trajectory Calculations

The complicated nature of a collision event between a vibrationally highly excited polyatomic molecule and a collider can be illustrated by

[4] M. Quack and J. Troe, *Int. Rev. Phys. Chem.*, 1981, **1**, 97.

classical trajectory calculations. Figures 1 and 2 show the results of Lim and Gilbert[5] on excited azulene ($C_{10}H_8$) colliding with Ar, at an initial azulene excitation energy corresponding to 35 455 cm^{-1} and a relative translational energy between the colliders of kT ($T = 630$ K). The internal energy E_i of azulene fluctuates during the collision time and, in addition, depending on the initial conditions either 'smoothly' changes from the initial to the final energy, or changes in a sequence of successive 'jumps'. The effective distance r_{eff} between the collision partners also reflects the 'single-' or 'multiple-' collision character of the events. The collision duration of ~500 fs demonstrates that a long-lived collision complex is not formed.

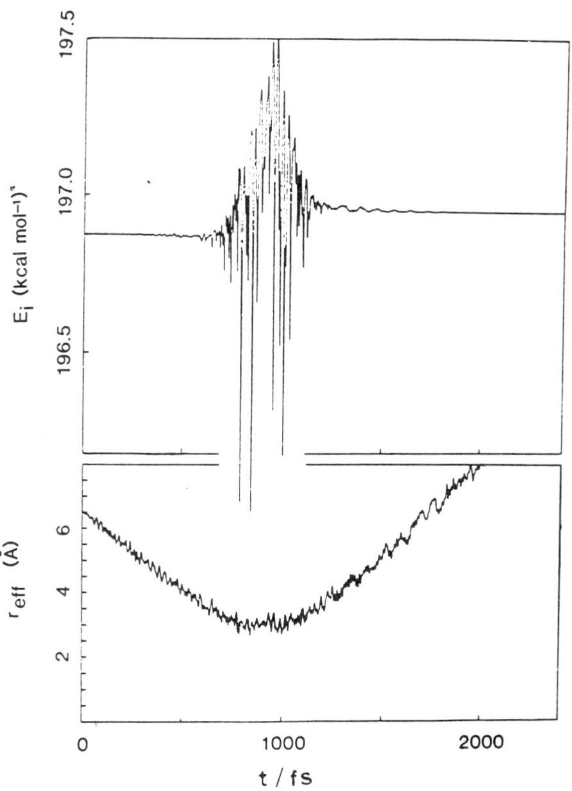

Figure 1 *Trajectory calculations of energy transfer in azulene* + Ar collisions. E_i = internal energy of azulene and r_{eff} = effective separation of the collision partners [initial energy (E) of azulene = 35455 cm^{-1}, temperature (T) of the bath gas = 630 K]*
(Reproduced with permission from *J. Chem. Phys.*, 1986, **84**, 6129)

[5] K. F. Lim and R. G. Gilbert, *J. Chem. Phys.*, 1986, **84**, 6129; R. G. Gilbert, *ibid.*, 1984, **80**, 5501.

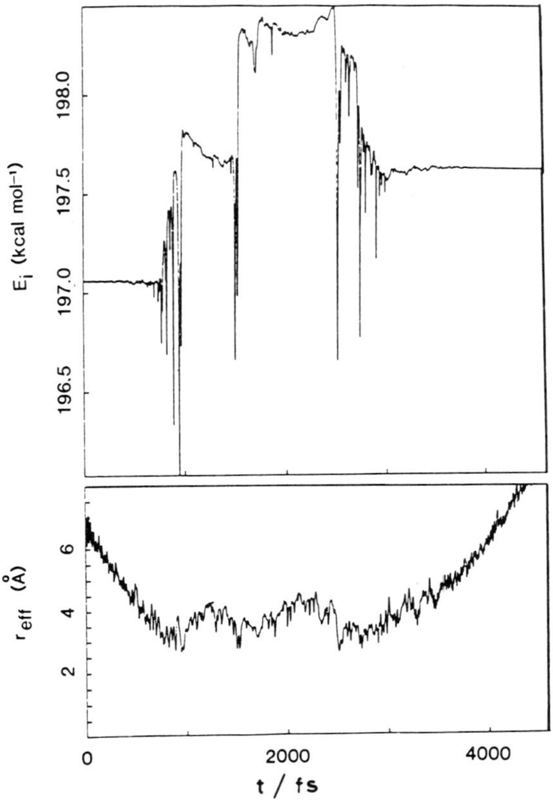

Figure 2 *As Figure 1; the same E and T but different initial conditions (positions and momenta)*
(Reproduced with permission from *J. Chem. Phys.*, 1986, **84**, 6129)

Since the azulene molecule at an internal energy of 35 400 cm^{-1} has a vibrational density of states, $\rho(E)$, of $2.6 \times 10^{21}/\text{cm}^{-1}$, the variety of collision patterns like Figures 1 and 2 will be immense, such that only the average behaviour is of interest. This requires the calculation of a sufficiently large number of trajectories to permit the construction of a histogram in which the fraction of collisions leading to final energies between E' and $E' + \delta E'$ is indicated. Histograms of this type have now been calculated for various small polyatomic molecules colliding with noble gases (see, for example, refs. 6—12). Figure 3 shows an

[6] H. Hippler, H. W. Schranz, and J. Troe, *J. Phys. Chem.*, 1986, **90**, 6158.
[7] H. W. Schranz and J. Troe, *J. Phys. Chem.*, 1986, **90**, 6168.
[8] A. J. Stace and J. N. Murrell, *J. Chem. Phys.*, 1978, **68**, 3028.
[9] N. J. Brown and J. A. Miller, *J. Chem. Phys.*, 1984, **80**, 5568.
[10] A. Gelb, *J. Chem. Phys.*, 1985, **89**, 4189.
[11] M. B. Grinchak, A. A. Levitzky, L. S. Polak, and S. Ya. Umanski, *Chem. Phys.*, 1984, **88**, 365.
[12] W. L. Hase, N. Date, L. B. Bhuiyan, and D. G. Buckowski, *J. Chem. Phys.*, 1985, **89**, 2502.

example for SO_2, at an energy of 100 kcal mol^{-1}, colliding with Ar in a thermal gas at 1000 K. The histogram defines a collisional transition probability distribution $P(E',E)$ for transitions from the initial energy E to the final energy E'. The results seem to be representable by an exponential collisional model:[13–15]

$$P(E', E) \approx \frac{1}{\alpha + \gamma} \exp\left(-\frac{E - E'}{\alpha}\right) \qquad \text{for } E' \leq E \tag{1}$$

$$P(E', E) \approx \frac{1}{\alpha + \gamma} \exp\left(-\frac{E' - E}{\gamma}\right) \qquad \text{for } E' \geq E \tag{2}$$

where the average energy γ transferred in activating collisions is smaller than the average energy α transferred in deactivating collisions. However, one may also prefer[9] a biexponential representation on the deactivation side with a dominant 'weak collision' component and a minor underlying 'strong collision' tail removing large amounts of internal energy.

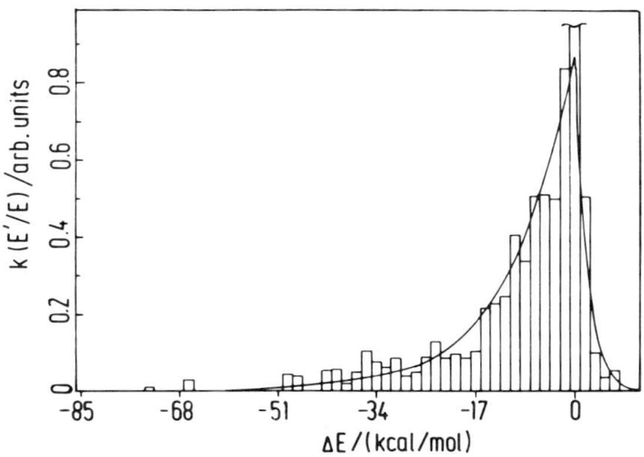

Figure 3 *Trajectory calculations of energy transfer in* SO_2^* + Ar *collisions. Histogram of (state E)-to-(state E') rate coefficients k(E',E) with initial internal energy (E) of* SO_2 = 100 kcal mol^{-1} *and bath-gas temperature (T)* = 1000 K; *the central peak is truncated*
(Reproduced with permission from *J. Phys. Chem.*, 1986, **90**, 6158)

The relation between a histogram like Figure 3 and a state-to-state cross-section for energy transfer is not easy to determine. The random selection procedure for initial conditions leads to an averaging over a sample of initial states (which appears to be unavoidable with a dense

[13] J. Troe and H. Gg. Wagner, in 'Recent Advances in Aerothermochemistry', AGARD, Oslo, 1966, p.21; *Ber. Bunsenges. Phys. Chem.*, 1967, **71**, 937.
[14] B. S. Rabinovitch and D. C. Tardy, *J. Chem. Phys.*, 1966, **45**, 3720; 1968, **48**, 1282.
[15] J. Troe, *J. Chem. Phys.* 1977, **66**, 4745.

quasi-continuum of initial states). Obviously, there is a problem with the central grain around the elastic peak: which events should be rejected as being 'elastic'?; how narrow has one to choose the central grains?; to which range should one limit the collision parameter b? The calculated average energy transferred per collision

$$\langle \Delta E(E) \rangle = \int_0^\infty (E' - E) P(E',E) \, dE' \tag{3}$$

obviously depends on the convention about the collision frequency Z, which is calculated according to the choice of the collision parameter b.

Starting from state-to-state cross-sections for collisional energy transfer, averaging these over narrow samples of initial- and final-state energies of the excited molecule, and averaging over a thermal distribution of the colliding species, one obtains an energy transfer rate coefficient $k(E',E)$. From this comes the collision frequency Z, the collisional transition probability $P(E',E)$, and the energy transfer moments $\langle \Delta E^n \rangle$ in an unambiguous way via

$$Z = \int_0^\infty k(E',E) \, dE' \tag{4}$$

$$P(E',E) = k(E',E)/Z \tag{5}$$

$$\langle \Delta E^n \rangle = \int_0^\infty (E' - E)^n P(E',E) \, dE' \tag{6}$$

Since in classical trajectory calculations one generally has to limit the calculations to a range of collision parameters, this defines Z in a different way from that corresponding to equation (4). Disagreements between experiments and trajectory calculations for HO_2, CS_2, and SO_2 collisions[7,9,16–18] may well be explicable in terms of the aforementioned problems with the central grains and the limitations of the collision parameter range in relation to the choice of the collision frequency Z. In order to compare trajectory calculations with experiments, the same collision frequency Z has to be employed since in most cases only the product $Z\langle \Delta E \rangle$ is measurable experimentally (see below). The problems with the truncation of the 'elastic' peak are more difficult to solve. By this truncation one rejects small collision parameters b, for elastic collisions. This contribution can be identified by varying b.[6,8,19] The contribution to the elastic peak from collisions with small b, however, has to be included in $\langle \Delta E \rangle$ calculations. Omitting this part will result in $\langle \Delta E \rangle$ values, that are too large.

[16] C. R. Gallucci and G. C. Schatz, *J. Chem. Phys.*, 1982, **86**, 2352.
[17] M. Bruehl and G. Schatz, *J. Phys. Chem.*, 1988, **42**, 7223.
[18] M. Heymann, H. Hippler, D. Nahr, H. J. Plach, and J. Troe, *J. Phys. Chem.*, 1988, **92**, 5507.
[19] J. Troe, *J. Phys. Chem.*, 1979, **83**, 114.

There is a further problem with histograms such as that shown in Figure 3 if one starts with vibrationally highly excited molecules which are rotationally cold. In this case the internal energy of the molecule before and after the collision has to be specified with respect to its vibrational and rotational parts (or its total energy E and its total angular momentum J). While the importance of angular momentum has often been realized,[9,19,16] a systematic investigation has only been made by Schranz and Troe[7] for SO_2 + Ar collisions. Figure 4 illustrates average vibrational and rotational energies transferred per collision for various initial conditions in the vibrational–rotational energy plane. Sequences of collisions obviously find pathways in the vibrational–rotational plane which are governed by the fine details of rovibrational energy transfer. Two-dimensional master equations using collision transition probabilities $P(E',E;J',J)$ are then required (see below).

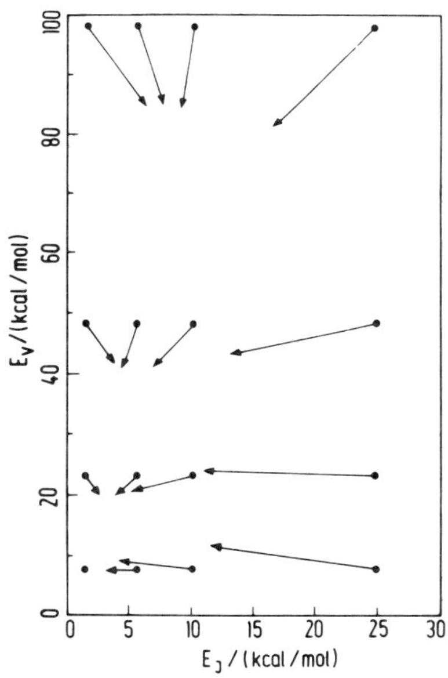

Figure 4 *Average vibrational (E_V) and rotational (E_J) energy flow in* SO_2^* + Ar *collisions at* 1000 K. *The length of the energy transfer arrows starting from the initial points* ● *are tenfold magnified*
(Reproduced with permission from *J. Phys. Chem.*, 1986, **90**, 6168)

There have been recent attempts to circumvent the graining problem of the histograms. Considering sequences of successive collisions with complete intramolecular redistribution between the collisions, Bruehl and Schatz[17] have simulated energy transfer of highly excited CS_2 in comparison with the experiments from refs. 20 and 21. In this procedure there is no need for a truncation of the elastic peak. The problem with the limitation of the range of collision parameters and the choice of a collision frequency again requires careful consideration. The problem is minimized when experimental and theoretical results on the product $Z\langle \Delta E \rangle$ are considered. A more serious problem, finally, concerns the classical character of the calculation in situations where the vibrational quasi-continuum is still sparse. This may be responsible for the disagreement between the measured[18,20,21] energy dependences of $\langle \Delta E \rangle \propto E^2$ for SO_2 and CS_2 and the calculated dependence, $\langle \Delta E \rangle \propto E$ (see below).

While the calculation of the energy- and angular momentum-resolved collisional transition probabilities for highly excited triatomic molecules involves large computational effort, trajectory calculations for molecules of the size of azulene may appear too time-consuming to construct histograms or to obtain average energies transferred per collision. An interesting attempt to reduce the required calculation time has been proposed recently by Lim and Gilbert.[5] Analysing a small number of smooth trajectories such as Figure 1 in terms of stochastic transport theories, an 'energy diffusion' coefficient was extracted. From only the few (10—50) trajectories considered, average energies transferred per collision of highly excited azulene were obtained of the order of the experimental results. In the quantitative comparison there is the problem of whether fine details of the $P(E',E)$ curves are really recovered from so small a number of trajectories. In particular, one has to ask whether a narrow weak collision peak near $E' \approx E$ can indeed be identified by this technique. More fundamentally, the multiple-collision character of some collision effects (Figure 2) may prevent the application of a simple transport approach.

3 Collision Frequencies

The essential quantities for collisional energy transfer are state-to-state rate coefficients $k(E',E)$ such as discussed in Section 2 and those considered in Section 4 below. In practical applications it is often convenient to separate the rate coefficient into an overall collision frequency Z and a transition probability $P(E',E)$, although only products such as $Z\langle \Delta E \rangle$ are accessible experimentally. The question arises, which collision frequency Z would provide the most reasonable choice? For a number of reasons, such as those discussed in refs. 2, 3, and 19, the use of a Lennard–Jones collision frequency is recommended. This is defined in molecular units by the relationship:

$$Z_{LJ} = \pi\sigma_{AB}^2(8RT/\pi\mu_{AB})^{1/2}\,\Omega_{AB}^{(2,2)*} \qquad (7)$$

[20] J. E. Dove, H. Hippler, and J. Troe, *J. Chem. Phys.*, 1985, **82**, 1907.
[21] M. Heymann, H. Hippler, H. J. Plach, and J. Troe, *J. Chem. Phys.*, 1987, **87**, 3867.

where the reduced collision integral $\Omega_{AB}^{(2,2)^*}$ is calculated from the Lennard–Jones well depth ε_{AB}. μ_{AB} denotes the reduced mass of the collision pair. There exist various methods for estimating the Lennard–Jones parameters σ_{AB} and ε_{AB} when no measurements are available. Also, the collision integrals $\Omega_{AB}^{(2,2)^*}$ are easily derived from ε_{AB} by approximate analytical expressions. Such techniques are demonstrated in the practical applications of Sections 11 and 12.

One might ask for the reasons for using Lennard–Jones collision frequencies Z_{LJ} as reference collision frequencies Z in energy transfer. These are indeed found in the classical trajectory calculations discussed in Section 2. If one calculates average energies $\langle \Delta E \rangle$ transferred per collision as a function of a maximum collision parameter b_{max} of the selected trajectories, one generally finds results like Figure 5. The calculations shown in this figure[8] correspond to O_3 + Xe collisions at 300 and 2500 K. A comparison[19] of b_{max} values, where $\langle |\Delta E| \rangle$ drops off, with the Lennard–Jones collision diameters $\sigma_{AB}\sqrt{\Omega_{AB}^{(2,2)^*}}$ showed that the absolute values and the temperature coefficients of the optimum choice of b_{max} are reproduced well with Lennard–Jones collision frequencies. For this reason, the general use of the Lennard–Jones collision frequency as the reference collision frequency is strongly recommended.

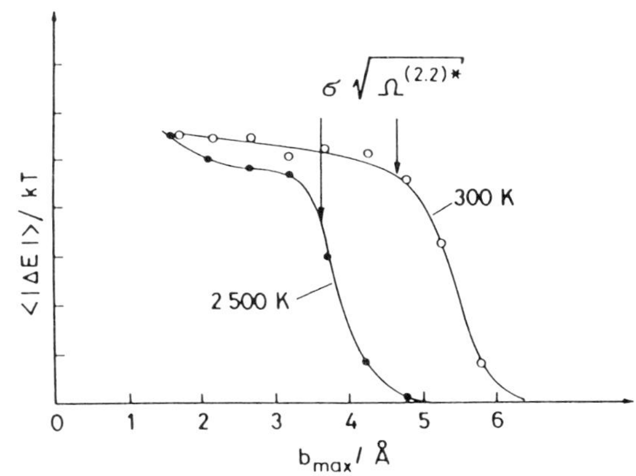

Figure 5 *Average energies* $\langle \Delta E \rangle$ *transferred in* O_3^* + Xe *collisions at* 300 *and* 2500 K *as a function of the cut-off collision parameter* b_{max}. *Classical trajectory calculations from ref.* 8, *Lennard–Jones diameters from ref.* 19

4 From State-to-State Cross-sections to Energy Transfer Averages: Quantum Scattering Calculations

Although classical trajectory calculations appear to reproduce well the experimental energy transfer properties for high excitation energies, one may expect problems in the sparse vibrational manifold at low excitation energies. It is known[1] that collisional energy transfer in this range selects specific pathways. Quantum calculations, therefore, appear most appropriate. One has to ask about the role of 'propensity rules' and, in particular, up to which energies specific quantum propensity rules survive averaging.

When the level manifold is sparse, state-to-state cross-sections may be calculated by approximate quantum scattering methods. Averaging over thermal distributions of translational velocities in the bath gas leads to (state i)-to-(state j) rate coefficients k_{ji}. Figure 6 schematically illustrates the results. With increasing numbers of levels and possible transitions, it may be necessary to smooth k_{ji} by a continuous function $k(E',E)$ such as used in Section 2. This is done by averaging over energy ranges δE_j and δE_i around the energies E' and E, respectively, *i.e.*

$$k(E', E) = \left\langle \sum_{\delta E_j} k_{ji}/\delta E_j \right\rangle_{\delta E_i} \tag{8}$$

This procedure becomes useful only when the averaging intervals δE_j and δE_i contain a sufficient number of levels and when the function $k(E',E)$ no longer depends on the size of δE_j and δE_i. In this way one moves from Figure 6 to Figure 7. One also may derive continuous collision transition probabilities $P(E',E)$, total collision frequencies Z, and average energy transfer moments $\langle \Delta E^n \rangle$ such as defined in equations (4)—(6).

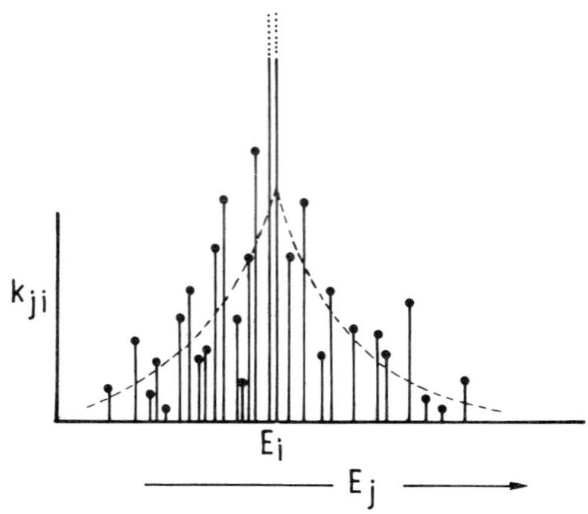

Figure 6 *Schematic representation of (state* i*)-to-(state* j*) collisional energy transfer rate coefficients*

The results from quantum scattering calculations on state-to-state energy transfer are illustrated here by the vibrational coupled channel–

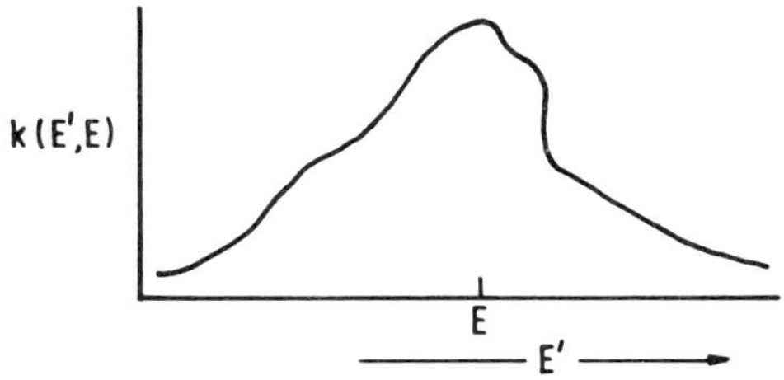

Figure 7 *As Figure 6; smoothed continuous rate coefficients*

infinite order sudden approximation (VCC–IOSA) such as applied to cyclopropane + He collisions by Clary.[22] Figure 8 gives rate coefficients for transitions from the various first excited normal vibrational modes to the ground state. The rate coefficients for deactivation increase more stongly with temperature than the gas-kinetic collision frequency Z, *i.e.* the state-to-state collision probability $P_{ji} = k_{ji}/Z$ is an increasing function of the temperature.

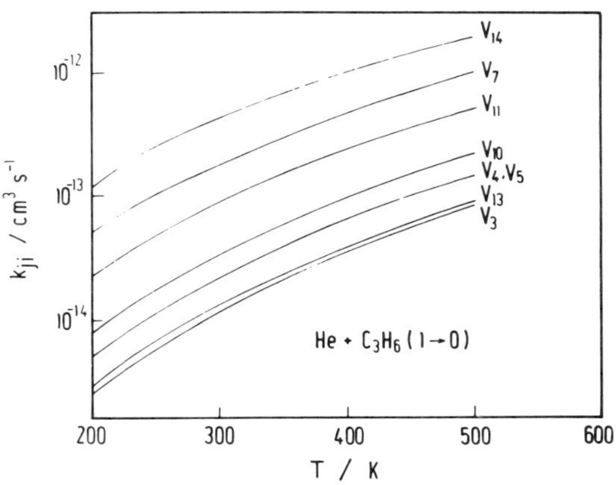

Figure 8 *Vibrational close coupling rotational infinite order sudden approximation of cyclopropane* + He *collisional energy transfer.*[22] *Rate cofficients for transitions from the first excited levels of various vibrational modes to the ground state*

In order to interpret the results on C_3H_6 + He collisions from Figure 8 in the way suggested by Figures 6 and 7, the rate coefficients are plotted as a function of E_j for $E_i = 0$. For this purpose all deactivating transition rate coefficients are converted into activation rate coefficients using the principle of detailed balance:

$$k_{ji} = k_{ij} \exp[-(E_j - E_i)/k_B T] \tag{9}$$

[22] D. C. Clary, *J. Am. Chem. Soc.*, 1984, **106**, 970.

Figure 9 shows the results. Apart from some state specific fluctuations around a 'smoothed' curve, an exponential collision model is observed, *i.e.* for the activating collisions shown one has (with $E = 0$):

$$k(E',E) = Z\exp\left(-\frac{E'-E}{\alpha} - \frac{E'-E}{k_{\mathrm{B}}T}\right) \tag{10}$$

where $Z \approx 4 \times 10^{-10}$ cm^3 molecule^{-1} s^{-1} and $\alpha \approx 110$, 120, and 140 cm^{-1} for $T = 200$, 300, and 500 K, respectively. Extending the smoothed expression (10) to higher energies and estimating an average energy transferred per collision $\langle \Delta E \rangle$ via

$$\langle \Delta E \rangle \approx -\alpha^2/(\alpha + k_{\mathrm{B}}T) \tag{11}$$

one would obtain $\langle \Delta E \rangle \approx -45$ cm^{-1} independent of the temperature. This is of the same order of magnitude as obtained at high excitation energies by classical trajectory calculations, and by experiments for highly excited triatomic (as well as for much larger polyatomic) molecules. However, the real $\langle \Delta E \rangle$ from the discrete transitions in Figure 9 would be much smaller because of the absence of low-frequency modes and the large gap between $E = 0$ and the smallest $E' > 0$.

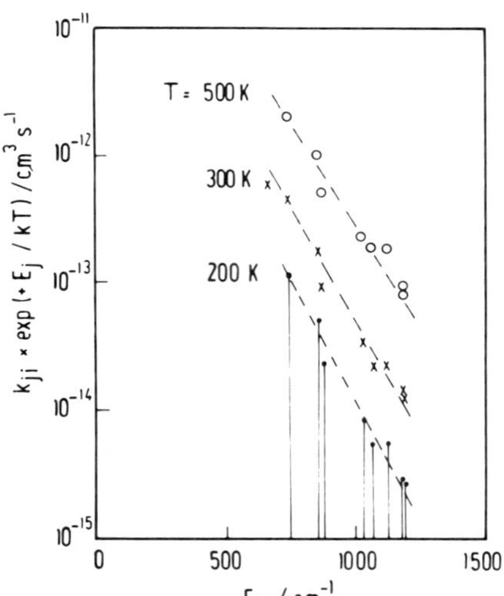

Figure 9 *As Figure 8; converted into activating collisions from the ground state to the excited state energy* E_j

One may be tempted to draw a number of conclusions from the foregoing discussion:

(*a*) Classical trajectory and quantum scattering calculations lead to energy transfer rate coefficients which are more or less equivalent in a sufficiently dense manifold of vibrational states.
(*b*) The $\langle \Delta E \rangle$ values obtained in the dense vibrational quasi-continuum are only weakly dependent on the temperature and on the size of the highly excited molecule.
(*c*) However, the vibrational quasi-continuum results become less and less adequate when the vibrational manifold becomes more sparse. $\langle \Delta E \rangle$ decreases dramatically when there are large gaps between the energy levels. Problems with the energy dependence of $\langle \Delta E \rangle$ in classical trajectory calculations of triatomic molecules[6,17] may be due to this 'dilution of the quasi-continuum'.

VCC–IOSA calculations by Clary on He + CO_2 and He + *p*-difluorobenzene collisions[23] confirm the conclusions drawn from the He + C_3H_6 results. There are state-specific vibrational and rotational effects. In *p*-difluorobenzene some low-frequency modes are transferring energy particularly well, others are less efficient. Very specific pathways are indeed observed at the lowest energies. Experiments have shown[1] that such 'propensities' survive up to energies where the vibrational quasi-continuum is already moderately dense. At which energies (or densities of states in the quasi-continuum) the low-energy propensities are becoming completely 'lost' within the large number of possible other transitions (if they are lost at all) remains an open question.

As for classical trajectories, only very few quantum calculations have been extended to the energy transfer of larger polyatomic molecules. The VCC–IOSA approach obviously gives more satisfactory results than the simple SSHT (Schwartz–Slawsky–Herzfeld–Tanczos) model. Alternative extensions of the SSHT model have been discussed in ref. 1. The approach by Parmenter and Tang[24] and, particularly, the modification by Knight and co-workers[25] express propensities in an increasingly satisfactory way.

5 Master Equations for Multiple Collision Systems

Single state-to-state resolved collisions of highly excited molecules as discussed in Sections 2 and 4 are observed only in exceptional cases. The experimental conditions more frequently correspond to processes involving sequences of many collisions. In this case the master equation has to be formulated either with state-to-state rate coefficients k_{ji} or with smoothed rate coefficient distributions $k(E',E; J', J)$. In the presence of activating and deactivating collisions only, the master equation

[23] D. C. Clary, *J. Chem. Phys.*, 1987, **86**, 802 and 813.
[24] K. Y. Tang and C. S. Parmenter, *J. Chem. Phys.*, 1983, **78**, 3922.
[25] S. H. Kable, J. W. Thoman, and A. E. W. Knight, *J. Chem. Phys.* 1988, **88**, 4748.

can be written as:

$$\frac{\mathrm{d}n(E,J,t)}{\mathrm{d}t} = \int_0^\infty \int_0^\infty \{-k(E',E;\, J',J)[\mathrm{M}]n(E,J,t) + k(E,E';\, J,J')[\mathrm{M}]n(E',J',t)\}\,\mathrm{d}E'\,\mathrm{d}J' \tag{12}$$

([M] is the bath gas concentration, n is the concentration of excited species). With a given set of rate coefficients $k(E',E;\, J',J)$ and initial conditions $n(E,J,t = 0)$, the solution of equation (12) in principle presents no difficulties. There are various powerful algorithms for simplifying the numerical solution of the one-dimensional master equation (see, for example, ref. 26). Numerical solutions of the two-dimensional master equation are much rarer. Solution procedures of the master equation are not described in the present review.

Two important experimental situations are described by equation (12): the stepwise collisional deactivation of an initially prepared narrow distribution of highly excited molecules (*i.e.* the 'direct' study of energy transfer to be discussed below) and the stepwise collisional activation in a thermal unimolecular reaction at low pressures. The latter situation is characterized by a generally short pre-steady-state 'incubation period' and a steady state with a time-independent limiting low-pressure rate coefficient k_0.

For the one-dimensional master equation in the diffusion limit of very weak collisions ($-\langle \Delta E \rangle \ll k_\mathrm{B}T$), k_0 is given by:

$$k_0 = Z[\mathrm{M}]\beta_\mathrm{c} \int_{E_0}^\infty f(E)\,\mathrm{d}E \tag{13}$$

where $f(E)$ is the equilibrium population of excited states, E_0 is the reaction threshold energy, β_c is a 'collision efficiency' which in the weak collision limit is given by:[15]

$$\beta_\mathrm{c} \approx \frac{-\langle \Delta E \rangle}{F_E k_\mathrm{B} T} \tag{14}$$

with

$$F_E = \int_{E_0}^\infty f(E)\,\mathrm{d}E / f(E_0) k_\mathrm{B} T \tag{15}$$

k_0 depends on the product $Z\langle \Delta E \rangle$, which was discussed above in Sections 2 and 4. For strong collisions, *i.e.* $-\langle \Delta E \rangle \gg F_E k_\mathrm{B} T$, β_c approaches unity. For intermediate cases, β_c not only depends on $\langle \Delta E \rangle$ but also on the shape of the collisional transition probability (see, for example, ref. 15). An analytical solution of the master equation is

[26] H. W. Schranz and S. Nordholm, *Chem. Phys.*, 1983, **74**, 365.

possible[13] with the exponential collision model of equation (1), leading to:[13,15]

$$\frac{\beta_c}{1-\sqrt{\beta_c}} \approx \frac{-\langle \Delta E \rangle}{F_E k_B T} \tag{16}$$

The transition of β_c from the weak collision expression (14) to the strong collision limit of unity was shown by Snider[27] to depend essentially only on the second moment $\langle \Delta E^2 \rangle$ of the average energy transfer. Since the strong collision limit is difficult to access by experiments or theoretical calculations, the fine details do not matter too much in practice. For this reason, equation (16) has found a quite general application.

Equations (13)—(16) provide the possibility of a 'direct' determination of the product $Z\langle \Delta E \rangle$, either via relative collision efficiency measurements or the absolute analysis of individual rate coefficients.[28] Results from this approach[2,3] will be referred to below in comparison with other 'direct' determinations when they are available.

The major difficulty with the described approach to $\langle \Delta E \rangle$ via k_0 lies in the 'two-dimensional character' of the energy transfer, *i.e.* its dependence on energy, E and angular momentum, J. Early analytical solutions of the two-dimensional master equation[15,29] have been discussed recently in detail.[29-32] In the weak collision limit for vibrational and rotational energy transfer (and in the absence of centrifugal barriers), one has:

$$\beta_c \approx -\frac{\langle \Delta E_{vib} \rangle}{(F_E)_{vib} k_B T} - \frac{\langle \Delta E_{rot} \rangle}{(F_E)_{rot} k_B T} \tag{17}$$

analogous to equation (16). Little is known about the partitioning of $\langle \Delta E \rangle$ into $\langle \Delta E_{vib} \rangle$ and $\langle \Delta E_{rot} \rangle$, and so it appears premature to proceed with a further interpretation of the solution of the two-dimensional master equation. The results from trajectory calculations such as those shown in Figure 4 may provide an understanding, although the conclusions are limited by the use of a truncation of the 'elastic peak', which may have contained a major contribution from weak collisions.

The second situation described by equation (12), *i.e.* the stepwise energy loss from highly excited states, may be seriously influenced by vibrational–rotational energy transfer. The 'flow' pattern in Figure 4

[27] N. Snider, *J. Phys. Chem.*, 1986, **90**, 4366.
[28] J. Troe, *J. Chem. Phys.*, 1977, **66**, 4758.
[29] J. Troe, *Z. Phys. Chem. Neue Folge*, 1987, **154**, 73.
[30] C. Knesse, M. Mangel, B. J. Matkowsky, Z. Schuss, and C. Tier, *J. Chem. Phys.*, 1984, **81**, 1285.
[31] M. Borkovec and B. J. Berne, *J. Chem. Phys.*, 1986, **84**, 4327.
[32] N. Snider, *J. Chem. Phys.*, 1986, **85**, 4207.

suggests that there is some rotational heating during the loss of vibrational energy. Neglecting this effect in the following, only solutions of the one-dimensional master equations for vibrational de-excitation are considered. The problem has been treated frequently in the past (see, for example, refs. 33, 34, and earlier references cited therein). If the collisional transition probability $P(E',E)$ is a narrowly confined function around $E' \approx E$, the energy loss in the presence of different bath gases i can be approximated well by:

$$\frac{\mathrm{d}\langle E\rangle}{\mathrm{d}t} = \sum_{i=1}^{n} Z_i \langle \Delta E\rangle_i [\mathrm{M}]_i \tag{18}$$

Non-linear mixing laws have not so far been observed experimentally. Experiments measuring the average energy $\langle E\rangle$ of the excited molecules as a function of time therefore provide a direct access to the product $Z\langle \Delta E\rangle$. Whenever more details of the distribution of excited states during the de-excitation sequence become available, an access to the second moment $\langle \Delta E^2\rangle$ may also be possible (see below). A number of 'energy-loss profiles' $\langle E\rangle$ as a function of $Z[\mathrm{M}]t$ will be illustrated in Section 11.

6 Master Equation for Collisions Competing with Other Processes

Information on collisional energy transfer of highly excited molecules has often been obtained from systems in which collisions compete with intramolecular rearrangements or other processes such as emission of light. Only a brief discussion of these 'indirect' experiments will be given in the following.

Starting with a cold distribution at $t = 0$, thermal unimolecular reactions in the fall-off range are governed by the competition between collisional energy transfer and unimolecular reaction. In this case the term $-k(E,J)n(E,J,t)$ has to be added to the right-hand side of equation (12). Here $k(E,J)$ denotes the specific rate constant of the unimolecular reaction. The influence of energy transfer properties on the shape of the fall-off curve has been discussed in refs. 2 and 35. The effects are generally not easily identified experimentally and are not further considered here.

Starting with narrow initial distributions of highly excited states after chemical or photochemical activation, one may either proceed to a steady-state measurement or perform a time-resolved measurement of the concentration of highly excited reactants. In both cases, the ex-

[33] J. Troe, *J. Chem. Phys.*, 1982, **77**, 3485.
[34] W. Forst, *J. Chem. Phys.*, 1984, **80**, 2504.
[35] R. G. Gilbert, K. Luther, and J. Troe, *Ber. Bunsenges. Phys. Chem.*, 1983, **87**, 169.

perimental observables are determined by the ratios of collisional energy transfer and unimolecular reaction rates. In this sense these experiments are 'indirect'. However, they can be made 'direct' when a time-resolved measurement of the unimolecular reaction rate is performed under collision-free conditions. Such complementary studies have been made for excited cycloheptatriene molecules.[36–39] An independent measurement of the thermal average of $k(E,J)$ was also possible for this system by thermal isomerization experiments.[40]

The result from the competition between collisional energy transfer and unimolecular reaction, after an initial excitation to the energy $\langle E_{ac} \rangle$, can be expressed by the ratio of the product yield D to the yield S of stabilized reactants or by a quantum yield Φ. The observable D/S ratio in steady-state chemical activation experiments can be expressed by:

$$D/S \approx k(\langle E_{ac} \rangle)/\gamma_c Z[\mathrm{M}] \tag{19}$$

where γ_c denotes a 'collision efficiency' to be derived from the solution of the master equation. Likewise, the quantum yield of steady-state photochemical activation experiments is expressed by:

$$\Phi^{-1} \approx 1 + \frac{\gamma_c Z[\mathrm{M}]}{k(\langle E_{ac} \rangle)} \tag{20}$$

The time-dependent concentration $c(t)$ of reactant molecules after laser excitation can be approximated in an analogous manner by:[41]

$$c(t) \approx c(t = \infty) + \{c(t = 0) - c(t = \infty)\} \exp[-\{k(\langle E_{ac} \rangle) + \gamma_c Z_{LJ}[\mathrm{M}]\}t] \tag{21}$$

where $c(t = \infty)/c(t = 0) = 1 - \Phi$. In general, γ_c will depend on [M], $k(E)$, and on energy transfer properties.

It should be emphasized that the collision efficiencies β_c for thermal and γ_c for chemical and photochemical activation systems are quite different. The two quantities have both to be derived from solutions of the master equation. However, they correspond to quite different starting conditions for $n(E,t)$ at time $t = 0$. Nevertheless, β_c and γ_c are both related to the average energy transferred per collision, $\langle \Delta E \rangle$. Whereas equations (13)—(16) provide useful approximations for β_c in

[36] H. Hippler, K. Luther, J. Troe, and H. J. Wendelken, *J. Chem. Phys.*, 1983, **79**, 239.
[37] H. Hippler, K. Luther, and J. Troe, *J. Chem. Phys.*, 1983, **78**, 6718.
[38] S. H. Luu and J. Troe, *Ber. Bunsenges. Phys. Chem.*, 1974, **78**, 766.
[39] J. Troe and W. Wieters, *J. Chem. Phys.*, 1979, **71**, 3931.
[40] D. C. Astholz, J. Troe, and W. Wieters, *J. Chem. Phys.*, 1979, **70**, 5107.
[41] J. Troe, *J. Phys. Chem.*, 1983, **87**, 1800.

the weak collision limit $-\langle \Delta E \rangle \ll (\langle E_{ac} \rangle - E_0)$, for chemical or photochemical activation the following is found:[27,41]

$$\gamma_c \approx \frac{-\langle \Delta E \rangle s^*}{\langle E_{ac} \rangle - E_0} \tag{22}$$

The coefficient s^* relates in a simple fashion the energy dependence of the specific rate constant $k(E)$ of the unimolecular reaction near $E = \langle E_{ac} \rangle$ according to:

$$k(E) \propto (E - E_0)^{s^*-1} \tag{23}$$

If $\langle E_{ac} \rangle$ approaches E_0, collisions become 'strong' for this experiment. It may then be possible[41] to replace equation (22) by:

$$\frac{\gamma_c}{1 - \gamma_c^2} \approx \frac{-\langle \Delta E \rangle s^*}{\langle E_{ac} \rangle - E_0} \tag{24}$$

However, the detailed shape of the collisional transition probability has to be taken into account more rigorously (see, for example, ref. 42). Photochemical and chemical activation experiments allow for the variation of the initial excitation energy $\langle E_{ac} \rangle$ over a certain range. However, energy dependences of $\langle \Delta E \rangle$ are not easily determined in this way, although the energy dependence of $k(E)$ was clearly revealed by early 'indirect' experiments (see, for example, refs. 2, 38, and 39).

The important question of the energy dependence of $\langle \Delta E \rangle$ can be addressed in a much better way when $\langle \Delta E \rangle$ is measured directly and followed during a complete collisional deactivation sequence. Such experiments start with rapid initial production of highly excited molecules. Then either the energy loss of the excited species or the energy gain of the surrounding heat bath is followed in a time-resolved manner. Such experiments will be considered in detail in the following sections.

7 Rapid Preparation of Vibrationally Highly Excited Molecules in the Electronic Ground State

Several pulsed laser methods have been used for the preparation of vibrationally highly excited molecules in their electronic ground state. The subsequent dynamics of these highly excited molecules were observed using various time-resolved techniques. Apparently, the first studies of this type were described in refs. 43 and 44. Like many later studies, (see, for example, refs. 45—50 and other work cited subsequently), these workers used the 'internal conversion (IC) technique', *i.e.* the preparation of highly excited electronic ground states via light

42 J. E. Baggott, *Chem. Phys. Lett.*, 1985, **119**, 47.
43 H. Hippler, K. Luther, J. Troe, and R. Walsh, *J. Chem. Phys.*, 1978, **68**, 323.
44 H. Hippler, K. Luther, and J. Troe, *Faraday Discuss. Chem. Soc.*, 1979, **67**, 173.

absorption into excited states followed by fast internal conversion. This technique has the great advantage that very large concentrations of excited species can be prepared with a narrow energy distribution. The large concentrations obtained simplify the detection problem. They also allow further excitation by multiphoton techniques in a particularly well-defined manner.[51]

An alternative pathway for the preparation of vibrationally highly excited electronic ground-state molecules, which uses the intermediate population of excited electronic states, is used in stimulated emission pumping (SEP) experiments. In principle, after laser excitation, high vibrational states in the electronic ground state can also be reached by spontaneous emission. However, fluorescence is a slow process that also may populate broad distributions of ground-state levels. The use of SEP has the advantage of a faster preparation rate. Particularly good state-selectivity can be achieved allowing the observation of the population of single states in sparse manifolds (see refs. 1, 52—54, and references cited therein). However, owing to the small populations of excited states achieved, the method requires sensitive state detection.

Besides the IC and SEP methods, which use the intermediate participation of electronically excited states, higher vibrational levels can also be populated directly within the electronic ground state. Single-photon excitation using low-power i.r. lasers allows the study of collisional energy transfer of low-lying vibrational states (see, for example, refs. 55—59 and work cited therein). This work is not discussed in the

[45] H. Hippler, V. Schubert, J. Troe, and H. J. Wendelken, *Chem. Phys. Lett.*, 1981, **84**, 253.

[46] H. Hippler, L. Lindemann, and J. Troe, *Ber. Bunsenges. Phys. Chem.*, 1988, **92**, 440.

[47] U. Brand, H. Hippler, L. Lindemann, and J. Troe, *J. Phys. Chem.*, 1989, in press.

[48] H. Hippler, *Ber. Bunsenges. Phys. Chem.*, 1985, **89**, 303.

[49] H. Hippler, in 'Time-Resolved Vibrational Spectroscopy', eds. A. Laubereau and M. Stockburger (Springer Proceedings in Physics), Springer Verlag, Berlin, 1985, Vol. 4, p. 45.

[50] J. R. Barker, *J. Phys. Chem.*, 1985, **88**, 11.

[51] M. Damm, H. Hippler, and J. Troe, *J. Chem. Phys.*, 1988, **88**, 3564.

[52] J. A. O'Neill, J. Y. Cai, G. W. Flynn, and R. E. Weston, jun., *J. Chem. Phys.*, 1986, **84**, 50; T. H. McGee, R. E. Weston, and G. W. Flynn, *ibid.*, 1985, **82**, 3607.

[53] J. W. Thoman, jun., S. H. Kable, A. B. Rock, and A. E. W. Knight, *J. Chem. Phys.*, 1986, **85**, 6234; D. J. Muller, W. D. Lawrance, and A. E. W. Knight, *J. Phys. Chem.*, 1983, **87**, 4952; W. D. Lawrance and A. E. W. Knight, *ibid.*, 1983, **87**, 389; *J. Chem. Phys.*, 1983, **79**, 6030.

[54] F. Temps, S. Halle, P. H. Vaccaro, R. W. Field, and J. L. Kinsey, *J. Chem. Phys.*, 1987, **87**, 1895; *J. Chem. Soc. Faraday Trans. 2*, 1988, **84**, 1457; E. Abramson, R. W. Field, D. Imre, K. K. Innes, and J. L. Kinsey, *J. Chem. Phys.*, 1984, **80**, 2298; C. Kittrell, E. Abramson, J. L. Kinsey, S. A. McDonald, D. E. Raisner, R. W. Field, and D. H. Katayama, *ibid.*, 1981, **75**, 2056.

[55] I. W. M. Smith, in 'Lasers as Reactants and Probes in Chemistry', eds. W. M. Jackson and A. B. Harvey, Howard University, Washington, D. C., 1985, p. 373.

[56] E. Weitz and G. Flynn, *Adv. Chem. Phys.*, 1981, **47**, 185.

[57] G. W. Flynn, *Acc. Chem. Res.*, 1981, **14**, 334.

[58] J. T. Yardley, 'Introduction to Molecular Energy Transfer', Academic Press, New York, 1980.

[59] J. D. Lambert, 'Vibrational and Rotational Relaxation in Gases', Oxford University Press, Oxford, 1977.

present review. I.r. multiple-photon excitation, mostly using intense pulsed CO_2 lasers (i.r.-*nhν*), can produce large populations of higher excited states. Therefore, this technique provides a third important access to the phenomena discussed here, although the distributions of excited states produced are relatively broad and not easily characterized. Thus, in energy transfer studies the determination of the initial pre-collision population distribution requires particular care (see, for example, refs. 60—62).

A much better defined energy distribution, using an excitation technique that stays within the electronic ground state, can be achieved by direct single-photon overtone excitation in the visible spectral range (VIS-*hν*). Excitations up to the ninth overtone of the C—H stretching vibration have been realized.[63–65] The concentrations of these excited molecules are relatively small and are therefore only detectable with the most sensitive techniques.

With the IC, SEP, i.r.-*nhν*, and VIS-*hν* methods, various pulsed excitation techniques are currently available which are often complementary. Each of the methods has advantages and limitations, such that one can choose among several alternatives in order to answer particular questions.

8 Detection of Collisional Energy Transfer of Vibrationally Highly Excited Molecules

The detection of collisional energy transfer of vibrationally highly excited molecules after pulsed preparation can be achieved by various time-resolved methods. These techniques are basically different to the comparative rate measurements used in earlier indirect studies. For this reason, the applied methods will be reviewed in some detail in the following. There are two possibilities: either one monitors the energy loss or the change of individual level populations in the excited molecules, or one follows the total energy gain or the change of population of individual levels of the accepting heat-bath molecules.

60 Y. S. Dolzhikov, V. S. Letokhov, A. A. Makarov, A. L., Malininski, and E. A. Ryabov, *Sov. Phys. JETP.*, 1986, **63**, 1161; V. M. Apatin and G. N. Makarov, *Sov. J. Quantum Electron.*, 1983, **13**, 932; *Appl. Phys. B.*, 1982, **28**, 367.

61 B. Abel, L. Brouwer, B. Herzog, H. Hippler, and J. Troe, *Chem. Phys. Lett.*, 1986, **127**, 541; B. Herzog, H. Hippler, L. Kang, and J. Troe, *ibid.*, 1985, **120**, 124.

62 J. M. Zellweger, T. C. Brown, and J. R. Barker, *J. Chem. Phys.*, 1985, **83**, 6251.

63 T. M. Ticich, T. R. Rizzo, H. R. Dübal, and F. F. Crim, *J. Chem. Phys.*, 1986, **84**, 1508; L. J. Butler, T. M. Ticich, M. D. Likar, and F. F. Crim, *ibid.*, 1986, **85**, 2231; H. R. Dübal and F. F. Crim, *ibid.*, 1985, **83**, 3863; K. V. Reddy, D. F. Heller, and M. J. Berry, *ibid.*, 1982, **76**, 2814; K. V. Reddy and M. J. Berry, *Faraday Discuss. Chem. Soc.*, 1979, **67**, 118.

64 M. C. Chuang and R. N. Zare, *J. Chem. Phys.*, 1985, **81**, 4501; D. L. Snavely, R. N. Zare, J. A. Miller, and D. W. Chandler, *J. Phys. Chem.*, 1986, **90**, 3544.

65 J. E. Baggott and D. W. Law, *J. Chem. Phys.*, 1986, **85**, 6475.

In principle, nearly all spectroscopic methods are suited for monitoring the energy loss of vibrationally highly excited molecules. u.v. absorption and i.r. emission play a particularly important role. Other types of absorption and emission, laser-induced fluorescence (LIF), and multiphoton ionization (MPI) have been used. The energy increase in the heat bath has been monitored using thermal lensing, optoacoustic spectroscopy, u.v. absorption of tracer gases, and interferometry. To date, the most widely applied technique providing a direct access to $\langle \Delta E \rangle$ values is time-resolved u.v. absorption spectroscopy.

U.v. absorption spectra of vibrationally highly excited molecules often appear continuous. Either they are truly continuous or the overlapping of many hot bands results in highly congested spectra. In general, the analysis of these 'hot u.v. absorption spectra' is not easy. A theoretical simulation as a function of specific excited-state populations rapidly becomes very difficult. Therefore, 'experimental calibrations' of hot u.v. absorption spectra as a function of the excited-state populations constitutes a central part of many direct energy transfer investigations (such as performed by our own group). Examples of this procedure will be demonstrated below.

Direct emission of visible or u.v. light from vibrational overtones is generally very weak and therefore not suitable as a detection method. However, in the case of strong mixing of the electronic ground state with electronically excited states (*e.g.* in NO_2,[66] SO_2,[67—72] and CS_2[73—80]) the temporal evolution of the u.v. emission from electronically excited states can be used as a measure for rates of collisional deactivation of vibrationally excited molecules in the ground electronic state.

I.r. spectra of vibrationally highly excited molecules are also highly congested. However, they provide a direct access to the average energy $\langle \Delta E \rangle$ of the excited molecules. The relation between excitation and i.r. emission intensity again has to be obtained either by experimental calibration or by theoretical modelling. I.r. emission has been applied in

[66] L. F. Keyser, S. Z. Levine, and F. Kaufman, *J. Chem. Phys.*, 1971, **54**, 335.
[67] S. J. Strickler and R. D. Ito, *J. Phys. Chem.*, 1985, **89**, 2366; S. J. Strickler and R. N. Rudolph, *J. Am. Chem. Soc.*, 1978, **100**, 3326.
[68] J. L. Ahl and G. W. Flynn, *J. Phys. Chem.*, 1983, **87**, 2172.
[69] C. Lalo-Kourilsky and C. Vermeil, *J. Photochem.*, 1982, **19**, 109.
[70] R. Shaw, J. E. Kent, and M. F. O'Dwyer, *J. Mol. Spectrosc.*, 1980, **82**, 1.
[71] J. Heicklen, N. Kelly, and K. Partymiller, *Rev. Chem. Intermediates*, 1980, **3**, 315.
[72] F. Su, F. B. Wampler, J. W. Bottenheim, D. L. Thorsell, J. G. Calvert, and E. K. Damon, *Chem. Phys. Lett.*, 1977, **51**, 150.
[73] G. W. Loge, J. J. Tiee, and F. B. Wampler, *J. Chem. Phys.*, 1986, **84**, 3624.
[74] J. A. Dueñas and A. Cabello, *J. Photochem.*, 1982, **19**, 13.
[75] S. J. Silvers and M. R. McKeever, *Chem. Phys.* 1978, **27**, 27; 1976, **18**, 333.
[76] A. Matsuzaki and S. Nagakura, *Bull. Chem. Soc. Jpn.*, 1976, **49**, 359.
[77] C. Lambert and G. H. Kimbell, *Can. J. Chem.*, 1973, **51**, 2601.
[78] L. E. Brus, *Chem. Phys. Lett.*, 1971, **12**, 116.
[79] J. Heicklen, *J. Am. Chem. Soc.*, 1963, **85**, 3562.
[80] F. Rosenkranz and H. Gg. Wagner, *Z. Phys. Chem. Neue Folge*, 1968, **61**, 302.

energy transfer studies of highly excited species by Barker and co-workers (see, for example, refs. 50, 81—85). I.r. absorption has been employed by Molinari and co-workers.[86]

LIF[55-57] allows the detection of populations of single rovibrational levels. However, this technique is applicable only to small molecules or to low levels of larger molecules.

MPI can be used in collisional energy transfer studies in a particularly detailed way.[87] By application of the kinetically controlled selective ionization (KCSI) technique it is possible to derive not only $\langle \Delta E \rangle$ but also $\langle \Delta E^2 \rangle$ for collisional energy transfer. Some additional information on the shape of the distribution of collisional transition probabilities is obtainable. This approach will be reviewed in more detail below.

The loss of internal energy of vibrationally highly excited molecules in the course of collisional deactivation leads to an energy increase of the heat bath. Temperature, refractive index, and density of the gas mixture change, and all have been used in studying collisional deactivation of highly excited molecules. Braun and co-workers[88] analysed the temperature increase by measuring the translational temperature of the heat bath via the Doppler broadening of a resonance absorption line of added traces of mercury. The temporal evolution of the refractive index has been monitored via time-dependent thermal lensing.[89] Density variations were recorded with time-dependent photoacoustic spectroscopy.[90-92]

9 U.V. Absorption Spectra of Vibrationally Highly Excited Molecules

Since most of the data on energy transfer have been obtained by time-resolved u.v. absorption spectroscopy, it seems appropriate to

[81] M. Rossi, J. R. Pladziewicz, and J. Barker, *J. Chem. Phys.*, 1983, **78**, 6695.
[82] J. R. Barker and R. E. Golden, *J. Phys. Chem.*, 1984, **88**, 1012.
[83] J. -M. Zellweger, T. C. Brown, and J. R. Barker, *J. Phys. Chem.*, 1985, **90**, 461; *J. Chem. Phys.*, 1985, **83**, 6261.
[84] J. H. Shi, D. Bernfeld, and J. R. Barker, *J. Chem. Phys.*, 1988, **88**, 6211.
[85] J. H. Shi and J. R. Barker, *J. Chem. Phys.*, 1988, **88**, 6219.
[86] A. Mele, F. Salvetti, E. Molinari, and M. L. Terranova, *J. Photochem.*, 1986, **32**, 265; M. Lenzi, E. Molinari, G. Piciacchia, V. Sessa, and M. L. Terranova, *Chem. Phys.*, 1986, **108**, 167; *Spectrochim. Acta, Part A*, 1987, **43**, 137.
[87] H. G. Löhmannsröben and K. Luther, *Chem. Phys. Lett.*, 1988, **144**, 473; K. Luther and K. Reihs, *Ber. Bunsenges. Phys. Chem.*, 1988, **92**, 442; K. Luther and K. Reihs, private communication.
[88] W. Braun, M. D. Scheer, and R. J. Cvetanovic, *J. Chem. Phys.*, 1988, **88**, 3715; W. Braun, M. D. Scheer, and V. Kaufman, *J. Res. Natl. Bur. Stand.*, 1986, **91**, 313.
[89] J. R. Barker and T. Rothem, *Chem. Phys.*, 1982, **68**, 331; P. L. Trevor, T. Rothem, and J. R. Barker, *ibid.*, 1982, **68**, 341.
[90] A. Karbach and P. Hess, *J. Chem. Phys.*, 1986, **84**, 2945.
[91] K. M. Beck and R. J. Gordon, *J. Chem. Phys.*, 1987, **87**, 5681.
[92] R. J. Gordon, K. M. Beck, M. Koshi, and Y. P. Vlahoyannis, in 'Multiphoton Processes', eds. S. J. Smith and P. L. Knight, Cambridge University Press, 1988, p. 193.

discuss the essential features of this detection technique in detail. U.v. absorption spectra often show pronounced temperature dependences. The broadening of the continuous spectrum of Cl_2[93] is a classic example. Dramatic changes of the u.v. absorption of CS_2[94,95] have been documented over a wide temperature range, from cold jets up to shock-wave conditions. Marked excitation dependences of u.v. spectra have also been reported for molecules such as SO_2,[96–100] N_2O,[101] O_3,[102] CF_3I,[103,104] benzene,[105] hexafluorobenzene,[106,107] toluene,[105,108] cycloheptatrienes,[109] and azulene.[110] Excitation dependences of radical spectra were reported for phenyl,[111] phenoxy,[112] and benzyl.[113,114] When absorption cross-sections show pronounced temperature dependences they can be used as fast thermometers[115–117] of the internal excitation. The response time of such thermometers is very fast, as shown in picosecond experiments.[118]

93 G. Herzberg, 'Molecular Spectra and Molecular Structure', Van Nostrand, Princeton, NJ, 1950, Vol. I.
94 R. J. Hemley, D. G. Leopold, J. L. Roebber, and V. Vaida, *J. Chem. Phys.*, 1983, **79**, 5219.
95 J. E. Dove, H. Hippler, H. J. Plach, and J. Troe, *J. Chem. Phys.*, 1984, **81**, 1209.
96 P. C. Sung and C. L. Chiu, *J. Mol. Spectrosc.*, 1980, **80**, 459.
97 C. Y. R. Wu and D. L. Judge, *Geophys. Res. Lett.*, 1981, **8**, 769.
98 H. Watanabe, Y. Hyodo, and S. Tsuchiya, *J. Phys. Chem.*, 1982, **86**, 685.
99 A. G. Gaydon, G. H. Kimbell, and H. B. Palmer, *Proc. R. Soc. London, Ser. A*, 1963, **276**, 462.
100 H. J. Plach and J. Troe, *Int. J. Chem. Kinet.*, 1984, **80**, 1853; H. Hippler, D. Nahr, H. J. Plach, and J. Troe, *J. Phys. Chem.*, 1988, **92**, 5503.
101 W. Jost, K. Michel, J. Troe, and H. Gg. Wagner, *Z. Naturforsch., A*, 1964, **19**, 59; M. G. Holliday and B. G. Reuben, *Trans. Faraday Soc.*, 1968, **64**, 1735.
102 D. C. Astholz, A. E. Croce de Cobos, and J. Troe, *J. Phys. Chem.*, 1982, **86**, 696; J. W. Simons, R. J. Paur, H. A. Webster, and E. J. Bair, *J. Chem. Phys.*, 1973, **59**, 1203.
103 L. Brouwer and J. Troe, *Chem. Phys. Lett.*, 1981, **82**, 1.
104 H. Hippler, H. A. Olschewski, J. Troe, and J. Willner, in preparation.
105 W. R. Richardson, S. H. Lin, and D. L. Evans, *J. Chem. Soc., Faraday Trans. 2*, 1982, **78**, 1.
106 T. Ichimura, M. Takahashi, and Y. Muri, *Chem. Phys.*, 1987, **114**, 116; T. Ichimura, Y. Muri, M. Nakashima, and K. Yoshihara, *J. Chem. Phys.*, 1985, **83**, 117.
107 H. Hippler, H. A. Olschewski, J. Troe, and J. Willner, in preparation.
108 H. Hippler, J. Troe, and H. J. Wendelken, *J. Chem. Phys.*, 1983, **78**, 5351.
109 D. C. Astholz, L. Brouwer, and J. Troe, *Ber. Bunsenges. Phys. Chem.*, 1981, **78**, 559.
110 L. Brouwer, H. Hippler, L. Lindemann, and J. Troe, *J. Phys. Chem.*, 1985, **89**, 4608.
111 N. Ikeda, N. Nakashima, and K. Yoshihara, *J. Am. Chem. Soc.*, 1985, **107**, 3318.
112 Y. Kajii, K. Obi, N. Nakashima, and K. Yoshihara, *J. Chem. Phys.*, 1987, **87**, 5059.
113 N. Ikeda, N. Nakashima, and K. Yoshihara, *J. Phys. Chem.*, 1984, **88**, 5803.
114 W. Müller-Markgraf and J. Troe, *J. Phys. Chem.*, 1988, **92**, 4899.
115 M. Heymann, H. Hippler, H. J. Plach, and J. Troe, *J. Chem. Phys.* 1987, **87**, 3867.
116 N. Selamoglu and C. Steel, *J. Phys. Chem.*, 1983, **87**, 1133; P. M. Mui and E. Grunwald, *J. Am. Chem. Soc.*, 1982, **104**, 6562; E. Grunwald, *J. Phys. Chem.*, 1981, **85**, 3409.
117 A. Seilmeier, P. O. J. Scherer, and W. Kaiser, *Chem. Phys. Lett.*, 1984, **105**, 140; W. Wild, A. Seilmeier, N. H. Gottfried, and W. Kaiser, *ibid.*, 1985, **119**, 259.
118 W. Kaiser, *Ber. Bunsenges. Phys. Chem.*, 1985, **89**, 213; W. Kaiser and A. Seilmeier, in 'Time-Resolved Vibrational Spectroscopy', eds. A. Laubereau and M. Stockburger (Springer Proceedings in Physics) Springer Verlag, Berlin, 1985, Vol. 4, p. 42.

Theoretical predictions of the temperature dependence of u.v. absorption spectra of polyatomic molecules are rare. The continuous spectra of diatomic halogens[93,119] have been analysed by simple applications of the Franck–Condon principle. With increasing temperature, the population of rovibrationally excited states increases; the absorption coefficient ε then decreases at the maximum of the absorption band while there is an increase of ε at the wings. For diatomic molecules, this behaviour can be described by a simple analytical formula due to Sulzer and Wieland:[120]

$$\varepsilon^{sw}(\tilde{\nu}, T) \approx \varepsilon_0^{max}\sqrt{\tanh(\theta_0/2T)}\,\exp\{-\tanh(\theta_0/2T)[(\tilde{\nu} - \tilde{\nu}_0)/\Delta\tilde{\nu}_0]^2\} \tag{25}$$

where ε_0^{max} is the maximum absorption coefficient at 0 K, $\tilde{\nu}_0$ denotes the origin of the transition in cm^{-1}, $\Delta\tilde{\nu}_0$ is the spectral width at 0 K, and θ_0 is the characteristic temperature $h\nu/k_B$ of the vibration. Each electronic transition is characterized by one set of the parameters ε_0^{max}, θ_0, $\tilde{\nu}_0$, and $\Delta\tilde{\nu}_0$. A generalization of equation (25) was proposed in ref. 108 which takes into account the shift of the maximum of the absorption coefficient with increasing temperature:

$$\varepsilon(\tilde{\nu},T) \approx \varepsilon^{sw}(\tilde{\nu},T)(\tilde{\nu}/\tilde{\nu}_0)(1 + Ax + Bx^3) \tag{26}$$

where

$$x = (\tilde{\nu} - \tilde{\nu}_0)/\Delta\tilde{\nu}_0$$

$$A \approx 1.9Z^{0.75}(1 - Z)^{0.6}w_0/\Delta\tilde{\nu}_0$$

$$B \approx -1.7Z^{0.68}(1 - Z)^{1.76}w_0/\Delta\tilde{\nu}_0$$

$$Z = \exp(-\theta_0/T)$$

$$w_0 = 0.695\theta_0\ \text{cm}^{-1}/\text{K}$$

It was found that continuous spectra of polyatomic molecules show a behaviour similar to the spectra of diatomic molecules and can be described by the Sulzer–Wieland formula.[103–109,121] Even more surprising is the fact that highly structured spectra (for example, CS_2,[95] SO_2[100]) at high temperatures can also be approximated by a Sulzer–Wieland expression. Some rationalization of this observation was given in the appendix of ref. 109 by a multi-dimensional application of the Franck–Condon principle. A more detailed discussion was provided in refs. 102,

[119] J. Le Roy, R. G. MacDonald, and G. Burns, *J. Chem. Phys.*, 1976, **65**, 1485.
[120] P. Sulzer and K. Weiland, *Helv. Phys. Acta.*, 1952, **25**, 653.
[121] G. Herzberg, 'Molecular Spectra and Molecular Structure', Van Nostrand, Princeton, NJ, 1966, Vol. III.

106, and 122—124 by theoretical calculations for specific molecules. In some cases, quantum-chemical calculations of upper electronic states were available.

Although theoretical predictions of the temperature dependence of u.v. spectra are qualitatively correct in general, the lack of precise information on electronically excited states in most cases still prevents quantitative predictions. For this reason, experimental investigations of the high-temperature spectra are necessary prior to the study of energy transfer processes. Modified Sulzer–Wieland expressions then help to interpolate or extrapolate the results to conditions outside the experimental range (see, for example, ref. 51).

The marked temperature dependences of u.v. absorption spectra are illustrated in Figures 10 and 11 for the molecules CS_2 and azulene. It

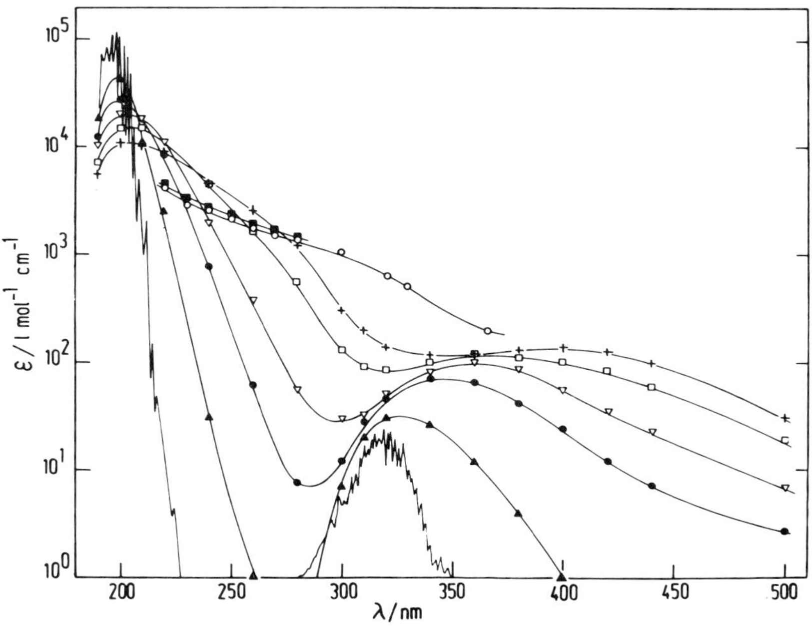

Figure 10 *'Hot' u.v. absorption spectra of* CS_2. *Highly structured spectrum at* 300 K, *curves with increasing absorption coefficients from shock waves at* T = 800, 1500, 2000, 3000, *and* 4000 K *from ref. 95; intersecting curves with highest absorption coefficients at* 300 nm *from laser excitation at* 351 *and* 308 nm, *from ref. 20*
(Reproduced with permission from *J. Chem. Phys.*, 1985, **82**, 1907)

[122] S. E. Adler-Golden, *J. Quant. Spectrosc. Radiat. Transfer*, 1983, **30**, 175; J. A. Joens and E. J. Blair, *J. Chem. Phys.*, 1983, **79**, 5780; M. G. Sheppard and R. B. Walker, *ibid.*, 1983, **78**, 7191.

[123] S. R. Langhoff and R. L. Jaffe, *J. Chem. Phys.*, 1979, **71**, 1475.

[124] Y. Fujimura, M. Arai, N. Nakashima, and K. Yoshihara, *Bull. Chem. Soc. Jpn.*, 1984, **57**, 2947.

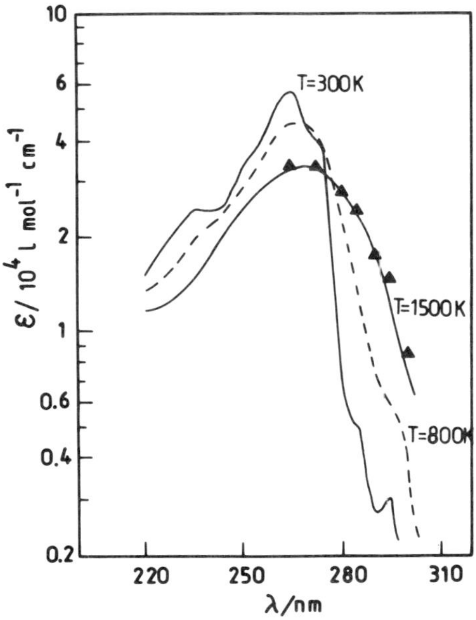

Figure 11 *'Hot' u.v absorption spectra of azulene. Lines depict thermal spectra at room temperature and from shock waves;* ▲ = *spectrum after laser excitation at* 337 nm, *after ref.* 110.

can be seen for CS_2 that there are wavelengths where the absorption coefficient ε monotonically decreases (*e.g.* 200 nm) or increases (*e.g.* 350 nm) with increasing temperature. However, there are also wavelengths where ε increases at low temperatures and decreases at high temperatures (*e.g.* 240 nm). For azulene, the absorption at the centre of the band near 265 nm decreases whereas it increases in the wings near 290 nm. In the following section we discuss how a knowledge of the temperature coefficients of the absorption spectrum allows the measurement of changes in the average internal energy of the molecules.

10 From 'Hot' U.V. Absorption Spectra to Average Energies of Non-canonical Ensembles

The u.v. absorption spectra of thermally excited polyatomic molecules may not necessarily agree with the spectra of vibrationally highly excited molecules after laser excitation. Thermal and photochemical excitation will generally result in quite different distribution functions of the total internal energy in the molecules. Figure 12 illustrates this difference for excited azulene molecules following thermal shock-wave excitation (600, 1000, and 1580 K) and laser-flash excitation (337 nm) conditions. Nevertheless, the u.v. absorption spectra of vibrationally highly excited azulene, such as recorded after pulsed laser excitation, agree very well[110] with the corresponding thermal spectra from shock-wave experiments, provided that a temperature is chosen for which the average energy $\langle E \rangle$ agrees with the laser excitation energy. Figure 11 shows this agreement between 'microcanonically hot' and 'canonically hot' spectra.

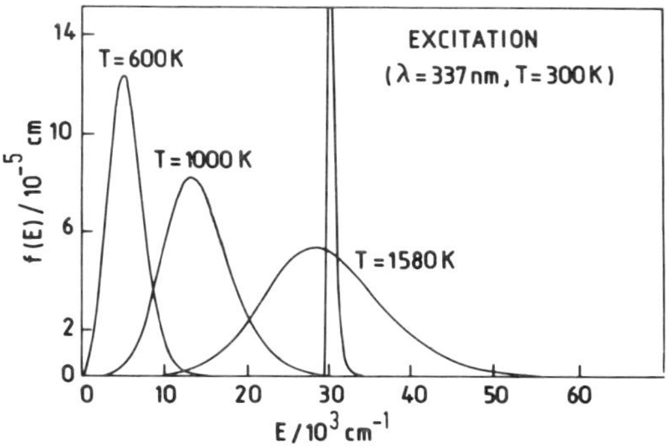

Figure 12 *Energy distributions in excited azulene. Thermal canonical distributions at* 600, 1000, *and* 1580 K; *nearly microcanonical distribution after laser excitation at* 337 nm *and* 300 K
(Reproduced with permission from *J. Phys. Chem.*, 1985, **89**, 4608)

For a triatomic molecule like CS_2, the comparison between the spectra of the two ensembles cannot be made: with laser excitation, stable molecules with energies close to the dissociation limit can be prepared. Thermal ensembles with an average energy corresponding to this energy would dissociate extremely rapidly, such that their u.v. absorption spectra cannot be measured. Nevertheless, Figure 10 shows that the u.v. spectra of laser-excited molecules follow the trend observed with thermal ensembles at much lower excitation energy.

The described observations suggest that u.v. absorption spectra can indeed be used as an 'average internal energy thermometer' that is essentially independent of the kind of energy distribution of the ensemble. In the stepwise collisional deactivation, the population probably evolves through a sequence of nearly Gaussian energy distributions.[33] If fully random intramolecular energy partitioning is realized at each total internal energy E of these distributions, the polyatomic molecule 'acts as its own heat bath'; *i.e.* each oscillator can be characterized by an intramolecular energy distribution which is not too different in various ensembles of the same average total energy $\langle E \rangle$.

Using classical statistics for a molecule like azulene, the energy distributions for one selected oscillator in a canonical and in a microcanonical ensemble of the same total energy $\langle E \rangle$ are compared in Figure 13. The distributions differ only in their high-energy tails. This is still approximately true for a triatomic molecule like CS_2 (see ref. 95). Applying quantum statistics, the conclusions are not very different.[125] Since the excitation dependence of the u.v. absorption spectra is governed by the energy distributions in the various oscillators of the molecule, the agreement between canonically hot and microcanonically hot spectra at identical average energies $\langle E \rangle$ of the ensembles becomes understandable.

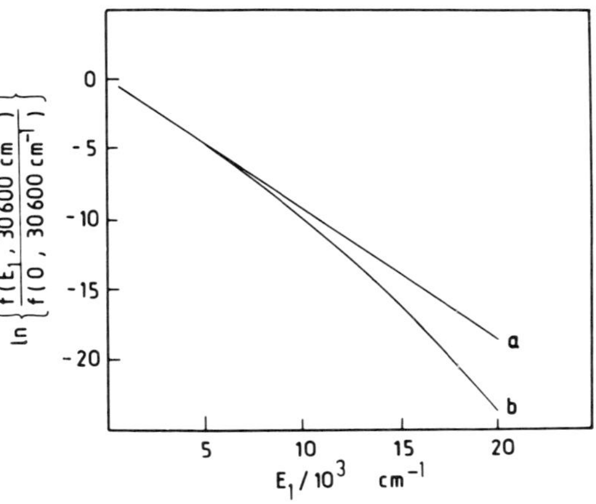

Figure 13 *Energy distribution in a single oscillator in excited azulene. (Classical calculations from ref. 110.) a = Canonical ensemble; b = microcanonical ensemble, equal average energy of the molecule* $\langle E \rangle = 30\,600\ \mathrm{cm}^{-1}$
(Reproduced with permission from *J. Phys. Chem.*, 1985, **89**, 4608)

Combining the available microcanonical and canonical experimental spectra of the excited molecules and bridging inaccessible energy ranges by Sulzer–Wieland-type models, calibration curves of the absorption coefficients as a function of the average energy of the excited molecules can be constructed. These curves serve for the interpretation of time-resolved absorption signals for collisional deactivation sequences (see below). Examples of such calibration curves are shown in Figures 14 (for azulene) and 15 (for CS_2).

11 Collisional Energy Transfer in Highly Excited Triatomic Molecules

The u.v. absorption technique discussed in the previous section has been used to measure the stepwise energy loss of highly excited SO_2 and CS_2 molecules after laser excitation followed by strong electronic mixing. The following illustrates the energy-loss profiles obtained and their analysis.

[125] M. Quack, *Nuovo Cimento*, 1981, **B63**, 358.

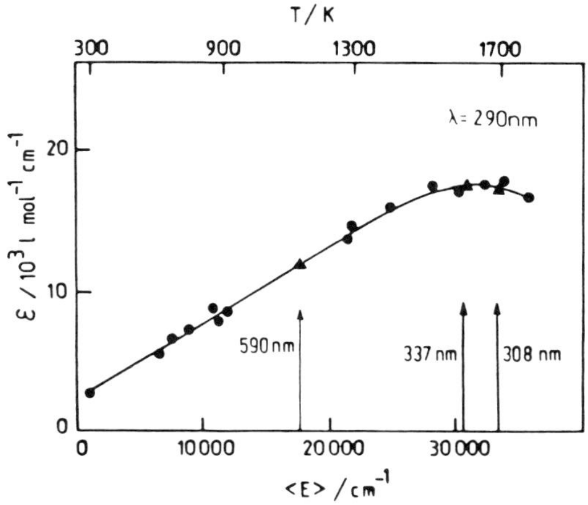

Figure 14 *'Calibration curve' of the azulene absorption coefficient at* 290 nm *as a function of the average excitation energy.* ●, *Canonical ensemble from shock-wave experiments;* ▲ *and arrows, nearly microcanonical ensembles from laser excitation (see refs. 110 and 152)*

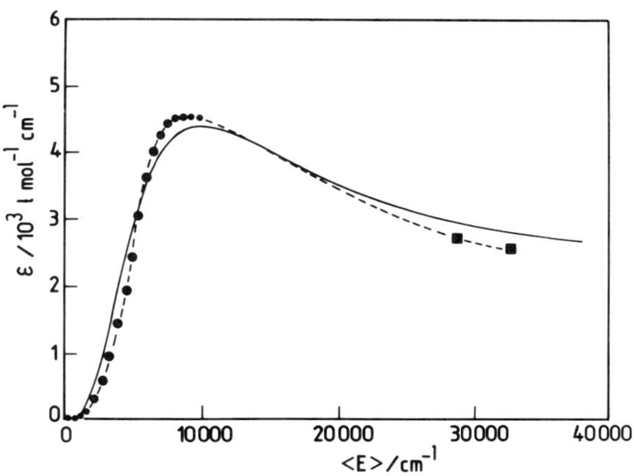

Figure 15 *As Figure 14; for excited* CS_2 *at* 240 nm. ●, *from shock-wave experiments;* ■, *from laser excitation;* —, *Sulzer–Wieland fit;* – – –, *modified Sulzer–Wieland fit*

(Reproduced with permission from *J. Chem. Phys.*, 1985, **82**, 1907)

Recorded absorption–time profiles can have quite different shapes. Figure 16 shows observations at 240 and 365 nm. The two profiles correspond to the same energy-loss process. However, the different dependence of the absorption coefficients on the excitation energy, such as derived from Figure 10, results in different absorption–time profiles.[20]

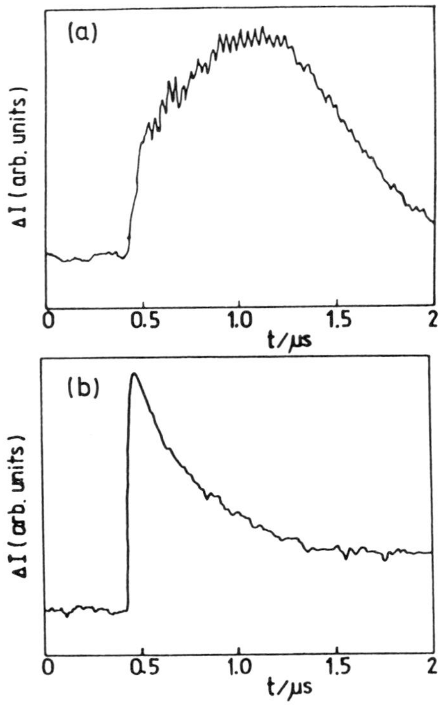

Figure 16 *Absorption–time profiles for collisional deactivation of excited* CS_2. *Excitation at* 308 nm, *observation at: (a),* 240 nm; *(b),* 365 nm; 8.5 Torr CS_2, 300 K
(Reproduced with permission from *J. Chem. Phys.*, 1985, **82**, 1907)

Using the respective absorption signal–average energy $\langle E \rangle$ calibration curves,[95] these profiles are converted into 'energy–time profiles' where $\langle E \rangle$ is plotted as a function of number of collisions $Z_{LJ}[M]t$ (or ZPt). The Lennard–Jones collision frequency is calculated using equation (7) and either measured or estimated Lennard–Jones parameters. One may use tables and the methods elaborated in refs. 126—129.

Examples of energy-loss profiles of $CS_2^* + M$[20,21] and $SO_2^* + M$[18,100] collisions are presented in Figures 17 and 18 for a range of gas temperatures, and for one monoatomic, one triatomic, and one large polyatomic bath-gas molecule. Laser-flash excitation experiments have been made at room temperature, in continuous wave CO_2 laser heated reactors[18,21,130] and behind reflected shock waves.[18,21] The final average

[126] J. O. Hirschfelder, C. F. Curtis, and R. B. Byrd, 'Molecular Theory of Gases and Liquids', Wiley, New York, 1964.
[127] R. C. Reid, J. M. Prausnitz, and T. K. Sherwood, 'The Properties of Gases and Liquids', 3rd Ed. McGraw-Hill, New York, 1977.
[128] L. S. Tee, S. Gotoh, and W. E. Stewart, *Ind. Eng. Chem., Fundam.*, 1966, **5**, 356.
[129] S. H. Fishtine, *Z. Phys. Chem. Neue Folge*, 1980, **123**, 39.
[130] M. Heymann, H. Hippler, and J. Troe, *J. Chem. Phys.*, 1984, **80**, 1853.

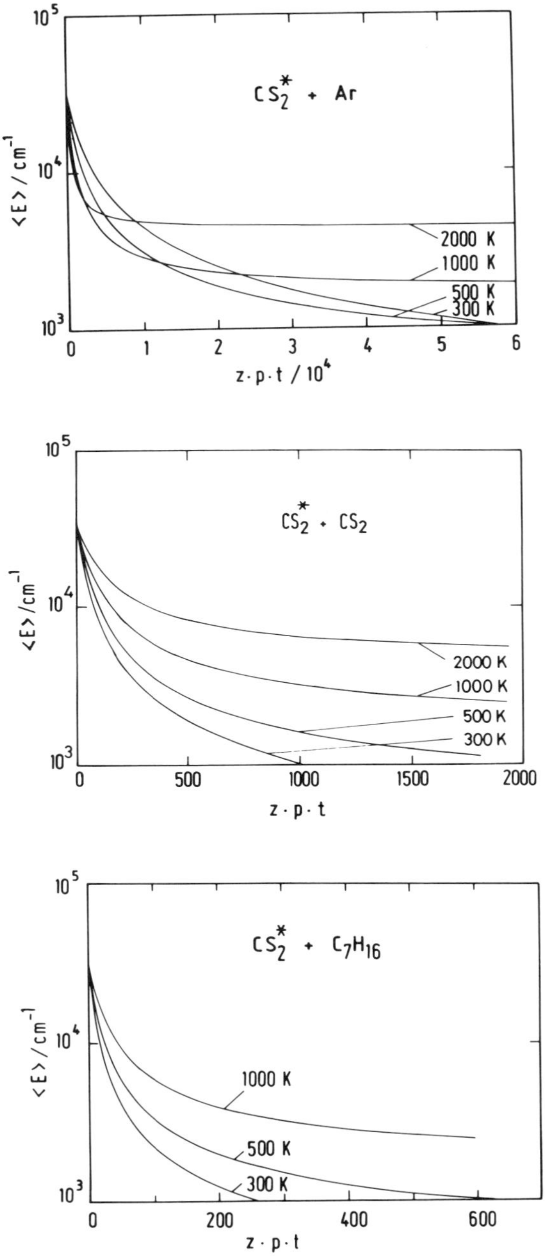

Figure 17 *Energy-loss profiles for* CS_2^* + M *collisions. Average energy* $\langle E \rangle$ *of* CS_2 *as a function of the number of collisions* $ZPt = Z_{LJ}[M]t$
(Reproduced with permission from *J. Chem. Phys.*, 1987, **87**, 3867)

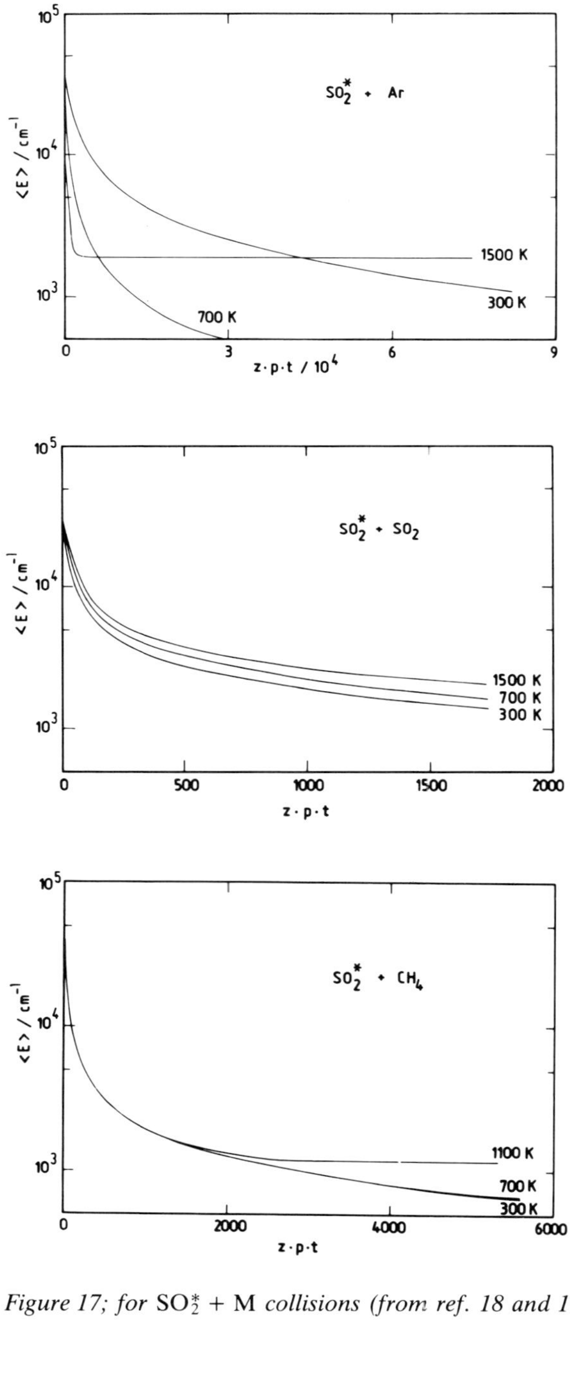

Figure 18 *As Figure 17; for* SO$_2^*$ + M *collisions (from ref. 18 and 100)*

thermal energy reached by the energy-loss processes obviousiy increases with temperature, as seen in the experiments. There are very large differences between the different colliders, in the time scales, and in the temperature effects.

Using equation (18), the slopes of the $\langle E \rangle$ vs. $Z_{LJ}[M]t$ profiles lead directly to the average energies $\langle \Delta E \rangle$ transferred per collision. This evaluation is valid as long as the collisional transition probability functions $P(E',E)$ are relatively narrow around $E' = E$. Furthermore, rotational effects, such as suggested by the flow pattern in Figure 4, should be kept in mind. Double-logarithmic representations of $\langle \Delta E \rangle$ as a function of $\langle E \rangle$ are shown in Figure 19 for various bath-gases at 300 K. The temperature dependences for selected collision pairs are demonstrated in Figures 20 and 21. The energy dependences of $\langle \Delta E \rangle$ most frequently correspond more or less to quadratic dependence:

$$\langle \Delta E \rangle \propto \langle E \rangle^2 \tag{27}$$

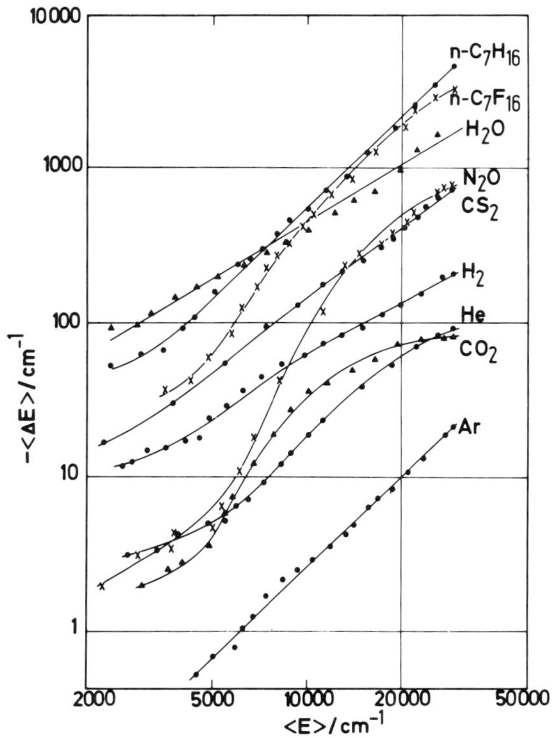

Figure 19 *Average energies $\langle \Delta E \rangle$ transferred per collision for excited* CS_2 *with various collision partners, as a function of the average excitation energy $\langle E \rangle$ (T = 300 K, experiments from ref. 20)*
(Reproduced with permission from *J. Chem. Phys.*, 1985, **82**, 1907)

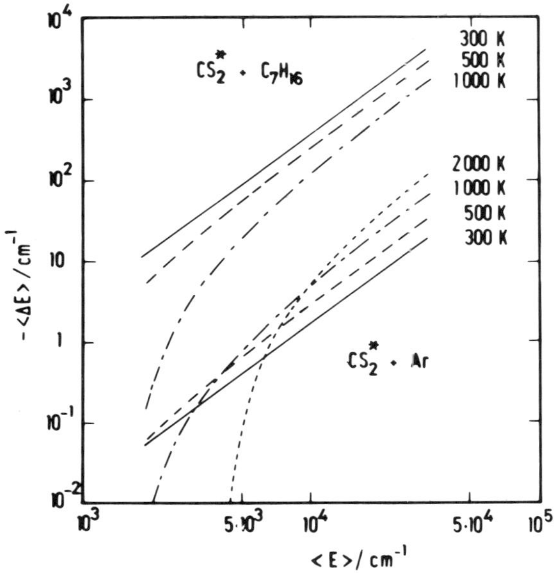

Figure 20 *As Figure 19; temperature dependence of* $\langle \Delta E(\langle E \rangle) \rangle$ (Reproduced with permission from *J. Chem. Phys.*, 1987, **87**, 3867)

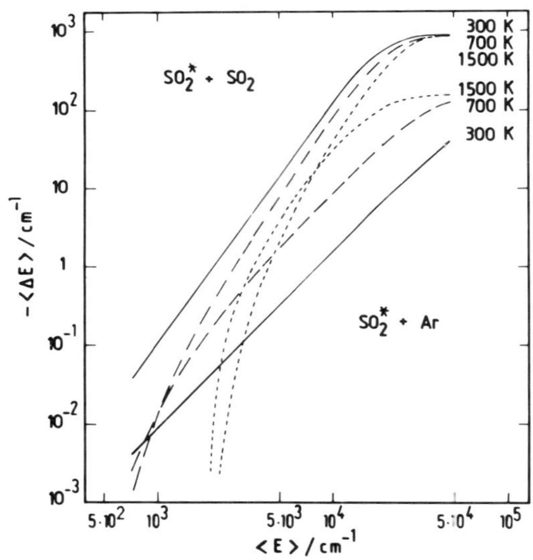

Figure 21 *As Figure 20; for excited* SO_2, *from ref. 18*

At low energies, of course, $\langle \Delta E \rangle$ approaches zero when $\langle E \rangle$ approaches the average thermal energy of the bath-gas. At high energies there appears to be a levelling-off of the energy dependence in some cases. However, this effect is not too clearly established for all collision gases (see below).

The dependence of $\langle \Delta E \rangle$ values of the size of the collision partners is pronounced. Changes over several orders of magnitude are observed. The dependence on the temperature, on the other hand, is only weak. Some bath-gases seem to show a specific behaviour which is not following the general trends of the majority of colliders. Simple analytical expressions representing the behaviour of $\langle \Delta E \rangle$ have been proposed in ref. 20. With these expressions the experimental observations can be satisfactorily reconstructed.

The 'direct' energy transfer results described so far for low energies $\langle E \rangle$ can be compared with energy transfer studies at low vibrational levels from conventional vibrational relaxation[131-133] or single-level laser excitation experiments.[68,134,135] A quantitative comparison is difficult since it is not known which specific relaxation pathways are accessed when the energy distribution approaches low energies from an initially high excitation. Nevertheless, if low-energy vibrational relaxation results are expressed in terms of a product $Z\langle \Delta E \rangle$, the same order of magnitude is obtained for SO_2 as derived from the energy-loss profiles starting at high energies. For CS_2, there was a gap between the accessible range of the energy-loss measurements and vibrational relaxation experiments. However, here again a consistent picture seems to arise. In any case, if the energy-loss sequence starts at high energies with the given $\langle \Delta E \rangle$ values, then only with the strong $\langle E \rangle$ dependence of equation (27) can one arrive at the very small $\langle \Delta E \rangle$ values characterizing energy transfer at low levels of excitation.

The 'direct' energy transfer results described above, at high energies $\langle E \rangle$ close to the dissociation energy, should be compared with results from thermal dissociation and recombination reactions. The analysis of absolute values of the low-pressure rate coefficients k_0 for dissociation and recombination in the CS_2 and SO_2 systems is not possible at present. There are large contributions from excited electronic states which are difficult to take into account.[12] However, relative collision efficiency measurements in the recombination reaction $O + SO + M \rightarrow$

[131] T. L. Cottrell and J. C. McCourby, 'Molecular Energy Transfer in Gases', Butterworths, London, 1961.

[132] B. Stevens, 'Collisional Activation in Gases', Pergamon Press, Oxford, 1967; J. D. Lambert, ref. 59. p. 90.

[133] V. V. N. Kishore, S. V. Babu, and V. S. Rao, *Chem. Phys.*, 1980, **46**, 296.

[134] J. K. Hancock, D. F. Starr, and W. H. Green, *J. Chem. Phys.*, 1974, **61**, 3071; D. C. Richman and R. C. Millikan, *ibid.*, 1974, **61**, 4263.

[135] G. A. West, R. E. Weston, and G. W. Flynn, *J. Chem. Phys.*, 1977, **67**, 4873; D. R. Siebert, and G. W. Flynn, *ibid.*, 1975, **62**, 1212; D. R. Siebert, F. R. Grabiner, and G. W. Flynn, *ibid.*, 1974, **60**, 1564.

SO_2 + M, via equation (16) and an estimated collision efficiency for a relatively strong collider such as C_3F_8, allow for a determination of $\langle \Delta E \rangle$ values near to the dissociation threshold. For the majority of bath gases, the $\langle \Delta E \rangle$ results from representations such as Figures 19—21 can readily be extrapolated[18] up to the measured recombination results for $\langle \Delta E \rangle$. In a few cases, however, there are marked discrepancies which are not well understood as yet.

An inspection of the SO_2 and CS_2 results, and a comparison of many indirect results of $\langle \Delta E \rangle$ from thermal dissociation and recombination studies,[28] confirms in general the position illustrated in Figure 19. $\langle \Delta E \rangle$ values are determined principally by the number of atoms in the collision partner. However, specific effects for some collision partners occur. This appears to happen more frequently with excited triatomic molecules than with excited larger polyatomic molecules.

A point of particular importance, especially for applications to thermal dissociation rate coefficients, is the question of the temperature dependence of $\langle \Delta E \rangle$. The direct results for CS_2 and SO_2, after an extrapolation to their respective dissociation energies, are shown in Figures 22 and 23. For inefficient collision partners there appears to be a weakly positive temperature coefficient of $\langle \Delta E \rangle$; for the most efficient colliders it is weakly negative. This is between the limits:

$$\langle \Delta E \rangle \propto T^{0 \pm 1} \tag{28}$$

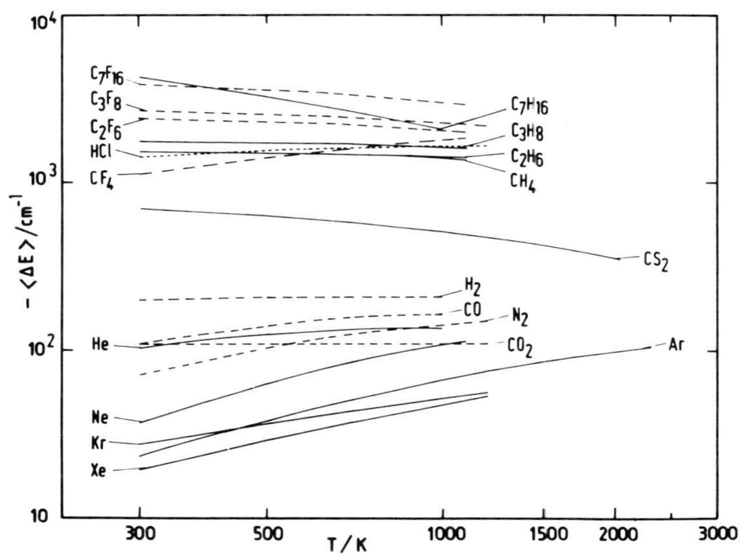

Figure 22 *Temperature dependence of* $\langle \Delta E \rangle$ *for collisions of excited* CS_2 *at its dissociation energy with various collision partners, from ref. 21*
(Reproduced with permission from *J. Chem. Phys.*, 1987, **87**, 3867)

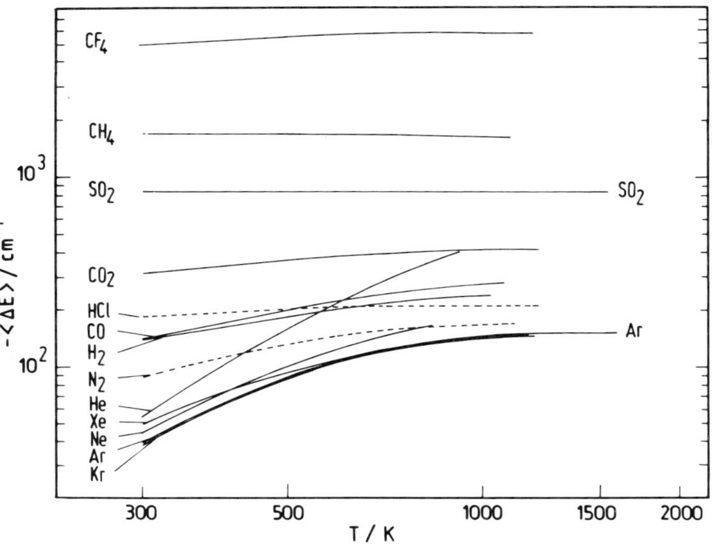

Figure 23 *As Figure 22; for excited* SO_2, *from ref. 18*

This general behaviour is in agreement with conclusions from thermal dissociation experiments. Relative collision efficiencies for several dissociation reactions of triatomic molecules have been measured in shock waves.[136] The analysis of absolute values of rate coefficients for H_2O[19] and SO_2[137,138] dissociation in Ar confirms the validity of equation (28) over a very wide temperature range (from 300 K up to 5000 K). For the $SO_2 + Ar \rightleftharpoons SO + O + Ar$ system, one may even identify[137] a small positive value of the temperature coefficient of $\langle \Delta E \rangle$ up to 3000 K and a change to a negative coefficient above this temperature.

A meaningful comparison with classical trajectory calculations can so far only be made for collisional energy transfer of highly excited triatomic molecules in collisions with noble gases. The trajectory results for SO_2 + Ar collisions from refs. 6 and 7 lead to $\langle \Delta E \rangle$ values about 10 times larger than those obtained experimentally.[18,137] It may well be that this problem is encountered quite generally when $P(E',E)$ histograms are constructed with a too coarsely grained $\delta E'$ scale around the truncated elastic peak (see Section 2). Apart from this problem, the derived energy dependence of $\langle \Delta E \rangle \propto \langle E \rangle$ must be too weak, since it does not lead to the correct order of magnitude of $\langle \Delta E \rangle$ at lower excitation energies. On the other hand, the calculations confirm a weak temperature dependence of $\langle \Delta E \rangle$.

136 H. Endo, K. Glänzer, and J. Troe, *J. Phys. Chem.*, 1979, **83**, 2083.
137 C. J. Cobos, H. Hippler, and J. Troe, *J. Phys. Chem.*, 1985, **89**, 1778.
138 H. J. Plach and J. Troe, *Int. J. Chem. Kinet.*, 1984, **16**, 1531; J. H. Kiefer, and R. Ramaprabhu, *Chem. Phys. Lett.*, 1982, **86**, 499; Th. Just and B. Rimpel, Proc. 11th Int. Symp. on Shock Tubes and Waves, 1977 (Seattle 1978), p. 226.

The trajectory modelling of a sequence of successive collisions in ref. 17 did not have the graining problem encountered in the construction of histograms. Figure 24 shows an example of calculated energy-loss profiles for $CS_2^* + He$ collisions. The agreement with the experimental observations from ref. 20 was quite satisfactory, although again a too weak energy dependence of $\langle \Delta E \rangle$ was obtained. Figure 25 shows a comparison of the measured $\langle \Delta E \rangle$ values with the calculations. As in the SO_2 + Ar trajectory calculations, $\langle \Delta E \rangle \propto \langle E \rangle^2$ was measured. As stated in Section 2, the stronger decrease of $-\langle \Delta E \rangle$ with decreasing energy of the experiments might be explained by the 'dilution' of the $k(E',E)$ curve in the sparse manifold at low energies. In this sense, quantum effects may limit the application of classical trajectory calculations for the collisional energy transfer of excited triatomic molecules. At which densities of levels this problem disappears remains an open question.

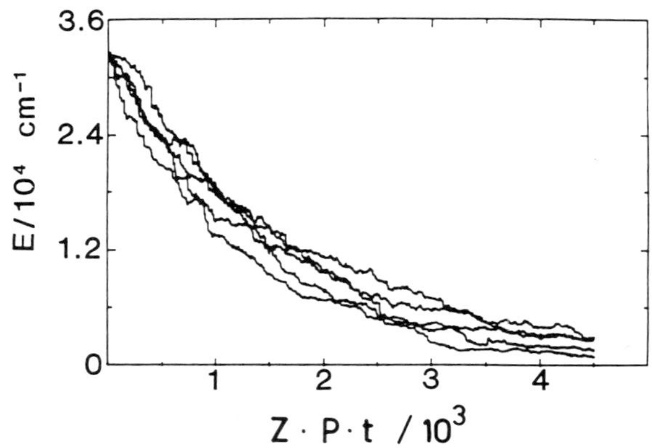

Figure 24 *Energy-loss profiles for sequences of* $CS_2^* + He$ *collisions (classical trajectory calculations from ref. 17)*

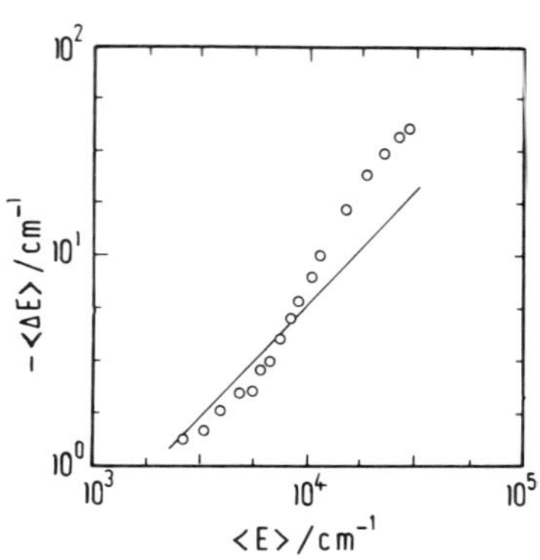

Figure 25 *Average energy* $\langle \Delta E \rangle$ *transferred per collision of* CS_2^*. —, *Classical trajectory calculations from ref. 17;* ○, *experiments from ref. 20*

12 Collisional Energy Transfer in Highly Excited Large Molecules

Direct studies of collisional energy transfer of vibrationally highly excited states have now been performed for a number of large polyatomic molecules. The best documented case is the azulene system. Several other larger cyclic molecules have also been investigated, although the range between triatomic and large cyclic molecules has still been somewhat neglected. In the following, the possibilities and limitations of current direct experiments with the azulene system are illustrated. A brief summary of results for other polyatomic molecules is then given.

Azulene

The spectral and photochemical properties of azulene[139–148] make this molecule a particularly suitable candidate for direct collisional energy transfer studies. The molecule has a series of low-lying electronically excited states which can be reached from the electronic ground state with sufficiently large absorption coefficients. Internal conversions from these states to the electronic ground state are sufficiently fast. There appear to be no perturbating reactions from highly excited molecules. At sufficiently high energies only the isomerization to naphthalene[149,150] has to be taken into account. This well-characterized process[150] can even be exploited at very high excitation energies, as a competing reference reaction.[51]

The excitation dependence of the azulene u.v. absorption spectrum was described in Sections 9 and 10. On the basis of these results, one can conveniently study collisional energy transfer by monitoring average energies $\langle E \rangle$ of excited molecules during the collisional deactivation process. Such studies have been performed by our own group.[48,151,152]

139 G. D. Gillespie and E. C. Lim, *J. Chem. Phys.*, 1978, **68**, 4578.
140 R. Pariser, *J. Chem. Phys.*, 1956, **25**, 1112.
141 J. B. Birks, *Chem. Phys. Lett.*, 1972, **17**, 370.
142 N. J. Turro, 'Modern Molecular Photochemistry', Benjamin/Cummings Publishing Co., Menlo Park, CA, 1978, p. 148.
143 'DMS UV Spektrenatlas organischer Verbindungen', Weinheim, London, 1967, Vol. III, F1/1.
144 H. J. Kray and B. Nickel, *Chem. Phys.*, 1980, **53**, 235.
145 A. Amirav and J. Jortner, *J. Chem. Phys.*, 1984, **81**, 4200.
146 Y. Hirata and E. C. Lim, *J. Chem. Phys.*, 1978, **69**, 3292.
147 D. Huppert, J. Jortner, and P. M. Rentzepis, *J. Chem. Phys.*, 1972, **56**, 4826.
148 M. Fujii, T. Ebata, N. Mikami, and M. Ito, *Chem. Phys.*, 1983, **77**, 191.
149 E. Heilbronner, in 'Non-Benzenoid Aromatic Compounds', ed. D. Ginsgurg, Interscience Publishers, New York, 1959, p. 263.
150 L. Brouwer and J. Troe, *Int. J. Chem. Kinet.*, 1988, **20**, 379.
151 H. Hippler, L. Lindemann, and J. Troe, *J. Chem. Phys.*, 1985, **89**, 1698.
152 H. Hippler, B. Otto, and J. Troe, *Ber. Bunsenges. Phys. Chem.*, 1989, in press; B. Otto, Ph.D. Thesis, University of Göttingen, 1987.

Previously, collisional energy transfer of excited azulene had been investigated in detail by Barker and co-workers,[50,81–85,153] who detected excited states by their i.r. emission. By using CO_2 as a bath-gas, the increase in the vibrational energy of CO_2 during the deactivation of azulene could also be monitored.[153,154] The use of i.r. emission as a detection method gives rise to a signal calibration problem analogous to that of the hot u.v. absorption method discussed above. Initially, a theoretical modelling of the i.r. emission intensity of excited states was applied. An experimental check of this calibration curve of the signal was performed[84] by comparing signals after laser excitation in the S_1 band (origin near 28 000 cm^{-1}). The comparison of the signal–average energy $\langle E \rangle$ calibration curves in Figure 26 for the i.r. emission and u.v. absorption detection techniques clearly shows some characteristic differences. The energy dependence of the signals is much stronger with the i.r. emission than that associated with the u.v. absorption. For this reason, the signal-to-noise ratio of energy-loss profiles decays more strongly with time in the i.r. emission experiments. The two techniques certainly should give identical results for $\langle \Delta E \rangle$. Differences would be due to problems with the calibration curves (see below).

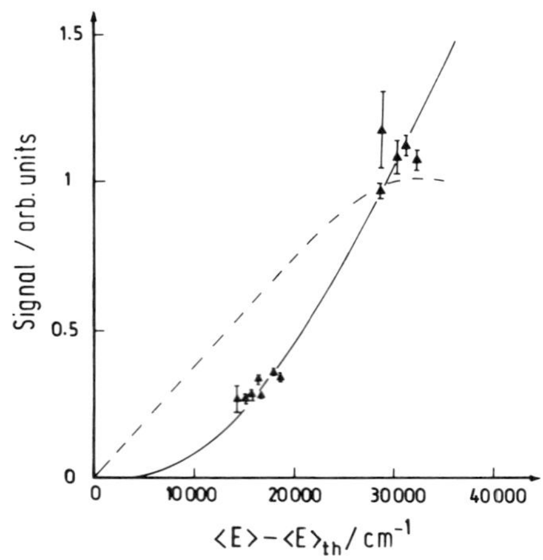

Figure 26 *Comparison of the signal–average energy* $\langle E \rangle$ *calibration curves for u.v. absorption (– – –, from Figure 11) and i.r. emission (— and ▲, from refs. 84 and 85) of hot azulene* ($\langle E \rangle_{th}$ = *average thermal energy of azulene at* 300 K)

[153] A. C. Baldwin, M. J. Rossi, J. R. Pladziewicz, and J. R. Barker, *Chem. Phys. Lett.*, 1982, **90**, 99.
[154] W. Jalenak, R. E. Weston, jun., T. J. Sears, and G. W. Flynn, *J. Chem. Phys.*, 1985, **83**, 6049.

Figures 27 and 28 show u.v. absorption and i.r. emission signals recorded during the collisional deactivation of azulene after laser excitation into the S_1 band. Using the signal calibration curves, energy-loss profiles such as Figure 29 are obtained. The slopes of these profiles give the $\langle \Delta E \rangle$ values as a function of the energy $\langle E \rangle$, as shown in Figure 30. The resulting $\langle \Delta E \rangle$ values are essentially identical, differences being explained by the unavoidable uncertainties of the calibration curves. While there is no doubt that $\langle \Delta E \rangle$ is energy dependent ($\langle \Delta E \rangle \propto \langle E \rangle$) for energies below 16 000 cm^{-1}, the energy dependence of $\langle \Delta E \rangle$ at higher excitation energies is still a matter of some dispute. While the u.v. absorption data indicate a levelling-off of $\langle \Delta E \rangle$ at high energies, this was much less apparent in the i.r. emission experiments, as shown in Figure 30. The reasons for this discrepancy are again found in the uncertainty of the calibration curve. Since the calibration curve for the u.v. absorption is much less steep than that for the i.r. emission, a single energy-loss profile from u.v. absorption probably reveals the energy dependence of $\langle \Delta E \rangle$ in a slightly more direct way.

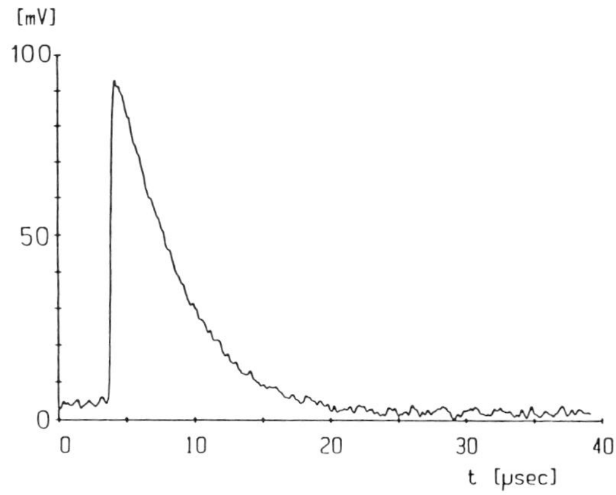

Figure 27 *Transient 'hot' u.v. absorption of azulene after laser excitation at* 590 nm *(absorption wavelength* 290 nm, 1.6 mbar He, 300 K*), from ref. 152*

The question of energy dependence of $\langle \Delta E \rangle$ at high excitation energies can probably only be solved rigorously when much wider energy ranges are covered. Such experiments have become possible by applying the novel u.v. multiple-photon excitation mechanism demonstrated in ref. 51 and illustrated in Figure 31. In the presence of fast internal conversion, several u.v. photons can be deposited in the molecule in the form of vibrational excitation. In this way one can reach vibrational excitation energies that exceed the ionization energy of the molecule. Ultimately, unimolecular reactions of the excited species take place faster than optical pumping and collisional energy transfer.

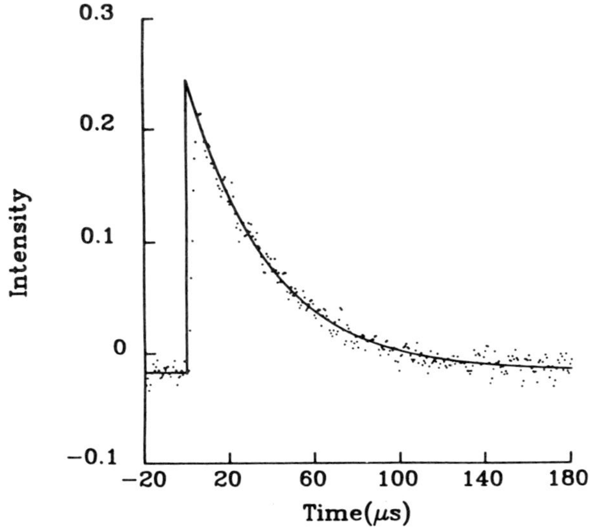

Figure 28 *Transient 'hot' i.r. emission of azulene after laser excitation at* 635 nm (6.2 mTorr *azulene*, 300 K)
(Reproduced with permission from *J. Chem. Phys.*, 1988, **88**, 6219)

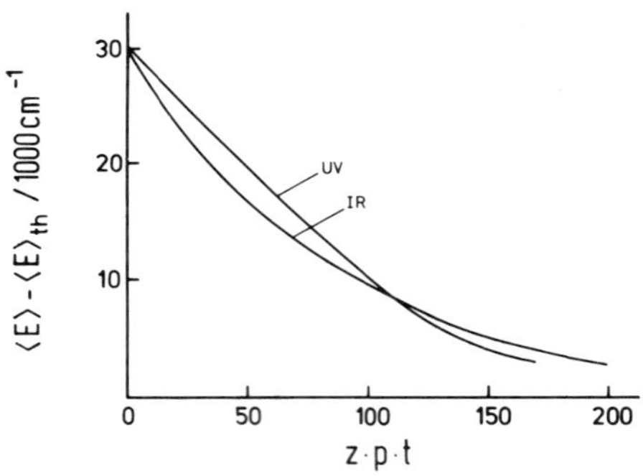

Figure 29 *Energy-loss profiles of azulene in collisions with* Ar *after laser excitation at* 337 nm *and* 300 K. *Comparison of u.v. absorption and i.r. emission results after conversion by the respective calibration curves (measurements from refs. 151, 152, and 85)*

However, with intense and short pumping pulses, energy transfer studies can be extended up to very high energies. Azulene experiments[51] up to $\langle E\rangle \approx 60\,000\ \mathrm{cm}^{-1}$ appear to confirm only a weak energy dependence of $\langle \Delta E\rangle$ between 30 000 and $60\,000\ \mathrm{cm}^{-1}$. However, so far these conclusions are based only on the evaluation of pressure- and pumping intensity-dependent reaction yields. Time-resolved experiments are underway to substantiate these conclusions.

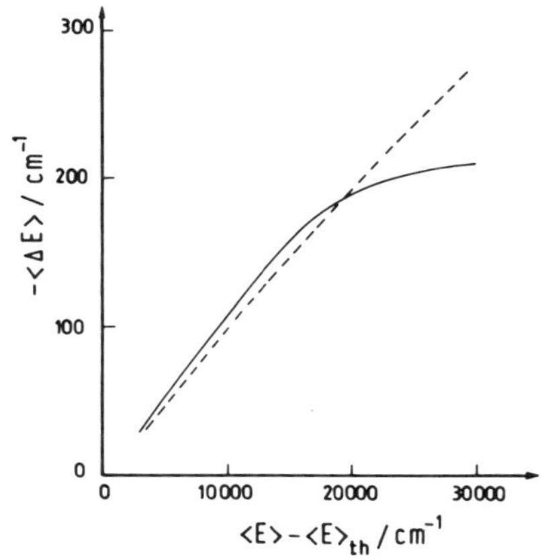

Figure 30 *Average energies* $\langle \Delta E \rangle$ *transferred per collision between excited azulene and* Ar. —, *u.v. absorption experiments from refs. 151 and 152;* – – –, *i.r. emission experiments from ref. 85;* $\langle E \rangle_{\text{th}}$ = *average thermal energy of azulene at* 300 K

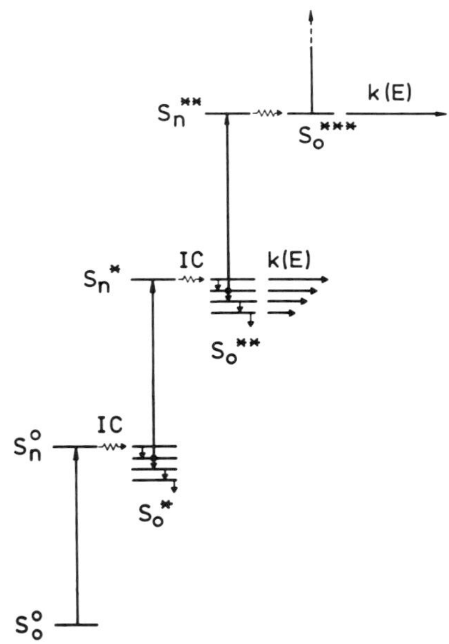

Figure 31 *Schematic representation of u.v. multiple-photon excitation of azulene, with internal conversion* (IC), *collisional energy transfer, and unimolecular isomerization to naphthalene (k(E))*
(Reproduced with permission from *J. Chem. Phys.*, 1988, **88**, 3564)

The generally good agreement between the $\langle \Delta E \rangle$ values from i.r. emission and u.v. absorption studies for various bath-gases is demonstrated in Table 1. The results differ by a factor of 2 only in rare cases:

usually the agreement is much closer. The experiments also agree in the observation of a weak temperature dependence of $\langle \Delta E \rangle$. A more systematic study of the temperature dependence of $\langle \Delta E \rangle$ will be described below for excited toluene.

Table 1 *Comparison of experimental average energies transferred per collision $-\langle \Delta E \rangle$ of azulene at different average excitation energies. All $\langle \Delta E \rangle$ values with respect to Lennard–Jones collision frequencies from ref. 151 and 152*[a]

Excitation at	600 nm	590 nm	337 nm	337 nm
$\langle E \rangle - \langle E \rangle_{th}$[b]	13 940	13 940	24 020	24 020
Method	*i.r.*	*u.v.*	*i.r.*	*u.v.*
He	44	75	97	79
Ne	90	120	175	179
Ar	134	160	235	205
Kr	140	154	205	225
Xe	110	160	220	225
H_2	110	95	160	140
CO	170	220	410	310
N_2	135	175	330	195
O_2	150	160	345	
CO_2	300	340	555	350
SF_6	380	515	645	700
H_2O	330	430	865	435
NH_3	300	300	954	320
CH_4	190		485	355
C_4H_{10}	750		1285	1005

[a] Energies in cm^{-1}; T = 300 K; i.r. emission from ref. 85; u.v. absorption from ref. 152.
[b] $\langle E \rangle_{th}$ = average thermal energy at 300 K.

Toluene

Collisional energy transfer in highly excited toluene has similarly received extensive study. Since there is no low-lying excited electronic state, excitation via internal conversion necessarily leads to initial energies above 40 000 cm^{-1}. A convenient second excitation mechanism starts with the isomer cycloheptatriene and passes through its internal conversion and subsequent fast unimolecular isomerization. Internal excitation energies of ~52 000 cm^{-1} are reached in this way. The evaluation of the energy-loss profiles from u.v. absorption detection suggested essentially energy-independent $\langle \Delta E \rangle$. When the energy decreases, of course, as in the azulene system, there must be a change in this behaviour: $-\langle \Delta E \rangle$ must decrease, pass through the low-energy vibrational relaxation range, and approach zero when $\langle E \rangle$ approaches the thermal equilibrium value. However, this change must occur only below 20 000 cm^{-1} where the observed absorption–time profiles are increasingly influenced by poor signal-to-noise.

The dependence of $\langle \Delta E \rangle$ on the nature of the bath-gases at room temperature has been studied systematically.[155] Since the results obtained are completely consistent with the high-energy results for azulene (excitation at 337 nm; see Table 1), they are not tabulated here. A systematic study of the temperature dependence of $\langle \Delta E \rangle$ between 300 and ~900 K has been made for various colliders:[130] Figure 32 illustrates the results. The temperature dependences of $\langle \Delta E \rangle$ are again quite weak. The results hardly differ from the high-energy values of $\langle \Delta E \rangle$ for excited triatomic molecules (see Figures 22 and 23). However, a slightly larger range of temperature coefficients $\langle \Delta E \rangle \propto T^{0 \pm 1}$ is seen for the excited triatomic molecules than for the excited polyatomic molecules, where $\langle \Delta E \rangle \propto T^{0 \pm 0.5}$ is found.

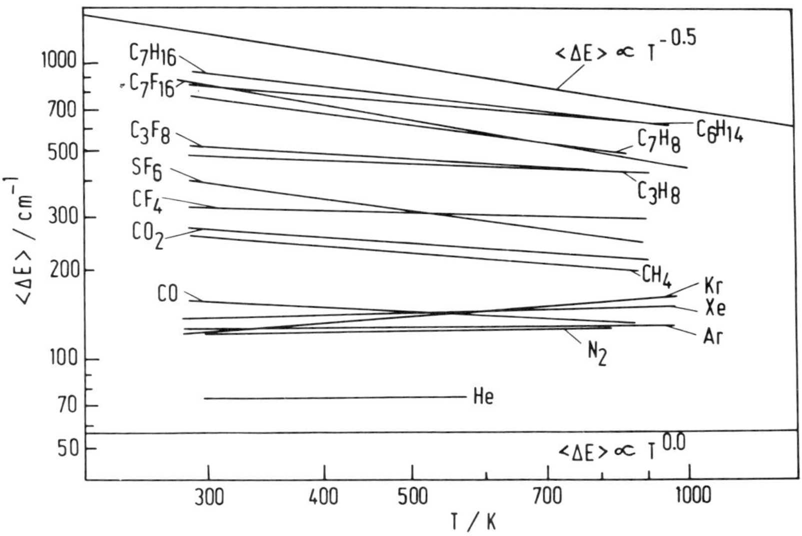

Figure 32 *Temperature dependence of $\langle \Delta E \rangle$ for excited toluene in various collider gases*
(Reproduced with permission from *J. Chem. Phys.*, 1984, **80**, 1853)

The surprisingly small dependence of $\langle \Delta E \rangle$ on the nature of the highly excited molecule (for the same bath-gas molecule) considerably facilitates the estimate of $\langle \Delta E \rangle$ values for arbitrary reaction systems. The values for azulene (Table 1) or toluene (refs. 130 and 155) are therefore recommended for *a priori* guesses.

[155] H. Hippler, J. Troe, and H. J. Wendelken, *J. Chem. Phys.*, 1983, **78**, 6709.

The toluene system has been studied further with respect to $\langle \Delta E \rangle$, the energy dependence of $\langle \Delta E \rangle$, $\langle \Delta E^2 \rangle$, and $P(E',E)$ by the MPI technique of ref. 87. Since these experiments go beyond the results reviewed previously, they will be discussed separately in the following section.

Cycloheptatrienes

Energy transfer in excited cycloheptatriene and in substituted cycloheptatrienes has been studied by activation via internal conversion and detection by u.v. absorption.[37] Since the results are quite comparable to the azulene and toluene systems, they are not given here. It appears worth mentioning that substitution of the cycloheptatriene has no identifiable effect on $\langle \Delta E \rangle$.

The cycloheptatriene experiments are of particular interest since direct[37] and indirect[38,39] measurements have been made independently and can be compared. In the indirect measurements, pressure dependences of photoisomerization quantum yields were evaluated. These energy transfer results can additionally be converted into 'direct' information since specific rate constants $k(E)$ for the competing unimolecular reaction were also measured directly.[36] Furthermore, thermal isomerization rate constants were determined.[40] It is useful to recall the history of these investigations. At first, the quantum yield measurements showed a very clear picture. However, when direct measurements of $\langle \Delta E \rangle$ became available, a discrepancy of about a factor of three was found. This discrepancy disappeared only when direct measurements of $k(E)$ replaced the data from thermal isomerization experiments in the evaluation. It seems plausible to attribute some discrepancies between earlier indirect[2] and new direct measurements of $\langle \Delta E \rangle$ to problems with the reference rates of the competing unimolecular reactions.

Collisional energy transfer of highly excited fluoroethylcycloheptatriene was also studied by u.v. absorption detection.[61] However, in this case vibrational excitation was achieved by i.r. multiple-photon excitation with a CO_2 laser. After identification of the average initial energy of excited molecules via their isomerization rate under low-pressure conditions, $\langle \Delta E \rangle$ values were derived in agreement with the u.v. single-photon excitation experiments. Certainly the initial excitation in the i.r. multiple-photon experiment is less well characterized than in the u.v. single-photon experiment. However, as long as one starts at sufficiently high energy, where $\langle \Delta E \rangle$ does not depend on the energy, this is not too significant a handicap.

Benzene and Hexafluorobenzene

Benzene, hexafluorobenzene, and other aromatic molecules show internal conversion in their first excited electronic states. Thus, they can be prepared in high vibrational levels of the electronic ground state by the internal conversion technique. The detection of energy transfer is then possible by hot u.v. absorption spectra. Experiments of this type have been performed in refs. 106 and 156. So far the calibration curves for the interpretation have been constructed on the basis of limited thermal spectra and simplified theoretical models. Unfortunately, experimental checks of these calibration curves in shock waves and directly after laser excitation[107] (such as described in Sections 9 and 10) did not confirm the predicted curves. Therefore a revision of the derived $\langle \Delta E \rangle$ values will be necessary. It appears improbable that, for these systems, $\langle \Delta E \rangle$ values different from the azulene, toluene, and cycloheptatriene results will be obtained.

CF_3I *and* SF_6

Energy transfer of highly excited molecules of intermediate size – between the triatomics and the larger cyclic molecules discussed above – has not yet been studied by the internal conversion technique. I.r. multiple-photon excitation was applied, instead, to CF_3I and SF_6. In the case of CF_3I, collisional deactivation could be studied[157] by hot u.v. absorption. A calibration of the spectrum by shock-wave experiments[158] and a careful analysis of the u.v. spectrum under truly collision-free conditions after the CO_2 laser pulse[61,157] are required in order to characterize the initial energy reached by the multiple-photon excitation pulse. At too large gas pressures, transient u.v. absorption spectra[159] are difficult to understand. Figure 33 shows the results from ref. 157 for energy transfer of CF_3I in three bath-gases. For the more efficient colliders propane and octane, $\langle \Delta E \rangle$ is found to depend linearly on $\langle E \rangle$ up to energies of about 17 000 cm^{-1}. For Ar, a stronger energy dependence is observed below $\langle E \rangle \approx 5000\ cm^{-1}$.

A transition from a stronger than linear energy dependence to a linear energy dependence of $\langle \Delta E \rangle$ with increasing energy was also observed with SF_6 collisions after CO_2 laser excitation. Experiments using i.r. absorption,[86] optoacoustic detection,[91] and the Hg tracer method[88] all put this transition close to an energy around 5000 cm^{-1}. Figure 34 demonstrates the $\langle \Delta E \rangle \propto \langle E \rangle^{3/2}$ dependence for $SF_6^* + SF_6$ collisions observed[88] between 1500 and 5000 cm^{-1}.

The results discussed for CF_3I and SF_6 put the energy transfer of these molecules 'half-way' between the excited triatomic and larger polyatomic molecules. The energy dependence of $\langle \Delta E \rangle$ apparently decreases with increasing energy. The larger the excited molecule, the lower is the energy where the strong energy dependence of $\langle \Delta E \rangle$

156 N. Nakashima and K. Yoshihara, *J. Chem. Phys.*, 1983, **79**, 2727.
157 B. Abel, B. Herzog, H. Hippler, and J. Troe, *J. Chem. Phys.* 1989, in press.
158 L. Brouwer and J. Troe, *Chem. Phys. Lett.*, 1981, **82**, 1.
159 Yu. A. Kudriavtsev and V. S. Letokhov, *Chem. Phys.*, 1980, **50**, 353.

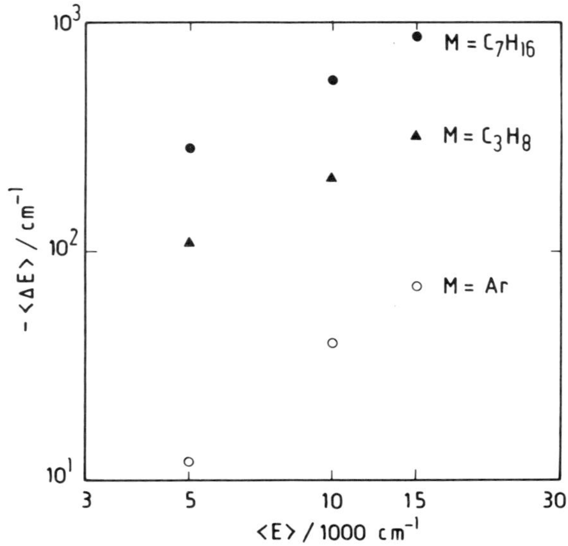

Figure 33 *Average energy* $\langle \Delta E \rangle$ *transferred in collisions of* CF_3I^* *with* M. *Time-resolved i.r. multiple-photon excitation experiments from ref.* 157 *with u.v. absorption detection technique*

ceases. This observation appears to support the discussion of the relation between the sparsity of the vibrational manifold and the energy transfer which was given in Section 4. Too sparse a vibrational quasi-continuum decreases $\langle \Delta E \rangle$ below the values obtained from the classical calculations with a true continuum of states.

A further discussion of energy transfer of SF_6 and several other examples is given in ref. 160. These examples will not be further elaborated here.

13 Detection by Multiphoton Ionization: An Access to $\langle \Delta E(E) \rangle$ and $\langle \Delta E^2(E) \rangle$

A new detection scheme for collisional energy transfer in highly vibrationally excited molecules has been developed[87] which offers opportunities beyond those provided by the other methods like u.v. absorption and i.r. emission. In this technique, a 'window' is placed in a suitable energy range, and the population of excited states passing through this window is monitored by MPI. According to the solution of the master equation for sequential collisional deactivation[33] from a narrow initial distribution of highly excited states, the population passing through a selected energy window, (which is chosen to be sufficiently far away

[160] R. J. Gordon, *Comments on Atomic and Molecular Physics*, 1988, **21**, 123.

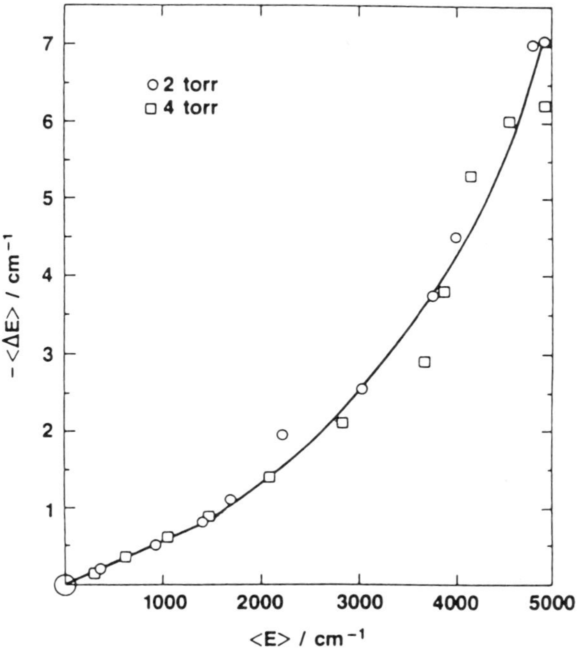

Figure 34 *Average energy* $\langle \Delta E \rangle$ *transferred in collisions of* $SF_6^* + SF_6$. *Time-resolved i.r multiple-photon excitation experiments from ref. 88 with* Hg *tracer detection technique*

from the initial excitation) depends only on the first and second moments $Z\langle \Delta E \rangle$ and $Z\langle \Delta E^2 \rangle$ of the energy transfer. If $\langle \Delta E \rangle$ is energy-dependent then, of course, the time evolution of the distribution observed at the detection window will reflect this as well. Furthermore, the presence of a strong collision component in the collision transition probability $P(E',E)$ would be revealed as a rapid precursor to the main arrival of population at the window.

In principle, there are various ways in which to set detection windows. In ref. 87, MPI was shown to provide the required properties. Figure 35 demonstrates the detection scheme for energy transfer in excited cyclic molecules like toluene. As before, preparation of highly excited electronic ground state molecules can be achieved by the internal conversion technique which exploits the onset of 'channel three' $S_1^* \rightarrow S_0^{**}$ transition at wavelengths below a certain threshold or, for aromatic molecules like toluene, uses the isomerization cycloheptatriene $\rightarrow$ toluene. The detection is achieved by a two-colour, two-photon ionization. If the energy of the first photon (wavelength λ') is too large, essentially no ions are formed since the fast $S_1^* \rightarrow S_0^{**}$ internal conversion depletes the intermediate S_1 state from which absorption of a second photon (wavelength λ'') would lead into the ionization continuum. If the energy

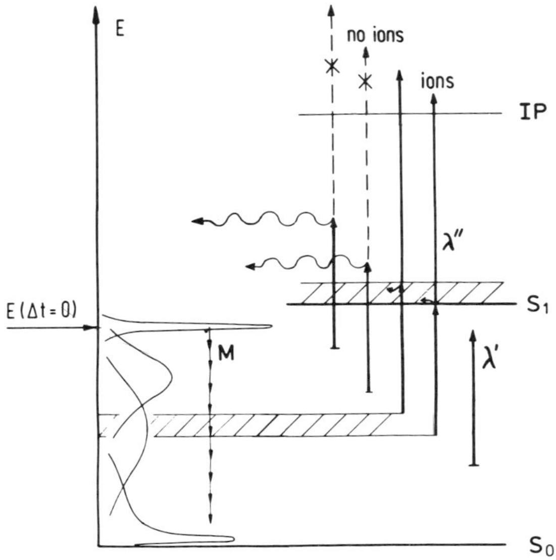

Figure 35 *Schematic representation of MPI detection of energy transfer after initial excitation at* $E(\Delta t = 0)$ (see text)
(Reproduced with permission from *Ber. Bunsenges. Phys. Chem.*, 1988, **92**, 442)

of the first photon is too small, the intermediate electronic state S_1 is not reached and again essentially no ions are formed. If, however, the energy of the first photon corresponds to the difference of the window energy and the lowest vibrational levels of S_1, there is resonant two-photon ionization. By varying the wavelength λ' of the first photon, the position of the detection window can be moved.

Figure 36 shows the dependence of the ion signal on the pressure of the bath gas heptane in the energy transfer of highly excited xylene, after preparation via fast isomerization of methylcycloheptatriene (C_8H_{10}, *i.e.* approximately the same size as azulene $C_{10}H_8$). Without added heptane, it takes about 20 μs before the distribution of deactivating excited molecules reaches the detection window. With increasing pressure the deactivation sequence is accelerated and the arrival time at the window is reduced. Figures 37 and 38 show the effect of the position of the detection window in experiments with excited toluene formed from cycloheptatriene. Figure 37 corresponds to a higher energy, where the distribution arrives and then passes out of the window; Figure 38 corresponds to a window close to the bottom of the energy scale where only an arrival is seen. The shape of the ion signals as a function of the window energy is so detailed that $\langle \Delta E \rangle$, $\langle \Delta E^2 \rangle$, and their energy dependences can all be derived. The fitted ion signals in Figures 37 and 38 agree well with the experimental signals. However, with cycloheptatriene as the bath gas there is a short time precursor signal superimposed

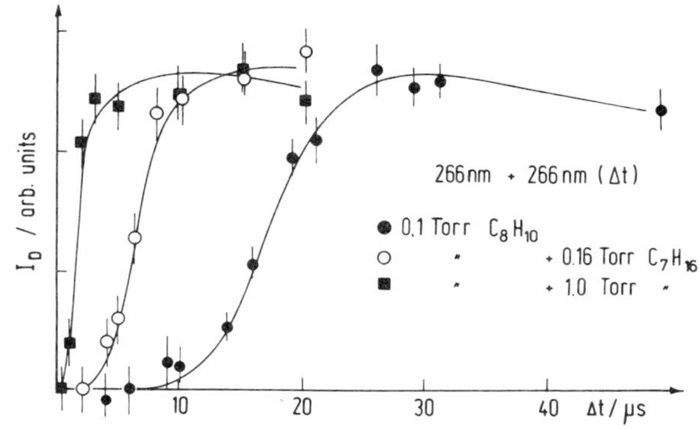

Figure 36 *Pressure dependence of transient two-photon ionization signals after laser excitation of methylcycloheptatriene* (C_8H_{10}) *and collisions with* C_7H_{16} (*ionization wavelength* = 266 nm)
(Reproduced with permission from *Chem. Phys. Lett.*, 1988, **144**, 473)

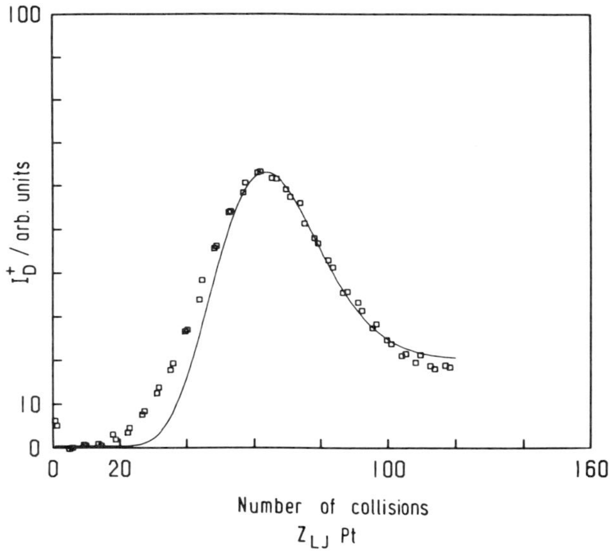

Figure 37 *Transient two-photon ionization signals from excited toluene with cycloheptatriene as collision partner (ionization wavelength* = 280 nm)

on the calculated profile. This may be tentatively attributed to a small strong collision component in the collisional transition probability (see below). Figure 39 shows the derived energy dependences of $\langle \Delta E \rangle$ and $\langle \Delta E^2 \rangle$ for the fitted signals (omitting the strong collision precursor). The signal shape and its dependence on the window position responds very sensitively to the details of $\langle \Delta E(E) \rangle$ such that the derived results appear particularly reliable. Modelling calculations with $\langle \Delta E \rangle$ either

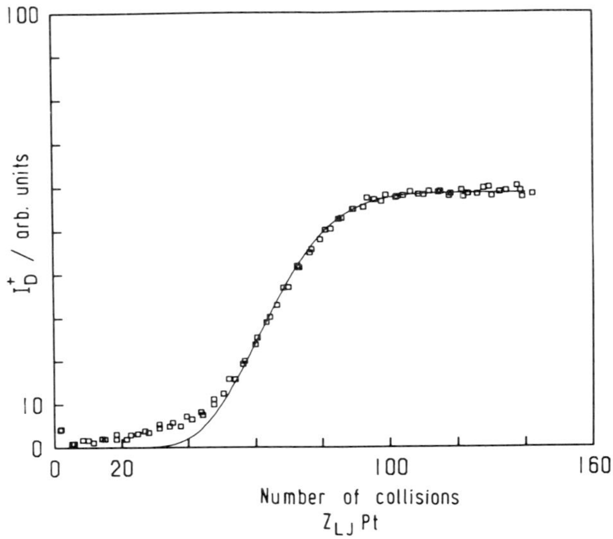

Figure 38 *As Figure 37; ionization wavelength* = 266 nm

independent of energy or proportional to $\langle E \rangle$ cannot reproduce the observed signals. Figure 39 confirms our earlier conclusions about the energy dependence of $\langle \Delta E \rangle$, *i.e.* a stronger than $\langle \Delta E \rangle \propto \langle E \rangle$ dependence at very low energies, a transition to a $\langle \Delta E \rangle \propto \langle E \rangle$ dependence at increasing energy (here above about 2000 cm^{-1}), and a levelling-off to a nearly energy-independent $\langle \Delta E \rangle$ at $\langle E \rangle$ larger than 15 000 cm^{-1}.

The profiles in Figures 37 and 38 suggest a minor contribution from strong collisions transferring large amounts of vibrational energy, exceeding the amounts expected from the tails of (for example) a weak collision exponential collision model. To date this precursor has only been observed with the bath gas cycloheptatriene; it was not detected with the bath-gas Ar.[87] The observation of a minor strong collision contribution is in accord with the results from classical trajectory calculations[6,9] which yielded a 'biexponential' $P(E',E)$ distribution, reflecting the superposition of narrow weak and broad strong collision contributions. An alternative means of identifying a strong collision contribution is provided by a chemical method. If the bath-gas molecules can undergo a chemical transformation characterized by a certain threshold energy, the yield of the reaction products is a measure of the number of strong collisions transferring more than the reaction threshold energy. Experiments with excited azulene in the bath gas quadricyclane isomerizing to norbornadiene[161] and with excited hexafluorobenzene in the bath gas cyclobutene[162] have demonstrated the presence of small fractions of strong collisions.

[161] S. Hasson, I. Oref, and C. Steel, *J. Chem. Phys.*, 1988, **89**, 1743.
[162] J. M. Margulis, S. S. Spares, C. Steel, and I. Oref, to be published.

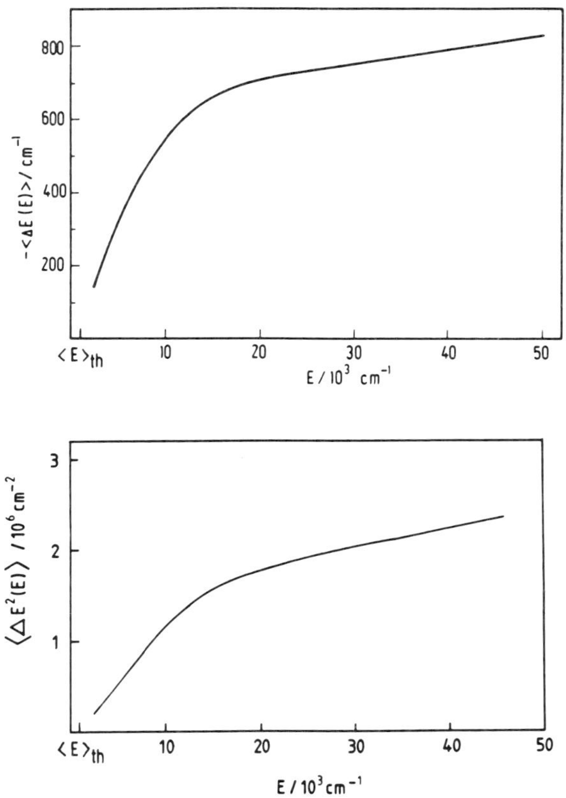

Figure 39 *$\langle \Delta E(E) \rangle$ and $\langle \Delta E^2(E) \rangle$ for collisions of toluene* + cycloheptatriene from the two-photon ionization experiments of ref. 87. $\langle E \rangle_{th}$ = thermal energy at* 300 K

14 Conclusions

Direct investigations of energy transfer phenomena now reveal a fairly complete picture of the collisional de-excitation of vibrationally highly excited molecules. The results are in general agreement with earlier indirect measurements once the basic problem of the calibration of reference rates in the indirect experiments has been solved. Average energies transferred per collision $\langle \Delta E \rangle$ appear to depend strongly on the energy $\langle E \rangle$ of the molecules. Typically, $\langle \Delta E \rangle \propto \langle E \rangle^2$ at low energies where the vibrational densities of states are small. With increasing energy there is a transition to a weaker energy dependence ($\langle \Delta E \rangle \propto \langle E \rangle$). At even higher energies there appears to be a levelling-off to a range with nearly energy-independent $\langle \Delta E \rangle$. The first measurements of $\langle \Delta E^2 \rangle$ and its energy dependence have also been reported. The transition between the various ranges of the energy dependence of

$\langle \Delta E \rangle$ depends on the size of the excited molecules. Triatomic molecules remain completely in the $\langle \Delta E \rangle \propto \langle E \rangle^2$ range up to their dissociation energies. Large polyatomic molecules show a switch to $\langle \Delta E \rangle \propto \langle E \rangle$ at energies of typically 2000 cm^{-1} and show a levelling-off to energy-independent $\langle \Delta E \rangle$ near 15 000 cm^{-1}. Molecules of intermediate size show a behaviour in between these limiting cases.

The strong energy dependence of $\langle \Delta E \rangle$ at low excitation energies can be attributed to the sparsity of the vibrational manifold. Classical trajectory calculations with a continuous vibrational energy scale reproduce the $\langle \Delta E \rangle \propto \langle E \rangle$ behaviour. The high-energy levelling-off of $\langle \Delta E \rangle$ has not yet been demonstrated theoretically, but some preliminary evidence for this behaviour has been found in trajectory calculations[163] for excited azulene.

Much still has to be done in this field both experimentally and theoretically. In particular, single state-to-state collision experiments with highly excited polyatomic molecules have yet to be performed. However, the data available from current experiments involving sequences of collisions allow one to implement the necessary energy transfer information into chemical reaction models in an increasingly meaningful manner.

Acknowledgement. Financial support of this work by the Deutsche Forschungsgemeinschaft (Sonderforschungsbereich 93 'Photochemie mit Lasern') as well as discussions and preprints from many colleagues are gratefully acknowledged.

[163] K. F. Lim and R. G. Gilbert, private communication.

CHAPTER 6

Bimolecular Collisions Involving Electronically Excited Alkaline Earth Atoms

D. HUSAIN and G. ROBERTS

1 Introduction

Collisional processes involving alkaline earth atoms in electronically excited states have constituted a major theme in kinetic investigations of atomic reactions during the past fifteen or more years, and continue to be studied in increasing detail as experimental techniques for their investigation improve. The importance of tunable laser sources as a tool for elucidating the details of molecular photodissociation dynamics has been stressed in the first volume of 'Advances in Gas-Phase Photochemistry and Kinetics'.[1] Similarly, their impact on the study of bimolecular processes involving excited species cannot be underestimated, and reference can be made to a variety of general review articles and monographs that have appeared in recent years outlining some of their uses in this research area.[2–10] For the study of elementary reactions involving alkaline earth atoms, lasers have been married to such

[1] 'Molecular Photodissociation Dynamics', Advances in Gas-Phase Photochemistry and Kinetics, Vol. 1, eds. M. N. R. Ashfold and J. E. Baggott, The Royal Society of Chemistry, London, 1987.
[2] S. R. Leone, *Science*, 1985, **227**, 889.
[3] F. Kaufman, *Science*, 1985, **230**, 393.
[4] F. Kaufman, *J. Phys. Chem.*, 1984, **88**, 4909.
[5] 'Laser Applications in Chemistry', N.A.T.O. ASI Series, eds. K. L. Kompa and J. Wanner, Plenum Press, New York, 1984, Series B, Vol. 103.
[6] I. W. M. Smith, *Ber. Bunsenges. Phys. Chem.*, 1982, **86**, 395.
[7] A. Ben-Shaul, Y. Haas, K. L. Kompa, and R. D. Levine, 'Lasers and Chemical Change', Springer Series in Chemical Physics, Vol. 10, Springer-Verlag, Berlin, 1981.
[8] R. N. Zare and R. B. Bernstein, *Physics Today*, 1980, **33**, 43.
[9] Y. T. Lee and Y. R. Shen, *Physics Today*, 1980, **33**, 52.
[10] R. W. Solarz, E. F. Warden, and J. A. Paisner, *Opt. Eng.*, 1980, **19**, 85.

well-known techniques as molecular beams, low-pressure flames, and flash photolysis, both as a means of preparing desired metal–atom reactants in different electronic states and as a state-resolved probe for detection of the nascent molecular products of such processes in specific electronic, vibrational, and rotational quantum levels.

That reactions involving electronically excited reactants and/or products involve more than one potential energy hypersurface has long been recognized.[11–25] However, precise and accurate theoretical treatments of these systems are difficult to obtain, primarily because their description is inherently complicated by the need to treat several potential energy hypersurfaces simultaneously while at the same time taking into account such effects as changes in spin multiplicity and non-Born–Oppenheimer perturbations. While adiabatic atomic recombination can be analysed theoretically with almost ultimate rigour,[21,23,26] in general only semi-quantitative arguments can be advanced for the interpretation of atom–molecule exchange reactions in which formation of product states involves an energy-dependent probability of transition between different potential energy hypersurfaces.[16,17,21,23,26] For simple triatomic cases, however, dynamical problems governed by multiple potential energy surfaces may be amenable to (nearly) rigorous computational treatment by trajectory surface-hopping, path integral methods and other semi-classical techniques.[27–30] A full quantum

[11] R. D. Levine and R. B. Bernstein, 'Molecular Reaction Dynamics and Chemical Reactivity', Oxford University Press, New York, 1987.

[12] R. B. Bernstein, 'Chemical Dynamics via Molecular Beam and Laser Techniques', The Hinshelwood Lectures, University of Oxford, 1980, Oxford University Press, New York, 1982.

[13] I. V. Hertel, *Adv. Chem. Phys.*, 1982, **50**, 475.

[14] D. M. Hirst, *Adv. Chem. Phys.*, 1982, **50**, 517.

[15] I. V. Hertel, *Adv. Chem. Phys.*, 1981, **45**, 341.

[16] M. Menzinger, *Adv. Chem. Phys.*, 1980, **42**, 1.

[17] C. Bottcher, *Adv. Chem. Phys.*, 1980, **42**, 169.

[18] I. W. M. Smith, 'Kinetics and Dynamics of Elementary Gas Reactions', Butterworths, London, 1980.

[19] R. J. Donovan, *Prog. React. Kinet.*, 1979, **10**, 252.

[20] D. L. King and D. W. Setser, *Annu. Rev. Phys. Chem.*, 1976, **27**, 407.

[21] M. F. Golde and B. A. Thrush, *Adv. Atom. Mol. Phys.*, 1975, **11**, 361.

[22] E. E. Nikitin, *Adv. Chem. Phys.*, 1975, **28**, 317.

[23] E. E. Nikitin, 'Theory of Elementary Atomic and Molecular Processes in Gases', Clarendon Press, Oxford, 1974, p. 99.

[24] R. D. Levine and R. B. Bernstein, 'Molecular Reaction Dynamics', Oxford University Press, New York, 1974.

[25] B. A. Thrush, *Annu. Rev. Phys. Chem.*, 1973, **19**, 371.

[26] T. Carrington, in 'Chemiluminescence and Bioluminescence', eds. M. J. Cormier, D. M. Hercules, and J. Lee, Plenum Press, New York, 1973, p. 1.

[27] J. C. Tully, in 'Dynamics of Molecular Collisions', Part B, ed. W. H. Miller, Plenum Press, New York, 1976, p. 217.

[28] J. N. Murrell, in 'Gas Kinetics and Energy Transfer', eds. P. G. Ashmore and R. J. Donovan (Specialist Periodical Reports), The Chemical Society, London, 1978, Vol. 3, p. 82.

[29] M. S. Child, in 'Atom–Molecule Collision Theory: A Guide for the Experimentalist', ed. R. B. Bernstein, Plenum Press, New York, 1979, p. 427.

[30] D. G. Truhlar and J. T. Muckerman, in ref. 29, p. 505.

mechanical treatment of electronically non-adiabatic processes therefore requires calculations in multi-dimensional space[17,27,30–33] together with an accurate knowledge of the topologies of the appropriate potential energy hypersurfaces in the region where they intersect.[34] For processes during which the total electron spin changes on going from reactants to products, the probability of the system undergoing a non-adiabatic transition is fundamentally linked to the magnitude of the appropriate spin–orbit operator, which is itself related to the mass of the atoms in the system, thus rendering its calculation a non-trivial task in all but the simplest of cases.[23,27,29] Thus, although *ab initio* calculations of excited-state potential energy hypersurfaces and their mutual couplings are becoming increasingly available for simple collisional processes involving electronically excited species, the conceptual framework of quantum dynamical scattering calculations on more than one hypersurface is still sufficiently complex at present[31,35,36] to be of limited practical use for investigating many of the reactive systems considered in this chapter.

A simpler and more convenient starting point for obtaining a rudimentary understanding of the dynamics of simple reactive atom–molecule collisions involving electronically excited species in the gas phase is afforded by a consideration of adiabatic potential energy surfaces correlating specific electronic states of reactants and products.[18,37–46] Donovan and Husain were the first to provide a systematic interpretation of the differing reactivities of a range of electronically excited atoms with diatomic and triatomic collisional partners in terms of adiabatically accessible reaction channels,[43–46] and several examples will be presented in this chapter. Although the use of state correlations as a reliable predictive tool has certain, well-known, shortcomings,[18] and can never be quantitatively successful in the absence of detailed knowledge of the nature of the potential energy hypersurfaces involved, they do provide a useful qualitative guide to the

31 J. C. Whitehead, in 'Comprehensive Chemical Kinetics', eds. C. H. Bamford and C. F. H. Tipper, Elsevier, Amsterdam, 1983, Vol. 24, p. 357.

32 R. B. Walker and J. C. Light, *Annu. Rev. Phys. Chem.*, 1980, **31**, 401.

33 M. S. Child, 'Molecular Collision Theory', Academic Press, London, 1974, p. 224.

34 T. Carrington, *Acc. Chem. Res.*, 1974, **7**, 20.

35 G. C. Schatz, in 'Potential Energy Surfaces and Reaction Dynamics Calculations', ed. D. G. Truhlar, Plenum Press, New York, 1981, p. 287.

36 'The Theory of Chemical Reaction Dynamics', N.A.T.O. ASI Series, ed. D. C. Clary, D. Reidel, Dordrecht, 1986, Series C. Vol. 170.

37 B. H. Mahan, *Acc. Chem. Res.*, 1975, **8**, 55.

38 R. S. Mulliken, *Rev. Mod. Phys.*, 1932, **4**, 1.

39 G. Herzberg, 'Molecular Spectra and Molecular Structure I: Spectra of Diatomic Molecules', 2nd Ed., Van Nostrand Reinhold, New York, 1950, p. 315.

40 K. E. Shuler, *J. Chem. Phys.*, 1953, **21**, 624.

41 K. J. Laidler and K. E. Shuler, *Chem. Rev.*, 1951, **48**, 153.

42 E. K. Gill and K. J. Laidler, *Can. J. Chem.*, 1958, **36**, 79.

43 R. J. Donovan and D. Husain, *Chem. Rev.*, 1970, **70**, 489.

44 R. J. Donovan and D. Husain, *Annu. Rep. Chem. Soc. A.*, 1972, **68**, 123.

45 R. J. Donovan, D. Husain, and L. J. Kirsch, *Annu. Rep. Chem. Soc. A.*, 1972, **69**, 19.

46 D. Husain, *Ber. Bunsenges. Phys. Chem.*, 1977, **81**, 168.

identity of product electronic states as well as an appreciation of the degree to which non-Born–Oppenheimer perturbations influence the outcome of reactive events. Ultimately, of course, the validity of adiabatic correlation rules is governed by the same approximations as the angular momentum coupling scheme upon which they are based.[46]

An alternative description of observed product-state distributions arising from reactions proceeding over more than one potential energy hypersurface is afforded by statistical rate theories, in which the evolution of the reacting system from initial to final states via an energetically accessible pathway is assumed to proceed with little or no dynamical bias. In such cases, the partitioning of atomic electronic energy may then be accounted for by statistical methods such as phase-space theory[47] or the information–theoretic approach.[7,11,12,24,48,49] Dynamical criteria can be included in statistical models by constraining product-state branching ratios to be consistent with certain key dynamical factors such as, for example, conservation of spin and/or electronic adiabaticity.[50] Indeed, the importance of non-adiabatic crossings between potential energy hypersurfaces and changes in electronic multiplicity that occur in many reactions of electronically excited alkaline earth atoms is such that models without any dynamical features may fail to give even an adequate qualitative interpretation of the observed product-state selectivity.

The collisional quenching of electronically excited alkaline earth atoms has been the subject of two review articles published by Breckenridge within the past six years.[51,52] A recent monograph on gas-phase chemiluminescence and chemi-ionization edited by Fontijn contains chapters on the reactions of ground-state and electronically excited alkaline earth atoms with halogen molecules.[53–55] The celebrated $Ba + N_2O$ reaction has been reviewed by several authors (see, for example, those by Menzinger,[16] Whitehead,[31] and Field[56]), while valuable compilations of the large number of articles devoted to this reaction may be found in papers by Cox and Dagdigian[57] and Reuther and Palmer.[58] The early review of Zare and Dagdigian[59] provides useful

[47] P. Pechukas, J. C. Light, and C. Rankin, *J. Chem. Phys.*, 1966, **44**, 794.
[48] R. B. Bernstein and R. D. Levine, *Adv. Atom. Mol. Phys.*, 1975, **11**, 216.
[49] R. D. Levine and A. Ben-Shaul, in 'Chemical and Biochemical Applications of Lasers', ed. C. B. Moore, Academic Press, London, 1977, Vol. 2, p. 145.
[50] M. H. Alexander and P. J. Dagdigian, *Chem. Phys.*, 1978, **33**, 13.
[51] W. H. Breckenridge and H. Umemoto, *Adv. Chem. Phys.*, 1982, **50**, 325.
[52] W. H. Breckenridge, in 'Reactions of Small Transient Species: Kinetics and Energetics', eds. A. Fontijn and M. A. A. Clyne, Academic Press, London, 1983, p. 157.
[53] M. Menzinger, in 'Gas-Phase Chemiluminescence and Chemi-ionization', ed. A. Fontijn, North-Holland, Amsterdam, 1985, p. 25.
[54] P. J. Dagdigian, in ref. 53, p. 203.
[55] M. H. Alexander, in ref. 53, p. 221.
[56] R. W. Field, in 'Molecular Spectroscopy: Modern Research', ed. K. N. Rao, Academic Press, New York, 1976, Vol. II, p. 261.
[57] J. W. Cox and P. J. Dagdigian, *J. Chem. Phys.*, 1983, **79**, 5351.
[58] J. J. Reuther and H. B. Palmer, *J. Chem. Phys.*, 1982, **77**, 83.
[59] R. N. Zare and P. J. Dagdigian, *Science*, 1974, **185**, 739.

information on laser-induced fluorescence detection of product molecules resulting from reactions of alkaline earth atoms with oxidants under molecular beam conditions.

The earlier of the two articles by Breckenridge[51] describes in some detail the identity and relative importance of major exit channels for the dissipation of electronic energy from excited metal atoms belonging to Groups IA, IIA, and IIB on collision with various molecular quenching agents, supported by the presentation of a large body of relevant rate data. The later review, though similar in choice and description of conceptual material, is extended to cover the efficiency of deactivation of excited states of Group VI atoms.[52] Both articles provide a critical evaluation of recent state-resolved measurements of internal energy distributions within a given exit channel, with particular emphasis given to the work of Breckenridge's own research group on the rotational population distributions of product molecules formed by reaction of electronically excited Group IIA and IIB metal atoms with hydrogen-atom-containing species. It is not our intention to provide another re-evaluation of the results of these investigations here. Rather, we shall describe in some detail the collisional removal of metastable excited states of atomic magnesium, calcium, and strontium in the presence of two classes of quenching partners.

The first of these classes consists of the isotopic molecules hydrogen and deuterium. Kinetic investigations of the collisional behaviour of electronically excited metallic states with these species are of fundamental concern to the study of the collision dynamics of excited atoms in general, and to the competing roles of chemical reaction and physical energy transfer in particular, since such investigations provide a basis for further detailed consideration of the primary collision processes involved in terms of appropriate electronic states of the metal dihydrides and dideuterides MH_2 and MD_2. The reviews of Breckenridge contain detailed discussions of experimental and theoretical attempts to elucidate the quenching mechanisms of the short-lived $3s3p(^1P_1)$ and lower-lying metastable $3s3p(^3P_J)$ state of atomic magnesium. In Section 3 we shall instead concentrate on the quenching of metastable states of calcium and strontium by H_2 and D_2, experimental and theoretical studies of which have been carried out since the publication of Breckenridge's articles in 1982 and 1983. Naturally, this will necessitate a comparison with, and hence some degree of repetition of earlier descriptions of, previous investigations of the $Mg(^3P_J) + H_2,D_2$ reactions.

The other class of quenching processes of interest here, to be described in Sections 4, 5, and 6, are the reactions of electronically excited Mg, Ca, and Sr atoms with simple molecular oxidants. These systems have received considerable attention over many years, from both experimental and theoretical points of view. Interest in these processes stems from the fact that there are usually a number of energetically accessible exit channels available, leading to several low-lying electronic states of the diatomic metal monoxide product whose formation and decay can be probed either by spectral analysis of chemiluminescence in the case of radiating states, or by laser fluorescence excitation in the case of 'dark' (*i.e.* non-emitting) states. Much of this work has therefore been undertaken in order to examine the possibilities for creation of a population inversion between electronic levels of the metal monoxide reaction products, leading to potential development of high-power chemical laser systems operating in the visible and ultraviolet regions.[16,60–62] The intense, highly coloured emission resulting from electronic transitions between states of the alkaline earth monoxides has also long been known to have applications in pyrotechnic devices.[63] The advent of experimental techniques capable of indentifying state-resolved products has opened up the possibility of characterization of such processes with a view to probing their collision dynamics and mechanisms in intimate detail. From a fundamental perspective, it is of paramount importance in attempting to obtain even a rudimentary understanding of the collision dynamics of excited metal atom plus oxidant systems to be able to predict branching ratios into different product electronic states which are energetically accessible and to identify the factors influencing the nascent formation of these states populated via processes involving multiple potential energy hypersurfaces. Hence oxidation reactions of electronically excited alkaline earth atoms afford an ideal opportunity to test current theories of molecular interactions which attempt to explain their many and varied characteristics, especially the number and type of product electronic states. Although mentioned by Breckenridge,[51,52] these systems were not discussed at great length, and so are considered in more detail here.

In this chapter, greatest importance is attached to experimental results and their interpretation using the simple theoretical concepts outlined above. The results of relevant *ab initio* calculations will be discussed where appropriate. Although we shall be concerned primarily with the collisional behaviour of metastable electronically excited Group IIA states, attention will also be given to corresponding processes involving the ground-state metal atoms in cases where there are sufficient data to merit useful discussion. We also stress that in this chapter we shall only consider net quenching (reactive and non-reactive) of excited metal-atom states. We shall not therefore discuss the reactivity of individual spin–orbit states of alkaline earth atoms, which has been the subject of a recent review,[64] or non-reactive collisional intramultiplet mixing processes, which have also been the subject of several experimental and

[60] C. R. Jones and H. P. Broida, *Laser Focus*, 1974, 37.

[61] D. J. Eckstrom, J. R. Barker, J. G. Hawley, and J. P. Reilly, *Appl. Opt.*, 1977, **16**, 2102.

[62] D. J. Benard, W. D. Slafer, P. J. Love, and P. H. Lee, *Appl. Opt.*, 1977, **16**, 2108.

[63] R. F. Barrow and E. F. Caldin, *Proc. Phys. Soc. B (London)*, 1949, **62**, 32.

theoretical investigations during the past few years.[65–70]

Before describing the collisional quenching processes mentioned above, we present in Section 2 atomic data that are pertinent to the excited states to be treated in Sections 3—6.

2 Data for Electronically Excited Alkaline Earth Atoms

Table 1 shows a compilation of data relating to some of the physical properties of the lowest electronic states of atomic magnesium, calcium, and strontium. Excitation energies and ionization potentials were taken from Moore.[71] Also listed are the wavelengths connecting a given excited state to the ground state for transitions in air. In addition, overall radiative lifetimes and Einstein coefficients for spontaneous emission via the transition to the ground state are given for these states. Einstein coefficients for spin-forbidden emission from the optically metastable $nsnp(^3P_1)$ levels of Mg, Ca, and Sr to their ground states (Table 1) are those values determined in our laboratory in recent years.[72–74] Mean radiative lifetimes of these excited states are given by the simple relationship $\tau_e = 1/A_{nm}$. Useful summaries of earlier measurements of τ_e for these triplet states have been compiled by Husain and Schifino[75–77] and by Swagel and Lurio.[78] For the higher-lying $ns(n-1)d(^1D_2)$ levels of Ca and Sr, which may decay radiatively to the $nsnp(^3P_{1,2})$ levels by spin–orbit mixing or to the ground state via electric-quadrupole-allowed emission, A_{nm} coefficients for the latter transitions were taken from recent experimental[79] and theoretical[80–82] work. Overall radiative lifetimes of Ca[$4s3d(^1D_2)$] and Sr[$5s4d(^1D_2)$] atoms are taken from time-resolved measurements.[73,83] The only electric-dipole-allowed transition out of the $3s3p(^1P_1)$ state of Mg is

[64] P. J. Dagdigian and M. L. Campbell, *Chem. Rev.*, 1987, **87**, 1.
[65] M. O. Hale and S. R. Leone, *J. Chem. Phys.*, 1983, **79**, 3352.
[66] H. -J. Yuh and P. J. Dagdigian, *Phys. Rev. A.*, 1983, **28**, 63.
[67] M. H. Alexander, T. Orlikowski, and J. E. Straub, *Phys. Rev. A.*, 1983, **28**, 73.
[68] T. Orlikowski and M. H. Alexander, *J. Phys. B*, 1984, **17**, 2269.
[69] B. Pouilly, B. H. Lengsfield, and D. R. Yarkony, *J. Chem. Phys.*, 1984, **80**, 5089.
[70] B. Pouilly, T. Orlikowski, and M. H. Alexander, *J. Phys. B*, 1985, **18**, 1953.
[71] C. E. Moore, 'Atomic Energy Levels', Natl. Bur. Stand. Ref. Data, Ser. 35, US Government Printing Office, Washington, D.C., 1971, Vols. I and II.
[72] D. Husain and G. Roberts, *J. Chem. Soc., Faraday Trans. 2*, 1986, **82**, 21.
[73] D. Husain and G. Roberts, *J. Chem. Soc., Faraday Trans. 2*, 1986, **82**, 1921.
[74] D. Husain and G. Roberts, *J. Chem. Soc., Faraday Trans. 2*, 1985, **81**, 1085.
[75] D. Husain and J. Schifino, *J. Chem. Soc., Faraday Trans. 2*, 1982, **78**, 2083.
[76] D. Husain and J. Schifino, *J. Chem. Soc., Faraday Trans. 2*, 1983, **79**, 1265.
[77] D. Husain and J. Schifino, *J. Chem. Soc., Faraday Trans. 2*, 1984, **80**, 321.
[78] M. W. Swagel and A. Lurio, *Phys. Rev.*, 1968, **169**, 114.
[79] K. Fukuda and K. Ueda, *J. Phys. Chem.*, 1982, **86**, 676.
[80] L. Pasternack, D. M. Silver, D. R. Yarkony, and P. J. Dagdigian, *J. Phys. B*, 1980, **13**, 2231.
[81] R. N. Diffenderfer, P. J. Dagdigian, and D. R. Yarkony, *J. Phys. B*, 1981, **14**, 21.
[82] C. W. Bauschlicher, jun., S. R. Langhoff, and H. Partridge, *J. Phys. B*, 1985, **18**, 1523.
[83] D. Husain and G. Roberts, unpublished results.

Table 1 *Selected properties of the ground and low-lying electronically excited states of atomic magnesium, calcium, and strontium*

Atom	*Level*	*Energy* /eV[a]	*Ionization potential* eV[a]	λ/nm[b]	A_{nm}	τ_e/s
Mg	$3s^2(^1S_0)$	0	7.644			
	$3s3p(^3P_0)$	2.709				
	$3s3p(^3P_1)$	2.712		457.1	435^{+41}_{-35} [c]	$(2.3 \pm 0.2) \times 10^{-3}$ [c]
	$3s3p(^3P_2)$	2.717				
	$3s3p(^1P_1)$	4.346		285.2	5.00×10^8 [d]	2.0×10^{-9} [d]
Ca	$4s^2(^1S_0)$	0	6.111			
	$4s4p(^3P_0)$	1.879				
	$4s4p(^3P_1)$	1.886		657.3	3021^{+75}_{-71} [e]	$(331 \pm 8) \times 10^{-6}$ [e]
	$4s4p(^3P_2)$	1.889				
	$4s3d(^3D_1)$	2.521				
	$4s3d(^3D_2)$	2.523				
	$4s3d(^3D_3)$	2.526				
	$4s3d(^1D_2)$	2.709		457.5	40[f,g] 80—90[h]	$(1.71 \pm 0.03) \times 10^{-3}$ [e]
	$4s4p(^1P_1)$	2.932		422.7	2.13×10^8 [d]	4.7×10^{-9} [d]
Sr	$5s^2(^1S_0)$	0	5.692			
	$5s5p(^3P_0)$	1.775				
	$5s5p(^3P_1)$	1.798		689.3	52400 ± 1100 [i]	$(19.1 \pm 0.4) \times 10^{-6}$ [i]
	$5s5p(^3P_2)$	1.847				
	$5s4d(^3D_1)$	2.251				
	$5s4d(^3D_2)$	2.259				
	$5s4d(^3D_3)$	2.271				
	$5s4d(^1D_2)$	2.498		496.1	45[g]	$(412 \pm 9) \times 10^{-6}$ [j]
	$5s5p(^1P_1)$	2.690		460.7	2.04×10^8 [d,k]	4.9 ± 10^{-9} [d,k]

[a] From ref. 71. (1 eV $\equiv$ 96.485 kJ mol^{-1} $\equiv$ 8065.54 cm^{-1}). [b] λ = wavelength in air of the radiative transition to the ground state. [c] From ref. 72. [d] From ref. 84. [e] From ref. 73. [f] From ref. 79. [g] From ref. 82. [h] From refs. 80 and 81. [i] From ref. 74. [j] From ref. 83. [k] From ref. 85.

that to the ground state. For Ca and Sr atoms, however, an alternative dipole-allowed radiative pathway exists connecting the $nsnp(^1P_1)$ and $ns(n-1)d(^1D_2)$ levels. For the transitions to the ground state, A_{nm} coefficients given in Table 1 have been calculated from the experimental oscillator strengths listed in a recent paper by Hunter and Berry[84] in the case of Mg and Ca, and taken from the work of Hunter *et al.*[85] for Sr. Using a Stark interference technique, this last group of workers[85] have also measured the Einstein coefficient for the $5s5p(^1P_1) \rightarrow 5s4d(^1D_2)$ transition in Sr as $A_{nm}(^1P\text{–}^1D) = 3.85 \times 10^3\ \text{s}^{-1}$. That this value is some five orders of magnitude smaller than $A_{nm}(^1P\text{–}^1S)$ enables us to estimate the observed radiative lifetime of the $5s5p(^1P_1)$ state in Sr as

[84] J. E. Hunter III and R. S. Berry, *Phys. Rev. Lett.*, 1987, **59**, 2959.
[85] L. R. Hunter, W. A. Walker, and D. S. Weiss, *Phys. Rev. Lett.*, 1986, **56**, 823.

$\tau_e^{obs} = 1/A_{nm}(^1P\text{–}^1S)$. Analogous measurements do not appear to have been carried out for the corresponding transition in atomic Ca, but inspection of the theoretical oscillator strengths calculated by Hunter and Berry[84] indicates that, to a good approximation, $\tau_e^{obs} = 1/A_{nm}(^1P\text{–}^1S)$ in this case also.

3 Collisional Quenching of Metastable Alkaline Earth Atoms by H_2 and D_2

It was noted in Section 1 that kinetic investigations of the collisional behaviour of electronically excited metal atoms with the isotopes H_2 and D_2 can provide both a quantitative and qualitative description of the appropriate potential energy hypersurfaces involving different electronic states of the triatomic metal dihydrides and dideuterides, MH_2 and MD_2. Of primary concern here are measurements of the temperature variation of the thermal bimolecular rate constants for such processes, since these allow activation energies for collisional removal to be elucidated. In cases where the metal atom M belongs to the first or second rows of the Periodic Table, the $MH(D)_2$ molecule has a sufficiently small number of electrons to be amenable to accurate theoretical study by *ab initio* MCSCF techniques.

The collisional quenching of electronically excited Group IIA and IIB metal atoms by H_2 and D_2 has been discussed previously by Breckenridge in his two review articles.[51,52] The earlier of these two publications[51] describes in detail the results of investigations carried out by the author and collaborators on the collisional removal of $Mg[3s3p(^3P_J)]$ atoms by H_2 and D_2.[86,87] More recent laser 'pump-and-probe' measurements of the initial rotational distribution of $MgH(X^2\Sigma^+)$ in its ground vibrational state produced by reaction of $Mg[3s3p(^1P_1)]$ atoms with H_2 are outlined in the later review,[52] since which time nascent rotational quantum-state distributions in $MgH(X^2\Sigma^+, v = 1)$ and $MgD(X^2\Sigma^+, v = 0)$ resulting from the $Mg[3s3p(^1P_1)] + H_2,D_2$ reactions have also been studied.[88–90] It is not our intention to offer another description of these studies: rather we shall describe results of quenching experiments involving metastable Ca and Sr atoms that have been completed since these reviews were presented, together with a brief recapitulation of earlier work on the $Mg(^3P_J) + H_2,D_2$ system.

Table 2 shows the energetics of diatomic metal hydride and deuteride formation for quenching of the metastable alkaline earth states listed in Table 1 by H_2 and D_2. Reaction ergicities, ΔE are calculated for

[86] W. H. Breckenridge and W. L. Nikolai, *J. Chem. Phys.*, 1980, **73**, 2763.
[87] W. H. Breckenridge and J. Stewart, *J. Chem. Phys.*, 1982, **77**, 4469.
[88] W. H. Breckenridge and H. Umemoto, *J. Chem. Phys.*, 1984, **80**, 4168.
[89] W. H. Breckenridge and J. -H. Wang, *Chem. Phys. Lett.*, 1987, **137**, 195.
[90] W. H. Breckenridge and J. -H. Wang, *Chem. Phys. Lett.*, 1987, **139**, 28.

Table 2 *Energetics of ground-state metal hydride and metal deuteride formation arising from the quenching of metastable electronically excited alkaline earth states by hydrogen and deuterium*

Atomic state	Ergicity (ΔE) of ground-state product formation[a] ΔE/eV		ΔE/kJ mol^{-1}	
	$M^* + H_2$	$M^* + D_2$	$M^* + H_2$	$M^* + D_2$
$Mg[3s3p(^3P_J)]$	0.43	0.49	41.5	47.3
$Ca[4s4p(^3P_J)]$	≥0.89	≥0.95	≥85.9	≥91.7
$Ca[4s3d(^1D_2)]$	≥0.07	≥0.13	≥6.8	≥12.5
$Sr[5s5p(^3P_J)]$	≥1.01	≥1.05	≥97.5	≥101.3
$Sr[5s4d(^1D_2)]$	≥0.32	≥0.36	≥30.8	≥34.7

[a] Relevant bond dissociation energies are taken from ref. 91; electronic excitation energies of M* states are taken from Table 1

formation of diatomic products in their ground electronic and vibrational states via the process:

$$M^* + H_2/D_2(X^1\,\Sigma_g^+, v = 0) \rightarrow MH/D(X^2\,\Sigma^+, v = 0) + H/D[1s(^2S_{1/2})] \quad (1)$$

where M* is an alkaline earth metal atom in its $nsnp(^3P_J)$ or $ns(n-1)d(^1D_2)$ excited state. Bond dissociation energies D_0^0 of MH and MD, together with those for H_2 and D_2, have been taken from the compilation of Huber and Herzberg.[91]

$Mg[3s3p(^3P_J)]$

Breckenridge and co-workers have monitored the decay of fluorescence from $Mg(^3P_1)$ atoms in the presence of known concentrations of H_2 and D_2 by time-resolved atomic resonance emission.[86,87] These workers have utilized the temperature dependence of the rate constants $k_Q(T)$ obtained for collisional removal of $Mg(^3P_J)$ atoms by H_2 and D_2 to probe chemical reaction and physical energy transfer in overall removal of the excited state.[51,52,86,87] The limited curvature observed in Arrhenius plots for $Mg(^3P_J) + H_2,D_2$ was rationalized by Breckenridge and co-workers using a simple model based on chemical exit-channel control at temperatures greater than ~625 K and E–(V,R) energy transfer at temperatures below this. Thus the temperature dependence of the bimolecular quenching rate constants was expressed in the form:

$$k_Q(T) = A\exp(-E_{act}/RT) + B \quad (2)$$

[91] K. P. Huber and G. Herzberg, 'Molecular Spectra and Molecular Structure IV: Constants of Diatomic Molecules', Van Nostrand Reinhold, New York, 1979.

The temperature-independent term B was ascribed by Breckenridge and co-workers to physical energy transfer and the exponential 'Arrhenius' term to chemical reaction to yield MgH/D products in their ground electronic and vibrational states.[86,87] Values of the Arrhenius parameters A and E_{act} for the $Mg(^3P_J) + H_2,D_2$ reactions are displayed in Table 3. These are consistent with chemical removal of the excited atomic state in which there is no appreciable activation barrier beyond the reaction endoergicities shown in Table 2. Further, production of ground-state $MgH(X^2\Sigma^+, v = 0)$ resulting from reaction of $Mg(^3P_J)$ atoms with H_2 at 800 K has been detected by laser-induced fluorescence (LIF).[86] The low values of $B(= 3.8 \times 10^{-14}$ and 1.5×10^{-14} cm^3 molecule^{-1} s^{-1}) relative to the exponential term in equation (2) for the $Mg(^3P_J) + H_2$ and $Mg(^3P_J) + D_2$ reactions, obtained from a fit of temperature-dependent rate data[87] to the form of this expression, enabled Breckenridge and collaborators to conclude that E–(V,R) energy transfer is a minor exit channel for removal of $Mg(^3P_J)$ atoms by H_2 and D_2 over the temperature range investigated.[86,87]

Table 3 *Optimized values of Arrhenius parameters A and E_{act} for collisional quenching of electronically excited alkaline earth atoms by hydrogen and deuterium*

	$A/10^{-9}$ cm^3 molecule^{-1} s^{-1}	E_{act}/kJ mol^{-1}	*References*
$Mg[3s3p(^3P_J)] + H_2$	1.1	47.3 ± 2.9	86, 87
$Mg[3s3p(^3P_J)] + D_2$	0.86	52.3 ± 5.0	86, 87
$Ca[4s4p(^3P_J)] + H_2$	$7.2^{+5.0}_{-3.0}$	88.3 ± 4.2	106
$Ca[4s4p(^3P_J)] + D_2$	$3.9^{+3.5}_{-1.9}$	92.9 ± 5.0	106
$Ca[4s4p(^3P_J)] + H_2$	$3.7^{+79.3}_{-3.5}$	79.2 ± 24.9	*a*
$Ca[4s4p(^3P_J)] + D_2$	$4.4^{+3.6}_{-2.0}$	87.2 ± 5.0	*a*
$Ca[4s3d(^1D_2)] + H_2$	$0.55^{+0.10}_{-0.09}$	14.4 ± 1.3	108
$Ca[4s3d(^1D_2)] + D_2$	0.39 ± 0.04	22.2 ± 0.8	108
$Sr[5s5p(^3P_J)] + H_2$	$1.4^{+0.3}_{-0.2}$	49.4 ± 1.3	74
$Sr[5s5p(^3P_J)] + D_2$	$0.94^{+0.14}_{-0.12}$	50.6 ± 1.1	74

[a] Values obtained from fluorescence measurements on $Ca(^3P_1)$ following pulsed laser excitation of $Ca(^1D_2)$ atoms; D. Husain and G. Roberts, unpublished results

The experimental results outlined above are in excellent agreement with *ab initio* theoretical calculations of portions of the lowest potential energy hypersurface corresponding to the lowest-lying triplet state of MgH_2 over which $Mg(^3P_J) + H_2$ reactants can be converted adiabatically into $MgH(X^2\Sigma^+) + H(^2S_{1/2})$ products.[92–95] A portion of an adiabatic

[92] N. Adams, W. H. Breckenridge, and J. Simons, *Chem. Phys.*, 1981, **56**, 327.
[93] R. P. Blickensderfer, K. D. Jordan, N. Adams, and W. H. Breckenridge, *J. Phys. Chem.*, 1982, **86**, 1930.
[94] K. D. Jordan, in ref. 53, p. 239.
[95] P. Chaquin, A. Serin, and H. Yu, *J. Phys. Chem.*, 1985, **89**, 2813.

correlation diagram connecting defined electronic states of $Mg + H_2$ with those of $MgH + H$, constructed for arbitrary approach of reactants (C_s geometry) and assuming LS coupling and spin conservation, is presented in Figure 1. While a planar configuration has been assumed for the purposes of labelling the different potential energy surfaces, the same correlations apply for an arbitrary approach configuration. In general, spectroscopic data for the Group IIA deuterides are not as comparably established as those for the hydrides,[91] so the corresponding diagrams for the D_2 reactions are not presented. For C_{2v} geometries [*i.e.* side-on approach of $Mg(^3P_J)$ atoms to H_2] this potential energy hypersurface has $^3B_2$ symmetry (and $^3A'$ symmetry for C_s geometries). The calculations of Chaquin and co-workers,[95] carried out at the SCF level, indicate that C_{2v} approach of $Mg(^3P_J)$ atoms to H_2 provides a pathway via which chemical reaction can occur to yield $MgH(X^2\Sigma^+, v = 0) + H(^2S_{1/2})$ products with no substantial activation barrier ($\leq 7.5\ kJ\,mol^{-1}$) over and above the reaction endoergicity of

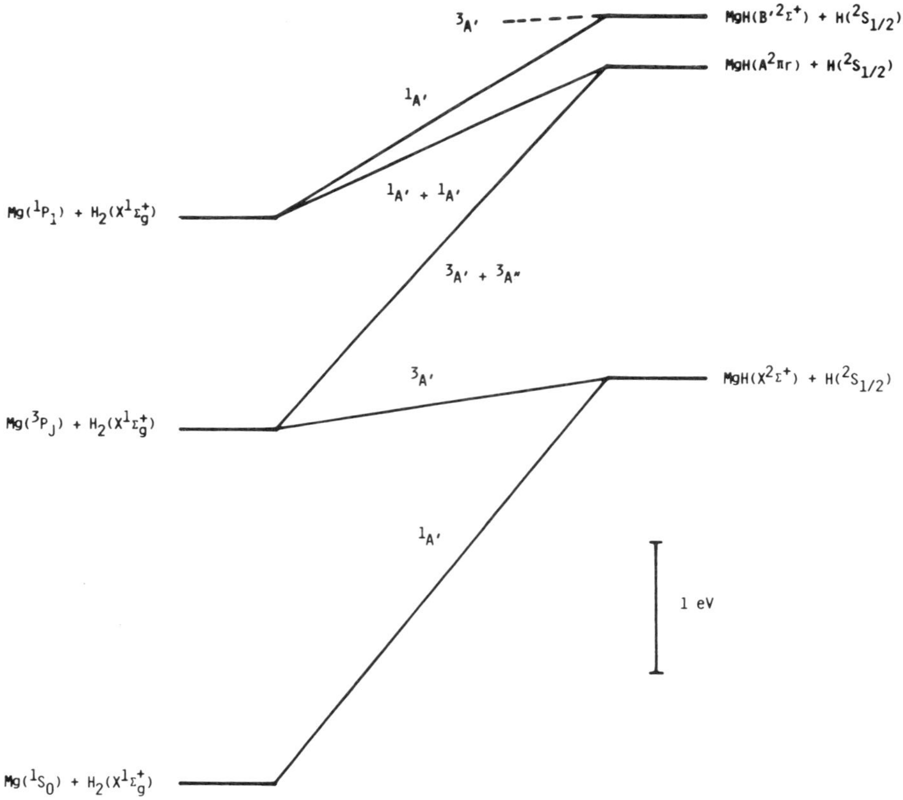

Figure 1 *Correlation diagram connecting the states of* $Mg + H_2$ *and* $MgH + H$ *constructed on the basis of the weak spin–orbit coupling approximation*

41.5 kJ mol^{1}. SCF computations also predict that end-on approach of $Mg(^3P_J)$ atoms to H_2 generates an activation barrier of ~59.8 kJ mol^{-1} beyond the endoergicity.[92,95]

The theoretical studies of Breckenridge's group,[92,93] later confirmed by Chaquin and co-workers,[95] failed to locate a bound or metastable MgH_2 complex on the $^3B_2$ potential hypersurface. Breckenridge has utilized this result to rationalize the low rate of E–(V,R) energy transfer in the $Mg(^3P_J) + H_2$ system.[52] Interactions of an attractive upper-state hypersurface with a repulsive hypersurface of lower energy are known to facilitate energy transfer, for example in $Li[2p(^2P_J)] + H_2$ and $Na[3p(^2P_J)] + H_2$ collisions.[13,96,97] In the $Mg(^3P_J) + H_2$ case, however, even though the ground-state $MgH_2(^1A_1)$ surface is stongly repulsive, so is the excited $MgH_2(^3B_2)$ surface (see figure 8 of ref. 95). The MCSCF–CI calculations of Blickensderfer *et al.*[93] indicate that the crossing seam of the $MgH_2(^3B_2)$ and $MgH_2(^1A_1)$ surfaces occurs very close to the saddle point of the former, where $MgH(X^2\Sigma^+) + H(^2S_{1/2})$ products can form adiabatically with no further potential energy barrier. Because $Mg(^3P_J) + H_2$ reagents do not form a bound intermediate, vibronic resonance to form ground-state products can only occur far along the chemical reaction co-ordinate.[52] Collisional transformation of the initial atomic electronic energy into vibrational and rotational excitation of the molecular product has also been predicted[51,52] to occur with a low cross-section in this system using a charge-transfer model.[98–101] Conservation of spin angular momentum in the E–(V,R) process could also play a restrictive role for $Mg(^3P_J)$ where spin–orbit coupling may be expected to be small. Adiabatic chemical reaction is therefore the dominant exit channel even though it is endoergic. Earlier measurements of the rate constants $k_{H_2}(T)$ and $k_{D_2}(T)$, carried out at single temperatures, are not susceptible to sensible analysis in Arrhenius, or other, form.[102,103]

Kleiber *et al.*[104,105] have studied the reaction dynamics of $Mg[3s^2(^1S_0)] + H_2$ collisions using a far-wing laser scattering technique. Although substantially endoergic, the MgH_2 collision system can be pumped in the presence of a laser radiation field into an excited

[96] J. Elward-Berry and M. J. Berry, *J. Chem. Phys.*, 1980, **72**, 4510.
[97] P. Botschwina, W. Meyer, I. V. Hertel, and W. Reiland, *J. Chem. Phys.*, 1981, **75**, 5438.
[98] E. Bauer, E. R. Fisher, and F. R. Gilmore, *J. Chem. Phys.*, 1969, **51**, 4173.
[99] E. R. Fisher and G. K. Smith, *Chem. Phys. Lett.*, 1970, **6**, 438.
[100] E. R. Fisher and G. K. Smith, *Chem. Phys. Lett.*, 1972, **133**, 448.
[101] J. R. Barker, *Chem. Phys.*, 1976, **18**, 175.
[102] R. P. Blickensderfer, W. H. Breckenridge, and D. S. Moore, *J. Chem. Phys.*, 1975, **63**, 3681.
[103] G. Taieb and H. P. Broida, *J. Chem. Phys.*, 1976, **65**, 2914.
[104] P. D. Kleiber, A. M. Lyyra, K. M. Sando, S. P. Heneghan, and W. C. Stwalley, *Phys. Rev. Lett.*, 1985, **54**, 2003.
[105] P. D. Kleiber, A. M. Lyyra, K. M. Sando, V. Zafiropulos, and W. C. Stwalley, *J. Chem. Phys.*, 1986, **85**, 5493.

potential energy hypersurface of the transient MgH_2 complex over which chemical reaction can occur. These workers have also developed a quasi-static theoretical model to explain semi-quantitatively the overall shape of their absorption profiles and wing intensities for both non-reactive formation of $Mg[3s3p(^1P_1)]$ atoms and reactive formation of two distinct rotational states of the diatomic product $MgH(X^2\Sigma^+, v = 0, J = 6, 23)$ following laser pumping of MgH_2.[104,105]

$Ca[4s4p(^3P_J)]$

Husain and Roberts have studied the collisional quenching of $Ca[4s4p(^3P_J)]$ atoms by H_2 and D_2 as a function of temperature across the range 850—1075 K using the technique of time-resolved atomic resonance emission.[106] Analysis of the rate data obtained by these workers in simple Arrhenius form according to the equation

$$k_Q(T) = A\exp(-E_{act}/RT) \tag{3}$$

yields values of the pre-exponential factor and activation energy shown in Table 3. An attempt[106] to fit the rate data to the form of equation (2), following the earlier work of Breckenridge and co-workers on $Mg(^3P_J) + H_2D_2$,[86,87] yielded values of A, E_{act}, and B from which the ratios $A\exp(-E_{act}/RT)/B = 2310$ and 29500 for $Ca(^3P_J) + H_2$ at $T = 850$ and 1075 K, respectively, and $A\exp(-E_{act}/RT)/B = 533$ and 9600 for $Ca(^3P_J) + D_2$ at $T = 850$ and 1075 K may be calculated. The Arrhenius expression is seen to provide a good description of the rate data over the entire temperature range.

Appropriate reaction thermochemistries for production of diatomic calcium hydride and deuteride in their electronic and vibrational ground states are displayed in Table 2. These are clearly in good agreement with the measured activation energies.[106] Husain and Roberts concluded that there is no basis for invoking any specific role for vibrational energy either as a means of promoting the $Ca(^3P_J) + H_2,D_2$ reactions or as product excitation energy.[106] Thus the kinetic data reported by these workers are fully consistent with chemical removal of $Ca(^3P_J)$ atoms by H_2 and D_2 to yield $CaH(X^2\Sigma^+, v = 0)$ and $CaD(X^2\Sigma^+, v = 0)$ products at every appropriately energized collision. Any contribution to deactivation of the excited atomic state by physical energy transfer appears to be negligible across the temperature range studied.[106] Figure 2 shows an adiabatic correlation diagram for the $Ca + H_2 \rightarrow CaH + H$ system. It can be seen that trajectories originating from $Ca(^3P_J) + H_2$ starting materials can proceed adiabatically via an endoergic reaction channel to ground-state $CaH(X^2\Sigma^+) + H(^2S_{1/2})$ products over a potential energy surface of $^3A'$ symmetry.

[106] D. Husain and G. Roberts, *J. Chem. Soc., Faraday Trans. 2*, 1985, **81**, 527.

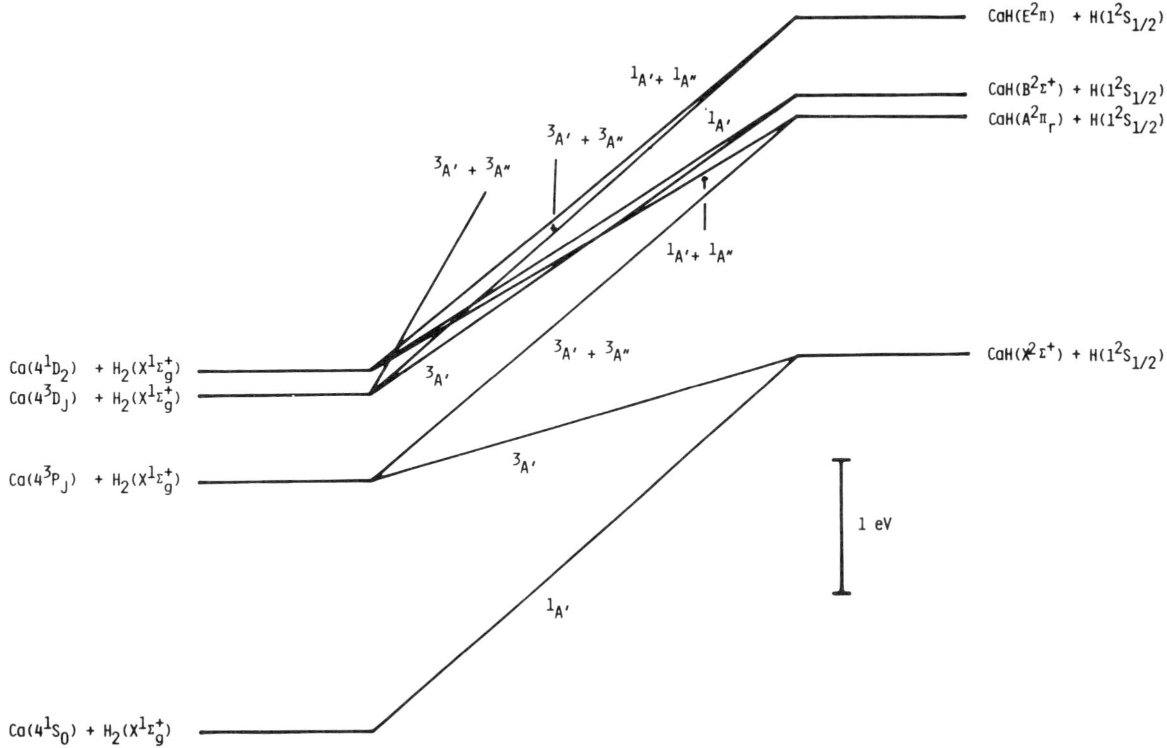

Figure 2 *Correlation diagram constructed on the basis of the weak spin–orbit coupling approximation connecting low-lying states of* Ca + H_2 *with those of* CaH + H (*cf. ref.* 108, *Fig.* 5)

Finally, we may note that the observed large pre-exponential factors for the $Ca(^3P_J) + H_2,D_2$ reactions are in reasonable agreement with gas-kinetic collision numbers, calculated from simple collision theory,[106] of 1.9×10^{-9} and 1.3×10^{-9} cm^3 molecule^{-1} s^{-1} for $Ca(^3P_J) + H_2,D_2$ collisions at $T = 1075$ K and 1.7×10^{-9} and 1.2×10^{-9} cm^3 molecule^{-1} s^{-1} at $T = 850$ K.

Particularly relevant to the experimental work described above is the theoretical investigation of the $Ca(^3P_J) + H_2$ insertion reaction reported by Simandiras and Handy.[107] *Ab initio* calculations at both SCF and CEPA levels indicate that for C_{2v} approach of reactants, groundstate products $CaH(X^2\Sigma^+) + H(^2S_{1/2})$ may be formed adiabatically over the lowest-lying triplet potential energy hypersurface (of $^3B_2$ symmetry for a C_{2v} collision geometry) with an activation barrier for chemical reaction less than 5 kJ mol^{-1} greater than the endoergicity of 85.9 kJ mol^{-1}. At both SCF and CEPA levels of theory, the $^3B_2$ potential energy hypersurface is totally repulsive and no local minima corresponding to bound $CaH_2(^3B_2)$ complexes can be located. A complete profile for the $Ca(^3P_J) + H_2$ reaction calculated at the SCF level indicates that rapid dissociation from $CaH_2(^3B_2)$ to form ground-state products along a C_s pathway proceeds without any additional energy barrier following the initial C_{2v} approach of $Ca(^3P_J)$ atoms to H_2 over the repulsive $^3B_2$ surface. End-on attack of H_2 by $Ca(^3P_J)$ atoms in $C_{\infty v}$ symmetry, leading to abstraction of a hydrogen atom, generates an additional energy barrier of ~40 kJ mol^{-1} in excess of the reaction endoergicity.[107] Of course, a bulk kinetic experiment such as that carried out by Husain and Roberts[106] cannot select between C_{2v} and $C_{\infty v}$ approach geometries for $Ca(^3P_J) + H_2$ reagents, amd consequenctly only a thermally averaged activation energy is measured, representing a distribution of possible collision configurations.

$Ca[4s3d(^1D_2)]$

Collisional removal of the higher-lying $4s3d(^1D_2)$ metastable state of atomic calcium by H_2 and D_2 has also been investigated[108] in the time-resolved mode using the same technique as that for study of the $Ca(^3P_J) + H_2,D_2$ reactions described above. For quenching of $Ca[4s3d(^1D_2)]$ atoms, however, observed rates of removal are some three orders of magnitude greater than those for the corresponding $Ca(^3P_J)$ processes.[106,108] Arrhenius parameters A and E_{act} for the $Ca(^1D_2) + H_2,D_2$ reactions are shown in Table 3. Comparison of the measured energy barriers with the reaction thermochemistries for these processes listed in Table 2 indicates the presence of a small activation energy over and above the endoergicity of production of ground-state $CaH(X^2\Sigma^+, v = 0)$ and $CaD(X^2\Sigma^+, v = 0)$.

[107] E. D. Simandiras and N. C. Handy, *J. Chem. Soc., Faraday Trans. 2*, 1986, **82**, 269.
[108] D. Husain and G. Roberts, *J. Chem. Soc., Faraday Trans. 2*, 1986, **82**, 1935.

The Arrhenius pre-exponential factors for the $Ca(^1D_2) + H_2,D_2$ reactions, shown in Table 3, provide some, but less detailed, quantitative support for chemical removal of the metastable state. In the absence of any information concerning the size of $Ca(^1D_2)$ atoms, it can only be presumed that given the greater degree of electronic excitation of this state, gas-kinetic collision cross-sections for the quenching of $Ca(^1D_2)$ atoms by H_2 and D_2 will be marginally higher than the values reported above for collisions between the lower-lying ^3P_J state with these isotopes.[108] The lower pre-exponential factors displayed in Table 3 for $Ca(^1D_2) + H_2,D_2$ quenching with respect to the estimated collision numbers of 1.9×10^{-9} and 1.3×10^{-9} cm^3 molecule^{-1} s^{-1} at $T = 1075$ K for the $Ca(^3P_J) + H_2,D_2$ reactions have been considered[108] to reflect the role of non-adiabatic transitions between potential hypersurfaces along the reaction co-ordinate leading to products. Inspection of the adiabatic correlation diagram presented as Figure 2 indicates that $Ca(^1D_2) + H_2$ reactants do not correlate adiabatically with ground-state $CaH(X^2\Sigma^+) + H(^2S_{1/2})$ products. Furthermore, there will presumably be some degree of breakdown in the weak spin–orbit coupling approximation employed to construct this diagram, with an approach to (J,Ω) coupling for a molecule such as CaH. Nevertheless, the observed Arrhenius pre-exponential factors for the $Ca(^1D_2) + H_2,D_2$ reactions, as well as their measured activation energies, lend support to the hypothesis of removal of the excited atomic state by chemical reaction with both quenching partners.

The measurements of Husain and Roberts[108] do not permit isolation of any role, however minor, due to physical deactivation of $Ca(^1D_2)$ atoms by H_2 and D_2 to the $^1S_0$ ground state, though a tentative estimate of the branching ratio for collision-induced (radiationless) intersystem crossing to yield the lower-lying $Ca[4s4p(^3P_J)]$ levels may be extracted from measurements of the intensity of fluorescence emitted from the $^3P_1$ state under different experimental conditions. While $Ca(^3P_J)$ atoms could be detected in emission via the spin-forbidden transition to the ground state at $\lambda = 657.3$ nm following pulsed laser generation of $Ca(^1D_2)$ atoms, observations of atomic fluorescence at this wavelength were confined to measurements in the absence and presence of low concentrations of H_2 and D_2 quenching gases. First-order decay coefficients for loss of $Ca(^3P_J)$ atoms were obtained assuming that this species is generated by radiative depopulation of the higher-lying $^1D_2$ state created directly by laser pumping,[80–82,109–111] or by collisional quenching via process (4) (see below).

Second-order rate constants for collisional removal of $Ca(^3P_J)$ atoms by H_2 and D_2 following their generation by radiative emission from, or

[109] D. Husain and G. Roberts, *J. Chem. Soc., Faraday Trans. 2*, 1986, **82**, 1921.
[110] D. Husain and G. Roberts, *J. Chem. Soc., Faraday Trans. 2*, 1988, **84**, 1213.
[111] D. Husain and G. Roberts, *J. Phys. Chem.*, 1988, **92**, 2079.

collisional quenching of, the $Ca(^1D_2)$ state were determined as a function of temperature, leading to the Arrhenius parameters listed in Table 3. These values may be compared, not too unfavourably, with the Arrhenius parameters for the $Ca(^3P_J) + H_2,D_2$ reactions derived from rate data obtained following direct optical generation of $Ca(^3P_J)$ atoms as described earlier in this section.

The branching ratio for collisional production of $Ca(^3P_J)$ atoms according to the spin–forbidden quenching process

$$Ca[4s3d(^1D_2)] + Q \rightarrow Ca[4s4p(^3P_J)] + Q^{(\ddagger)} \qquad (4)$$

was determined for those experimental conditions under which $Ca(^3P_J)$ atoms could be readily monitored. In this case, Q is H_2 or D_2, and the superscript ‡ in parentheses indicates possible excitation of vibrational degrees of freedom in the quenching molecule on collision with a $Ca(^1D_2)$ atom. The fraction of $Ca(^1D_2) + Q$ collisions resulting in formation of $Ca(^3P_J)$ atoms was calculated using a procedure, *mutatis mutandis*, analogous to that described by Breckenridge and co-workers for production of the lower-lying $5s5p(^3P_J)$ state of atomic cadmium arising from collisions between the higher-lying $Cd[5s5p(^1P_1)]$ state and quenching molecules Q.[112–114] Branching ratios so obtained formally represent a lower limit to the formation of $Ca(^3P_J)$ atoms by collisional quenching of the $Ca(^1D_2)$ state since it is assumed that there will be little or no concomitant net quenching of the $Ca(^3P_J)$ state by gases Q. For those temperatures and concentrations of H_2 and D_2 at which $Ca(^3P_J)$ atoms could be detected, branching ratios for production of this state via process (4) are less than ~0.4, indicating that chemically reactive pathways yielding $CaH(X^2\Sigma^+, v = 0)$ and $CaD(X^2\Sigma^+, v = 0)$ probably constitute the predominant routes for removal of $Ca(^1D_2)$ atoms in collisions with H_2 and D_2. Confirmation of this conclusion of course awaits a direct determination of the branching ratio for production of $CaH(D)(X^2\Sigma^+, v = 0)$. Such a determination should be possible by measurement of the ratio of total LIF intensities for selected transitions following laser pumping of the $Ca(^3P_J)$ and $CaH(D)(X^2\Sigma^+, v = 0)$ states at a given pressure of H_2 and D_2 coupled with known transition probabilities for the radiative processes of interest.

$Sr[5s5p(^3P_J)]$

The collisional quenching of $Sr[5s5p(^3P_J)]$ atoms by H_2 and D_2 has also been studied in the time-resolved mode across the temperature range

[112] W. H. Breckenridge, R. J. Donovan, and O. K. Malmin, *Chem. Phys. Lett.*, 1979, **62**, 608.
[113] W. H. Breckenridge and O. K. Malmin, *J. Chem. Phys.*, 1981, **74**, 3307.
[114] W. H. Breckenridge, W. L. Nikolai, and D. Oba, *J. Phys. Chem.*, 1986, **90**, 5724.

725—1100 K, yielding the Arrhenius parameters shown in Table 3. Attempts to fit the rate data to the functional form of equation (2) result in values of the temperature-independent term B that are negligibly small or even negative.[74]

The observed Arrhenius pre-exponential factors are in sensible accord with estimates based on simple collision theory.[74] These yield gas-kinetic pre-exponential factors for $Sr(^3P_J) + H_2$ and $Sr(^3P_J) + D_2$ collisions of 2.2×10^{-9} and 1.6×10^{-9} cm^3 molecule^{-1} s^{-1} at $T = 725$ K and 2.6×10^{-9} and 1.8×10^{-9} cm^3 molecule^{-1} s^{-1} at $t = 1100$ K. Considerably more important are the experimental activation energies, which may be compared with the spectroscopic reaction endoergicities listed in Table 2 for production of SrH and SrD molecules in their ground electronic and vibrational states. It can be seen that in these two cases, the magnitude of the observed activation energies are approximately half the endoergicities for chemical reaction to yield ground-state products. Husain and Roberts[74] have therefore postulated that collisional deactivation of excited $Sr(^3P_J)$ atoms by H_2 and D_2 occurs by a physical energy transfer mechanism to give ground-state $Sr[5s^2(^1S_0)]$ atoms and H_2 and D_2 in one or more vibrationally excited states:

$$Sr[5s5p(^3P_J)] + H_2(D_2)(X^1\,\Sigma_g^+, v = 0) \rightarrow Sr[5s^2(^1S_0)] + H_2(D_2)(X^1\,\Sigma_g^+, v \geq 1) \quad (5)$$

Multi-vibrational quantum jumps in both H_2 and D_2 must be involved in the electronic-to-vibrational energy transfer process (5) in order to take up the 174.3 kJ mol^{-1} of electronic excitation energy[71] of $Sr(^3P_J)$. For example, the energy difference $\Delta E_{1,0}$ between the ground and first excited vibrational levels of H_2 and D_2 may be calculated as 49.7 and 35.8 kJ mol^1, respectively.[91] We may note that the observed activation barrier for collisional removal of $Sr(^3P_J)$ atoms by H_2 is identical within experimental error to the energy required to populate the lowest-lying vibrationally excited state of the quenching molecule, but that this is clearly not the case for collisions between $Sr(^3P_J)$ atoms and D_2, where E_{act} is considerably larger than $\Delta E_{1,0}$.

An adiabatic correlation diagram connecting low-lying reactant and product states in the $Sr + H_2 \rightarrow SrH + H$ reaction is presented in Figure 3. For an atom as heavy as strontium, however, it may reasonably be expected that the weak spin–orbit coupling approximation upon which Figure 3 is based is not strictly valid and that Hund's coupling case (c) will describe the interaction of atom and molecule. Correlations between reactant and product states should therefore be considered in the context of the (J,Ω) coupling scheme[46] using the method of extended groups.[115] For the purposes of this discussion it is sufficient to point out that violations of spin and spatial symmetry rules may occur to an appreciable extent for collisions between $Sr(^3P_J)$ atoms and H_2 and D_2, thus permitting efficient non-adiabatic transitions between potential energy hypersurfaces.

While the experimental measurements[74] reported above do not permit the specific roles played by particular vibrational states of H_2 and D_2 to be elucidated, including the distribution of product vibrational energy as a function of temperature, the broad conclusions are clear. Physical deactivation of $Sr(^3P_J)$ atoms by H_2 and D_2 constitutes the dominant mode of removal of the excited state over the temperature range investigated[74] and is a highly temperature-dependent process, in marked contrast to the assumption of a temperature independence for collisional quenching of $Mg(^3P_J)$ atoms by these species via physical energy transfer routes.[51,52,86,87]

4 Oxidation Reactions of Electronically Excited Magnesium Atoms

During the past fifteen or more years there have been a considerable number of kinetic and dynamical investigations of the reactions of atomic magnesium, calcium, and strontium with oxygen-atom-containing species. These include a number of measurements performed under single-collision conditions in molecular beams and the description of a large body of data obtained from low-pressure flames and time-resolved techniques. Concurrent with these collisional studies have been spectroscopic investigations and *ab initio* computations whose purpose has been to fix the locations of low-lying electronic states of the metal monoxide products of these processes. Indeed, it is through the spectroscopy of the metal monoxide products of these reactions that their inherent complexity becomes first apparent, thus rendering the determination of primary product-state distributions a non-trivial and on-going task. Today chemiluminescence and LIF from product states resulting from collisions of metal atoms with an oxidizing gas provide the major tools for elucidating the roles of particular reactive pathways.

Sections 4, 5, and 6 are devoted primarily to describing kinetic and dynamical features of oxidation reactions of Mg, Ca, and Sr atoms as revealed by recent experimental and theoretical investigations. At the beginning of each section, a brief account of the ordering of electronic terms of the relevant metal monoxide molecule is given; it is obvious that an understanding of the complex state manifolds and radiative properties of the excited states of alkaline earth monoxides is an important prerequisite for elucidating the mechanisms of the chemiluminescent reactions in which these species are formed. Similarly, a

[115] G. Herzberg, 'Molecular Spectra and Molecular Structure III: Electronic Structure of Polyatomic Molecules', Van Nostrand Reinhold, New York, 1966, p. 15.

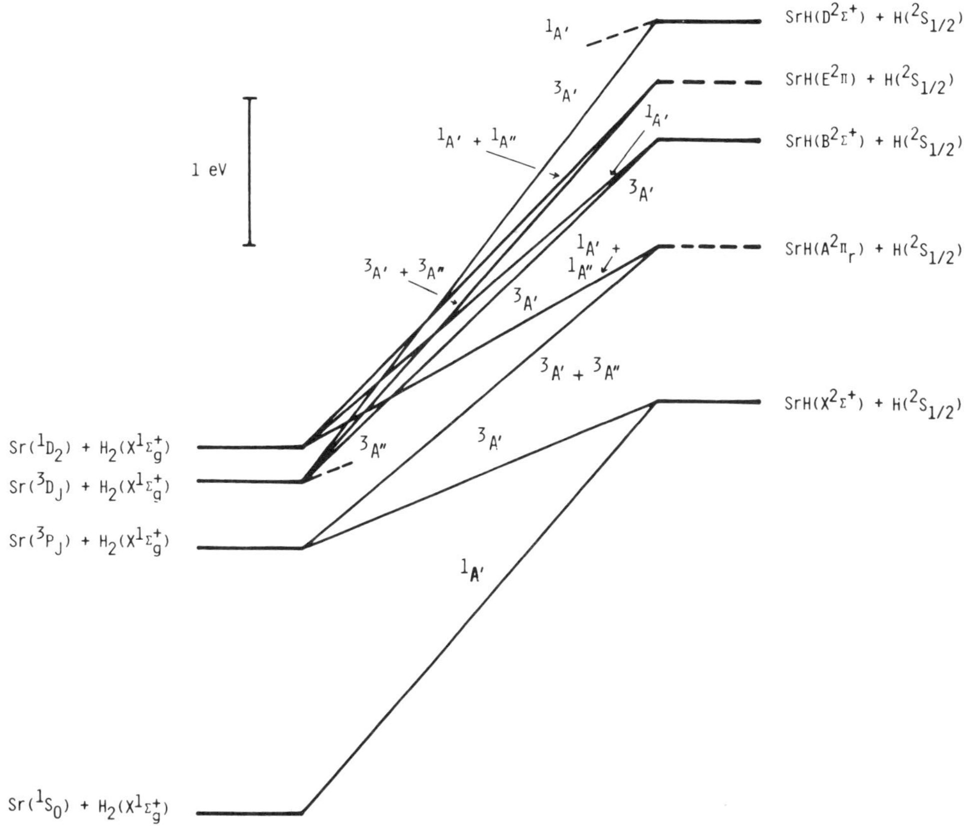

Figure 3 *Correlation diagram connecting the states of* Sr + H_2 *and* SrH + H *constructed on the basis of the weak spin–orbit coupling approximation*

short introductory discussion of the appropriate reaction energetics is also presented, as this forms the basis for subsequent consideration of the collision dynamics and kinetics of these processes.

Electronic States of Magnesium Oxide

Although a viable description of the electronic properties of the alkaline earth metal monoxide molecules has taken many years to build up, a combination of recent experimental observations and theoretical calculations has enabled a more or less complete picture of the low-lying electronic states of MgO to be constructed. The valuable compilations of Huber and Herzberg,[91] Pearse and Gaydon,[116] and Rosen[117] together provide a comprehensive evaluation and interpretation of the many earlier spectroscopic investigations on MgO (and also CaO and SrO) carried out between the late 1920s and the mid 1970s. Molecular term energies for known electronic states of diatomic MgO are shown in Table 4, which compares the experimentally determined values compiled by Huber and Herzberg[91] and that for MgO($a^3\Pi_i$) determined by Ikeda *et al.*[118] with the theoretical estimates reported by Bauschlicher *et al.*[119] as a result of their *ab initio* MCSCF–CI calculations.

Table 4 *Term values of electronic states of magnesium oxide below* 5.0 eV

MgO *electronic state*	*Experimental* T_e/eV[a]	Ab initio T_e/eV[b]
$X^1\Sigma^+$	0.000	0.000
$a^3\Pi_i$	0.326[c]	0.298
$A^1\Pi$	0.442	0.432
$(b)^3\Sigma^+$		1.026
$B^1\Sigma^+$	2.478	2.584
$c^3\Sigma^+$	3.356[d]	3.429
$d^3\Delta$	3.656[d]	3.562
$D^1\Delta$	3.701	3.638
$e^3\Sigma^-$	Repulsive or weakly bound	Repulsive
$C^1\Sigma^-$	3.729	3.674
$E^1\Sigma^+$	4.677	
$F^1\Pi$	4.702	
$G^1\Pi$	4.992	

[a] From ref. 91. [b] From ref. 119. [c] From ref. 118. [d] Changed from ref. 91 to reflect the value for the $a^3\Pi_i$ state reported in ref. 118

[116] R. W. B. Pearse and A. G. Gaydon, 'The Identification of Molecular Spectra', 4th Ed., Chapman & Hall, London, 1976.

[117] B. Rosen, 'Spectroscopic Data Relative to Diatomic Molecules', Pergamon Press, New York, 1970.

[118] T. Ikeda, N. B. Wong, D. O. Harris, and R. W. Field, *J. Mol. Spectrosc.*, 1977, **68**, 452.

[119] C. W. Bauschlicher, jun., B. H. Lengsfield III, D. M. Silver, and D. R. Yarkony, *J. Chem. Phys.*, 1981, **74**, 2379.

Energetics of Mg$[3s^2(^1S_0), 3s3p(^3P_J), 3s3p(^1P_1)]$ + *Oxidant Reactions*

Basic to any discussion of the kinetics and dynamics of bimolecular collisions is the energetics, calculated here for the reactions of the three lowest electronic states of atomic magnesium with N_2O, CO_2, and O_2 to yield diatomic MgO. For thermal collisions between reactants, conservation of energy yields the equality:

$$\Delta E = D_0^0(\text{Oxidant}) + (T_0^{\text{p}} + G_v^{\text{p}}) - D_0^0[\text{MgO}(X^1\Sigma^+)] - E_{\text{Mg}} - (T_0^{\text{r}} + G_v^{\text{r}}) \quad (6)$$

where D_0^0(Oxidant) and $D_0^0[\text{MgO}(X^1\Sigma^+)]$ are thermodynamic dissociation energies at 0 K and $(T_0^{\text{r}} + G_v^{\text{r}})$ and $(T_0^{\text{p}} + G_v^{\text{p}})$ are the internal energies (electronic and vibrational) of molecular reactants and products. For simplicity, rotational excitation of the molecular reactant and products(s) is neglected. E_{Mg} is the electronic excitation energy of atomic magnesium. In order to evaluate reaction ergicities ΔE given by equation (6), we consider here formation of products in their ground electronic and vibrational states in an experiment in which the molecular oxidant is also in its electronic and vibrational ground state. Thus both $(T_0^{\text{p}} + G_v^{\text{p}})$ and $(T_0^{\text{r}} + G_v^{\text{r}})$ are zero in this case. The energies of the atomic reactant states of interest are taken from the data listed in Table 1, while values of the oxidant bond dissociation energies were taken from Herzberg's books.[91,115] The dissociation energy of MgO is much less precisely known than those of the oxidant molecules, and indeed has thus far eluded definitive determination. Here we employ the value $D_0^0[\text{MgO}(X^1\Sigma^+)] = (3.47 \pm 0.26)$ eV recommended by the 1978 Supplement to the JANAF Tables;[120] use of an *ab initio* result derived by Bauschlicher *et al.*[121] would lower available product energies by some 0.82 eV.

Figure 4 displays term energies for the known electronic states of MgO below 5.0 eV and the available reaction energies for Mg$[3s^2(^1S_0), 3s3p(^3P_J), 3s3p(^1P_1)] + N_2O$, CO_2, and O_2 collisions as calculated using equation (6). Experimentally determined term values from Table 4 are preferred wherever possible; the results of the theoretical calculations of Bauschlicher *et al.*[119] are employed to locate the $(b)^3\Sigma^+$ state whose spectroscopy remains as yet unknown.

[120] M. W. Chase, jun., J. L. Curnutt, R. A. McDonald, and A. N. Syverud, *J. Phys. Chem. Ref. Data*, 1978, **7**, 793.
[121] C. W. Bauschlicher, jun., B. H. Lengsfield III, and B. Liu, *J. Chem. Phys.*, 1982, **77**, 4084.

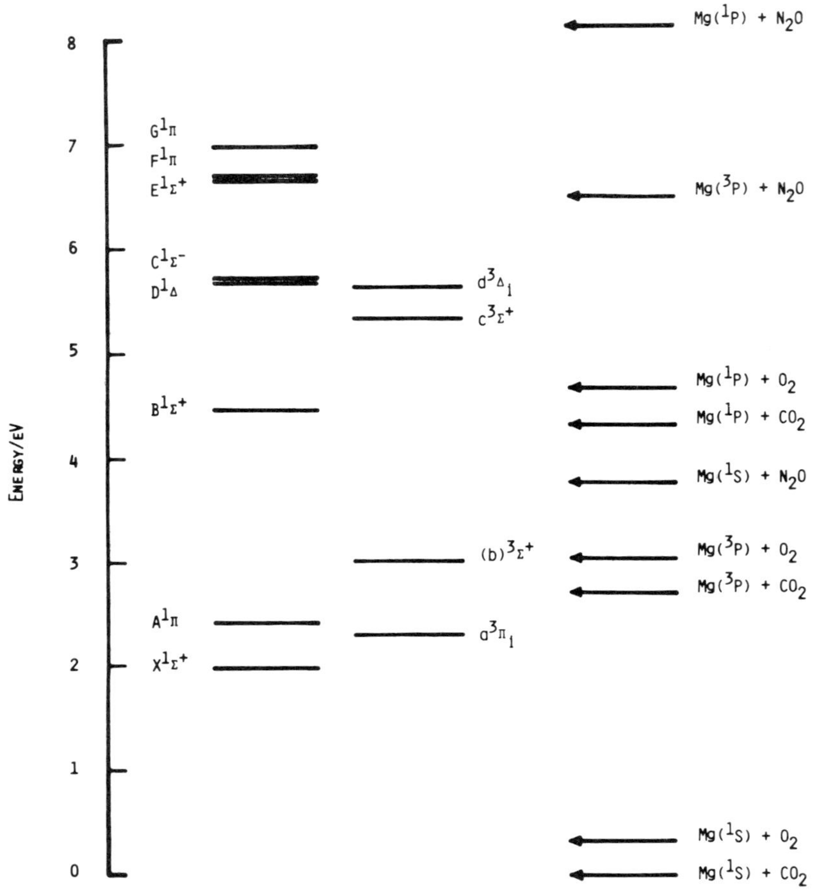

Figure 4 *Energy level diagram showing states of* MgO *below* 5.0 eV *and available energies for reaction of* Mg(^1S,^3P,^1P) *atoms with* N_2O, CO_2, *and* O_2

Reactions with N_2O

Inspection of Figure 4 shows that for the very exoergic reaction of Mg[$3s3p(^3P_J)$] atoms with N_2O, a large number of MgO product states are available up to and including the upper states giving rise to the well-known arc band transitions[91,116,117] situated near 3.5—3.8 eV. It is somewhat surprising, therefore, that no chemiluminescence has been detected from this highly exoergic reaction to date,[122–129] even though high-lying emitting product states are energetically accessible. Dagdigian, alone[122] and with Cox,[123,124] has studied the reaction of Mg(^3P_J) atoms with N_2O employing a method in which an atomic beam containing the metastable species is scattered by a low-pressure oxidant gas and using both chemiluminescence[123] and LIF[122,124] techniques to map out the MgO product internal energy distribution. In their chemiluminescence study, Cox and Dagdigian determined an upper limit to the chemiluminescence cross-section of only 2×10^{-5} nm^2 for the $Mg(^3P_J) + N_2O$ reaction.[123] These workers employed LIF to detect the MgO product in the ground $X^1\Sigma^+$ (refs. 122 and 124) and low-lying excited $a^3\Pi_i$ and $A^1\Pi$ (ref. 122) electronic states via the $B^1\Sigma^+ \leftarrow X^1\Sigma^+$, $d^3\Delta \leftarrow a^3\Pi_i$, and $D^1\Delta \leftarrow A^1\Pi$ transitions. Comparison of integrated fluorescence intensities for these band systems, corrected for electronic transition moments, Franck–Condon factors, and the wavelength dependence of the photomultiplier sensitivity, has enabled Dagdigian to derive electronic state branching ratios for the low-lying states of MgO monitored by LIF. Normalized to unit production rate for $MgO(a^3\Pi_i)$, relative rates of formation are found to be $(0.77 \pm 0.16) : 1.00 : \geq(0.09 \pm 0.05)$ for $MgO(X^1\Sigma^+)$:$MgO(a^3\Pi_i)$:$MgO(A^1\Pi)$.[122] A smooth distribution of the relative populations in the six lowest vibrational levels of $MgO(X^1\Sigma^+)$ and a rotational temperature $T_{rot} = (700 \pm 50)$ K, characterizing the rotational distribution in the $v = 0$ level of this state, were also determined from $B^1\Sigma^+ \leftarrow X^1\Sigma^+$ excitation spectra.[122]

A lack of spectroscopic data pertaining to the triplet states of MgO precluded any reliable population analysis of the $a^3\Pi_i$ state by Dagdigian.[122] Recently, however, Bourguignon *et al.* have carried out a comprehensive spectroscopic analysis of the $d^3\Delta \leftarrow a^3\Pi_i$, $D^1\Delta \leftarrow a^3\Pi_i$, $D^1\Delta \leftarrow A^1\Pi$, and $C^1\Sigma^- \leftarrow A^1\Pi$ band systems[126,130–132] following the previous work of Schamps and Gandara,[133] who first located and rotationally assigned an intense grouping of lines in the ($v' = 0$, $v'' = 0$) band of the $d^3\Delta$–$a^3\Pi_i$ system, of Ikeda *et al.*,[118] who subsequently modified the earlier assignment,[133] and of Trajmar and co-workers[134–136] and Schamps *et al.*,[137] who identified and analysed the (0,0) and (1,1) bands of the $D^1\Delta$–$A^1\Pi$ and $C^1\Sigma^-$–$A^1\Pi$ systems some twenty or more

[122] P. J. Dagdigian, *J. Chem. Phys.*, 1982, **76**, 5375.
[123] J. W. Cox and P. J. Dagdigian, *J. Phys. Chem.*, 1982, **86**, 3738.
[124] J. W. Cox and P. J. Dagdigian, *J. Phys. Chem.*, 1984, **88**, 2455.
[125] B. Bourguignon, J. Rostas, and G. Taieb, *J. Chem. Phys.*, 1982, **77**, 2979.
[126] B. Bourguignon, J. Rostas, G. Taieb, M. -A. Gargoura, and J. McCombie, *Laser Chem.*, 1986, **6**, 15.
[127] B. Bourguignon, *Ann. Phys. (Paris)*, 1986, **11**, 23.
[128] W. H. Breckenridge and H. Umemoto, *J. Phys. Chem.*, 1983, **87**, 1804.
[129] D. Husain and G. Roberts, *J. Chem. Soc., Faraday Trans. 2*, 1986, **82**, 395.
[130] B. Bourguignon, J. McCombie, and J. Rostas, *Chem. Phys. Lett.*, 1985, **113**, 323.
[131] P. C. F. Ip, K. J. Cross, R. W. Field, J. Rostas, B. Bourguignon, and J. McCombie, *J. Mol. Spectrosc.*, submitted for publication.
[132] B. Bourguignon and J. Rostas, *Natl. Bur. Stand. Spec. Publ.*, 1986, **716**, 83.
[133] J. Schamps and G. Gandara, *J. Mol. Spectrosc.*, 1976, **62**, 80.
[134] L. Brewer, S. Trajmar, and R. A. Berg, *Astrophys. J.*, 1962, **135**, 955.
[135] S. Trajmar and G. E. Ewing, *J. Chem. Phys.*, 1964, **40**, 1170.
[136] S. Trajmar and G. E. Ewing, *Astrophys. J.*, 1965, **142**, 77.
[137] J. Schamps, G. Gandara, and M. Becart, *Can. Spectrosc.*, 1969, **14**, 13.

years ago. Improved molecular constants for the $d^3\Delta$ and $a^3\Pi_i$ states have been obtained by Bourguignon and colleagues from $d^3\Delta \leftarrow a^3\Pi_i$ and $D^1\Delta \leftarrow a^3\Pi_i$ laser excitation spectra by means of partial rotational analysis and band contour calculations.[127,130] Comparison of simulated and observed $D^1\Delta \leftarrow A^1\Pi$ and $C^1\Sigma^- \leftarrow A^1\Pi$ laser excitation spectra has enabled this same group to estimate a rotational temperature $T_{rot} \approx 300$ K and a vibrational temperature $T_{vib} \approx 1200$ K characterizing a Boltzmann internal energy distribution within the $A^1\Pi$ state of MgO.[130] These results show that under the room-temperature conditions of the discharge–flow experiment employed by these workers to generate MgO from $Mg(^3P_J) + N_2O$ reactants, the $MgO(A^1\Pi)$ rotational distribution is in thermal equilibrium while vibrational energy is not totally relaxed.[130] In a later study,[126] a semi-quantitative population analysis of the $\Delta v = 0$ sequence of the highly congested $d^3\Delta \leftarrow a^3\Pi_i$ band at $\lambda \approx 372$ nm permitted these workers to interpret the $d^3\Delta \leftarrow a^3\Pi_i$ laser excitation spectrum for $MgO(a^3\Pi_i)$ formed in the reaction of $Mg(^3P_J)$ atoms with N_2O and O_2 reported by Dagdigian.[122] In contrast to the internal energy distribution of the $X^1\Sigma^+$ ground state, it transpires that vibrational populations of $MgO(a^3\Pi_i)$ molecules generated by reaction of $Mg(^3P_J)$ atoms with N_2O shows a marked maximum at $v' = 2$—3 and that rotational excitation is very high, T_{rot} being ~2300 K.[126] Bourguignon *et al.* have also discovered that population of the $a^3\Pi_2$ component of the low-lying triplet state is anomalously high compared with those of the other $a^3\Pi_{0,1}$ spin–orbit levels.[126] That the rotvibrational population distributions of the $X^1\Sigma^+$ and $a^3\Pi_i$ states of MgO resulting from the $Mg(^3P_J) + N_2O$ reaction differ so markedly suggests that the mechanism of their formation is different.[126] This point will be discussed in more detail later in relation to *ab initio* potential energy surfaces for the $Mg + N_2O$ system calculated by Yarkony.[138] Overall, it has been found that only a small fraction [~10% and 19% for $MgO(X^1\Sigma^+)$ and $MgO(a^3\Pi_i)$] of the total energy available from $Mg(^3P_J) + N_2O$ starting materials is deposited as internal energy of nascent product states.[126,127] MgO product has not been detected in the as yet unidentified $(b)^1\Sigma^+$ state or in higher-lying chemiluminescent states, even though these are energetically accessible and, in the case of $MgO[(b)^1\Sigma^+]$ product, symmetry allowed.

Following earlier work on $CaO(A'^1\Pi) \rightarrow CaO(A^1\Sigma^+)$ collision-induced transitions,[192–194] Alexander and co-workers[195] have applied a dipolar model for collision-induced electronic energy transfer between closely-spaced rotational manifolds of different electronic states[192,194] to the study of collisions between low-lying excited states of MgO with N_2O:

$$MgO(a^3\Pi_i, A^1\Pi) + N_2O(\widetilde{X}^1\Sigma^+) \rightarrow MgO(X^1\Sigma^+) + N_2O(\widetilde{X}^1\Sigma^+) \quad (17)$$

[138] D. R. Yarkony, *J. Chem. Phys.*, 1983, **78**, 6768.

Calculation of cross-sections for inelastic energy transfer from rotational manifolds of the lower vibrational levels of $MgO(a^3\Pi_i, A^1\Pi)$ into rotational manifolds of the $X^1\Sigma^+$ state on collision with N_2O requires[192,194,195] knowledge of appropriate non-Born–Oppenheimer perturbations and Coriolis spin–orbit and orbit–rotation parameters[118] and the necessary dipole matrix elements.[179] Pouilly *et al.*[195] report substantial cross-sections of the order of 0.1—0.2 nm^2 for transitions between the rotational manifolds of selected vibrational levels of mixed states of predominantly different electronic character $[MgO(X^1\Sigma^+, a^3\Pi_{0e}, A^1\Pi_{e,f})]$ in the region where two ro vibrational manifolds of different Born–Oppenheimer states cross.

Ground-state $MgO(X^1\Sigma^+)$ product resulting from the reactions of $Mg(^3P_J)$ and $Mg(^1S_0)$ atoms with various diatomic and triatomic oxidants, including N_2O, at a temperature of 520 K has been detected by Breckenridge and Umemoto using a laser 'pump-and-probe' technique.[128] The rate of reaction of ground-state $Mg(^1S_0)$ atoms with N_2O was found by these workers to be very slow, consistent with the qualitative theoretical prediction of an activation barrier in the entrance channel portion of the controlling potential energy hypersurface.[138] No electronically excited states of MgO were monitored by Breckenridge and Umemoto in their study of the $Mg(^1S_0, ^3P_J) + N_2O$ reactions.[128] Indeed, $MgO(X^1\Sigma^+, a^3\Pi_i, A^1\Pi)$ are the only thermochemically accessible products from $Mg(^1S_0) + N_2O$ reactants as shown in Figure 4. The ground-state reaction has also been studied under single-collision conditions[124] and in a discharge–flow experiment:[125] in the former, $MgO(X^1\Sigma^+)$ product was detected by $B^1\Sigma^+ \leftarrow X^1\Sigma^+$ LIF; in the latter, no reaction of $Mg(^1S_0)$ atoms with N_2O was observed.

Oxidation reactions of magnesium atoms have also been studied in flames, where surprisingly complex kinetic behaviour is exhibited.[139–156]

[139] L. de Galan and J. D. Winefordner, *J. Quant. Spectrosc. Radiat. Transfer*, 1967, **7**, 251.

[140] D. J. Benard, *Chem. Phys. Lett.*, 1975, **35**, 167.

[141] D. J. Benard, W. D. Slafer, and P. H. Lee, *Chem. Phys. Lett.*, 1976, **43**, 69.

[142] D. J. Benard, W. D. Slafer, and J. Hecht, *J. Chem. Phys.*, 1977, **66**, 1012.

[143] D. J. Benard and W. D. Slafer, *J. Chem. Phys.*, 1977, **66**, 1017.

[144] H. H. Michels and R. A. Meinzer, *Chem. Phys. Lett.*, 1983, **98**, 6.

[145] G. Taieb, *J. Phys. (Paris)*, 1981, **42**, 537.

[146] O. E. Kashireninov, V. A. Kuznetsov, and G. B. Manelis, *Zh. Fiz. Khim.*, 1975, **49**, 888 [*Russ. J. Phys. Chem.*, 1975, **49**, 520].

[147] O. E. Kashireninov, V. A. Kuznetsov, and G. B. Manelis, *Zh. Fiz. Khim.*, 1975, **49**, 775 [*Russ. J. Phys. Chem.*, 1975, **49**, 454].

[148] V. A. Kuznetsov, A. V. Krestinin, O. E. Kashireninov, and G. B. Manelis, *Zh. Fiz. Khim.*, 1975, **49**, 2096 [*Russ. J. Phys. Chem.*, 1975, **49**, 1225].

[149] O. E. Kashireninov, V. A. Kuznetsov, and G. B. Manelis, *Zh. Fiz. Khim.*, 1977, **51**, 958 [*Russ. J. Phys. Chem.*, 1977, **51**, 566].

[150] O. E. Kashireninov, V. A. Kuznetsov, and G. B. Manelis, *AIAA J.*, 1977, **15**, 1035.

[151] V. A. Kuznetsov and O. E. Kashireninov, *Zh. Fiz. Khim.*, 1977, **51**, 3105 [*Russ. J. Phys. Chem.*, 1977, **51**, 1813].

[152] V. A. Kuznetsov and O. E. Kashireninov, *Zh. Fiz. Khim.*, 1977, **51**, 3109 [*Russ. J. Phys. Chem.*, 1977, **51**, 1815].

[153] O. E. Kashireninov and V. A. Kuznetsov, *Gorenie i Vzryv*, 1977, 521.

Benard and co-workers have observed efficient chemical production of the metastable $nsnp(^3P_J)$ states of Mg, Ca, and Sr where CO is introduced into a flame in which the ground-state metal atom M, entrained in helium carrier gas, is mixed with N_2O.[141–143] The basis of the production of excited metal atoms in such ternary flames has been proposed by Benard and co-workers as consisting of the two-step mechanism:

$$\mathrm{M}[ns^2(^1S_0)] + \mathrm{N_2O}(\widetilde{X}^1\,\Sigma^+) \rightarrow \mathrm{MO}^\dagger + \mathrm{N_2}(X^1\,\Sigma_g^+) \tag{8}$$

$$\mathrm{MO}^\dagger + \mathrm{CO}(X^1\,\Sigma^+) \rightarrow \mathrm{M}[nsnp(^3P_J)] + \mathrm{CO_2}(\widetilde{X}^1\,\Sigma_g^+) \tag{9}$$

where $MO^\dagger$ is a metal monoxide 'intermediate'.[141–143] It is not clear at present whether or not $MO^\dagger$ is a vibrationally excited ground state or an electronically excited state of the metal monoxide. In addition to the production of metastable metal atoms, a marked enhancement in flame chemiluminescence was observed by these workers,[141–143] which has since been verified,[123,157–161] in the case of $Ca(^3P_J)$ and $Sr(^3P_J)$ atoms, to result from subsequent reaction of the metastable states with N_2O:

$$\mathrm{M}[nsnp(^3P_J)] + \mathrm{N_2O}(\widetilde{X}^1\,\Sigma^+) \rightarrow \mathrm{MO}^* + \mathrm{N_2}(X^1\,\Sigma_g^+) \tag{10}$$

Here, MO^* is an electronically excited state of the product metal monoxide which decays rapidly by spontaneous emission of radiation. Taieb[145] has investigated reaction (10) in the case of M = Mg in a ternary Mg–N_2O–CO–He flame and has detected the low-lying $a^3\Pi_i$ state of MgO by exciting the (1,0) transition of the $\mathrm{MgO}(d^3\Delta \leftarrow a^3\Pi_i)$ band system with an ultraviolet line from an argon-ion laser and monitoring fluorescence from the (1,1) transition at $\lambda \approx 372$ nm. The absence of any detectable fluorescence from the very exoergic reaction of $Mg(^3P_J)$ atoms with N_2O implies that the sequence of events leading to production of electronically excited MO^* states in Mg–N_2O–CO–He flames is different, and inherently more complicated, than for the cases of M = Ca or Sr, though no satisfactory mechanism appears to have been postulated so far. The free-atom concentrations of alkaline earth

[154] O. E. Kashireninov, V. A. Kuznetsov, and L. F. Repka, *Zh. Fiz. Khim.*, 1978, **52**, 200 [*Russ. J. Phys. Chem.*, 1978, **52**, 107].
[155] L. F. Repka, A. D. Chervonnyi, A. B. Proudin, and O. E. Kashireninov, 'Kinet. Khim. Reakts. Mater. Vses. Simp. Goreniyu Vzryuvu', 6th, ed. G. B. Manelis, p. 57.
[156] O. E. Kashireninov and V. A. Kuznetsov, *Dokl. Akad. Nauk. SSSR*, 1981, **257**, 636 [Engl. Transl., 1981, **257**, 109].
[157] J. A. Irvin and P. J. Dagdigian, *J. Chem. Phys.*, 1981, **74**, 6178.
[158] P. J. Dagdigian, *Chem. Phys. Lett.*, 1978, **55**, 239.
[159] B. E. Wilcomb and P. J. Dagdigian, *J. Chem. Phys.*, 1978, **69**, 1779.
[160] H. Wang and M. Menzinger, *Can. J. Chem.*, 1983, **61**, 2703.
[161] D. Husain and G. Roberts, *Chem. Phys.*, 1986, **109**, 393.

metals in fuel-rich acetylene–air flames have been measured by de Galan and Winefordner.[139]

The kinetics of oxidation reactions of ground-state $Mg(^1S_0)$ atoms have been studied by Kashireninov *et al.* by probing the temperature pattern in a spherically symmetric reaction zone of a low-pressure diffusion flame, taking into account the large temperature gradients in the reaction zone, the occurrence of secondary reactions, the effect of condensation of gas-phase reaction products on measured temperature profiles in the diffusion flame, and other complicating factors.[146–156,162] To this end, general calculational methods necessary to extract Arrhenius parameters for the process of interest have been developed.[148,150,151,163,164] For reaction of $Mg(^1S_0)$ atoms with N_2O, Kashireninov *et al.* report an Arrhenius pre-exponential factor of $A = (2.2 \pm 0.5) \times 10^{-11}$ cm^3 molecule^{-1} s^{-1} and an activation energy of $E_{act} = (27.6 \pm 9.2)$ kJ mol^{-1},[149,156] with which we may compare analogous results for $Mg(^3P_J) + N_2O$[129] described in the following paragraph [see equation (11)]. These workers further find that the reaction of the alkaline earth metal atoms Mg, Ca, Sr, and Ba with N_2O satisfies the Polanyi–Semenov rule linking activation energy and heat of reaction.[156]

By direct monitoring of $Mg(^3P_1)$ fluorescence in the time-resolved mode following pulsed laser excitation of the triplet state, Husain and Roberts[129] have obtained absolute second-order rate constants for collisional removal of $Mg(^3P_J)$ atoms by N_2O over the temperature range 600—1100 K. Computerized curve-fitting of $k_{N_2O}(T)$ data as a bi-exponential function of temperature yielded the result:

$$k_{N_2O}(T) = (2.6^{+3.5}_{-1.5}) \times 10^{-10} \exp[-(41.3 \pm 5.3)\ \text{kJ mol}^{-1}/RT] + (2.73^{+0.92}_{-0.69}) \exp[-(232.8 \pm 32.5)\ \text{kJ mol}^{-1}/RT]\ \text{cm}^3\ \text{molecule}^{-1}\ \text{s}^{-1} \qquad (11)$$

The first term on the right-hand side of equation (11) comprises Arrhenius parameters for bimolecular collisions between $Mg(^3P_J)$ atoms and N_2O, while the second term reflects additional removal of the excited atomic state by molecular oxygen formed by decomposition of the primary oxidant at high temperatures.[129] In addition to monitoring the decay of metastable atomic species, the use of the time-domain as a diagnostic tool for probing molecular chemiluminescence in such systems has enabled these workers to investigate the different mechanistic routes via which emitting product states may be formed under thermal

[162] D. Garvin, V. P. Guinn, and G. B. Kistiakowsky, *Discuss. Faraday. Soc.*, 1954, **17**, 32.

[163] V. A. Kuznetsov, O. E. Kashireninov, and G. B. Manelis, *Zh. Fiz. Khim.*, 1977, **51**, 1377 [*Russ. J. Phys. Chem.*, 1977, **51**, 811].

[164] V. A. Kuznetsov, *Khim. Fiz. Protsessov Goreniya i Vzryva. Kinetika Khim. Reaktsii*, 1977, 106.

collisions.[129,161] In this case, Husain and Roberts were able to monitor molecular emission from the $B^1\Sigma^+$, $d^3\Delta$, and $D^1\Delta$ states of MgO in the time-resolved mode, and showed that the time dependence of decay traces so obtained was consistent with a mechanism in which the observed high-lying electronically excited product states arise from collisional E–(E,V) energy transfer between $Mg(^3P_J)$ atoms and long-lived dark electronic states of MgO subsequent to the initial oxidation process:[129]

$$Mg(^3P_J) + MgO[X^1\Sigma^+, a^3\Pi_i, A^1\Pi, (b)^3\Sigma^+] \rightarrow MgO(B^1\Sigma^+, d^3\Delta, D^1\Delta) + Mg(^1S_0) \quad (12)$$

Fractional yields of electronically excited product states MgO* were determined to be of the order of $[MgO^*]/[Mg(^3P_1)] \approx 10^{-6}$.[129] Vibrational temperatures in the $MgO(B^1\Sigma^+)$ state derived from progressions in the red $B^1\Sigma^+ \rightarrow A^1\Pi$ system were found to be some 2—5 times higher than the ambient temperature, suggesting that a certain degree of vibrational excitation accompanies the electronic exchange process (12).[129] Previous measurements of $k_{N_2O}(T)$ at single temperatures do not provide a consistent basis for estimation of Arrhenius parameters.[103,125,165]

In contrast to collisions between $Mg[3s^2(^1S_0)]$ and $Mg[3s3p(^3P_J)]$ atoms and N_2O, reaction of the higher-lying state $Mg[3s3p(^1P_1)]$ with the same oxidant results in copious chemiluminescence from electronically excited $MgO(B^1\Sigma^+)$ product,[125,127,128] formation of which is exoergic from these reactants (see Figure 4). Taieb and co-workers have succeeded in characterizing the initial vibrational and rotational quantum-state distributions of the nascent $MgO(B^1\Sigma^+)$ product of the $Mg(^1P_1) + N_2O$ reaction by means of a rotational temperature T_{rot} of ~1700 K and a vibrational temperature T_{vib} of ~5000 K from simulated spectra of the $MgO(B^1\Sigma^+ \rightarrow X^1\Sigma^+)$ chemiluminescence observed in their discharge–flow system at room temperature.[125] Although the nascent vibrational population decreases monotonically with increasing vibrational quantum number within the $MgO(B^1\Sigma^+)$ state, the distribution has been shown to be non-Boltzmann,[127] in contrast to the near-Boltzmann rotational population distribution monitored by these workers.[125] Ground-state $MgO(X^1\Sigma^+)$ and low-lying excited states $MgO(a^3\Pi_i, A^1\Pi)$ resulting from $Mg(^1P_1) + N_2O$ collisions do not appear to have been probed to date.

As a starting point for consideration of the dynamics of the $Mg(^1S_0, ^3P_J, ^1P_1) + N_2O$ reactions, a state correlation diagram is presented as Figure 5. Preliminary inspection of this diagram reveals that

[165] D. Husain and J. Schifino, *J. Chem. Soc., Faraday Trans. 2*, 1983, **79**, 919.

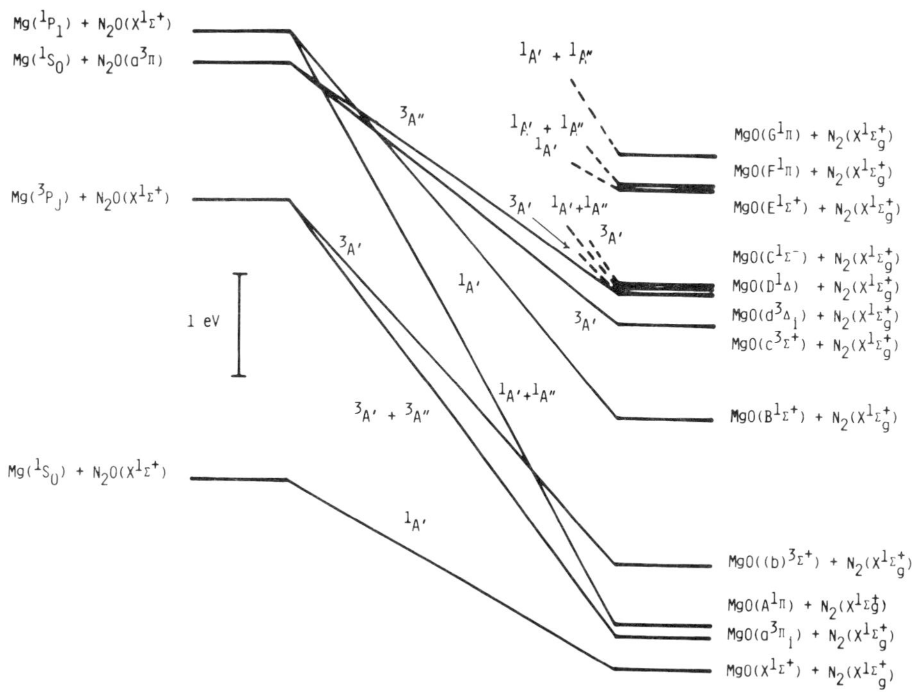

Figure 5 *Correlation diagram connecting the states of* Mg + N_2O *and* MgO + N_2 *constructed on the basis of the weak spin–orbit coupling approximation and* C_s *symmetry in the collision complex*

the absence of any chemiluminescence, and hence production of emitting MgO states, arising from the $Mg(^3P_J) + N_2O$ reaction, is consistent with this picture. Thus if the reaction were completely electronically adiabatic and spin-conserved, only the $a^3\Pi_i$ and hitherto unobserved $(b)^3\Sigma^+$ states of MgO product would be formed. Indeed, as mentioned earlier in this section, Dagdigian has found that the $MgO(a^3\Pi_i)$ state is the most highly populated product state resulting from $Mg(^3P_J) + N_2O$ reagents and that the yield of the adiabatically forbidden $MgO(A\,^1\Pi)$ state is low.[122] On the other hand, there is a rather substantial amount of ground-state $MgO(X^1\Sigma^+)$ formed in this reaction; the measured rate of formation of $MgO(X^1\Sigma^+)$ relative to that of $MgO(a^3\Pi_i)$[122] is somewhat greater than the ratio of 0.23 predicted from purely statistical considerations.[50] As Figure 5 suggests, formation of ground-state $MgO(X^1\Sigma^+)$ product can occur by nonadiabatic transitions from the lowest triplet surface to the lowest singlet surface. The correlation diagram also predicts correctly the formation of $MgO(X^1\Sigma^+)$ and $MgO(B^1\Sigma^+)$ products from $Mg(^1S_0) + N_2O$ and $Mg(^1P_1) + N_2O$ starting materials, respectively.

Yarkony has mapped out the lowest singlet ($^1A'$) and triplet ($^3A'$) potential energy hypersurfaces of the $Mg[3s^2(^1S_0),3s3p(^3P_J)] + N_2O(\tilde{X}^1\Sigma^+)$ system using *ab initio* SCF, two-configuration MCSCF, and CI wavefunctions.[138] According to these calculations, potential energy barriers are present on the $r_{MgO\text{–}N_2}$ reaction co-ordinate, corresponding to reaction of $Mg(^1S_0)$ and $Mg(^3P_J)$ atoms with linear N_2O. Furthermore, a colinear approach of either $Mg(^1S_O)$ or $Mg(^3P_J)$ atoms along the N—N—O internuclear axis is favoured since a prohibitively high potential barrier due to repulsive electronic interactions, the height of which has not been determined quantitatively, makes abstraction of an oxygen atom via attack perpendicular to the N—N—O internuclear axis unlikely.[138]

The reaction of both $Mg(^1S_0)$ and $Mg(^3P_J)$ atoms with N_2O is probably initiated by a short-range charge transfer from the magnesium atom to linear N_2O through a $Mg^+N_2O^-$ transition state. Such a mechanism is of course reminiscent of the celebrated 'harpoon' model employed so successfully to describe the dynamics of alkali metal atom plus halogen reactions.[11,12,24,166,167] Unlike the halogen molecules, however, N_2O has a negative vertical electron affinity[168–174] which becomes

[166] J. L. Magee, *J. Chem. Phys.*, 1940, **8**, 687.
[167] D. R. Herschbach, *Adv. Chem. Phys.*, 1966, **10**, 319.
[168] J. Balamuta and M. F. Golde, *J. Chem. Phys.*, 1982, **76**, 2430.
[169] D. G. Hopper, A. C. Wahl, R. L. C. Wu, and T. O. Tiernan, *J. Chem. Phys.*, 1976, **65**, 5474.
[170] W. E. Wentworth, E. Chen, and R. Freeman, *J. Chem. Phys.*, 1971, **55**, 2075.
[171] P. J. Chantry, *J. Chem. Phys.*, 1969, **51**, 3369.
[172] P. J. Chantry, *J. Chem. Phys.*, 1969, **51**, 3380.
[173] J. N. Bardsley, *J. Chem. Phys.*, 1969, **51**, 3384.
[174] E. E. Ferguson, F. C. Fehsenfeld, and A. L. Schmeltekopf, *J. Chem. Phys.*, 1967, **47**, 3085.

more positive as the molecule distorts from its near-equilibrium linear configuration.[169,170,173] The reaction mechanism therefore probably involves a 'collision-assisted' electron jump[50] in which a close encounter of the Mg atom with N_2O causes the molecule to deviate from linearity, thereby increasing its electron affinity. Owing to the negative electron affinity of N_2O, the electron jump occurs at much closer distances than in alkali atom plus halogen reactions. Thus, as the electron affinity of N_2O increases with excitation of the ν_2 in-plane bending vibration of the molecule ($\tilde{\nu}_2 = 589\ \mathrm{cm}^{-1}$),[175] we may expect the electron transfer distance, and with it the chemiluminescence cross-section, to increase also. Experimental data such as there are on collisional excitation of the ν_2 bending mode of N_2O (and CO_2) suggest that such vibrationally inelastic collision processes possess reasonably large cross-sections.[176] Collisional enhancement of the electron affinity of N_2O therefore facilitates an electron jump which then leads to reaction as the resulting Coulomb attraction draws the Mg^+ and N_2O^- ions together. Stabilization of highly excited N_2O^- in its linear configuration can be effected via the in-plane bending vibration which allows the system to bypass the potential energy barriers on the reaction co-ordinate.[138] In addition, the reacting system experiences a small local barrier of ~0.13 eV in the exit channel leading to $MgO(a^3\Pi_i) + N_2(X^1\Sigma_g^+)$ products, though no such barrier is encountered in the ground-state exit channel yielding $MgO(X^1\Sigma^+) + N_2(X^1\Sigma_g^+)$.[138] Experimental evidence for this type of model has been provided by Wren and Menzinger, who have reported an enhancement of chemiluminescence for the $Ba(^1S_0) + N_2O$ reaction with increasing in-plane bending vibration of N_2O.[16,177,178] Yarkony[138] has also noted that the $Mg^+N_2O^-$ complex correlates adiabatically with several states (non-$^1\Sigma^+$) in the product channel corresponding to a singly charged Mg^+O^- ion pair.[119,179–184]

Taieb and co-workers[126] have presented a qualitative argument to rationalize the observed[122] vibrational population distribution within $MgO(a^3\Pi_i)$ formed in collisions of $Mg(^3P_J)$ atoms with N_2O based on theoretical and experimental investigations[185] of the similar exoergic reaction of $C[2p^2(^3P_J)]$ atoms with N_2O and the characteristics of the

[175] G. Herzberg, 'Molecular Spectra and Molecular Structure II: Infra-Red and Raman Spectra of Polyatomic Molecules', Van Nostrand Reinhold, New York, 1945, p. 278.

[176] E. Weitz and G. Flynn, *Annu. Rev. Phys. Chem.*, 1974, **25**, 275.

[177] D. J. Wren and M. Menzinger, *J. Chem. Phys.*, 1975, **63**, 4557.

[178] D. J. Wren and M. Menzinger, *Faraday Discuss. Chem. Soc.*, 1979, **67**, 97.

[179] R. N. Diffenderfer, D. R. Yarkony, and P. J. Dagdigian, *J. Quant. Spectrosc. Radiat. Transfer*, 1983, **29**, 329.

[180] R. N. Diffenderfer and D. R. Yarkony, *J. Phys. Chem.*, 1982, **86**, 5098.

[181] C. W. Bauschlicher, jun., D. M. Silver, and D. R. Yarkony, *J. Chem. Phys.*, 1980, **73**, 2867.

[182] J. Schamps and H. Lefebvre-Brion, *J. Chem. Phys.*, 1972, **56**, 573.

[183] B. Huron and P. Rancurel, *Chem. Phys. Lett.*, 1972, **13**, 515.

[184] W. G. Richards, G. Verhaegen, and C. M. Moser, *J. Chem. Phys.*, 1966, **45**, 3226.

[185] G. Dorthe, M. Costes, C. Naulin, J. Joussot-Dubein, C. Vaucamps, and G. Nouch, *J. Chem. Phys.*, 1985, **83**, 3171.

calculated[138] $^3A'$ potential energy hypersurface. The vibrational population inversion found[122,126] for $MgO(a^3\Pi_i)$ is consistent with the highly exoergic nature of the reaction and an intermediate potential well located on the $^3A'$ hypersurface.[126] The observation that high rotational levels (up to $J \approx 130$) of the $a^3\Pi_i$ state[126] are populated is consistent with the reaction proceeding via a bent $Mg^+N_2O^-$ transition state, since it may be expected that transfer of angular momentum from a bending mode of the complex will result in a high degree of rotational energy in the $MgO(a^3\Pi_i)$ product.[126] More quantitative data on the vibrational and rotational population distributions in $MgO(a^3\Pi_i)$ can be obtained from appropriately selected laser fluorescence excitation spectra of the $B^1\Sigma^+ \leftarrow a^3\Pi_i$ and $D^1\Delta \leftarrow a^3\Pi_i$ intercombination transitions, which display relatively open ro vibrational structure.[126,130,131] Bourguignon *et al.* have suggested that such spectra would also be appropriate for measurement of individual rotational linewidths from which product translational energies may be deduced, and which, in turn, would provide a useful tool for estimation of the internal energy of the $N_2(X^1\Sigma_g^+)$ product.[126]

Unlike $MgO(a^3\Pi_i)$, the $X^1\Sigma^+$ ground state of MgO does not correlate adiabatically with $Mg(^3P_J) + N_2O$ reagents. Taieb and co-workers consider that non-adiabatic transitions from the lowest-lying triplet surface ($^3A'$) leading to $MgO(a^3\Pi_i)$ product to the lowest singlet surface ($^1A'$) may occur in the transition state or entrance channel regions.[126] In the first case, formation of the $X^1\Sigma^+$ state is enhanced by the close proximity of the potential surfaces in the transition-state region leading to $MgO(a^3\Pi_i)$ and $MgO(X^1\Sigma^+)$ products (see figure 6 of ref. 126) in which mixing between the two surfaces arising from the same configuration occurs via non-Born–Oppenheimer interactions between them.[118] Such a mechanism is similar to that previously put forward by Dagdigian for the $Mg(^3P_J) + N_2O$ and $Mg(^3P_J) + O_2$ reactions,[122] in which the substantial rate of formation of $MgO(X^1\Sigma^+)$ product was explained as resulting from non-adiabatic transitions in the exit channel from potential energy surfaces leading to adiabatically allowed $MgO(a^3\Pi_i)$ product to those giving rise to $MgO(X^1\Sigma^+)$. Potential energy curves for $MgO(a^3\Pi_i)$ and $MgO(X^1\Sigma^+)$ are known[118] to lie very close to each other and have significant non-Born–Oppenheimer interactions with each other. The absence of electronically excited $MgO(B^1\Sigma^+)$ as a product of the $Mg(^3P_J) + N_2O$ reaction has been explained by Cox and Dagdigian[123] as being due to the large energy difference between the asymptotic $B^1\Sigma^+$ and adiabatically accessible $a^3\Pi_i$ [and $(b)^3\Sigma^+$] states such that non-Born–Oppenheimer interactions between potential energy surfaces leading to these states will be small. While electronic matrix elements for perturbations between the $B^1\Sigma^+$ state and the $a^3\Pi_i$ and $A^1\Pi_i$ states of MgO are expected[118] to be of a similar magnitude to those between the $A^1\Sigma^+$ and $a^3\Pi_i$ and $A'^1\Pi$ states of CaO and SrO (see Sections 5 and 6), vibrational overlap between the potential energy surfaces for these MgO states will be significantly smaller than those for CaO and SrO, where the small energy splittings between the $A\,^1\Sigma^+$, $A'\,^1\Pi$, and $a\,^3\Pi_i$ asymptotic states facilitates non-adiabatic transitions in the exit channels of the $Ca(^3P_J) + N_2O$ and $Sr(^3P_J) + N_2O$ reactions.[123] Unlike the $A\,^1\Sigma^+$–$X\,^1\Sigma^+$ band spectra of CaO[186] and SrO,[187–189] no perturbations of the $MgO(B\,^1\Sigma^+)$ state have been identified.[190,191]

The different vibrational population distributions observed within the $X\,^1\Sigma^+$ and $a\,^3\Pi_i$ states of MgO arising from the $Mg(^3P_J) + N_2O$ reaction may be due to the absence of a potential well on the singlet surface, though as pointed out by Bourguignon *et al.*, such simple considerations cannot account for the very different rotational distributions in these two states.[126] A non-adiabatic transition in the entrance channel region of the reaction path of $C_{\infty v}$ symmetry would, according to these workers, be an indirect process going through an intermediate $[Mg^+(^2S) + N_2O^-(^2\Sigma)](^1\Sigma^+)$ surface to the ionic part of the lowest $^1\Sigma^+$ potential surface.[126] In this case, the evolution of the system in a linear configuration into the exit channel leading to $MgO(X\,^1\Sigma^+) + N_2(X\,^1\Sigma_g^+)$ products then depends upon the precise nature of the potential energy hypersurfaces involved and details of the hopping mechanism.[126] These quantities have not as yet been calculated, so definitive conclusions cannot be drawn about the sequence of events which results in the smooth vibrational population distribution within the $X\,^1\Sigma^+$ ground state which is actually observed.[122,126] However, in order to account for the relatively low rotational temperature of $MgO(X\,^1\Sigma^+)$ product, Bourguignon *et al.* consider that the entire reactive process would have to be rapid with respect to the bending vibrational period of the N_2O^- anion.[126]

Reaction of ground-state $Mg(^1S_0)$ atoms with N_2O is symmetry-allowed and may proceed adiabatically to $MgO(X\,^1\Sigma^+) + N_2(X\,^1\Sigma_g^+)$ ground-state products over the lowest singlet potential energy hypersurface of $^1A'$ symmetry, as shown in Figure 5. Approach of the $Mg(^1S_0)$ atom to N_2O is energetically unfavourable: the theoretical computations of Yarkony indicate that a barrier to charge transfer appears to be associated with the absence of accessible open-shell electronic configurations which correlate with excited singlet states in the isolated atom.[138] Breckenridge and Umemoto have estimated the magnitude of this barrier to lie in the range approximately 63—84 kJ mol^{-1} following their

[186] M. Hultin and A Lagerqvist, *Ark. Fys.*, 1951, **2**, 471.
[187] G. Almkvist and A Lagerqvist, *Ark. Fys.*, 1949, **1**, 477.
[188] G. Almkvist and A. Lagerqvist, *Ark. Fys.*, 1950, **2**, 233.
[189] A. Lagerqvist and L. -E. Selin, *Ark. Fys.*, 1959, **11**, 323.
[190] A. Lagerqvist, *Ark. Mat. Astronom. Fys. A*, 1943, **29**, 1.
[191] A. Lagerqvist and U. Uhler, *Ark. Fys.*, 1949, **1**, 459.
[192] M. H. Alexander, *J. Chem. Phys.*, 1982, **76**, 429.
[193] M. H. Alexander and M. G. Osmolovsky, *J. Chem. Phys.*, 1982, **77**, 839.
[194] M. H. Alexander and B. Pouilly, *J. Chem. Phys.*, 1983, **79**, 1545.
[195] B. Pouilly, J. -M. Robbe and M. H. Alexander, *J. Phys. Chem.*, 1984, **88**, 140.

LIF measurements on $MgO(X^1\Sigma^+)$ produced in the $Mg(^1S_0) + N_2O$ reaction.[128]

In the absence of theoretical calculations on the $Mg[3s3p(^1P_1)] + N_2O$ reaction, only broad qualitative features of the potential energy hypersurfaces controlling this process can be inferred[127] by comparison with the $Mg[3s3p(^3P_J)] + N_2O$ system. Bourguignon[127] has speculated on the factors governing the formation of $MgO(B^1\Sigma^+)$ in collisions of $Mg(^1P_1)$ atoms with N_2O on the basis of molecular orbital configurations of the $Mg^+N_2O^-$ transition state and by reference to the calculated[138] surfaces for the $Mg(^3P_J) + N_2O$ reaction. It is thought likely by Bourguignon that the first step of the $Mg(^1P_1) + N_2O$ reaction is charge transfer leading to formation of an $Mg^+N_2O^-$ ion pair.[127] Using a simple Coulombic potential, Bourguignon has estimated the Mg—ON_2 distance at which charge transfer takes place on potential energy surfaces of $^1\Pi$ and $^1\Sigma^+$ symmetry (see figure 4 on ref. 127). As the calculated separations for the $Mg(^1P_1) + N_2O$ reaction are somewhat larger than those for $Mg(^3P_J) + N_2O$ collisions, it has been postulated[127] that no potential barrier to charge transfer exists in the entrance channels of the $^1\Pi$ and $^1\Sigma^+$ potential energy hypersurfaces. Possible formation of the chemiluminescent $MgO(B^1\Sigma^+)$ state therefore depends on the height of the barrier encountered on the $3^1A'$ potential energy surface (see Figure 5) which arises from a change in electronic configuration of the $Mg^+N_2O^-$ transition state occurring between the entrance channel and transition state regions.[127] Since the maximum of this potential barrier probably lies below the asymptotic energy of $Mg(^1P_1) + N_2O$ reactants, $MgO(B^1\Sigma^+)$ product may be formed in this reaction.[127] The actual magnitude of this barrier has not, however, been quantified as yet.

Comparison of the internal energy distribution within $MgO(B^1\Sigma^+)$ resulting from the $Mg(^1P_1) + N_2O$ reaction with those of $MgO(a^3\Pi_i)$ and $MgO(X^1\Sigma^+)$ arising from the reaction of $Mg(^3P_J)$ atoms with the molecular oxidant reveals that rotational excitation in the $B^1\Sigma^+$ state is ~2.5 times larger than that of the $X^1\Sigma^+$ state[127] where an abstraction mechanism – leading to rotationally cold $MgO(X^1\Sigma^+)$ – has been postulated.[126] In view of the similar rotational temperatures of $MgO(B^1\Sigma^+)$ and $MgO(a^3\Pi_i)$ products, Bourguignon[127] has suggested that the $B^1\Sigma^+$ state is also formed via a bent $Mg^+N_2O^-$ intermediate with transfer of large amounts of angular momentum in the exit channel leading to products via a mechanism analogous to that described above for production of rotationally hot $MgO(a^3\Pi_i)$ from $Mg(^3P_J) + N_2O$ starting materials. The very different vibrational population distributions observed for $MgO(B^1\Sigma^+)$ and $MgO(a^3\Pi_i)$ are probably due to marked differences in the topologies of the potential energy hypersurfaces controlling the $Mg(^3P_J) + N_2O$ and $Mg(^1P_1) + N_2O$ reactions:[127] a late release of potential energy during formation of the $B^1\Sigma^+$ state is expected[127] to result in the smooth population distribution observed experimentally,[125] whereas an earlier release of potential energy leads preferentially to a peaked vibrational population distribution for $MgO(a^3\Pi_i)$.[126]

Reactions of Electronically Excited Magnesium Atoms with CO_2

The collisional behaviour of magnesium atoms with CO_2, like N_2O described above, is of particular interest on account of the linear sixteen-electron closed-shell structure of the triatomic molecule, which may be expected to give rise to significant energy barriers for oxygen-atom abstraction even though pathways to product states may exist which are both exoergic and symmetry-allowed. Unlike N_2O, however, the bond dissociation energy of CO_2 is relatively high and so the number of energetically accessible states of metal monoxide product is somewhat more limited, as shown in Figure 4. For this reason perhaps, the reactions of $Mg[3s^2(^1S_0)]$, $Mg[3s3p(^3P_J)]$, and $Mg[3s3p(^1P_1)]$ atoms with CO_2 have been less extensively studied that the corresponding processes involving N_2O. In fact, reaction of ground-state $Mg(^1S_0)$ atoms with CO_2 is (191.0 ± 25.1) kJ mol^{-1} endoergic[115,120] and does not appear to have been investigated by an experimental method at present. Also, it seems that no chemiluminescence from electronically excited states of MgO product arising from reaction of $Mg(^3P_J)$ and $Mg(^1P_1)$ atoms with this molecular oxidant has been reported, although ground-state $MgO(X^1\Sigma^+)$ resulting from reaction of both excited atoms has been detected by LIF.[124,128,196] Population of MgO product states up to 2.36 eV is feasible on energetic grounds from $Mg(^1P_1) + CO_2$ starting materials. Use of the JANAF recommended value of $D_0^0[MgO(X^1\Sigma^+)]$[120] indicates that production of $MgO(B^1\Sigma^+)$ in the $Mg(^1P_1) + CO_2$ reaction would be 11.1 kJ mol^{-1} endoergic, though this value increases to 90.2 kJ mol^{-1} when the calculated value of $D_0^0[MgO(X^1\Sigma^+)]$[121] is employed. Figure 6 shows a portion of an adiabatic correlation diagram connecting the low-lying electronic states of $Mg + CO_2$ with those of $MgO + CO$. The diagram includes the low-lying $^1A_1$ and $^3B_2$ electronic states of CO_2 whose energies have been calculated by England and co-workers.[197–199] From Figure 6 it can be seen that $Mg(^3P_J) + CO_2(\widetilde{X}\,^1\Sigma_g^+)$ reagents correlate with $MgO(a^3\Pi_i)$ and $MgO[(b)^3\Sigma^+]$ metal monoxide product states, formation of the second of which is endoergic by ~0.3 eV ($\approx$ 29 kJ mol^{-1}) with respect to these reactants. The long-lived product state $MgO(A\,^1\Pi)$ (τ_e = 80,[200,201]

[196] W. H. Breckenridge and H. Umemoto, *J. Phys. Chem.*, 1983, **87**, 476.
[197] W. B. England, B. J. Rosenberg, P. J. Fortune, and A. C. Wahl, *J. Chem. Phys.*, 1976, **65**, 684.
[198] W. B. England, W. C. Ermler, and A. C. Wahl, *J. Chem. Phys.*, 1977, **66**, 2336.
[199] W. B. England, D. Yeager, and A. C. Wahl, *J. Chem. Phys.*, 1977, **66**, 2344.
[200] L. Pasternack and P. J. Dagdigian, *Chem. Phys.*, 1978, **33**, 1.
[201] B. Huron, J. P. Malrieu, and P. Rancurel, *Chem. Phys.*, 1974, **3**, 277.

230,[179] and 400[118,201] μs) is also energetically accessible from $Mg(^3P_J) + CO_2$ starting materials, though it appears that no attempt has as yet been made to search for this product.[124,128]

Cox and Dagdigian have succeeded in probing the initial populations of the $v = 0$ and 1 quantum levels of $MgO(X^1\Sigma^+)$ produced from reaction of $Mg(^3P_J)$ atoms with CO_2 by laser fluorescence excitation of the $\Delta v = 0$ sequence of the $B^1\Sigma^+ \leftarrow X^1\Sigma^+$ band system under single-collision conditions in a molecular beam plus scattering gas arrangement.[124]

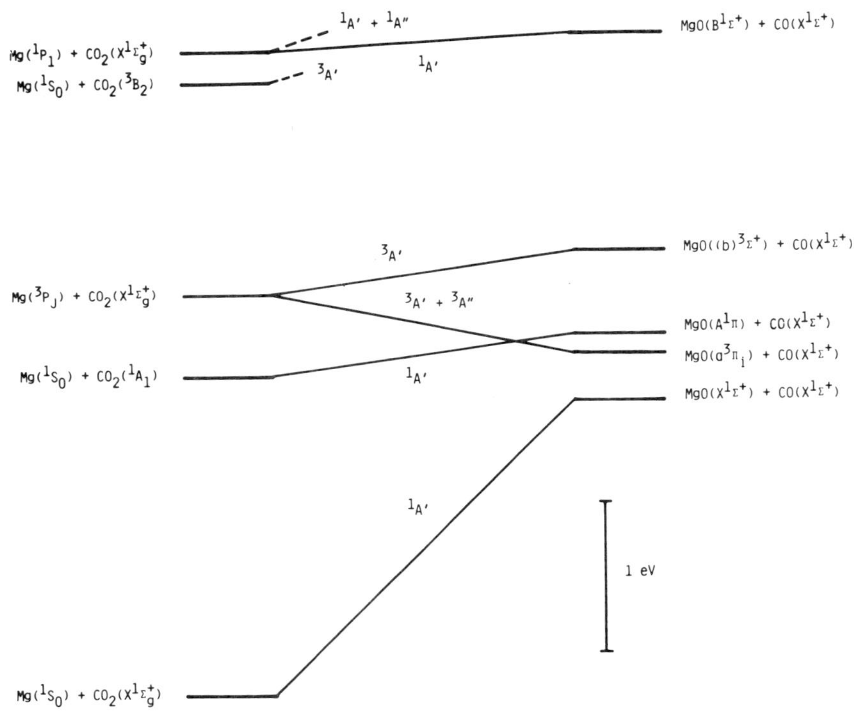

Figure 6 *Correlation diagram connecting the states of* $Mg + CO_2$ *and* $MgO + CO$ *constructed on the basis of the weak spin–orbit coupling approximation and* C_s *symmetry in the collision complex*

Husain and Roberts have studied the collisional behaviour of $Mg(^3P_J)$ atoms with CO_2 in the time-resolved mode over the temperature range 600—1100 K.[202] Absolute second-order rate constants for total collision-

[202] D. Husain and G. Roberts, *Ber. Bunsenges. Phys. Chem.*, 1986, **90**, 360.

al removal of this atomic state were expressed in the non-Arrhenius form:

$$k_{CO_2}(T) = (2.4^{+0.7}_{-0.5}) \times 10^{-10} \exp[-(47.6 \pm 5.5)\ \text{kJ mol}^{-1}/RT] + (8.1^{+2.9}_{-2.1}) \times 10^{-4} \exp[-(171.3 \pm 34.3)\ \text{kJ mol}^{-1}/RT]\ \text{cm}^3\ \text{molecule}^{-1}\ \text{s}^{-1} \quad (13)$$

Arrhenius parameters for $Mg(^3P_J) + CO_2$ collisions are given by the leading exponential in equation (13). Husain and Roberts presented a kinetic argument, based on the time dependence of $MgO(X^1\Sigma^+)$ formation, to account for non-Arrhenius behaviour at high temperatures in terms of additional removal of $Mg(^3P_J)$ atoms by ground-state $MgO(X^1\Sigma^+)$ product.[202] Molecular chemiluminescence from $MgO(B^1\Sigma^+)$ was also detected by these workers, the time dependence of which was shown to be in accord with the hypothesis that emitting vibronic states of $B^1\Sigma^+$ are formed by E–(E,V) energy transfer between $Mg(^3P_J)$, atoms and $MgO(X^1\Sigma^+)$ following the O-atom abstraction reaction [see equation (12)]. Vibrational temperatures in $MgO(B^1\Sigma^+)$ in the range 2346—3042 K reflect a degree of vibrational excitation during the electronic energy exchange process.[202] Bimolecular quenching rate constants for overall collisional removal of $Mg(^3P_J)$ atoms by CO_2 have also been reported at single temperatures by Taieb and Broida[103] and Husain and Schifino.[165]

Reaction of the higher-lying atomic state $Mg(^1P_1)$ with CO_2 has also been the object of an investigation by Breckenridge and Umemoto,[196] in which $MgO(X^1\Sigma^+)$ product was monitored by LIF via $\Delta v = 0$ vibrational sequence bands of the green $B^1\Sigma^+ \leftarrow X^1\Sigma^+$ system. These workers observed a smooth decrease in the relative populations of vibrational levels of $MgO(X^1\Sigma^+)$ from $v = 0$ and $v = 5$ and a distribution of nascent rotational quantum states within the $v = 0$ level for which a rotational temperature characterizing that distribution of $T_{rot} = 1600$ K was extracted assuming a Boltzmann functional form.[196] Comparison of the experimentally determined initial rotational state distribution with those expected from phase-space theory[47] and the information–theoretic approach,[7,11,12,24,48,49] both of which assume statistical partitioning of energy into all degrees of freedom, indicates that $MgO(X^1\Sigma^+)$ product is both rotationally and vibrationally colder than anticipated.[196] Breckenridge and Umemoto have postulated the existence of a late potential barrier located in the exit-channel portion of the ground-state $MgCO_2$ surface as providing a dynamical constraint leading to a colder than expected distribution of internal quantum states within $MgO(X^1\Sigma^+)$. Alexander and Dagdigian[50] have noted that orbital rearrangements necessary at exit-channel geometries could indeed lead to substantial barriers to formation of ground-state metal oxide products from metastable complexes such as, in this case, $Mg^+CO_2^-$.

Although no branching ratio could be determined by these workers, the measurements of Breckenridge and Umemoto[196] suggest that reaction of $Mg(^1P_1)$ atoms with CO_2 produces ground-state $MgO(X^1\Sigma^+)$ in good yield, even though $MgO(X^1\Sigma^+) + CO(X^1\Sigma^+)$ products do not correlate adiabatically with these reactants in C_s symmetry (*i.e.* for planar $MgCO_2$ geometries). Inspection of the state correlation diagram depicted as Figure 6 reveals that only formation of the $A\,^1\Pi$ and $B\,^1\Sigma^+$ excited states of MgO are symmetry-allowed from $Mg(^1P_1) + CO_2(\tilde{X}\,^1\Sigma_g^+)$ starting materials, although production of $MgO(B\,^1\Sigma^+)$ is endoergic. Indeed, Breckenridge and Umemoto were unable to detect any green $MgO(B\,^1\Sigma^+ \rightarrow X^1\Sigma^+)$ chemiluminescence resulting from the $Mg(^1P_1) + CO_2$ reaction. These workers have also estimated that the yield of $MgO(A\,^1\Pi)$ is less than 15% that of the ground-state product of this reaction, though no attempt has been made to detect the remaining energetically accessible product states $MgO(a\,^3\Pi_i)$ and $MgO[(b)^3\Sigma^+]$ by LIF.[196] It thus appears that ground-state $MgO(X^1\Sigma^+)$ is produced more or less to the exclusion of those final states – $MgO(a\,^3\Pi_i)$ and $MgO[(b)^3\Sigma^+]$ – which correlate with $Mg(^1P_1) + CO_2(\tilde{X}\,^1\Sigma_g^+)$ reactants.

Using the technique of time-resolved atomic resonance fluorescence spectroscopy, Breckenridge and Umemoto have reported efficient quenching of $Mg(^1P_1)$ atoms by CO_2.[203] In contrast to the analogous $Cd[5s5p(^1P_1)]$ level, production of $Mg[3s3p(^3P_J)]$ atoms by collision-induced intersystem crossing was found to be negligible for $Mg(^1P_1) + CO_2$ collisions.

Efficient production of adiabatically forbidden $MgO(X^1\Sigma^+)$ product from collisions of $Mg(^1P_1)$ atoms with CO_2 requires a crossing from the initial potential energy hypersurface of $^1A''$ symmetry, correlating with $Mg(^1P_1) + CO_2(\tilde{X}\,^1\Sigma_g^+)$ reactants, to the $^1A'$ hypersurface correlating with $MgO(X^1\Sigma^+) + CO(X^1\Sigma^+)$ products. As pointed out by Breckenridge and Umemoto,[196] it is possible that for a range of $MgCO_2$ geometries, there are direct crossings to the ground-state potential hypersurface from the hypersurface of $^1A''$ symmetry correlating with $MgO(A\,^1\Pi) + CO(X^1\Sigma^+)$ products providing that the initial hypersurface possesses regions of substantial attractive character and that in the crossing region the ground-state $^1A'$ hypersurface has repulsive character with respect to ground-state $Mg[3s^2(^1S_0)] + CO_2(\tilde{X}\,^1\Sigma_g^+)$. *Ab initio* calculations on the ground-state potential energy hypersurface of $MgCO_2$ indicate the existence of a metastable $Mg^{\delta+}CO_2^{\delta-}$ complex located at a minimum energy $\sim 126\ \mathrm{kJ\,mol^{-1}}$ above the asymptotic energy of $Mg(^1S_0) + CO_2(\tilde{X}\,^1\Sigma_g^+)$.[204] A well depth associated with this ionic complex sufficiently great so as to be able to support vibrational eigenstates is also expected[196] on the basis of these computations,[204]

[203] W. H. Breckenridge and H. Umemoto, *J. Chem. Phys.*, 1981, **75**, 698.

[204] K. D. Jordan, private communication cited in W. H. Breckenridge and H. Umemoto, *J. Phys. Chem.*, 1983, **87**, 476.

which further suggests that there must be a substantial potential barrier to formation of the $Mg^{\delta+}CO_2^{\delta-}$ ionic complex from ground-state $Mg(^1S_0) + CO_2(\tilde{X}\,^1\Sigma_g^+)$ species. If this barrier is sufficiently high, there may indeed be entrance-channel crossings from attractive $MgCO_2$ surfaces correlating with $Mg(^1P_1) + CO_2(\tilde{X}\,^1\Sigma_g^+)$ reactants to the ground-state ionic surface with eventual formation of $MgO(X^1\Sigma^+) + CO(X^1\Sigma^+)$ products.[196] However, Breckenridge and Umemoto consider that because there are several singlet and triplet potential energy hypersurfaces which may be involved, such a mechanism may be grossly over-simplified.[196]

Reactions with O_2

Dagdigian has detected MgO product in the energetically accessible $X^1\Sigma^+$ and $a^3\Pi_i$ states by LIF following reaction of $Mg(^3P_J)$ atoms with molecular oxygen under single-collision conditions.[122] By comparison of integrated fluorescence intensities for individual vibronic transitions within the $B^1\Sigma^+–X^1\Sigma^+$ and $d^3\Delta–a^3\Pi_i$ band systems, the relative rates of production of MgO in the $v = 0$ vibrational level of the $X^1\Sigma^+$ and $a^3\Pi_i$ states for the $Mg(^3P_J) + N_2O$ versus $Mg(^3P_J) + O_2$ reaction have been determined to be (0.12 ± 0.02) and (0.08 ± 0.02).[122] The branching ratio for production of $MgO(X^1\Sigma^+)$ and $MgO(a^3\Pi_i)$ states in the $Mg(^3P_J) + O_2$ reaction was measured as only $1 : (0.03 \pm 0.01)$,[122] which is some ten times smaller than that expected statistically. The measured[122] vibrational distribution with the $X^1\Sigma^+$ state has $v = 0$ as the most probable level, with population decreasing monotonically with increasing v up to $v = 4$, the highest level monitored. A rotational temperature characterizing the rotational distribution in the $v = 0$ level of $MgO(X^1\Sigma^+)$ was determined by Dagdigian as $T_{rot} = (1700 \pm 150)$ K.[122]

Dagdigian has also recorded a $d^3\Delta \leftarrow a^3\Pi_i$ laser excitation spectrum for MgO produced from the $Mg(^3P_J) + O_2$ reaction at a scattering gas pressure of 4 mTorr.[122] Bourguignon *et al.*[126] have pointed out that at this pressure vibrational and rotational relaxation will have occurred to some extent before laser excitation so that the $d^3\Delta–a^3\Pi_i$ spectrum reported by Dagdigian[122] does not truly represent the $MgO(a^3\Pi_i)$ nascent population. Bourguignon *et al.* find that relative populations in vibrational levels $v = 0$, 1, and 2 of $MgO(a^3\Pi_i)$ show a Boltzmann distribution, decreasing monotonically with increasing vibrational excitation.[126] Unlike the $Mg(^3P_J) + N_2O$ reaction, there appears to be no preference for population of a particular spin–orbit level of $MgO(a^3\Pi_i)$.[126] Rotational excitation of $MgO(a^3\Pi_i)$ produced from $Mg(^3P_J) + O_2$ reagents is also much lower than from $Mg(^3P_J) + N_2O$.

Breckenridge and Umemoto have also probed the initial ro vibrational state distribution within the $MgO(X^1\Sigma^+)$ product of the $Mg(^3P_J) + O_2$ reaction using their laser 'pump-and-probe' technique.[128] Because the laser bandwidth was not sufficient to resolve the rotational structure of the observed $\Delta v = 0$ sequences of the MgO($B^1\Sigma^+ \leftarrow X^1\Sigma^+$) excitation band, these workers determined the nascent population distribution of vibrational and rotational quantum states within the $X^1\Sigma^+$ state by an empirical procedure based upon the minimization of deviations between computer-simulated and experimental spectra.[128] Using this method, they report a rotational temperature characterizing the rotational distribution within the lowest vibrational state of MgO($X^1\Sigma^+$) of $T_{rot} = (1400 \pm 100)$ K. The initial vibrational distribution of MgO($X^1\Sigma^+$) resulting from the Mg(^3P_J) + O_2 reaction reported by Breckenridge and Umemoto[128] is consistent with that determined by Dagdigian[122] in that a monotonic decrease in population from $v = 0$ to $v = 4$ was observed. Rather surprisingly perhaps, these workers were unable to monitor MgO($X^1\Sigma^+$) and MgO($A\,^1\Pi$) products when Mg($^1P_1$) atoms were laser-excited in the presence of O_2, and neither could chemiluminescence from highly excited MgO($B^1\Sigma^+$) be observed, in contrast to the Mg($^1P_1$) + N_2O reaction.[128] Mg(^3P_J) atoms formed via an intersystem crossing mechanism were detected following Mg($^1P_1$) + O_2 collisions, however, by LIF of the Mg[$3s4s(^3S_1) \leftarrow 3s3p(^3P_J)$] transitions in the $\lambda = 516.7$—518.4 nm region.[128] Breckenridge and Umemoto have argued that production of Mg(^3P_J) atoms in the quenching of Mg($^1P_1$) by O_2 is a minor exit channel (with branching ratio ≤ 0.1[128]) even though such a process is spin-allowed.[128,203] From estimates of the Mg($^1P_1$) versus Mg(^3P_J) concentrations, these authors have further estimated that the yield of MgO($X^1\Sigma^+$) produced in the reaction of Mg($^1P_1$) atoms with O_2 is less than 10% of that observed in the Mg(^3P_J) + O_2 reaction.[128]

Kashireninov and collaborators have studied the kinetics of the reaction of ground-state Mg($^1S_0$) atoms with O_2 using their thermocouple variation of the diffusion flame method.[146–148,150,151,153,154,156,162–164] Activation energies for the Mg($^1S_0$) + O_2 reaction reported by these workers are $E_{act} = (66.9 \pm 8.8)$,[150,153] (64.0 ± 11.7),[146,147,156] and (58.8 ± 12.6)[146] kJ mol^{-1}. Arrhenius pre-exponential factors are approximately gas kinetic, being $A = (3.5 \pm 0.3) \times 10^{-10}$ cm^3 molecule^{-1} s^{-1} [150,153] and 2.8×10^{-10} cm^3 molecule^{-1} s^{-1}.[146,147,156] Like the reaction with N_2O described earlier in this section, adherence to the Polanyi–Semenov relationship is found for the reaction of Mg, Ca, Sr, and Ba atoms with O_2.[156] Bimolecular rate constants for overall removal of Mg(^3P_J) atoms by O_2 indicate quenching of the excited atomic state at a rate of about one in every ten collisions.[75,103]

In the absence of detailed *ab initio* calculations of the relevant potential energy hypersurfaces for the Mg + O_2 system, such as are available for Mg + N_2O,[138] Dagdigian[122] has suggested that the Mg(^3P_J) + O_2 reaction proceeds via a long-lived singly charged $Mg^+O_2^-$ intermediate by analogy with the Ca(^3P_J) + O_2 reaction.[50,200] Dagdigian has also pointed out that MgO_2 has been observed[205] using an infrared absorption technique in low-temperature matrices and was deduced to have a C_{2v} structure. This intermediate, the potential energy minimum of which is considered to lie some 2 eV below that of the $Mg(^3P_J) + O_2$ reactants,[206,207] can dissociate with little orbital rearrangement to the $a^3\Pi_i$ and $A^1\Pi$ states of MgO, which are reasonably described as singly-charged Mg^+O^- ion pairs.[119,179–184] However, the $X^1\Sigma^+$ ground state has a significantly different electronic structure,[119,179–184] being predominantly $Mg^{2+}O^{2-}$ in character, so that considerable orbital rearrangement would be required for dissociation of the $Mg^+O_2^-$ intermediate into ground-state products. This would, in turn, lead to the existence of a late potential energy barrier in the exit channel for $MgO(X^1\Sigma^+)$ formation.[122] Such a mechanism would also give rise to the observed cold internal energy distribution within the $X^1\Sigma^+$ state[122,128] and high product translational energy.

Experimentally observed[122,128] vibrational state population distributions of $MgO(X^1\Sigma^+)$ formed by reaction of $Mg(^3P_J)$ atoms with O_2 may be compared with the statistical expectations predicted by phase-space theory[47] and information theory.[7,11,12,24,48,49] Both Dagdigian[122] and Breckenridge and Umemoto[128] find that their experimental results exhibit considerably less excitation than the statistical models. Similar results have been observed for the $CaO(X^1\Sigma^+)$ product of the $Ca(^3P_J) + O_2$ reaction[200] and may again be rationalized[50] on the basis of a postulated potential barrier located in the exit channel leading to products in their electronic ground states.

The relatively low branching ratio for production of the $a^3\Pi_i$ state of MgO has been accounted for by Dagdigian[122] in terms of significant non-Born–Oppenheimer interactions[118] between the $a^3\Pi_i$ and $X^1\Sigma^+$ potential curves. Bourguignon *et al.*[126] have pointed out that non-adiabatic transitions in the exit channel from surfaces leading to $a^3\Pi_i$ product to those leading to $X^1\Sigma^+$ would allow efficient population transfer out of the $a^3\Pi_{0e}$ component.[194,195]

Reactions with Other Oxidants

Cox and Dagdigian have studied the reactions of $Mg(^1S_0)$ and $Mg(^3P_J)$ atoms with NO_2 under single-collision conditions in a molecular beam plus scattering gas system.[124] For the $Mg(^1S_0) + NO_2$ reaction, vibrational levels of $MgO(X^1\Sigma^+)$ product up to $v = 2$ were monitored by laser fluorescence excitation of $\Delta v = 0$ sequences of the $B^1\Sigma^+ \leftarrow X^1\Sigma^+$

[205] L. Andrews, E. S. Prochaska, and B. S. Ault, *J. Chem. Phys.*, 1978, **69**, 556.
[206] M. H. Alexander, *J. Chem. Phys.*, 1978, **69**, 3502.
[207] M. H. Alexander, unpublished work cited in P. J. Dagdigian, *J. Chem. Phys.*, 1982, **76**, 5375 based on an electrostatic model outlined in ref. 206.

band.[124] Herm and co-workers were able to detect the scattered metal monoxide product of the $Mg(^1S_0) + NO_2$ reaction mass spectrometrically in a crossed molecular beam experiment, though they were unable to derive MgO angular distributions because of interference from an Ar^+ peak also present.[208] Zare and co-workers were unable to detect any chemiluminescence from $Mg(^1S_0) + NO_2$ collisions.[209] This reaction has also been investigated in flame media,[152,155] from which work the rate constant for $Mg(^1S_0) + NO_2$ has been expressed in the form:[152]

$$k_{NO_2}(T) = (8.8 \pm 1.0) \times 10^{-14} \exp[(19.2 \pm 10.5)\ \mathrm{kJ\,mol^{-1}}/RT]\ \mathrm{cm^3\,molecule^{-1}\,s^{-1}} \quad (14)$$

The negative activation energy given in equation (14) has been rationalized[152,155] in terms of a mechanism involving a relatively long-lived $MgNO_2$ collision intermediate, though no adequate explanation of the surprisingly low Arrhenius pre-exponential factor appears to have been forthcoming. Cox and Dagdigian have observed a weak continuum emission extending over the wavelength range $\lambda = 440$—900 nm following reaction of $Mg(^3P_J)$ atoms with NO_2,[124] and have ascribed this chemiluminescence to transitions of the $\tilde{A}\,^2B_2 \rightarrow \tilde{X}\,^2A_1$ system of NO_2.[210] In view of the approximate resonance between the excited state energies of $Mg(^3P_J)$[71] and $NO_2(\tilde{A}\,^2B_2)$,[210] these authors[124] believe that $NO_2(\tilde{A}\,^2B_2)$ arises from the near-resonant electronic energy exchange process:

$$Mg(^3P_J) + NO_2(\tilde{X}\,^2A_1) \rightarrow Mg(^1S_0) + NO_2(\tilde{A}\,^2B_2) \quad (15)$$

Furthermore, from measurements of cross-sections for energy transfer and beam attenuation, they estimate that the energy transfer efficiency is $> 2\%$.[124] Although production of excited $MgO(B\,^1\Sigma^+)$ is energetically favourable from $Mg(^3P_J) + NO_2$ reactants ($\Delta E = -0.59$ eV[71,91,115,120]), the presence of continuum emission from $NO_2(\tilde{A}\,^2B_2)$ precluded any assignment of the well-known $B\,^1\Sigma^+ \rightarrow X\,^1\Sigma^+$ and $B\,^1\Sigma^+ \rightarrow A\,^1\Pi$ bands of MgO.[124]

Breckenridge and Umemoto failed to detect $MgO(X\,^1\Sigma^+)$ by LIF when $Mg(^3P_J)$ atoms were created in the presence of SO_2 using the laser 'pump-and-probe' method,[128] although Cox and Dagdigian did succeed in observing a weak excitation signal near the (0,0) band of the $MgO(B\,^1\Sigma^+ \leftarrow X\,^1\Sigma^+)$ system.[124] For the reaction of $Mg(^1P_1)$ atoms with SO_2, a vibrational and rotational quantum-state distribution within

[208] R. R. Herm, S. -M. Lin, and C. A. Mims, *J. Phys. Chem.*, 1973, **77**, 2931.
[209] C. D. Jonah, R. N. Zare, and Ch. Ottinger, *J. Chem. Phys.*, 1972, **56**, 263.
[210] D. Hsu, D. L. Monts, and R. N. Zare, 'The Spectral Atlas of Nitrogen Dioxide', Academic Press, New York, 1978.

$MgO(X^1\Sigma^+)$ has been observed[128] similar to that reported for $Mg(^1P_1) + CO_2$.[196]

Reaction of $Mg(^3P_J)$ atoms with ozone is sufficiently exoergic to populate electronic states of MgO up to $G^1\Pi$, located some 4.992 eV above the $X^1\Sigma^+$ ground state[91,211] ($\Delta E = -5.14$ eV[71,115,120]). Cox and Dagdigian have observed chemiluminescence bands at $\lambda = 373$ and 496 nm, corresponding to transitions within the $d^3\Delta \rightarrow a^3\Pi_i$ and $B^1\Sigma^+ \rightarrow X^1\Sigma^+$ band systems of MgO, when $Mg(^3P_J)$ atoms are reacted with O_3 under single-collision conditions.[124] These workers were not able to monitor any chemiluminescence from the ground-state $Mg(^1S_0) + O_3$ reaction, even though production of $MgO(B^1\Sigma^+)$ is slightly exoergic.[124]

Collisional quenching of $Mg(^3P_J)$ atoms by CO has been studied in bulk systems by time-resolved[75,102] and discharge–flow[103] methods, yielding rate constants of the order of $k_{CO}(T) = (1$—$2) \times 10^{-12}$ cm^3 molecule^{-1} s^{-1} at $T = 800$ K[75,102] and room temperature.[103] In addition, Breckenridge and Umemoto have measured a quenching cross-section of 0.55 nm^2 for collisional removal of $Mg(^1P_1)$ atoms by CO.[203] In these cases, production of ground-state monoxide is 4.91 and 3.28 eV endoergic[71,91,120] for reaction of $Mg(^3P_J)$ and $Mg(^1P_1)$ atoms, respectively. Using the same technique, these workers were unable to detect $MgO(X^1\Sigma^+)$ when $Mg(^3P_J)$ and $Mg(^1P_1)$ atoms were excited in the presence of NO,[128] values of ΔE being 0.315 and -1.319 eV, respectively.[71,91,120]

Engelke *et al.* have monitored chemiluminescence arising from the reaction of $Mg(^1S_0)$ atoms with ClO_2 in a crossed molecular beam experiment.[212] Molecular emission was observed only in the ultraviolet region and has been assigned to the $A^2\Pi \rightarrow X^2\Sigma^+$ band system of MgCl. Formation of $MgO(B^1\Sigma^+)$ is some 1.51 eV endoergic,[71,91,115,120] and although energetically accessible, the $MgO(A^1\Pi \rightarrow X^1\Sigma^+)$ band system lies in the infrared outside the wavelength range of the detection system employed by these workers.[212] The reaction dynamics of this system were discussed in terms of the familiar electron-jump model[11,12,24,166,167] with transfer of a valence electron of the metal atom to the half-filled antibonding orbital of ClO_2.[212]

5 Oxidation Reactions of Electronically Excited Calcium Atoms

Electronic States of Calcium Oxide

The visible and near-infrared spectrum of diatomic calcium oxide, constituting part of the chemiluminescence observed on reaction of $Ca[4s4p(^3P_J)]$ and $Ca[4s3d(^1D_2)]$ atoms with molecular oxidants, is

[211] M. Singh, *J. Phys. B*, 1973, **6**, 1917.
[212] F. Engelke, R. K. Sander, and R. N. Zare, *J. Chem. Phys.*, 1976, **65**, 1146.

exceedingly complex and indeed has not yet been completely assigned.[91,116,117] However, important progress in this area has been particularly forthcoming in recent years: Field and co-workers have employed the technique of sub-Doppler intermodulation spectroscopy to analyse partially the rotational fine-strucuture of the celebrated orange arc band of CaO,[213–215] the results of which have enabled these authors to identify with some certainty the upper and lower electronic states involved and also to speculate on the electronic transitions giving rise to the green arc band system.[214,215] Molecular parameters for the low-lying excited CaO($A'^1\Pi$) state have been obtained by Field *et al.* from observations of CaO($A'^1\Pi \rightarrow X^1\Sigma^+$) luminescence in Ca + N_2O and Ca + O_3 flames.[216] Field has also developed an analysis technique for assignment of perturbations in electronic spectra of the alkaline earth monoxides CaO, SrO, and BaO,[217] and has successfully utilized this method to assign perturbations of the CaO($A^1\Sigma^+ \rightarrow X^1\Sigma^+$) spectrum to the low-lying $a^3\Pi_i$ and $A'^1\Pi$ states using data for the $A^1\Sigma^+ \rightarrow X^1\Sigma^+$ band system reported by Hultin and Lagerqvist some thirty years ago.[186]

Complementary to these spectroscopic observations are various theoretical calculations on the properties of a number of low-lying electronic states of CaO,[218–222] the exact nature and ordering of which has been the subject of considerable controversy for some years. Molecular term energies of electronic states of CaO lying lower than about 29000 cm^{-1} above the ground state are presented in Table 5. Term energies have been taken directly from the compilations of Huber and Herzberg[91] and Rosen[117] for the well-known singlet states $X^1\Sigma^+$, $A^1\Sigma^+$, $B^1\Pi$, and $C^1\Sigma^+$; otherwise term values were estimated from recent spectroscopic[213–215,223] and theoretical[219,220] work. Limitations of space prevent us from describing how these estimations were made, but details may be found in a recent publication.[161] Table 6 summarizes current knowledge of the occupancies of the six valence electrons giving rise to the electronic states of CaO listed in Table 5.[213, 215,219–225] We may note that while *ab initio* studies[218–222] provide no adequate single-configuration description of the $X^1\Sigma^+$ ground state of CaO, it is occasionally better described[182,200,218,219,221,222] as $8\sigma 9\sigma 3\pi^4$ rather than the commonly invoked molecular orbital description[217,220] $8\sigma^2 3\pi^4$ (shown in

[213] R. F. Marks, R. A. Gottscho, and R. W. Field, *Phys. Scr.*, 1982, **25**, 312.
[214] R. F. Marks, H. S. Schweda, R. A. Gottscho, and R. W. Field, *J. Chem. Phys.*, 1982, **76**, 4689.
[215] R. F. Marks, H. S. Schweda, R. A. Gottscho, and R. W. Field, *J. Chem. Phys.*, 1982, **77**, 4795.
[216] R. W. Field, G. A. Capelle, and C. R. Jones, *J. Mol. Spectrosc.*, 1975, **54**, 156.
[217] R. W. Field, *J. Chem. Phys.*, 1974, **60**, 2400.
[218] K. D. Carlson, K. Kaiser, C. Moser, and A. C. Wahl, *J. Chem. Phys.*, 1970, **52**, 4678.
[219] C. W. Bauschlicher, jun., and D. R. Yarkony, *J. Chem. Phys.*, 1978, **68**, 3990.
[220] W. B. England, *Chem. Phys.*, 1980, **53**, 1.
[221] R. N. Diffenderfer and D. R. Yarkony, *J. Chem. Phys.*, 1982, **77**, 5573.
[222] R. N. Diffenderfer and D. R. Yarkony, *J. Chem. Phys.*, 1983, **78**, 7017.
[223] C. -S. Wei, S. -W. Guo, and Y. -P. Lee, *J. Chem. Phys.*, 1985, **82**, 2942.

Table 5 *Term values of electronic states of calcium oxide below* 3.6 eV

CaO *electronic state*	T_e/cm^{-1}	T_e/eV	*References*
$X^1\Sigma^+$	0	0.000	91, 117
$a^3\Pi_i$	8386	1.040	213—215
$A'^1\Pi$	8693	1.078	213—215
$(b)^3\Sigma^+$	9900^{+2000}_{-100}	$1.227^{+0.248}_{-0.012}$	213—215, 219, 220
$A^1\Sigma^+$	11555	1.433	91, 117
$(c)^3\Sigma^+$	24300^{+300}_{-500}	$3.013^{+0.037}_{-0.062}$	219, 223
$D^1\Delta$	24752	3.069	213—215
$d^3\Delta$	24817	3.077	213—215
$^{3,1}\Sigma^-$	24992	3.090	219
$e^3\Pi$	25931	3.215	219
$B^1\Pi$	25991	3.222	91, 117
$C^1\Sigma^+$	28858	3.578	91, 117

Table 6), corresponding to a doubly charged ion-pair configuration $Ca^{2+}O^{2-}$. With the former description, transitions connecting the manifold of states of configuration $8\sigma^2 3\pi^3 4\pi$ to the ground state correspond to a two-electron jump and hence should be of low intensity, while radiative decay out of the same configuration to the $a^3\Pi_i$ and $A'^1\Pi$ states of configuration $8\sigma^2 9\sigma 3\pi^3$ should give rise to strong fluorescence (the orange arc bands).[213–215] Bauschlicher and Yarkony find that all strongly bound non-$^1\Sigma^+$ valence states of CaO are well represented by single configurations that are nominally Ca^+O^-.[219]

Table 6 *Single-configuration occupancies of the six valence electrons constituting the low-lying electronic states of calcium oxide*[a]

	Outer electronic configuration[b]	CaO *electronic states*	*References*
1	$8\sigma^2 3\pi^4$	$X^1\Sigma^+$	219, 220
2	$8\sigma^2 9\sigma 3\pi^3$	$a^3\Pi_i, A'^1\Pi$	213, 214, 219, 220
3	$8\sigma 9\sigma 3\pi^4$	$(b)^3\Sigma^+$, $A^1\Sigma^+$	214, 219, 220
4	$8\sigma^2 3\pi^3 4\pi$	$(c)^3\Sigma^+$, $D^1\Delta$, $d^3\Delta$, $^{3,1}\Sigma^-$, $C^1\Sigma^+$	213, 214, 219
5	$8\sigma 3\pi^4 4\pi$	$e^3\Pi$, $B^1\Pi$	214, 219, 220
6	$8\sigma^2 10\sigma 3\pi^3$	$^3\Pi$, $^1\Pi$	214
7	$8\sigma 10\sigma 3\pi^4$	$^3\Sigma^+$	214

[a] Non-valence electrons $1\sigma^2 2\sigma^2 3\sigma^2 4\sigma^2 5\sigma^2 6\sigma^2 7\sigma^2 1\pi^4 2\pi^4$ are omitted for clarity. [b] Approximate orbital correspondences are: $8\sigma \equiv \mathrm{O}(2p\sigma)$; $9\sigma \equiv \mathrm{Ca}(4s\sigma)$; $10\sigma \equiv \mathrm{Ca}(4p\sigma)$; $3\pi \equiv \mathrm{O}(2p\pi)$; $4\pi \equiv \mathrm{Ca}(4p\pi)$

Energetics of Ca[$4s^2(^1S_0)$, $4s4p(^3P_J)$, $4s3d(^1D_2)$] + *Oxidant Reactions*

Figure 7 displays the term values of CaO electronic states of excitation less than 5 eV and the available reaction energies for the Ca[$4s^2(^1S_0)$,

$4s4p(^3P_J)$, $4s3d(^1D_2)$] + N_2O, CO_2, and O_2 reactions. Term energies for CaO are taken from Table 5. Reaction energetics were calculated via a relationship analogous to equation (6) for formation of CaO product in its electronic and vibrational ground state. Accurate evaluation of the dissociation energy of CaO($X^1\Sigma^+$), like that of MgO($X^1\Sigma^+$), has been bedevilled with uncertainty even though a large number of workers have addressed the problem of its determination using a wide variety of techniques. The value chosen here for construction of Figure 7 is $D_0^0[\text{CaO}(X^1\Sigma^+)] \geq (4.11 \pm 0.07)$ eV obtained from the CaO chemiluminescence spectrum arising from the Ca($^1D_2$) + $O_2(X^3\Sigma_g^-)$ reaction observed under single-collision conditions.[224] This appears to be the most accurate and precise experimental determination of $D_0^0[\text{CaO}(X^1\Sigma^+)]$ to date and is in total accord with a recent theoretical estimate.[225]

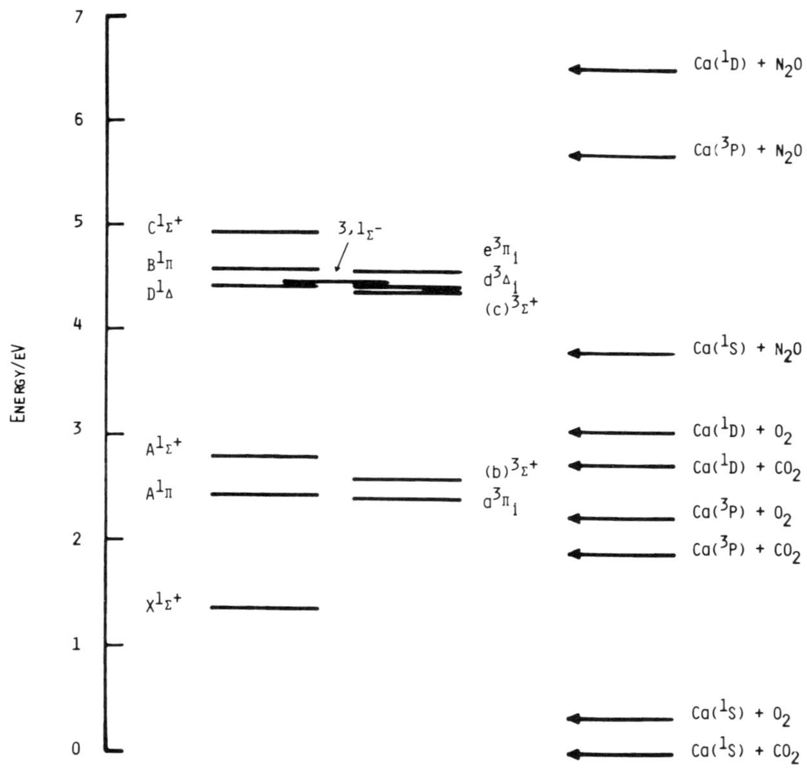

Figure 7 *Energy level diagram showing states of* CaO *below* 3.6 eV *and available energies for reaction of* Ca(^1S,^3P,^1D) *atoms with* N_2O, O_2, *and* CO_2

[224] J. A. Irvin and P. J. Dagdigian, *J. Chem. Phys.*, 1980, **73**, 176.
[225] C. W. Bauschlicher, jun., and H. Partridge, *Chem. Phys. Lett.*, 1983, **94**, 366.

Reactions with N_2O

The light-emitting reactions of Ca atoms with N_2O have been studied under single-collision conditions using an atomic beam plus scattering gas arrangement by Irvin and Dagdigian[157,158] and Wang and Menzinger[160] (electronically excited and ground-state atoms) and by Zare and co-workers[209,226] (ground-state atoms only). Examination of Figure 7 indicates that production of CaO states with electronic configurations $8\sigma^2 3\pi^3 4\pi$, $8\sigma 3\pi^4 4\pi$, $8\sigma^2 10\sigma 3\pi^3$, and $8\sigma 10\sigma 3\pi^4$ (see Table 6) located about 3.1–3.3 eV above the ground state is feasible from the reactions of both metastable states $Ca[4s4p(^3P_J)]$ and $Ca[4s3d(^1D_2)]$ with N_2O. Therefore, on energetic grounds alone, we may expect to observe orange and green electronic chemiluminescence characteristic of the CaO arc band transitions in both these cases, in addition to emission in the red and infrared regions resulting from transitions out of $CaO(A\,^1\Sigma^+)$ and $CaO(A'\,^1\Pi)$. For the $Ca[4s^2(^1S_0)]$ ground-state atom, however, production of the upper arc band states is endoergic, though low-lying electronically excited products are energetically accessible. Indeed, the orange and green arc bands of CaO constitute the prominent features of the observed chemiluminescence spectrum over the wavelength range $\lambda = 300$—900 nm arising from the reactions of $Ca(^3P_J)$ and $Ca(^1D_2)$ atoms with N_2O under molecular beam conditions.[157,158,160] A broad unresolved continuum emission is also a persistent spectral feature, the principal emitter of which has been shown to be $A'\,^1\Pi \rightarrow X\,^1\Sigma^+$ bands from rotationally hot $CaO(A'\,^1\Pi)$ [$\approx (69 \pm 25)$ %][157,158,160] with the remainder attributed to $CaO(A\,^1\Sigma^+ \rightarrow X\,^1\Sigma^+)$[157,160] and $CaO(B\,^1\Pi \rightarrow X\,^1\Sigma^+)$[160] emission. The assignment of chemiluminescence spectra reported by these two research groups[157,160] relies greatly on direct inference based on the technique of lifetime-separated spectroscopy[227,228] in order to distinguish between the relatively long-lived $A'\,^1\Pi \rightarrow X\,^1\Sigma^+$ emission[200,221,222] and shorter-lived $A\,^1\Sigma^+ \rightarrow X\,^1\Sigma^+$ chemiluminescence.[157,221,222] In contrast, only very little chemiluminescence has been detected from the analogous reaction of the ground-state atom.[157,158] State-resolved photon emission yields over the wavelength range $\lambda = 300$—900 nm of $>(0.15 \pm 0.09)$ % for $Ca(^1S_0) + N_2O$,[157] $>(7.2 \pm 2.6)$ %, $>(6.5 \pm 0.6)$ %, and (10 ± 3) % for $Ca(^3P_J) + N_2O$,[157,158,160] and $>(5.1 \pm 3.2)$ % for $Ca(^1D_2) + N_2O$[157] imply that collisions of the metastable triplet state with the molecular oxidant are most efficient at producing light. Dagdigian and co-workers[50,158] have further noted that regardless of the exact assignment of the upper states giving rise to the orange and green visible chemiluminescence characteristic of the $Ca(^3P_J) + N_2O$ reaction, the measured photon yields listed above are much too large to be explained by a

[226] Ch. Ottinger and R. N. Zare, *Chem. Phys. Lett.*, 1970, **5**, 243.
[227] A. Siegal and A. Schultz, *Chem. Phys.*, 1978, **28**, 265.
[228] J. G. Pruett and R. N. Zare, *J. Chem. Phys.*, 1975, **62**, 2050.

purely statistical decay of a $Ca^+N_2O^-$ ion-pair intermediate. Dagdigian and co-workers[57,123,157–159,224] and Telle and Brinkmann[299,230] have shown independently that reactions of metastable excited atoms lend themselves readily to measurements of chemiluminescence and quenching cross-sections without the need for inherently inaccurate calibrations based upon knowledge of absolute detection sensitivities. Qualitative inspection of the state-resolved chemiluminescence cross-sections derived from molecular beam measurements[157,160] reveals that for reaction of $Ca(^1D_2)$ atoms with N_2O, the reaction exoergicity is channelled mainly into production of electronically excited CaO states which give rise to emission of the orange and green arc band systems rather than the lower-lying $A\,^1\Sigma^+$ and $A'\,^1\Pi$ states; for reaction of $Ca(^3P_J)$ atoms, the available reaction energy is utilized in such a manner that results in preferential production of the $A'\,^1\Pi$ state – and to a lesser extent $A\,^1\Sigma^+$ – rather than the high-lying upper arc band states. This conjecture is supported by consideration of adiabatic reaction pathways linking reactants and products, as will be discussed later in this section. The relatively small magnitudes of reactive (beam attenuation) cross-sections for the $Ca(^1S_0, ^3P_J, ^1P_J) + N_2O$ reactions, measured under single-collision conditions,[157,158,160] provide evidence for a short-range electron jump as the rate-determining step during close-in encounters of the colliding particles, as has been established elsewhere[178] for the $Ba(^1S_0) + N_2O$ system. Such a mechanistic possibility will also be considered in more detail later in this section.

In addition, Irvin and Dagdigian have investigated collisional energy transfer between low-lying $CaO(A'\,^1\Pi)$ produced reactively in $Ca(^3P_J) + N_2O$ collisions and the molecular oxidant under a multiple-collision regime[157] where the chemiluminescence yield from $CaO(A\,^1\Sigma^+)$ was found to increase with scattering gas pressure:

$$CaO(A'\,^1\Pi) + N_2O(\widetilde{X}\,^1\Sigma^+) \rightarrow CaO(A\,^1\Sigma^+) + N_2O(\widetilde{X}\,^1\Sigma^+) \quad (16)$$

These workers[157] have proposed that process (16) occurs predominantly from high rotational levels of $CaO(A'\,^1\Pi)$, which are known[217] to be strongly mixed with levels of the $A\,^1\Sigma^+$ state, though some transfer may also occur from high vibrational levels of the $CaO(X^1\Sigma^+)$ ground state, which is mixed at high rotational levels with the $A'\,^1\Pi$ state, and hence also with $A\,^1\Sigma^+$. A similar mechanism has been invoked to explain collisional energy transfer at low pressures into $BaO(A\,^1\Sigma^+)$ in the $Ba(^1S_0) + N_2O$ reaction.[231] In a related study, a model for *E–E* collisional energy transfer induced by dipolar forces has been proposed by Alexander in an attempt to deduce the mechanism of the observed

229 H. Telle and U. Brinkmann, *Mol. Phys.*, 1980, **39**, 361.
230 U. Brinkmann, W. Hartig, H. Telle, and H. Walther, *Appl. Phys. (Berlin)*, 1974, **5**, 109.
231 R. W. Field, C. R. Jones, and H. P. Broida, *J. Chem. Phys.*, 1974, **60**, 4377.

electronic energy transfer.[192,193] This is based upon the long-range coupling of the permanent dipole of the polar collision partner – in this case N_2O – and a transition dipole between electronic states of the alkaline earth metal oxide of interest – in this case CaO.[192] The transition dipole arises primarily from the small non-Born–Oppenheimer coupling of the ro vibronic manifolds of individual dark ($a^3\Pi_i$, $A'^1\Pi$) and radiating ($A^1\Sigma^+$) excited electronic states of CaO. Rate constants for process (16) have been calculated[192,193] to be $>1 \times 10^{-10}$ cm^3 molecule^{-1} s^{-1} over a wide range of rotational energy levels in the neighbourhood of approximate degeneracy where, for a particular pair of vibrational quantum numbers, the rotational ladders of the two states become nearly isoenergetic. Employing these computed rate constants, Alexander and Osmolovsky have modelled the pressure dependence of CaO($A^1\Sigma^+ \rightarrow X^1\Sigma^+$) emission[193] under kinetic conditions typical of the beam–gas experiments of Dagdigian and co-workers.[157,158] The spectral variation and increase in intensity of fluorescence from CaO($A^1\Sigma^+$) predicted by Alexander and Osmolovsky are found[193] to be qualitatively similar to features in the $A^1\Sigma^+ \rightarrow X^1\Sigma^+$ chemiluminescence spectrum reported by Irvin and Dagdigian.[157]

In their study of chemiluminescence from the reaction of Ca($^1S_0$) with N_2O under molecular beam conditions,[157] Irvin and Dagdigian were able only to discern CaO($A'^1\Pi \rightarrow X^1\Sigma^+$) and CaO($A^1\Sigma^+ \rightarrow X^1\Sigma^+$) bands in the near infrared; no chemiluminescence was observed at wavelengths shorter than $\lambda = 600$ nm. Such a chemiluminescence spectrum obtained by these workers is in considerable disagreement with the single-collision results of Zare and co-workers,[209,226] who have reported a broad structureless emission over the wavelength range $\lambda = 400$—570 nm, peaking around $\lambda = 490$ nm. Zare and his collaborators have extracted a chemiluminescence cross-section for emission arising from the Ca($^1S_0$) + N_2O reaction of $\sigma_{chem} = 0.16$ nm^2 based on kinetic arguments,[209] which may be contrasted with the much smaller value of $\sigma_{chem} = (1.3 \pm 0.7) \times 10^{-3}$ nm^2 determined by Irvin and Dagdigian.[157] These latter workers noted that the spectral distribution of chemiluminescence reported by Zare's group[209,226] bears a remarkable resemblence to that for the reaction of ground-state Ba($^1S_0$) atoms with N_2O, also studied with the same apparatus, and for which a chemiluminescence cross-section of $\sigma_{chem} = 0.02$ nm^2 has been measured.[232] It now seems likely[157] that the chemiluminescence from the Ca($^1S_0$) + N_2O reaction reported by Zare and co-workers[209,226] arose from reaction with the molecular oxidant of ground-state barium atoms present as an impurity in the atomic beam source.

Chemiluminescence has also been reported from Ca–N_2O–He/Ar flames[62,233] and from ternary Ca–N_2O–CO–He/Ar flames.[61,62,141,142] The

[232] C. R. Dickson, S. M. George, and R. N. Zare, *J. Chem. Phys.*, 1977, **67**, 1024.
[233] G. A. Capelle, C. R. Jones, J. Zorskie, and H. P. Broida, *J. Chem. Phys.*, 1974, **61**, 4777.

spectrum determined by Capelle *et al.* from a Ca–N_2O–Ar flame is complex and exhibits the $A'^1\Pi \rightarrow X^1\Sigma^+$ and $A^1\Sigma^+ \rightarrow X^1\Sigma^+$ bands of CaO together with an unidentified band in the orange at $\lambda \approx 625$ nm and a continuum emission stretching to shorter wavelengths.[233] Chemiluminescence from CO-seeded flames has been studied in emission,[141,142] by modulated intracavity resonance spectroscopy,[62] and by intracavity dye-laser spectroscopy.[61] In these experiments, the orange and green arc band transitions of CaO and the Ca($^3P_1 \rightarrow {}^1S_0$) atomic line at $\lambda = 657.3$ nm are prominent features, indicating that secondary reaction of Ca(^3P_J) atoms with N_2O dominates the observed chemiluminescence spectrum. This explanation has subsequently been verified by the observations of Dagdigian,[157,158] who noted that chemiluminescence emission from the ground-state reaction is negligible while that from the excited-atom reaction is not, as discussed above. Again, an unresolved continuum emission extending into the blue region is present in the CO-seeded flames.[61,62,141,142] A mechanism for production of excited metal atoms in such flames, proposed by Benard and co-workers,[141,142] has been given as equations (8), (9), and (10) and is described in Section 4. Early flame photometric investigations of the alkaline earth monoxides CaO, SrO, and BaO have been discussed by van der Hurk *et al.*[234] These authors also report a method for identification of emitting monoxide and monohydroxide species responsible for some of the visible bands observed when a solution containing the appropriate alkaline earth metal is sprayed into C_2H_2–air, H_2–air, and moist CO–N_2–O_2 and CO–N_2O flames.[234]

The kinetics of oxidation processes involving ground-state atomic calcium have been investigated by probing the spatial temperature variation in the reaction zone of a diffusion flame of Ca atoms burning in the appropriate molecular oxidant.[146–151,153–156,162–164,235] For reaction of Ca($^1S_0$) atoms with N_2O, Kashireninov and co-workers have extracted an Arrhenius activation energy of $E_{act} = (23.4 \pm 10.0)$ kJ mol^{-1} and a pre-exponential factor of $A = (2.7\,^{+0.3}_{-0.4}) \times 10^{-11}$ cm^3 molecule^{-1} s^{-1}.[149,156] The time-resolved investigations of Husain and Roberts[161] have shed some light on the different routes leading to formation of different product electronic states of CaO from the reactions of Ca(^3P_J) and Ca($^1D_2$) atoms with N_2O under bulk conditions. A comparison of the time dependences of the molecular and atomic emission profiles enabled these workers to ascribe the production of emitting molecular states either to direct generation from chemical reaction between Ca(^3P_J) and Ca($^1D_2$) atoms and N_2O:

$$\mathrm{Ca}(^3P_J, {}^1D_2) + N_2O \rightarrow \mathrm{CaO}^* + N_2 \qquad (17)$$

[234] J. van der Hurk, Tj. Hollander, and C. Th. J. Alkemade, *J. Quant. Spectrosc. Radiat. Transfer*, 1973, **13**, 273.

[235] V. A. Kuznetsov, L. F. Repka, and O. E. Kashireninov, *Zh. Fiz. Khim.*, 1978, **52**, 430, [*Russ. J. Phys. Chem.*, 1978, **52**, 237].

or to *E–E* energy transfer between these two atomic states and ground-state $CaO(X^1\Sigma^+)$:

$$Ca(^3P_J,{}^1D_2) + CaO(X^1\Sigma^+) \rightarrow Ca(^1S_0) + CaO^* \quad (18)$$

Following laser generation of $Ca(^3P_J)$ atoms in the presence of N_2O, the highly excited $(c)^3\Sigma^+$, $D^1\Delta$, $d^3\Delta$, $^{3,1}\Sigma^-$, $e^3\Pi$, and $B^1\Pi$ states (responsible for orange arc band fluorescence[213—215]), the $^3\Pi$ and $^1\Pi$ states of electronic configuration $8\sigma^2 10\sigma 3\pi^3$ and the $^3\Sigma^+$ state of electronic configuration $8\sigma 10\sigma 3\pi^4$ (thought to be responsible for green arc band emission[214]), and the lower-lying $A'^1\Pi$ state in vibrational levels $v' = 13$–15 were shown to arise directly from collisions between the excited atom and molecular oxidant.[161] On the other hand, production of the vibronic states $CaO(A'^1\Pi,\ 8 \leq v' \leq 12)$ and CaO $(A^1\Sigma^+, v' \leq 4)$ was demostrated to occur via secondary electronic energy transfer collisions between $Ca(^3P_J)$ and $CaO(X^1\Sigma^+)$.[161] The time dependence of orange and green arc band chemiluminescence from excited CaO^*, monitored following laser excitation of $Ca(^1D_2)$ atoms, is also indicative of the production of the high-lying arc band states by direct reaction. Conversely, the lower-lying product states $CaO(A'^1\Pi)$ (in vibrational levels $v' = 8$—15) and $CaO(A^1\Sigma^+)$ (monitored via sequences $\Delta v = -2$ to -9 with $v' = 14$ as the energetically allowed maximum[71,91]) displayed chemiluminescence profiles consistent with bimolecular energy transfer leading to their formation.[161] Estimated relative populations in emitting product states showed, in agreement with the earlier work of Dagdigian[157,158] and Menzinger,[160] that reaction of the $^1D_2$ state with N_2O results in preferential population of the high-lying arc band states of CaO rather than the lower-lying $A'^1\Pi$ state.[161] In the case of molecular states arising from the $Ca(^3P_J)$ reaction, all product-state populations are of comparable magnitude.[161] Measurement of vibrational temperatures in $CaO(A'^1\Pi)$ indicates some measure of vibrational excitation accompanying both reaction and *E–E* energy transfer processes [equations (17) and (18)].[161] From their temperature-dependent rate data for overall removal of both metastable atomic states in the presence of N_2O, Husain and Roberts derived the equations:[161]

$$k_{N_2O}(T) = (3.5^{+8.5}_{-2.5}) \times 10^{-10} \exp[-(36.1 \pm 11.6)\ \mathrm{kJ\ mol^{-1}}/RT] + (7.2^{+5.7}_{-3.3}) \exp[-(219.2 \pm 36.9)\ \mathrm{kJ\ mol^{-1}}/RT]\ \mathrm{cm^3\ molecule^{-1}\ s^{-1}} \quad (19)$$

for loss of $Ca(^3P_J)$ atoms, and:

$$k_{N_2O}(T) = (3.8^{+2.9}_{-1.7}) \times 10^{-10} \exp[-(17.5 \pm 3.7)\ \mathrm{kJ\ mol^{-1}}/RT] + (25.7^{+10.9}_{-7.7}) \exp[-(240.4 \pm 54.4)\ \mathrm{kJ\ mol^{-1}}/RT]\ \mathrm{cm^3\ molecule^{-1}\ s^{-1}} \quad (20)$$

for removal of the $^1D_2$ state. In a manner analogous to that described in Section 4 for the $Mg(^3P_J) + N_2O$ reaction, the first term on the right-hand side of equations (19) and (20) represents a temperature-dependent rate of removal of both excited atomic states according to the simple Arrhenius expectation [see equation (3)]. The second term in both equations (19) and (20) reflects additional removal processes involving the atomic species due to quenching by molecular oxygen arising from thermal decomposition of N_2O. The only other measurement of a bimolecular rate constant for quenching of $Ca(^3P_J)$ atoms by N_2O appears to be that of Husain and Schifino,[236] who report $k_{N_2O}(T) = (6.5 \pm 0.5) \times 10^{-13}$ cm^3 molecule^{-1} s^{-1} at $T = 1000$ K using the same time-resolved technique as the later study.[161] Apart from the work described above,[161] no other measurements of rate constants for collisional removal of $Ca(^1D_2)$ atoms by N_2O appear to have been reported to date.

In the absence of any *ab initio* calculations of the potential energy hypersurfaces for the reactions of electronically excited calcium atoms with N_2O, discussion of the criteria governing the dynamical behaviour of such processes is necessarily more speculative than the description of the $Mg(^1S_0,^3P_J) + N_2O$ systems presented in Section 4. Because of the low ionization potentials of $Ca(^3P_J)$ and $Ca(^1D_2)$ atoms (listed in Table 1), their reaction with N_2O is probably initiated by the transfer of an electron from the excited atom to N_2O[11,12,24,166,167] to form a singly charged ion-pair intermediate $Ca^+N_2O^-$.[50,158,160] Addition of an electron to the 3π antibonding orbital (LUMO) of N_2O which transforms as the a' irreducible representation in C_s symmetry (the b_2 representation in C_{2v} symmetry)[169,170,174,237] stabilizes the incipient N_2O^- anion in a bent structure. If the $Ca^+N_2O^-$ intermediate has a planar geometry, its electronic state will be symmetric (A') on reflection through a molecular plane. It is improbable, however, that a planar geometry would be retained on dissociation, so that electronic-state symmetry arguments probably do not provide strict constraints on the accessible CaO product states formed (see later). Following the discussion of $Mg(^3P_J) + N_2O$ collisions given in Section 4, we may simply postulate here that a close encounter of excited $Ca(^3P_J)$ or $Ca(^1D_2)$ atoms with N_2O causes the molecule to distort from linearity, thereby increasing its electron affinity[169,170,173,177] and facilitating a 'collision-assisted' electron jump. By analogy with the results obtained by Wren and Menzinger for the $Ba[6s^2(^1S_0)] + N_2O$ system,[177,178] we may again expect the electron transfer distance and chemiluminescence cross-section to increase as the electron affinity of the molecule becomes more positive on excitation of the ν_2 in-plane bending mode.

An adiabatic correlation diagram for the reactions of various electronic states of atomic calcium with N_2O is portrayed in Figure 8. For

[236] D. Husain and J. Schifino, *J. Chem. Soc., Faraday Trans. 2*, 1983, **79**, 1677.
[237] G. V. Pfeiffer and L. C. Allen, *J. Chem. Phys.*, 1969, **51**, 190.

the $Ca(^3P_J) + N_2O$ reaction, this diagram indicates that there is no adiabatic pathway to the principal $CaO(A'^1\Pi)$ product.[157,160] Instead, these reactants correlate with the two lowest-lying triplet states $a^3\Pi_i$ and $(b)^3\Sigma^+$. It is for this reason that this reaction is a suitable candidate for locating the $(b)^3\Sigma^+$ state which has thus far not been observed; in BaO, it has been identified from LIF measurements.[238] Wang and Menzinger[160] have offered an explanation for efficient chemical production of $CaO(A'^1\Pi)$ from $Ca(^3P_J) + N_2O$ reactants in terms of a spin–orbit mechanism similar to that responsible for the quenching of $O[2p^4(^1D_2)]$ atoms by molecular nitrogen. This involves formation of a long-lived complex and multiple passage of trajectories through the coupling region.[239,240] It appears therefore that the $Ca^+N_2O^-$ ion-pair intermediate is able to rearrange with ease into the principal $CaO(A'^1\Pi)$ product regardless of the presence of spin restrictions. Indeed, observation of a significant yield of $CaO(A'^1\Pi)$[157,160] argues strongly against conservation of spin as a stringent selection rule in the $Ca(^3P_J) + N_2O$ reaction.

The spin- and symmetry-conserved picture displayed in Figure 8 reveals that formation of the minor products of the $Ca(^3P_J) + N_2O$ reaction[157,160] are adiabatically forbidden, even though production of $CaO(A^1\Sigma^+)$ and the upper arc band states at about 3.1 eV above the ground state are energetically accessible. The $A^1\Sigma^+$, $A'^1\Pi$, and $D^1\Delta$ states of CaO do, however, correlate with the more highly energized $Ca(^1D_2)$ atomic reagent. Indeed, the observed[157,160,161] yields of $CaO(A^1\Sigma^+)$ and of $Ca(4p\pi)O(2p\sigma)^2(2p\pi)^3$ and $Ca(4p\pi)O(2p\sigma)(2p\pi)^4$ states, which include $CaO(D^1\Delta)$ as indicated in Table 6, are not inconsistent with $Ca(^1D_2)$ as the major precursor atomic state. The trend noted initially by Irvin and Dagdigian,[157] namely that reaction of $Ca(^1D_2)$ atoms with N_2O produces arc band emitters about ten times more efficiently than the analogous reaction of $Ca(^3P_J)$ atoms, is supported by Figure 8 in terms of the greater number of surface crossings required for their formation for $Ca(^3P_J) + N_2O$ starting materials. Thus although both $Ca(^3P_J) + N_2O$ and $Ca(^1D_2) + N_2O$ reactions proceed over several adiabatic potential energy hypersurfaces, products which require only a limited number of curve crossings tend to be favoured.[157] It remains unclear, however, why the available energy from the $Ca(^1D_2) + N_2O$ reaction is channelled predominantly into the production of $CaO(D^1\Delta)$ and other upper arc band states with a correspondingly low yield of the lower-lying $CaO(A^1\Sigma^+)$ and $CaO(A'^1\Pi)$ products, even though formation of these last two states is adiabatically allowed.

Facile production of the manifold of states constituting the high-lying arc band emitters, with valence electronic configurations $CaO(8\sigma^2 3\pi^3 4\pi)$

[238] R. A. Gottscho, *J. Chem. Phys.*, 1979, **70**, 3554.
[239] J. C. Tully, *J. Chem. Phys.*, 1974, **61**, 61.
[240] J. C. Tully, *J. Chem. Phys.*, 1975, **62**, 1893.

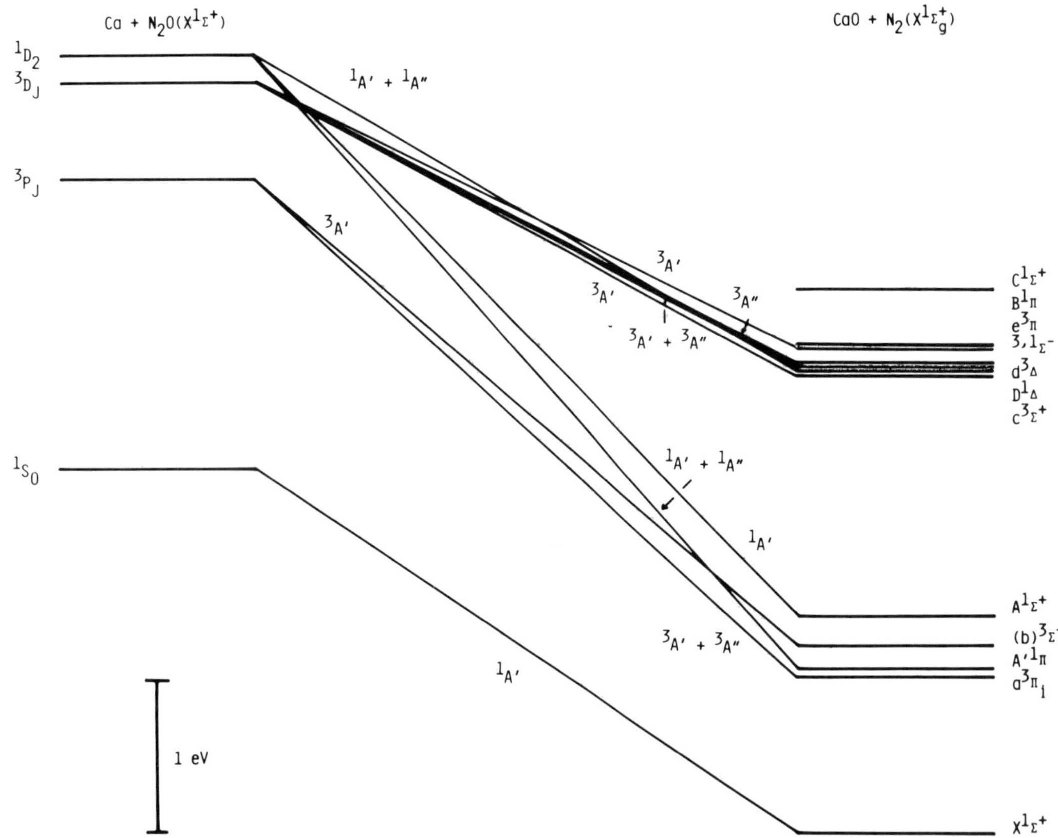

Figure 8 *Correlation diagram connecting the states of* Ca + N_2O *and* CaO + N_2 *constructed on the basis of the weak spin–orbit coupling approximation and* C_s *symmetry in the collision complex*. (Reproduced with permission from *Chem. Phys.*, 1986, **109**, 393)

and $CaO(8\sigma 3\pi^4 4\pi)$ (see Table 6), from the $Ca(^3P_J) + N_2O$ reaction can be understood by a mechanism which emphasizes transitions between electronic configurations.[50,158,160] The same mechanism has been independently proposed by Brinkmann and Telle for production of electronically excited $CaCl(A\,^2\Pi)$ from the reactions of $Ca(^3P_J)$ atoms with molecular chlorine and $Ca(^1D_2)$ atoms with both chlorine and hydrogen chloride.[229,241] Initially, it is supposed that collision-induced bending of the N_2O molecule will result in a co-planar tetra-atomic transient species. If the projection of the orbital angular momentum of the excited calcium atom on to the axis connecting the atom and colliding molecule is chosen to be non-zero following Brinkmann and Telle,[241] then for nearly co-planar (C_s) collisions in which the $4p$ orbital of the excited $Ca(^3P_J)$ state is orientated perpendicular to the molecular plane (*i.e.* a'' symmetry), transfer of the electron from this atomic orbital to the 3π antibonding LUMO of N_2O (a' symmetry) is symmetry-forbidden.[169,237] Consequently, the subjacent $Ca(4s)$ electron (a' symmetry), being closer to the colliding molecule in any case,[241] is transferred, as would be expected to occur in reaction of the ground-state atom. As the excited $4p$ electron remains localized at the calcium ion so generated within a lone $4p\pi$ orbital, the resulting excited ionic intermediate $Ca^+N_2O^-$ may eventually rearrange smoothly into products with electronic configurations $Ca(4p\pi)O(2p\sigma 2p\pi^4)$ and $Ca(4p\pi)O(2p\sigma^2 2p\pi^3)$,[150,158,160] from which the $B\,^1\Pi$ and other arc band states derive (see Table 6).

In view of the results of various theoretical treatments of the low-lying electronically excited states of CaO,[218–222] and by analogy with the situation for ground and electronically excited states of alkaline earth monohalides,[242,243] the ionic character of the low-lying excited states of CaO may be anticipated from the similarity of their potential energy curves to the ground-state potential, which is known to be ionic.[220] As a consequence of this, a reactive collision leaving the excited $4p$ electron localized at the incipient Ca ion while the second valence electron is switched to the oxygen atom of N_2O proceeds on a highly attractive potential energy hypersurface, any energy barrier being situated in the entrance valley.[241] Such reactions are known to release excess energy mainly into vibrational excitation of the newly created diatomic species.[244] The departing nitrogen molecule probably plays a 'spectator' role in such processes. Together with measurements of product internal energies, the use of LIF techniques of high spectral resolution[245] to

[241] U. Brinkmann and H. Telle, *J. Phys. B*, 1977, **10**, 133.

[242] L. B. Knight, jun., W. C. Eastley, W. Weltner, jun., and M. Wilson, *J. Chem. Phys.*, 1971, **54**, 322.

[243] P. J. Dagdigian, H. W. Cruse, and R. N. Zare, *J. Chem. Phys.*, 1974, **60**, 2330.

[244] J. C. Polanyi and J. L. Schreiber, in 'Physical Chemistry, An Advanced Treatise: Kinetics of Gas Reactions', ed. W. Jost, Academic Press, New York, 1974, Vol. VIA, p. 383.

[245] P. J. Dagdigian and R. N. Zare, *J. Chem. Phys.*, 1974, **61**, 2464.

measure recoil translational energies of separating products would permit these hypotheses to be tested. Alexander and Dagdigian[50] have also pointed out that if an entropy deficiency is assumed for the $CaO(X^1\Sigma^+)$ product of the $Ca(^3P_J) + N_2O$ reaction comparable in magnitude to the values reported for the reactions of $Ca(^3P_J)$ and $Ca(^1D_2)$ atoms wtih CO_2, the predicted branching ratios in the upper arc band states are still significantly smaller than the observed photon yields.[157,158,160]

Cox and Dagdigian[123] have suggested that CaO in its $A^1\Sigma^+$ and $A'^1\Pi$ states is formed in the $Ca(^3P_J) + N_2O$ reaction via a mechanism which involves exit-channel mixing of the potential energy surfaces leading to these states with surfaces leading to the spin-allowed $a^3\Pi_i$ and (possibly) $(b)^3\Sigma^+$ product states. It is further proposed[123] that such non-adiabatic processes are facilitated by the significant electronic matrix elements and small energy splittings[217] between the asymptotic product states. Indeed, extensive perturbations in the $A^1\Sigma^+$–$X^1\Sigma^+$ band spectrum of CaO were first noted some thirty-seven years ago,[186] being only more recently assigned[217] to mixing between the $A^1\Sigma^+$ state and the $a^3\Pi_i$ and $A'^1\Pi$ states. In addition to large electronic matrix elements, these perturbations are enhanced by large vibrational overlaps arising from the fact that potential energy curves for these states lie so close to each other in energy.[217] Marks *et al.* have also shown that the $a^3\Pi_i$ and $A'^1\Pi$ states of CaO are strongly mixed.[213]

An information–theoretic analysis[7,11,12,24,48,49] has been performed by Wang and Menzinger[160] to establish 'prior' branching ratios into electronically excited states of CaO product resulting from reactions of $Ca(^3P_J)$ and $Ca(^1D_2)$ atoms wtih N_2O. In neither of the $Ca(^3P_J) + N_2O$ and $Ca(^1D_2) + N_2O$ reactions do the calculated branching fractions decrease monotonically with electronic excitation energy, suggesting qualitatively some measure of non-statistical behaviour.[160] For both reactions, the principal chemiluminescence channel leading to $CaO(A'^1\Pi)$ product is characterized by relatively low surprisals, and phase-space theory[47] serves as a reasonable approximation to their dynamics.[50] The reactive pathways leading to $CaO(A^1\Sigma^+)$ product, on the other hand, exhibit substantial surprisals (*i.e.* under-population) for $Ca(^3P_J)$ as atomic reactant but a nearly statistical population for the $Ca(^1D_2)$ precursor state.[160] For both $Ca(^3P_J) + N_2O$ and $Ca(^1D_2) + NO_2$ reactions, moderate surprisals have been calculated[160] for the high-energy channels, particularly those leading to formation of the upper arc band states of electronic configurations $8\sigma^2 3\pi^3 4\pi$ and $8\sigma 3\pi^4 4\pi$ (see Table 6). It has been suggested by Wang and Menzinger that the conspicuously non-linear surprisals characterizing the $Ca(^3P_J, ^1D_2) + N_2O$ and NO_2 reactions probably reflect the specific nature of electronic transitions between pairs of potential energy hypersurfaces that are generally localized in different regions of configuration space.[160]

Reactions with CO_2

Objective analysis of the study of the $Ca(^3P_J) + CO_2$ reaction carried out by Pasternack and Dagdigian[200] under single-collision conditions in a molecular beam plus scattering gas arrangement is complicated by the fact that it now seems certain that the low-lying $a^3\Pi_i$ and $A'^1\Pi$ electronic states of CaO probed by laser excitation spectroscopy did not arise from reaction of $Ca(^3P_J)$ atoms with the molecular oxidant but were in fact produced by reaction with the target gas of a trace amount ($\approx 2\%$) of $Ca(^1D_2)$ atoms present in the atomic beam.[224] Based on their lower estimate of $D_0^0[CaO(X^1\Sigma^+)]$,[224] considered to be accurate,[246] Irvin and Dagdigian showed that electronically excited states of CaO could only be reached via endoergic reaction channels from $Ca(^3P_J) + CO_2$ starting materials. Figure 7 confirms ground-state $CaO(X^1\Sigma^+)$ as the only energetically allowed product of the reaction of $Ca(^3P_J)$ atoms with CO_2, and also indicates that for $Ca(^1D_2)$ as atomic reactant, production of electronically excited $CaO(a^3\Pi_i, A'^1\Pi, (b)^3\Sigma^+)$ as well as ground-state $CaO(X^1\Sigma^+)$ can occur via exoergic reaction pathways. Although no chemiluminescence spectrum (presumably consisting of $A'^1\Pi \rightarrow X^1\Sigma^+$ bands) appears to have been reported for collisions of an atomic beam of metastable $Ca(^3P_J)$ and $Ca(^1D_2)$ atoms with CO_2 target gas, in view of the foregoing, Irvin and Dagdigian expect on energetic grounds alone that it would be dominated by reaction of $Ca(^1D_2)$ atoms with the molecular oxidant, while the chemical pathway leading to ground-state products would be traversed mainly by trajectories originating from $Ca(^3P_J) + CO_2$ reactants.[224] Pasternack and Dagdigian[200] have correctly[224] derived internal energy distributions for the $CaO(X^1\Sigma^+)$ product of $Ca(^3P_J, ^1D_2) + CO_2$ reactions by comparison of their observed $CaO(B^1\Pi \leftarrow X^1\Sigma^+)$ LIF spectrum with simulated spectra which convolute calculated rotational lines over the laser bandwidth.[247] In particular, the relative population distribution in the first four vibrational quantum levels of $CaO(X^1\Sigma^+)$ together with a rotational temperature characterizing a Boltzmann distribution of rotational energy in the $v = 0$ level were determined from laser excitation spectra.[200] These authors found that most of the population resides in the lowest vibrational state with a sharp decline exhibited with increasing vibrational quantum number. A lower limit to the rotational temperature of the $v = 0$ level was determined to be $T_{rot} \geq (750 \pm 200)$ K, which decreased with increasing scattering gas pressure. Only a small fraction of the available reaction energy in the CaO ground-state channel ($\sim 10\%$) appears as internal excitation of the diatomic product. Of course, some of the remaining energy may be taken up as internal excitation of the CO molecule also formed on collision of the excited atoms with CO_2.

[246] J. B. Pedley and E. M. Marshall, *J. Phys. Chem. Ref. Data*, 1983, **12**, 967.
[247] P. J. Dagdigian, H. W. Cruse, A Schultz, and R. N. Zare, *J. Chem. Phys.*, 1974, **61**, 4450.

Branching ratios for the $X^1\Sigma^+$ and unresolved $a^3\Pi_i/A'^1\Pi$ product states of the $Ca(^3P_J) + CO_2$ reaction derived by Pasternack and Dagdigian from observed relative integrated fluorescence intensities of the $B^1\Pi \leftarrow X^1\Sigma^+$ and orange and green arc band systems strictly represent lower limits for electronic-state branching in the $Ca(^1D_2) + CO_2$ reaction.[224] The branching ratio for formation of CaO product in the excited $a^3\Pi_i$ and $A'^1\Pi$ states to that for production of $X^1\Sigma^+$ ground state was found to be 0.3,[200] with an experimental uncertainty of about a factor of two.

The temperature dependence of the rate of removal of $Ca(^3P_J)$ and $Ca(^1D_2)$ atoms by CO_2 has been studied by Husain and Roberts using the technique of time-resolved atomic resonance emission following pulsed laser generation of the metastable atomic states.[248] Absolute second-order rate constants were obtained by these workers across the temperature ranges 725–1100 K $[Ca(^3P_J) + CO_2]$ and 750–1100 K $[Ca(^1D_2) + CO_2]$, yielding the non-Arrhenius forms:

$$k_{CO_2}(T) = (2.2^{+4.1}_{-1.4}) \times 10^{-10} \exp[-(51.5 \pm 9.2)\ \mathrm{kJ\,mol^{-1}}/RT] + (1.5^{+2.1}_{-0.9}) \times 10^{-3} \exp[-(197.5 \pm 48.0)\ \mathrm{kJ\,mol^{-1}}/RT]\ \mathrm{cm^3\,molecule^{-1}\,s^{-1}} \quad (21)$$

and

$$k_{CO_2}(T) = (3.4^{+9.7}_{-2.5}) \times 10^{-10} \exp[-(29.7 \pm 7.3)\ \mathrm{kJ\,mol^{-1}}/RT] + (7.6^{+10.2}_{-4.5}) \times 10^{-3} \exp[-(182.8 \pm 51.5)\ \mathrm{kJ\,mol^{-1}}/RT]\ \mathrm{cm^3\,molecule^{-1}\,s^{-1}} \quad (22)$$

for collisional removal of $Ca(^3P_J)$ and $Ca(^1D_2)$ atoms, respectively.[248] Kinetic arguments similar to those derived for the $Mg(^3P_J) + CO_2$ system[202] were again employed by Husain and Roberts to explain observed deviations from the Arrhenius law at high temperatures ($T \geq 950$ K) for both reactions.[248] At these higher temperatures, the enhanced rate of removal of the metastable atomic states were considered by these workers as arising from additional removal processes involving ground-state $CaO(X^1\Sigma^+)$ product. Chemiluminescence profiles for various vibronic transitions out of excited $CaO(A'^1\Pi)$ and $CaO(A^1\Sigma^+)$ product states, monitored by Husain and Roberts following laser excitation of $Ca(^3P_J)$ and $Ca(^1D_2)$ atoms,[248] all displayed a bi-exponential dependence on time indicative of a collisional E–(E,V) energy transfer step following initial O-atom abstraction from CO_2 by the excited atomic state:

$$Ca(^3P_J,{}^1D_2) + CaO(X^1\Sigma^+) \rightarrow Ca(^1S_0) + CaO(A'^1\Pi, A^1\Sigma^+) \quad (23)$$

[248] D. Husain and G. Roberts, *Z. Phys. Chem., N. F.*, 1986, **150**, 133.

Vibrational temperatures in the range $T_{vib} = 1076$—1697 K indicate that vibrational excitation accompanying process (23) is not that significant. Although $CaO(A'^1\Pi)$ product may be formed in vibrational levels up to $v' = 3$ from the $Ca(^1D_2) + CO_2$ reaction,[71,115,216,224] no evidence was obtained in these time-resolved experiments[248] for direct generation of this state (in any case, transitions emanating from vibrational levels $v' \leq 7$ of the $A'^1\Pi$ state lie at infrared wavelengths outside the transmission range of the monochromator employed in that work[75,248]). Production of electronically excited $CaO(A^1\Sigma^+)$ in its lowest vibrational level is 0.05 eV (≈ 4.9 kJ mol^{-1}) endoergic[71,91,115,224] from $Ca(^1D_2) + CO_2$ reagents but may be formed from direct O-atom abstraction given the uncertainty associated with the value of $D_0^0[CaO(X^1\Sigma^+)]$.[224] No evidence for such a process was observed, however.[248] There appear to have been two other determinations of the thermal bimolecular rate constant $k_{CO_2}(T)$ for the reaction of $Ca(^3P_J)$ atoms with CO_2. This has been measured by Husain and Schifino,[236] who report a value of $k_{CO_2}(T) = (3.1 \pm 0.3) \times 10^{-13}$ cm^3 molecule^{-1} s^{-1} at $T = 1000$ K using the same time-resolved technique as that employed in the temperature-dependent investigation of Husain and Roberts.[248] Nicks *et al.* have employed the technique of modulated LIF with standard Stern–Volmer analyses to obtain a similar estimate of $k_{CO_2}(T) = 1.5 \times 10^{-13}$ cm^3 molecule^{-1} s^{-1} at $T \approx 1073$ K.[249] There do not appear to have been any other measurements of the rate constant for overall removal of $Ca(^1D_2)$ atoms by CO_2.

Alexander and Dagdigian have developed a statistical procedure for making *a priori* predictions of the branching ratios into various energetically allowed product electronic states following reaction of $Ca(^3P_J)$ atoms with several oxidant molecules.[50] This method is based solely upon the total volume of translational phase space associated with each electronic channel, and the branching ratios so obtained are constrained so as to be consistent with electronic adiabaticity and/or conservation of spin angular momentum. However, values of electronic-state branching ratios determined by these workers[50] for reactions of $Ca(^3P_J)$ atoms with CO_2, O_2, and N_2O have subsequently been shown[224] to be quantitatively incorrect owing to the erroneous value of $D_0^0[CaO(X^1\Sigma^+)]$[200,224] employed in their calculation. Nevertheless, it does appear that a purely statistical interpretation of both $Ca(^3P_J) + CO_2$ and $Ca(^3P_J) + N_2O$ reactions without any dynamical constraints consistently underestimates the fractional production of electronically excited CaO.[50,157,158,160,224]

By analogy with the oxidation reactions discussed earlier in this section and in Section 4, it seems reasonable to suppose that both these reactions proceed via an electron-jump mechanism[11,12,24,50,166,167] leading to formation of a singly charged ion pair intermediate $Ca^+CO_2^-$.

249 C. M. Nicks, R. J. Malins, and D. J. Benard, *Chem. Phys. Lett.*, 1982, **89**, 523.

Although the vertical electron affinity of CO_2 is also negative,[250–252] addition of an electron to the 3π antibonding orbital of CO_2[169,174,237] can stabilize the CO_2^- anion in a bent geometry in a manner analogous to that described earlier for the N_2O^- negative ion. Presumably the requisite distortion from collinear geometry is again accomplished by collision with the incoming metal atom. Any of these singly charged $Ca^+CO_2^-$ ion-pair states so formed can dissociate without drastic orbital rearrangement to yield electronically excited states of CaO product.[182,218,219] However, since the electronic distribution and individual orbitals of ground-state CaO differ significantly from those of the single-charged excited electronic states,[182,218–222] extensive orbital rearrangement will therefore accompany the formation of ground-state products, which could result in an energy barrier or other constriction in the exit channel of the appropriate adiabatic surface.[50] This type of feature could therefore be responsible for the very cold vibrational and rotational distributions within $CaO(X^1\Sigma^+)$ product resulting from collisions of $Ca(^3P_J,{}^1D_2)$ atoms with CO_2 observed by Pasternack and Dagdigian.[200] An alternative explanation concerning the puzzle of the extremely cold internal energy distribution within the $CaO(X^1\Sigma^+)$ product of the $Ca(^3P_J) + CO_2$ reaction, put forward by Alexander and Dagdigian,[50] involves the introduction of entropy deficiencies associated with formation of $CaO(X^1\Sigma^+)$ by analogy with previous studies of the $Ba(^1S_0) + CO_2,O_2$ reactions under single-collision conditions.[228,246] These investigations have indicated that the internal energy distribution of nascent energetically accessible $BaO(X^1\Sigma^+)$ reaction products can be modelled successfully using a statistical phase-space theory.[47] In these cases, a fit of the $BaO(X^1\Sigma^+)$ internal energy distribution to a linear surprisal model results, within the uncertainty associated with the experimental data, in negligibly small vibrational and rotational surprisals as well as entropy deficiencies.[228,246]

Figure 9 shows an adiabatic correlation diagram connecting states of $Ca + CO_2$ reactants with $CaO + CO$ products based upon the weak spin–orbit coupling approximation. We may immediately note that if reaction of $Ca(^3P_J)$ atoms with CO_2 were electronically adiabatic and spin-conserved, only the low-lying $a^3\Pi_i$ and as yet unidentified $(b)^3\Sigma^+$ states of CaO product would be populated. This constrasts sharply with the rather substantial amount of ground-state $CaO(X^1\Sigma^+)$ arising from this reaction observed by Pasternack and Dagdigian.[200] Thus although spin–orbit coupling in both Ca and O atoms is weak, the spin selection rule is significantly violated in the $Ca(^3P_J) + CO_2$ reaction and C_s symmetry may not apply. The experiments of Pasternack and Dagdigian clearly show that a significant reactive flux can occur through pathways which are forbidden by strict considerations of electronic adiabaticity,[200]

[250] M. Krauss and D. Neumann, *Chem. Phys. Lett.*, 1972, **14**, 26.
[251] S. Y. Tang, E. W. Rothe, and G. P. Reck, *J. Chem. Phys.*, 1974, **61**, 2592.
[252] J. Pacansky, U. Wahlgren, and P. S. Bagus, *J. Chem. Phys.*, 1975, **62**, 2740.

though simple considerations of the reaction energetics would appear to rule out electronically excited $CaO(a^3\Pi_i)$ and $CaO[(b)^3\Sigma^+]$ as major products. The comparable yields of $CaO(X^1\Sigma^+)$ and unresolved $CaO(a^3\Pi_i,A'^1\Pi)$ products observed by these workers[200] are probably indicative of the importance of non-adiabatic channels in governing the dynamics of the reaction of $Ca(^1D_2)$ atoms with CO_2. A similar violation of the conservation of spin angular momentum has been observed in the $Ba(^1S_0) + SO_2$ reaction[253,254] and, as originally suggested by Smith and Zare,[254] can be rationalized in terms of the ion-pair reaction mechanism discussed above. Thus the tightly bound ion-pair complex may exist over a sufficiently long time period such that non-Born–Oppenheimer terms can act for a time long enough to influence substantially the course of reactive events. Relatively weak spin-orbit perturbations may thus significantly enhance the probability of non-spin-conserving products in the reactions of $Ca(^3P_J)$ and $Ca(^1D_2)$ reactions with CO_2 and N_2O.[50]

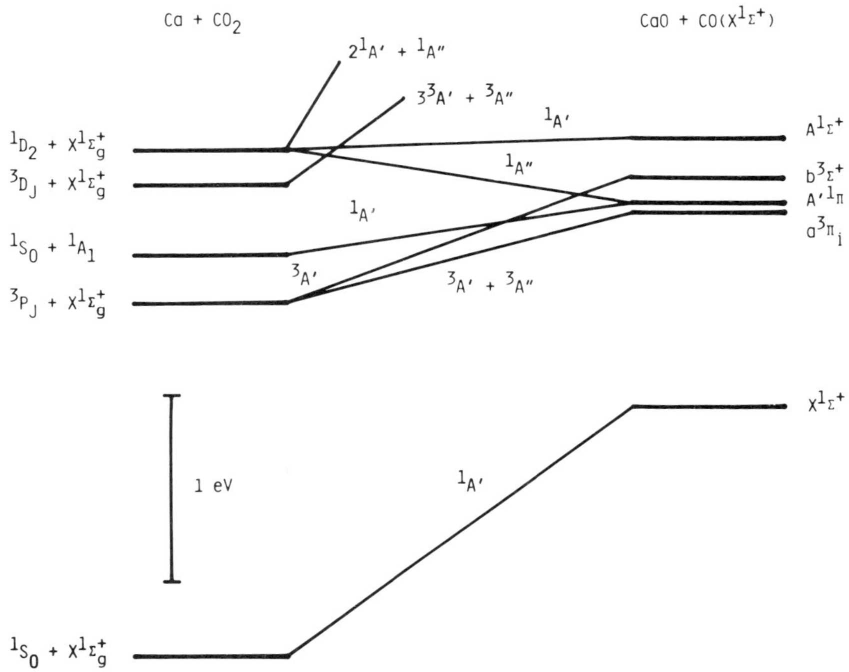

Figure 9 *Correlation diagram connecting the states of* $Ca + CO_2$ *and* $CaO + CO$ *constructed on the basis of the weak spin–orbit coupling approximation and* C_s *symmetry in the collision complex*
(Reproduced with permission from *Z. Phys. Chem., NF*, 1986, **150**, 133)

[253] R. Behrens, jun., A. Freedman, R. R. Herm, and T. P. Parr, *J. Am. Chem. Soc.*, 1976, **98**, 294.
[254] G. P. Smith and R. N. Zare, *J. Am. Chem. Soc.*, 1976, **97**, 1985.

Reactions with O_2

The reaction of $Ca(^1D_2)$ atoms with O_2 has been studied under molecular beam conditions by Irvin and Dagdigian,[224] yielding absolute chemiluminescence cross-sections of $(3.2 \pm 1.4) \times 10^{-16}$ and $(1.1 \pm 0.5) \times 10^{-16}$ cm^2 for emission from $CaO(A'^1\Pi)$ and $CaO(A^1\Sigma^+)$, respectively, a total photon yield of $\geq (5.2 \pm 2.5)\%$ and a dissociation energy of $CaO(X^1\Sigma^+)$ of $D_0^0 \geq (4.11 \pm 0.07)$ eV. Of the energetically accessible product states available from $Ca(^1D_2) + O_2$ starting materials, shown in Figure 7, only $A'^1\Pi$ and $A^1\Sigma^+$ are expected to give rise to chemiluminescence. $CaO(a^3\Pi_i, A'^1\Pi)$ product states of the $Ca(^3P_J) + O_2$ reaction, probed by LIF in an earlier molecular beam study,[200] are now known[224] to arise from $Ca(^1D_2) + O_2$ reactants. Figure 7 shows that reaction of $Ca(^3P_J)$ atoms with O_2 is endoergic for formation of all electronically excited states of the CaO product. In a manner analogous to that described earlier for $Ca(^3P_J, ^1D_2) + CO_2$, it is likely that for $Ca(^3P_J, ^1D_2) + O_2$ collisions, the reaction pathway leading to ground-state $CaO(X^1\Sigma^+)$ product will be dominated by the higher density of $Ca(^3P_J)$ atoms of the two metastable species present in the atomic beam.[200,224] Irvin and Dagdigian[224] have pointed out that the CaO product-state branching ratio $a^3\Pi_i, A'^1\Pi : X^1\Sigma^+$, determined earlier to be 2.0,[200] strictly represents a lower limit for the $Ca(^1D_2) + O_2$ reaction. The CaO ground-state vibrational distributions measured by Pasternack and Dagdigian following reaction of $Ca(^3P_J)$ atoms with O_2 show that the vibrational population falls off rapidly with increasing quantum number, the largest population residing in the $v = 0$ state.[200] A lower limit of $T_{rot} \geq (800 \pm 200)$ K for the rotational distribution in the $v = 0$ level of $CaO(X^1\Sigma^+)$ was also determined by these workers.[200]

The simplest dynamical model for the $Ca(^1D_2) + O_2(X^3\Sigma_g^-)$ reaction assumes conservation of spin so that the reaction proceeds adiabatically on potential energy hypersurfaces of triplet multiplicity. Both $Ca(^3P_J) + O_2$ and $Ca(^1D_2) + O_2$ reactions are probably initiated by an electron-jump mechanism.[11,12,24,50,166,167] By analogy with BaO_2[255] and the alkali metal superoxides,[256–261] Alexander and Dagdigian have considered the CaO_2 intermediate to be a singly charged planar ion pair which is well bound with respect to both reactant and product channels.[50] These authors note that such a long-lived complex would permit any relatively weak spin–orbit perturbations to act over a period of time sufficiently long so as to enhance significantly the formation of spin-forbidden product states.[50]

[255] B. S. Ault and L. Andrews, *J. Chem. Phys.*, 1975, **62**, 2312.
[256] D. M. Lindsay, D. R. Herschbach, and A. L. Kwiram, *Chem. Phys. Lett.*, 1974, **25**, 175.
[257] F. J. Adrian, E. L. Cochran, and V. A. Bowers, *J. Chem. Phys.*, 1973, **59**, 56.
[258] R. R. Smardzewski and L. Andrews, *J. Chem. Phys.*, 1972, **57**, 1327.
[259] L. Andrews, *J. Chem. Phys.*, 1971, **54**, 4935.
[260] L. Andrews, *J. Chem. Phys.*, 1969, **50**, 4288.
[261] L. Andrews, *J. Am. Chem. Soc.*, 1969, **73**, 3922.

As Dagdigian and collaborators have pointed out,[50,200] the ground $^{1,3}A_2$ and low-lying $^{1,3}B_2$ electronic states of CaO_2, in which the unpaired electron of the O_2^- entity occupies the $1\pi_g$ O_2 orbital parallel and perpendicular to the CaO_2 plane, can dissociate with only small orbital rearrangement to yield electronically excited CaO product states, which can also be described as singly charged ion pairs.[218–222] It has also been postulated,[50,200] in a manner analogous to that described earlier for $Ca(^3P_J, ^1D_2) + CO_2$ collisions, that the very cold vibrational and rotational energy distribution monitored[200] for the $CaO(X^1\Sigma^+)$ product of the $Ca(^3P_J) + O_2$ reaction is due to an activation barrier or other dynamical constraints in the exit channel leading to $CaO(X^1\Sigma^+)$ arising from extensive orbital rearrangement accompanying formation of ground-state products.

The observed chemiluminescent products of the $Ca(^1D_2) + O_2(X^3\Sigma_g^-)$ reaction, $CaO(A'^1\Pi)$ and $CaO(A^1\Sigma^+)$, are both spin-allowed and so may be formed via trajectories which remain on potential energy surfaces of triplet multiplicity along the entire reaction co-ordinate. However, while the $CaO(A'^1\Pi)$ state is adiabatically accessible from $Ca(^1D_2) + O_2$ reactants, $CaO(A^1\Sigma^+)$ product is not. Irvin and Dagdigian have proposed two mechanisms for production of $CaO(A^1\Sigma^+)$ in the $Ca(^1D_2) + O_2$ reaction.[224] In the first of these, $CaO(A^1\Sigma^+)$ is produced via a non-adiabatic spin-allowed curve crossing between triplet potential energy hypersurfaces which correlate separately with $Ca(^1D_2) + O_2(X^3\Sigma_g^-)$ reactants and $CaO(A^1\Sigma^+) + O(^3P_J)$ products.[224] Such a process may involve an activation barrier and hence could channel more energy into product translation. Irvin and Dagdigian have rationalized their non-observance of bands above $\Delta v = 2$ in the $CaO(A^1\Sigma^+ \rightarrow X^1\Sigma^+)$ chemiluminescence spectrum (although sufficient energy is available to populate the $v' = 4$ level of $A^1\Sigma^+$ [71,91,224]) and the colder rotational distribution of the $\Delta v = 2$ sequence compared with that for $\Delta v = 0$ in terms of the existence of such a barrier.[224] The other mechanistic suggestion of these workers for formation of $CaO(A^1\Sigma^+)$ involves multiple crossings from the triplet surfaces correlating with $Ca(^1D_2) + O_2(X^3\Sigma_g^-)$ reactants to surfaces of the same multiplicity leading to $CaO(A^1\Sigma^+) + O(^3P_J)$ products via a singlet surface that intersects both triplet surfaces. Such crossings could take place if the $Ca^+O_2^-$ intermediate is sufficiently long-lived for spin–orbit perturbations to take place,[224] though it does not account for the observed ro vibrational distribution of the $A^1\Sigma^+$ states.[224]

Red chemiluminescence arising from the well-known CaO $(A^1\Sigma^+ \rightarrow X^1\Sigma^+)$ bands and an intense orange band centred at $\lambda = 625$ nm have been observed on reaction of ground-state $Ca(^1S_0)$ atoms with O_2 in a low-pressure diffusion flame.[233] A photon yield of 0.002% was determined for $CaO(A^1\Sigma^+ \rightarrow X^1\Sigma^+)$ emission.[233] Using the same method as that employed for study of the $Mg(^1S_0) + O_2$ reaction,[146–148,150,151,153,154,156,162–164] Kashireninov *et al.* have obtained a thermal activation energy of $E_{act} = (60.2 \pm 8.8)\,\mathrm{kJ\,mol^{-1}}$ and a pre-exponential factor of $A = (4.2 \pm 0.5) \times 10^{10}\,\mathrm{cm^3\,molecule^{-1}\,s^{-1}}$ for reaction of Ca($^1S_0$) atoms with O_2.[147,150,153,156]

Reactions with Other Oxidants

Wang and Menzinger have carried out a detailed investigation of the reactions of Ca(^3P_J) and Ca($^1D_2$) atoms with NO_2 in a molecular beam experiment.[160] The chemiluminescence spectrum obtained by these workers for the Ca(^3P_J,$^1D_2$) + NO_2 reactions comprises the orange and green arc bands of CaO and the $A^1\Sigma^+ \rightarrow X^1\Sigma^+$ and $A'^1\Pi \rightarrow X^1\Sigma^+$ band systems at wavelengths greater than $\lambda \approx 700$ nm.[160] A broad continuum emission is also observed extending into the blue region and has been ascribed to the $A'^1\Pi \rightarrow X^1\Sigma^+$ band system as described above for the Ca(^3P_J,$^1D_2$) + N_2O reactions. A much lower population of the CaO($B^1\Pi$) state was monitored[160] following the Ca(^3P_J,$^1D_2$) + NO_2 reactions compared with Ca(^3P_J,$^1D_2$) + N_2O. A photon yield of $(5.95 \pm 1.30)\%$ for emission over the wavelength range $\lambda = 300$—900 nm for the Ca(^3P_J) + NO_2 reaction[160] indicates that its light-producing efficiency is similar to that of Ca($^1D_2$) + N_2O.[157] Inspection of product-state-resolved chemiluminescence cross-sections reported by Wang and Menzinger suggest that, like the Ca(^3P_J) + N_2O system, the exoergicity of the Ca(^3P_J) + NO_2 reaction is utilized in such a way that results in the predominant production of CaO($A'^1\Pi$) rather than CaO($A^1\Sigma^+$) or the high-lying electronically excited states that give rise to orange and green arc band emission.[160] Measured values of total beam attenuation (reactive) cross-sections for the reactions of Ca(^3P_J) and Ca($^1D_2$) atoms with NO_2 under single conditions are consistent with a long-range harpoon mechanism operating in the entrance channel.[160] Formation of the high-lying arc band emitters and the $B^1\Pi$ product state from Ca(^3P_J) + NO_2 collisions can be understood by a mechanism analogous to that for Ca(^3P_J) + N_2O.[50,158,160] Thus, transfer of the Ca(4*s*) electron (*a'* symmetry) to the LUMO of NO_2 during nearly co-planar collisions results in an excited ionic intermediate which rearranges smoothly into products of electronic configuration Ca($4p\pi$)O($2p\sigma 2p\pi^4$) and Ca($4p\pi$)O($2p\sigma^2 2p\pi^3$). Wang and Menzinger have suggested that the lower yield of upper arc band states from the Ca(^3P_J,$^1D_2$) + NO_2 reaction relative to the Ca(^3P_J,$^1D_2$) + N_2O system is consistent with a reduced probability of co-planar collisions due to the non-linearity of the NO_2.[160]

Calculated surprisals[7,11,12,24,48,49] for CaO states produced by the Ca(^3P_J) + NO_2 and Ca($^1D_2$) + NO_2 reactions are, like their N_2O counterparts, non-zero, non-linear, and erratic.[160] These results, coupled with the fact that branching fractions into electronically excited products of both Ca(^3P_J) and Ca($^1D_2$) reactions do not exhibit a monotonic decrease with increasing excitation energy, are indicative of some form of dynamical control of the product electronic state distribution.[160] Again, the surprisal characterizing the principal chemiluminescence pathway leading to CaO($A'^1\Pi$) product is low while the channel leading to CaO($A^1\Sigma^+$) product is found to be under-populated for Ca(^3P_J) reactant and statistically populated for Ca($^1D_2$) as the atomic precursor.[160]

Measured product laboratory angular distributions and derived centre-of-mass recoil distributions for the Ca($^1S_0$) + NO_2 reaction have been presented by Herm and co-workers as a result of their crossed molecular beam study.[208] Centre-of-mass angular distributions are found to be asymmetric with scattering of product into the forward hemisphere favoured. The observed large reaction cross-sections and forward product scattering suggest that the Ca($^1S_0$) + NO_2 reaction is direct and proceeds via a long-range electron-jump mechanism.[208] Zare and co-workers have also investigated the Ca($^1S_0$) + NO_2 reaction,[209,226] observing a broad continuum emission extending from $\lambda \approx 400$—575 nm. Total beam attenuation cross-sections of $\sigma_{tot} = 1.22$ nm^2 [226] and 0.93 nm^2 [209] are found to be consistent with a value of $\sigma_{tot} = 0.70$ nm^2 [209] calculated on the basis of the harpoon mechanism.[11,12,24,166,167] The reaction of Ca($^1S_0$) atoms with NO_2 has also been studied in low-pressure diffusion flames.[155,233,235] A photon yield of 4×10^{-4} was obtained by Capelle *et al.* for the CaO($A^1\Sigma^+ \rightarrow X^1\Sigma^+$) band system.[233] An approximate estimate of the activation energy for the Ca($^1S_0$) + NO_2 reaction has been determined as $E_{act} = -16.7$ kJ mol^{-1}.[235] The unusual temperature dependence of the thermal rate constant for this process has been discussed by Repka *et al.*, who have put forward a mechanism involving concurrent reactions of the metal atom with the oxidant.[155]

Chemiluminescence arising from the reactions of Ca($^1S_0$) atoms with O_3 and NO have been studied in dilute diffusion flames.[233] For both oxidants, CaO($A^1\Sigma^+ \rightarrow X^1\Sigma^+$) emission has been recorded. In addition, especially prominent lines of the CaO($A'^1\Pi \rightarrow X^1\Sigma^+$) band system are observed in a flame containing Ca atoms burning in O_3.[216,233] Photon yields of 0.03—0.06% and 0.001% have been determined for the Ca($^1S_0$) + O_3 and Ca($^1S_0$) + NO reactions, respectively. Amongst other quenching partners, Husain and Schifino have reported $k_{NO}(T) = (3.2 \pm 0.3) \times 10^{-13}$ cm^3 molecule^{-1} s^{-1} and $k_{CO}(T) = (3.7 \pm 0.5) \times 10^{-13}$ cm^3 molecule^{-1} s^{-1} from their time-resolved investigations of the collisional removal of Ca(^3P_J) atoms by these molecules.[236] Chemiluminescence arising from the Ca($^1S_0$) + ClO_2 reaction has been observed in crossed molecular beams.[212] Molecular emission was found to extend over the entire visible region, unlike the Mg($^1S_0$) + ClO_2 system. Both $A^1\Sigma^+ \rightarrow X^1\Sigma^+$ and $A'^1\Pi \rightarrow X^1\Sigma^+$ band systems of CaO are present in the chemiluminescence spectrum reported by Zare and co-workers,[212] in addition to the $A^2\Pi \rightarrow X^2\Sigma^+$,

$B^2\Sigma^+ \rightarrow X^2\Sigma^+$, and $C^2\Pi \rightarrow X^2\Sigma^+$ bands of the monochloride. The product electronic state distribution reported by these authors shows that CaO($A'^1\Pi$) and CaCl($A^2\Pi$) are the most highly populated oxide and chloride products, these two being in a ratio of about 4:3. Assignment of individual vibronic transitions within the CaO($A'^1\Pi \rightarrow X^1\Sigma^+$) band system has enabled Engelke *et al.* to determine relative vibrational population distributions from observed relative emission intensities.[212] These results show that the vibrational population distribution with the CaO($A'^1\Pi$) state is distinctly non-Boltzmann in character, peaking at about $v' = 7$. The dynamics of the Ca($^1S_0$) + ClO_2 reaction have been discussed[212] in terms of the electron-jump model,[11,12,24,166,167] whereby a valence electron of the metal atom is transferred to the half-filled antibonding orbital of ClO_2.

6 Oxidation Reactions of Electronically Excited Strontium Atoms

Electronic States of Strontium Oxide

The spectroscopy of strontium monoxide has yet to receive a measure of attention comparable to that afforded in recent years to calcium monoxide. Thus although the energies of the low-lying $a^3\Pi_i$, $A'^1\Pi$, and $A^1\Sigma^+$ states and the higher-lying $B^1\Pi$ and $C^1\Sigma^+$ states are well established,[91,116,117] the locations of states lying between $A^1\Sigma^+$ and $B^1\Pi$, which are expected[61,62,142] to give rise to the well-known red and orange arc band systems, still await definitive spectroscopic assignment. Benard and co-workers have suggested that the shorter wavelength orange band arises from a $^3\Delta \rightarrow a^3\Pi_i$ transition of the diatomic oxide, while the longer wavelength red system consists of transitions from one or more upper singlet states to the lower-lying $A'^1\Pi$ state, $^1\Delta, ^1\Sigma^- \rightarrow A'^1\Pi$.[61,62,142] By analogy with the rotational analysis and assignments of the CaO orange and green arc bands carried out by Field and co-workers,[213–215] Cox and Dagdigian have suggested that the SrO arc bands are due to transitions between similar states.[123] Band heads that belong to the SrO($A'^1\Pi \rightarrow X^1\Sigma^+$) system were first observed and vibrationally assigned by Field and co-workers,[262,263] who recorded the $A'^1\Pi \rightarrow X^1\Sigma^+$ spectrum in emission following reaction of Sr atoms with N_2O and other oxidizing species in a low-pressure flow system. The locations of the lowest-lying electronically excited states, $a^3\Pi_i$ and $A'^1\Pi$, have been determined by Field[217] from perturbations of the $A^1\Sigma^+$–$X^1\Sigma^+$ bands of SrO reported by Lagerqvist and co-workers.[187–189,264] Table 7 shows term energies of the known excited states of SrO taken from the compilation of Huber and Herzberg.[91]

[262] G. A. Capelle, H. P. Broida, and R. W. Field, *J. Chem. Phys.*, 1975, **62**, 3131.
[263] J. Hecht, *J. Chem. Phys.*, 1976, **65**, 5026.
[264] G. Almkvist and A. Lagerqvist, *Nature (London)*, 1949, **164**, 665.

Table 7 *Term values of electronic states of strontium oxide below* 3.6 eV

SrO electronic state	T_e/cm^{-1} [a]	T_e/eV
$X\,^1\Sigma^+$	0	0.000
$a\,^3\Pi_i$	9149	1.134
$A'\,^1\Pi$	9400	1.165
$A\,^1\Sigma^+$	10886.59	1.350
$B\,^1\Pi$	24701.0	3.062
$C\,^1\Sigma^+$	28632.7	3.550

[a] From ref. 91

Energetics of Sr[$5s^2(^1S_0)$, $5s5p(^3P_J)$, $5s4d(^1D_2)$] + *Oxidant Reactions*

Term values of the well-established electronic states of SrO are displayed in Figure 10, which also shows the energetics of the Sr[$5s^2(^1S_0)$,

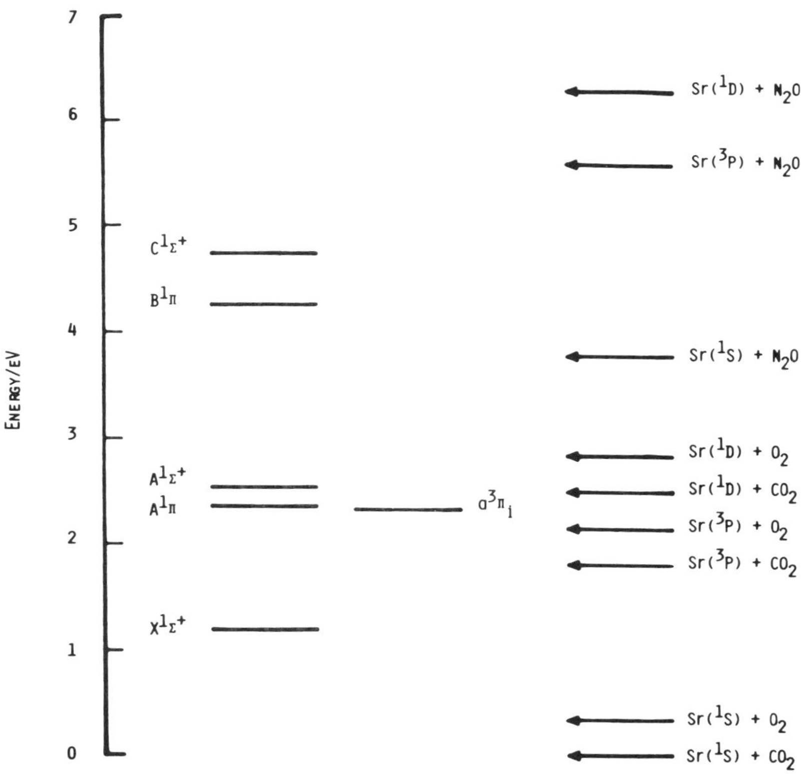

Figure 10 *Energy level diagram showing states of* SrO *below* 3.6 eV *and available energies for reaction of* Sr(^1S, ^3P, ^1D) *atoms with* N_2O, O_2, *and* CO_2

$5s5p(^3P_J)$, $5s4d(^1D_2)] + N_2O$, CO_2, and O_2 reactions. These were calculated for production of SrO in its electronic and vibrational ground state using a relationship analogous to equation (6). The value of $D_0^0[SrO(X^1\Sigma^+)]$ required for input into this expression is taken from a revised[265] lower limit of $D_0^0[SrO(X^1\Sigma^+)] \geq (4.27 \pm 0.15)$ eV obtained from a reinterpretation of the chemiluminescence spectrum of the $Sr(^1S_0) + ClO_2$ reaction recorded under single-collision conditions by Engelke *et al.*[212] This result is in good agreement with a recently computed *ab initio* value.[265]

Reactions with N_2O

Of the oxidation reactions of ground-state and electronically excited strontium atoms, that with N_2O appears to have received most experimental attention (refs. 61, 62, 123, 141, 142, 149, 156, 159, 209, 234, 235, 262, 266). The $Sr(^1S_0) + N_2O$[123,209] and $Sr(^3P_J) + N_2$[123,159] reactions have been investigated under single-collision conditions in molecular beam experiments and exhibit attributes that are qualitatively very similar to their Ca-atom counterparts discussed in Section 5. For the $Sr(^1S_0) + N_2O$ reaction, a number of band heads corresponding to the $A^1\Sigma^+ \rightarrow X^1\Sigma^+$ and $A'^1\Pi \rightarrow X^1\Sigma^+$ systems can be observed,[123] the lack of sharpness of which has been attributed to substantial rotational excitation of the electronically excited SrO molecules, since these bands heads are formed at low J values. Cox and Dagdigian were unable to identify individual band heads of the $A^1\Sigma^+ \rightarrow X^1\Sigma^+$ and $A'^1\Pi \rightarrow X^1\Sigma^+$ systems in their chemiluminescence spectrum arising from the $Sr(^3P_J) + N_2O$ reaction, though the orange and red arc band systems at $\lambda \approx 580$—600 and 640—685 nm are clearly discernable.[123,159] The well-established ultraviolet $C^1\Sigma^+ \rightarrow X^1\Sigma^+$ and blue $B^1\Pi \rightarrow X^1\Sigma^+$ band systems of SrO do not, however, constitute part of the $Sr(^3P_J) + N_2O$ chemiluminescence spectrum.[123,159] In contrast, Zare and collaborators have reported a highly structured chemiluminescence system extending over the entire visible region.[209] Dagdigian and co-workers have reported photon yields of $(40 \pm 10)\%$ and $> (6.8 \pm 3.8)\%$ for the $Sr(^3P_J) + N_2O$ reaction[123,159] and $> (0.24 \pm 0.15)\%$ for $Sr(^1S_0) + N_2O$.[123]

An adiabatic correlation diagram for the $Sr + N_2O$ system (although incomplete due to the lack of spectroscopic information pertaining to the upper arc band states of SrO) is presented in Figure 11. Qualitatively, it may be expected that the dynamics of the $Sr(^3P_J) + N_2O$ reaction are similar to the analogous $Ca(^3P_J) + N_2O$ system. Thus easy production of the upper arc band states of SrO (whose orbital occupancies are

[265] H. Partridge, S. R. Langhoff, and C. W. Bauschlicher, jun., *Chem. Phys. Lett.*, 1985, **121**, 163.

[266] D. Husain and J. Schifino, *J. Chem. Soc., Faraday, Trans. 2*, 1984, **80**, 647.

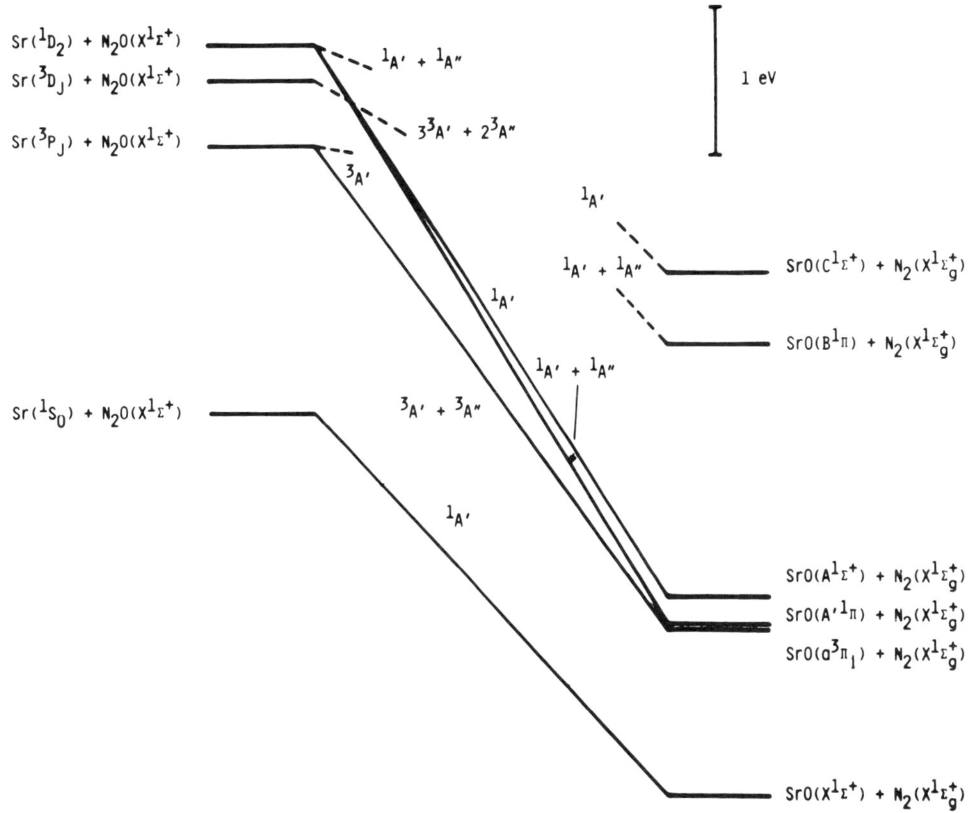

Figure 11 *Correlation diagram connecting the states of* Sr + N_2O *and* SrO + N_2 *constructed on the basis of the weak spin–orbit coupling approximation and* C_s *symmetry in the collision complex*

as yet unknown) may be postulated, by transfer of the subjacent 5*s* electron of $Sr(^3P_J)$ to the 3π antibonding LUMO of N_2O (see Section 5). The excited $Sr^+N_2O^-$ ionic complex so formed, with the 5*p* electron of Sr located within a $5p\pi$ orbital localized on the Sr ion, then rearranges into products with electron configurations from which the upper arc band states are derived. A mechanism analogous to that for production of $CaO(A'^1\Pi, A^1\Sigma^+)$ from $Ca(^3P_J) + N_2O$ reagents[123] and $MgO(X^1\Sigma^+)$ from $Mg(^3P_J) + N_2O$,[122] described in Sections 4 and 5, has been invoked to explain formation of the corresponding states of SrO as well.[123] Thus $SrO(A'^1\Pi, A^1\Sigma^+)$ states are considered to arise from non-Born–Oppenheimer interactions between potential energy surfaces leading to these states and that for the adiabatically allowed $SrO(a^3\Pi_i)$ product.[123] Again, such processes are enhanced by the small energy spacings between these surfaces.[217] Large electronic perturbation matrix elements for mixing of the $A^1\Sigma^+$ state with $a^3\Pi_i$ and $A'^1\Pi$ have been determined by Field[217] from previously reported[187–189] partially assigned perturbations of the $SrO(A^1\Sigma^+–X^1\Sigma^+)$ spectrum. Unlike the CaO case,[213] however, perturbations between the $A'^1\Pi$ and $a^3\Pi_i$ states of SrO have not been shown to exist, but it is likely that they are strongly mixed.[123]

There have been a number of investigations of the chemiluminescence emitted from flames containing Sr atoms.[61,62,141,142,149,156,234,262] Capelle *et al.*[262] have reported both $SrO(A^1\Sigma^+ \rightarrow X^1\Sigma^+)$ and $SrO(A'^1\Pi \rightarrow X^1\Sigma^+)$ emission from $Sr–N_2O–Ar$ 'metal-lean' low-pressure flames, this work constituting the first observation of the $A'^1\Pi \rightarrow X^1\Sigma^+$ band system of SrO. Photon yields for $SrO(A^1\Sigma^+ \rightarrow X^1\Sigma^+)$ chemiluminescence were determined to be 6.6 and 2.2% at total pressures of 1.3 and 9.5 Torr, respectively.[262] Intense chemiluminescence from $Sr–N_2O–CO–He$ flames has been observed by Benard and co-workers,[62,141,142] in which the orange and red SrO arc bands constitute prominent features. Addition of CO to $Sr–N_2O–He$ flames has been shown by these authors to enhance substantially the photon yield for $SrO(A^1\Sigma^+ \rightarrow X^1\Sigma^+)$ emission,[62,142] increasing from 6.4 to 18.0% at total pressures in the Torr range.[142] The well-known spin-forbidden atomic line at $\lambda = 689.3$ nm, corresponding to the $Sr(^3P_1 \rightarrow {}^1S_0)$ transition, is also a feature of the emission spectrum derived from ternary flames involving Sr.[62,141,142] The mechanism of production of visible chemiluminescence in such flames, namely the reaction with N_2O of metastable Sr atoms produced by the metal-catalysed oxidation of CO by N_2O,[141,142] has been discussed in Section 4, and verified in the case of Sr by the molecular beam experiments of Dagdigian and co-workers.[123,159] For the reaction of $Sr(^1S_0)$ atoms with N_2O, an activation energy of $E_{act} = (25.1 \pm 10.5)$ kJ mol^{-1} and pre-exponential factor of $A = (3.2 \pm 0.3) \times 10^{-11}$ cm^3 molecule^{-1} s^{-1} have been obtained from temperature measurements in diffusion flames.[148–151,156,163,164] Husain and Schifino have obtained a rate constant of $k_{N_2O}(T) =$

$(1.5 \pm 0.2) \times 10^{-11}$ cm^3 molecule^{-1} s^{-1} at $T = 950$ K for collisional quenching of Sr(^3P_J) atoms by N_2O using the method of time-resolved atomic resonance emission.[266] No activation energy for the Sr(^3P_J) + N_2O reaction appears to have been determined to date.

Reactions with Other Oxidants

A crossed molecular beam study of the reaction of Sr($^1S_0$) atoms with O_2 has been performed[267] from which an estimate of the dissociation energy of SrO($X^1\Sigma^+$) was obtained from the dependence of the reaction cross-section on relative collisional energy. Photon yields have been determined and chemiluminescence spectra recorded by Capelle *et al.* with a low-pressure flame apparatus for the reactions of Sr($^1S_0$) atoms with O_2, O_3, and NO in addition to N_2O discussed above.[262] Thermal rate parameters for the reaction of Sr($^1S_0$) atoms with O_2 have been determined as $E_{act} = (54.4 \pm 7.5)$ kJ mol^{-1} and $A = (4.0 \pm 0.3) \times 10^{-10}$ cm^3 molecule^{-1} s^{-1} [147,150,153,154,156] by Kashireninov and co-workers from measurements of the temperature profile inside the reaction zone of a dilute diffusion flame.[148,150,151,162–164]

The reaction of Sr($^1S_0$) atoms with NO_2 has also been the subject of a molecular beam study.[209] The chemiluminescence spectrum observed by Jonah *et al.*[209] was reported to be similar to that obtained from the Ca($^1S_0$) + NO_2 reaction; *i.e.* a broad featureless emission was recorded over the visible regions from $\lambda \approx 400$ to 575 nm. A total beam attenuation cross-section of $\sigma_{tot} = 1.27$ nm^2 is again consistent with a value of $\sigma_{tot} = 1.00$ nm^2 derived from a simple electrostatic harpoon model.[209] Wide-angle reactive scattering of Sr atoms from NO_2, observed by Herm and co-workers,[208] is in qualitative agreement with the large reaction cross-section for this process measured by Zare's group.[209] The forward-peaked angular distribution of SrO product in the centre-of-mass frame is indicative of a long-range electron transfer mechanism operating during the collisional event, resulting in an $Sr^+NO_2^-$ ion pair which immediately breaks up in the force field provided by the metal cation Sr^+ to form SrO + NO products.[208] Such reaction features are qualitatively similar to those discussed in Section 5 for Ca($^1S_0$) + NO_2 collisions. Kuznetsov *et al.*[235] have obtained a value of $E_{act} = -12.6$ kJ mol^{-1} for the activation energy of the Sr($^1S_0$) + NO_2 reaction from a kinetic investigation using the method of diffusion flames.[148,150,151,162–164] An explanation of the observed negative temperature dependence in terms of a mechanism involving competing reaction channels for Sr($^1S_0$) + NO_2 has been postulated analogous to that for Ca($^1S_0$) + NO_2.[155]

Husain and Schifino have reported absolute second-order rate constants, determined by time-resolved resonance emission measurements,

[267] C. Batalli-Cosmovici and K. -W. Michel, *Chem. Phys. Lett.*, 1972, **16**, 77.

for total removal of $Sr(^3P_J)$ atoms by various quenching partners including oxidant molecules.[266] Apart from their result for $Sr(^3P_J) + N_2O$ mentioned earlier, these authors report $k_{CO_2}(T) = (1.3 \pm 0.1) \times 10^{-11}$ and $k_{CO}(T) = (2.0 \pm 0.1) \times 10^{-11}$ cm^3 molecule^{-1} s^{-1} at $T = 950$ K for quenching of $Sr(^3P_J)$ atoms by CO_2 and CO. In the former case, the relatively rapid quenching rate has been rationalized in terms of exoergic symmetry-allowed routes for chemical removal of $Sr(^3P_J)$ atoms leading to electronically excited $SrO(a\,^3\Pi_i)$. Jansen *et al.* have determined quenching cross-sections of $\sigma_{tot}/nm^2 = (1.52 \pm 0.20)$, (0.30 ± 0.05), and (0.49 ± 0.08) for collisional removal of the higher-lying $5s5p(^1P_1)$ state of Sr by O_2, CO_2, and CO, respectively.[268]

The $Sr(^1S_0) + ClO_2$ reaction has been studied under single-collision conditions using crossed molecular beams.[212] The chemiluminescence spectrum from $\lambda \approx 350$ to 850 nm exhibits vibronic bands of the $SrO(A'\,^1\Pi \rightarrow X^1\Sigma^+)$ system and bands of the $A\,^2\Pi \rightarrow X^2\Sigma^+$, $B\,^2\Sigma^+ \rightarrow X^2\Sigma^+$, and $C\,^2\Pi \rightarrow X^2\Sigma^+$ systems of SrCl.[212] Nascent electronic state distributions reported by Zare and co-workers for the $Sr(^1S_0) + ClO_2$ reaction show that the partitioning of available reaction energy amongst $SrO(A'\,^1\Pi)$, $SrCl(A\,^2\Pi)$, $SrCl(B\,^2\Sigma^+)$, and $SrCl(C\,^2\Pi)$ products occurs in ratios of approximately 9 : 10 : 4 : 1. Like the analogous reaction of $Ca(^1S_0)$ atoms therefore, the $A'\,^1\Pi$ state of the monoxide and $A\,^2\Pi$ state of the monochloride are the major products.[212] The existence of a vibrational population inversion within the $SrO(A'\,^1\Pi)$ state (maximum population at about $v' = 7$) has also been clearly established. A harpoon mechanism[11,12,24,166,167] has been invoked[212] to describe the dynamics of the $Sr(^1S_0) + ClO_2$ reaction, as has been discussed in Sections 4 and 5 for the $Mg(^1S_0) + ClO_2$ and $Ca(^1S_0) + ClO_2$ reactions.

Acknowledgement. The authors would like to thank the SERC for support of their own work in this area by way of an equipment grant for the purchase of a dye laser system.

[268] B. J. Jansen, Tj. Hollander, and H. Van Helvoort, *J. Quant. Spectrosc. Radiat. Transfer*, 1977, **17**, 193.

CHAPTER 7

Studies of the Reactions of Metal Clusters

M. F. JARROLD

1 Introduction

A metal (or semiconductor) cluster is a small particle of bulk material containing between two and several thousand atoms. Unlike van der Waals clusters or ion–molecule clusters, these species are usually bound by chemical bonds with cohesive energies greater than 1 eV per atom. Metal clusters are important technologically in such diverse areas as catalysis[1] and materials processing.[2] From a fundamental point of view, metal clusters are interesting because of the position they occupy between molecules and bulk material. It is clearly appropriate to think of a dimer as a molecule, but is a metal cluster with 50 atoms a molecule or a small piece of bulk?

Over the past few years various techniques have been used to investigate the properties of bare metal clusters. Since these clusters are not covered with ligands they are often highly reactive and usually studied either in the gas phase or isolated in an inert matrix. Some of the techniques that have recently been used to study metal clusters include: studies of the cluster size distribution using mass spectrometry to determine if any cluster sizes are particularly favoured ('magic numbers')[3–7]; measurement of cluster ionization potentials[8–11] and electron affinities;[12] studies of magnetic properties;[13,14] investigations of the

[1] 'Metal Clusters in Catalysis', eds. B. C. Gates, L. Gucz, and H. Knozinger, Elsevier, Amsterdam, 1986.

[2] I. Yamada, H. Usui, and T. Takagi, in 'Metal Clusters', eds. F. Trager and G. zu Putlitz, Springer Verlag, Berlin, 1986, p. 37; E. Kay, *ibid.*, p. 151; T. Araya, Y. Ibaraki, S. Hioka, R. Okada, and M. Kanamaru, in 'Physics and Chemistry of Small Clusters', eds. P. Jena, B. K. Rao, and S. N. Khanna, Plenum, New York, 1987, p. 25.

[3] E. Recknagel, *Ber. Bunsenges. Phys. Chem.*, 1984, **88,** 201.

[4] M. M. Kappes, R. W. Kunz, and E. Schumacher, *Chem. Phys. Lett.*, 1982, **91,** 413.

[5] W. D. Knight, K. Clemenger, W. A. de Heer, W. A. Saunders, M. Y. Chou, and M. L. Cohen, *Phys. Rev. Lett.*, 1984, **52,** 2141.

collision-induced dissociation[15–18] and photodissociation[19,20] of clusters; photoelectron spectroscopy of size-selected cluster anions[21,22] and size-selected clusters deposited on a surface;[23] and studies of the chemical properties of metal clusters. The recent studies of the chemical reactions of isolated gas-phase metal clusters is the subject of this chapter.

There are several reasons for studying the reactions of metal clusters. We would like to see how the reactivity changes with cluster size and determine if there are any particularly reactive or particularly inert clusters. We would like to know if the reactivity correlates with any other physical property and ultimately understand the physical basis for the observed reactivity trends. Finally, we would like to determine at what cluster size the chemistry approaches that of the bulk surface, and perhaps use clusters as models for surface phenomena.

It is useful to explore some of the important differences expected between reactions on surfaces and reactions on clusters. The simplest chemical process that occurs on surfaces is chemisorption, and the heat of chemisorption is accommodated by the bulk lattice. The same sort of process cannot occur with a dimer since there is no lattice, and the transient collision complex that forms will quickly dissociate. With larger clusters, the lifetime of the collision complex increases (as the heat of chemisorption is randomized among the internal degrees of freedom in the cluster) and it will be possible to observe a metastable adduct directly. With even larger clusters, the lifetime of the adduct will exceed the lifetime for infrared emission. As is described below, running the

[6] H. W. Kroto, J. R. Heath, S. C. O'Brian, R. F. Curl, and R. E. Smalley, *Nature (London)*, 1985, **318,** 162.

[7] J. C. Phillips, *Chem. Rev.*, 1986, **86,** 619.

[8] M. Kappes, M. Schar, P. Radi, and E. Schumacher, *J. Chem. Phys.*, 1986, **84,** 1863, and references cited therein.

[9] R. E. Walstedt and R. F. Bell, *Phys. Rev. A*, 1986, **33,** 2830.

[10] W. D. Knight, W. A. de Heer, and W. A. Saunders, *Z. Physik. D*, 1986, **3,** 109.

[11] E. A. Rohlfing, D. M. Cox, A. Kaldor, and J. H. Johnson, *J. Chem. Phys.*, 1984, **81,** 3846.

[12] L. -S. Zheng, C. M. Karmer, P. J. Brucat, S. H. Yang, C. L. Pettiette, M. J. Craycraft, and R. E. Smalley, *J. Chem. Phys.*, 1986, **85,** 1681.

[13] D. M. Cox, D. J. Trevor, R. L. Whetten, E. A. Rohlfing, and A. Kaldor, *Phys. Rev. B*, 1986, **32,** 7290.

[14] D. M. Cox, D. J. Trevor, R. L. Whetten, E. A. Rohlfing, and A. Kaldor, *J. Chem. Phys.*, 1986, **84,** 4651.

[15] K. Ervin, S. K. Loh, N. Aristov, and P. B. Armentrout, *J. Phys. Chem.*, 1983, **87,** 3593.

[16] W. Begemann, S. Dreihofer, K. H. Meiwes-Broer, and H. O. Lutz, *Z. Physik. D*, 1986, **3,** 183.

[17] M. F. Jarrold, J. E. Bower, and J. S. Kraus, *J. Chem. Phys.*, 1987, **86,** 3876.

[18] L. Hanley, S. A. Ruatta, and S. L. Anderson, *J. Chem. Phys.*, 1987, **87,** 260.

[19] L. A. Bloomfield, R. R. Freeman, and W. L. Brown, *Phys. Rev. Lett.*, 1985, **54,** 2246.

[20] P. J. Brucat, L. -S. Zheng, C. L. Pettiette, S. Yang, and R. E. Smalley, *J. Chem. Phys.*, 1986, **84,** 3078.

[21] D. G. Leopold, J. Ho, and W. C. Lineberger, *J. Chem. Phys.*, 1987, **86,** 1715.

[22] S. H. Yang, C. L. Pettiette, J. Conceicao, O. Cheshnovsky, and R. E. Smalley, *Chem. Phys. Lett.*, 1987, **139,** 233.

[23] S. B. DiCenzo, S. D. Berry, and E. H. Hartford, *Phys. Rev. B.*, in press.

chemisorption reactions in the presence of a buffer gas provides a way to stabilize the chemisorption adducts of smaller clusters so that they can be directly observed.

Another important difference between metal clusters and surfaces is that even in quite large clusters most of the atoms are surface atoms. For example, in a 10-atom cluster essentially all the atoms are surface atoms, in a 100-atom cluster 72 are surface atoms (assuming structureless packing),[24] and even in a 1000-atom cluster around 40% are surface atoms. Owing to their small dimensions, the surfaces of clusters are strongly curved with high step densities. It is known that steps and other defects on surfaces often enhance reactivity, so we might predict that bare metal clusters are highly reactive.

On attempting to study the reactivity of metal clusters one is faced with several problems. Generating the clusters is no easy task, and virtually all cluster sources produce a range of cluster sizes, so there is the problem of identifying reactants and products. In all the approaches currently used to study the reactivity of clusters in the gas phase, mass spectrometry is used to identify the products. When studying the reactions of neutral clusters, this means that the products must be ionized and there is no selection of the reactant cluster size before the reaction. By studying the reactions of ionized clusters it is possible to select a particular cluster size before reaction and avoid the uncertainties (such as unknown ionization cross-sections and fragmentation) associated with ionizing neutral clusters for subsequent mass analysis. Three approaches have been used to study the reactivity of metal clusters: the flow reactor, Fourier transform ion cyclotron resonance, and low-energy ion beam techniques. We will start by considering the flow reactor.

2 Cluster Reactions Studied using a Flow Reactor

Experimental Aspects

The cluster flow reactor is similar to conventional flow-tube techniques where reactions are studied in the presence of an inert buffer gas. This method can be used to study the reactions both of neutral clusters and of cluster ions, although as described below, there are much more elegant ways to study the reactions of cluster ions. The cluster flow reactor has been employed principally by three groups, at Argonne,[25] Rice,[26] and Exxon.[27] There are two variations of the basic approach

[24] E. K. Parks, K. Liu, S. C. Richtsmeier, L. G. Pobo, and S. J. Riley, *J. Chem. Phys.*, 1985, **82,** 5470.

[25] S. J. Riley, E. K. Parks, L. G. Pobo, and S. Wexler, *Ber. Bunsenges. Phys. Chem.*, 1984, **88,** 287.

[26] M. E. Geusic, M. D. Morse, and R. E. Smalley, *J. Chem. Phys.*, 1985, **82,** 590.

[27] D. J. Trevor, R. L. Whetten, D. M. Cox, and A. Kaldor, *J. Am. Chem. Soc.*, 1985, **107,** 518.

with continuous[28] and pulsed[29] gas flow. The continuous flow version of Riley and co-workers[28] will be used here as an example to illustrate the important features of this technique. Figure 1 shows a schematic diagram of a cluster flow reactor experiment, which can be conveniently broken into several sections: generating the clusters, the reactor, and the analysis and detection of the products.

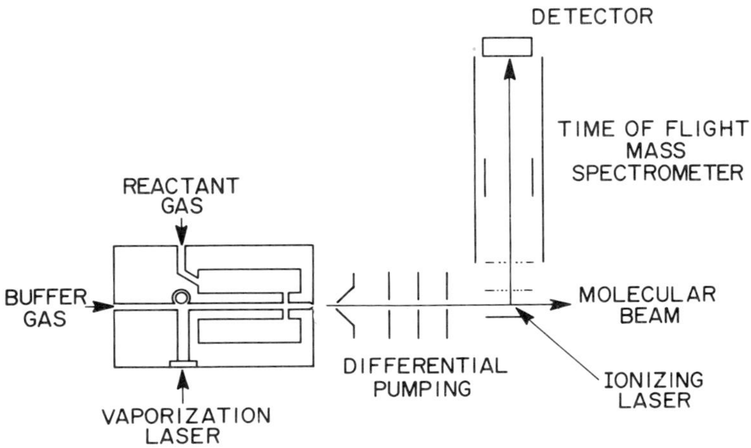

Figure 1 *Schematic diagram of a cluster flow reactor experiment*

Generating the Clusters. Clusters are generated by laser vaporization.[30,31] A pulsed laser is focused on to a target. Material is evaporated and entrained in a flowing buffer gas, where it cools and clusters grow. Using laser vaporization it is possible to make large clusters of even the most refractory materials. In the experiment of Riley and co-workers,[28] shown schematically in Figure 1, clusters are generated with an exciplex laser operating at 308 nm. The target rod is continuously rotated and translated to prevent a hole being bored in the target, which would result in low and wildly fluctuating cluster signals. The pressure in the vaporization region is ~30 Torr. For metals, most of the material is removed from the rod as atoms and it is believed that clusters grow predominantly by monomer addition. Cluster growth is terminated by diffusion of atoms to the walls. This is estimated to occur a few centimeters downstream from the point of vaporization.[32]

[28] S. C. Richtsmeier, E. K. Parks, K. Liu, L. G. Pobo, and S. J. Riley, *J. Chem. Phys.*, 1985, **82,** 3659.

[29] M. E. Geusic, M. D. Morse, S. C. O'Brian, and R. E. Smalley, *Rev. Sci. Instrum.*, 1985, **56,** 2124.

[30] T. D. Dietz, M. A. Duncan, D. E. Powers, and R. E. Smalley, *J. Chem. Phys.*, 1981, **74,** 6511.

[31] V. E. Bondybey and J. H. English, *J. Chem. Phys.*, 1981, **74,** 6978.

[32] S. J. Riley and E. K. Parks, in 'Physics and Chemistry of Small Clusters', eds. P. Jena, B. K. Rao, and S. N. Khanna, Plenum, New York, 1987, p. 727.

The Reactor. The reactor is a simple extension of the cluster growth channel where reagent gases can be added. The temperature of the clusters in the reaction region is of course a critical parameter, particularly since residual heat from the vaporization laser could be a factor. Riley and co-workers have presented evidence suggesting the clusters are close to the temperature of the source block.[28] With pulsed flow reactors the clusters may be slightly hotter.[27,29] After travelling through the reactor tube, the clusters and products undergo a supersonic expansion. The resulting molecular beam is skimmed and passes through several regions of differential pumping before entering the analysis section of the experiment.

Mass Analysis and Detection of Products. The products are analysed and detected using photoionization and time-of-flight mass spectrometry. Photoionization is accomplished using a pulsed laser. Unfortunately, attempts at multiphoton ionization often result in extensive cluster fragmentation,[33,34] so the best approach is to work with a short-wavelength laser using low fluence and attempt single-photon ionization. However, even then one must be careful not to use a photon energy too much above the ionization threshold or fragmentation could occur. With readily available laser sources the maximum photon energy that can be obtained is 7.87 eV (an exciplex laser operating with F_2). This is generally enough to single-photon ionize the bare metal cluster, but in some cases chemisorption results in an increase in the cluster ionization potential[35] with the result that the products cannot be one-photon ionized.

Interpretation of Results. To determine absolute rate constants, accurate values for the residence time in the reactor, the reagent pressure, and the buffer-gas (helium) pressure are required. More accurate values for these parameters can be estimated with continuous flow reactors. The reaction kinetics are generally studied by monitoring the depletion of the reactant metal cluster. Since a range of cluster sizes are present in the flow reactor, the only reactions that can be studied are reactions resulting in adduct formation (chemisorption):

$$M_n + AB \rightarrow M_nAB \tag{1}$$

or addition reactions:

$$M_n + AB \rightarrow M_nA + B \tag{2}$$

Cluster fragmentation, if it occurred, would lead to erroneous conclusions. The majority of the reactions studied so far have been simple

[33] K. Liu, E. K. Parks, S. C. Richtsmeier, L. G. Pobo, and S. J. Riley, *J. Chem. Phys.*, 1985, **83,** 2882.

[34] G. C. Nieman, E. K. Parks, S. C. Richtsmeier, K. Liu, L. G. Pobo, and S. J. Riley, *High Temp. Sci.*, 1986, **22,** 115.

adduct formation or chemisorption reactions.

Invoking the standard mechanism for association reactions:

$$M_n + AB \underset{k_b}{\overset{k_f}{\rightleftharpoons}} M_nAB^* \tag{3}$$

$$M_nAB^* + He \overset{k_s}{\rightarrow} M_nAB + He \tag{4}$$

assuming [AB] and [He] $\gg$ [M_n], then a simple steady-state analysis yields a pseudo first-order rate constant:

$$k = k_f[AB]\left(\frac{k_s[He]}{k_s[He] + k_b}\right) \tag{5}$$

with

$$\ln\frac{[M_n]_f}{[M_n]_i} = kt \tag{6}$$

In the high-pressure limit, $k_s[He] \gg k_b$,

$$k = k_f[AB] \tag{7}$$

and the observed kinetics can be related to the chemisorption step. Based on RRK estimates,[36] this limit is expected to be approached for medium-sized clusters ($n > 5$) with quite strongly bound adducts (>1 eV), but this assumption has not been confirmed experimentally. If k_s[He] and k_b are comparable, the observed kinetics are a convolution of the forward and reverse steps which can only be unravelled by studies as a function of buffer-gas pressure. Even with large clusters, however, thermally activated desorption will influence the results if the adduct is weakly bound. Estimates using an Arrhenius expression[37] suggest that this will not be significant for adducts with a desorption energy greater than 0.8 eV.

Reactions of Neutral Transition Metal Clusters

Hydrogen. The reactions of transition-metal clusters with hydrogen (and deuterium) have been extensively studied because of their fundamental importance. Studies have been performed with V,[38,39] Fe,[28,37,40]

[35] M. R. Zakin, D. M. Cox, R. L. Whetten, D. J. Trevor, and A. Kaldor, *Chem. Phys. Lett.*, 1987, **135**, 223.

[36] D. M. Cox, K. C. Reichmann, D. J. Trevor, and A. Kaldor, *J. Chem. Phys.*, 1988, **88**, 111.

[37] M. D. Morse, M. E. Geusic, J. R. Heath, and R. E. Smalley, *J. Chem. Phys.*, 1985, **83**, 2293.

Co,[37] Ni,[37,41] Cu,[37] Nb,[37,42] and Pt[43] clusters. The reactions of Fe clusters with hydrogen have been the most widely studied. Figure 2 shows a plot of the rate constants measured by Riley and co-workers[28] for chemisorption of hydrogen on Fe clusters with up to 30 atoms. Similar results have been obtained by Smalley and co-workers[37] and Kaldor and co-workers[40] using pulsed flow reactors. The rate constants in Figure 2 are for the addition of the first H_2 on to the cluster:

$$Fe_n + H_2 \rightarrow Fe_nH_2 \tag{8}$$

As discussed below, further H_2 molecules add up to a saturation value. With a mixture of D_2 and H_2, isotope exchange is observed. This provides convincing evidence that chemisorption of H_2 on Fe clusters is dissociative rather than molecular. As can be seen from Figure 2, wide swings in reactivity (over three orders of magnitude) are observed. There is a maximum in the reactivity for clusters with 10—12 atoms and

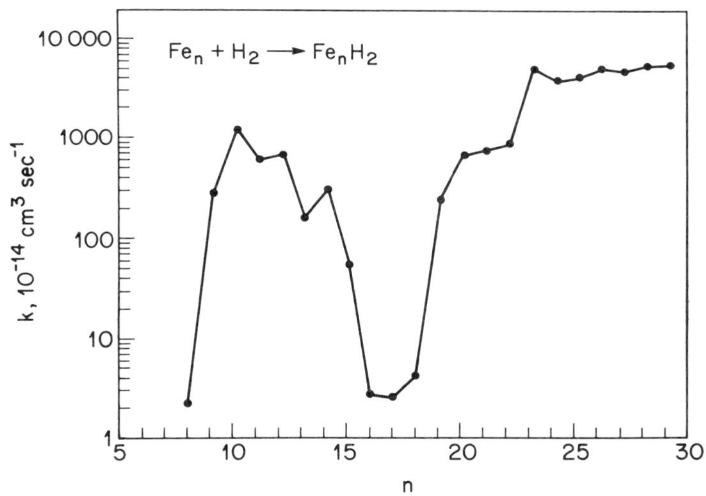

Figure 2 *Rate constants measured by Riley and co-workers for the reactions of iron clusters with hydrogen*
(Adapted from *J. Chem. Phys.*, 1988, **88,** 1622

[38] D. M. Cox, M. R. Zakin, and A. Kaldor, in ref. 32, p. 741.

[39] D. J. Trevor and A. Kaldor, *ACS Symp. Ser.*, 1987, **333,** 43.

[40] R. L. Whetten, D. M. Cox, D. J. Trevor, and A. Kaldor, *Phys. Rev. Lett.*, 1985, **54,** 1494.

[41] W. F. Hoffman, E. K. Parks, G. C. Nieman, L. G. Pobo, and S. J. Riley, *Z. Phys. D*, 1987, **7,** 83.

[42] R. L. Whetten, M. R. Zakin, D. M. Cox, D. J. Trevor, and A. Kaldor, *J. Chem. Phys.*, 1986, **85,** 1697.

[43] A. Kaldor, D. M. Cox, D. J. Trevor, and M. R. Zakin, in 'Metal Clusters', eds. F. Trager and G. zu Putlitz, Springer Verlag, Berlin, 1986, p. 95.

a minimum for clusters with 16—18 atoms. Using an Arrhenius expression, an activation barrier of 0.19 eV was estimated for Fe_{17}. For clusters with more than 23 atoms, the reactivity appears to have levelled off and only changes slowly. This trend has been found to continue for clusters with up to 70 atoms.[28] The rate constants for clusters with more than 23 atoms are quite close to estimates based on the sticking probability of hydrogen on iron surfaces.

It was soon recognized that there is a correlation between the reactivity of iron clusters towards hydrogen and the ionization potentials of the bare clusters.[40] This is illustrated in Figure 3 by the rate constant data from the Exxon group[40] (the correlation is slightly less convincing when the rate constants of the Argonne or Rice groups are considered). This correlation was interpreted as indicating that charge transfer from the metal cluster's highest occupied molecular orbital (HOMO) to the H_2 σ^* orbital plays an important role in weakening the H_2 bond. There have been several theoretical studies of hydrogen chemisorption on organometallic complexes, clusters, and metal surfaces.[44–48] These studies agree that M(HOMO) $\rightarrow$ $H_2(\sigma^*)$ and $H_2(\sigma)$ $\rightarrow$ M(LUMO) donation

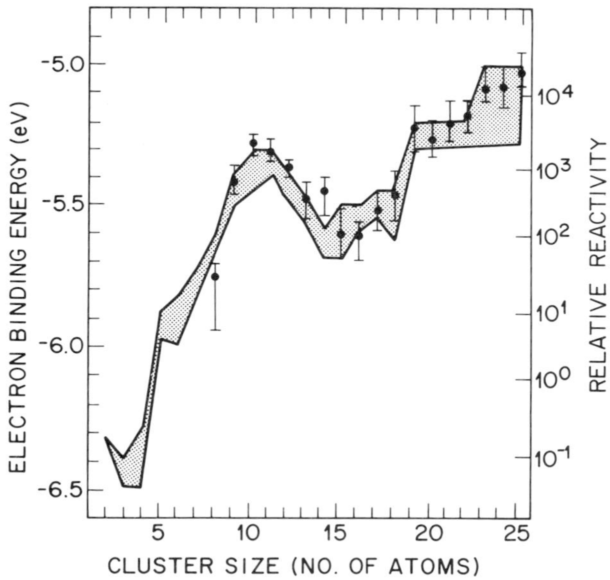

Figure 3 *Comparison of measured ionization thresholds* (*the grey shaded band reflects the uncertainties*) *and the relative reactivity of iron clusters with hydrogen* (*points*)
(Reproduced with permission from *Phys. Rev. Lett.*, 1985, **54,** 1494)

[44] P. K. Johansson, *Surf. Sci.*, 1981, **104,** 510.
[45] T. H. Upton, *J. Am. Chem. Soc.*, 1984, **106,** 1561.
[46] J. -Y. Saillard and R. Hoffman, *J. Am. Chem. Soc.*, 1984, **106,** 2006.
[47] J. Harris and S. Anderson, *Phys. Rev. Lett.*, 1985, **55,** 1583.
[48] T. H. Upton, D. M. Cox, and A. Kaldor, in ref. 32, p. 755.

may play a role in activating the H_2 bond and stabilizing the transition state, but they also suggest that reorganization of the molecular orbitals on M to accommodate the new M—H bonds is important. A strong correlation between reactivity and ionization potential implies that charge transfer from the cluster to the H_2 σ^* orbital is the dominant factor in determining the size of the activation barrier, and that other factors such as cluster structure are less important.

A low ionization potential alone is not enough, however, to cause efficient chemisorption of hydrogen. Smalley and co-workers[37] found no evidence for chemisorption of hydrogen on copper clusters, even though some of them have ionization potentials as low as the highly reactive iron clusters. Thus the electronic structure of the metal (in this case the presence of an open *d* shell) is also important, as it is for bulk surfaces.

Measurements of cluster ionization potentials and rates for hydrogen chemisorption have also been performed for vanadium[38,39] and niobium[37,42] clusters. The rates for hydrogen chemisorption on both V and Nb clusters show a strong cluster-size dependence. For vanadium there appears to be an odd–even alternation in the reactivity: the clusters with an even number of atoms are less reactive than their neighbours, and V_6 is particularly inert. With niobium, clusters with 8, 10, and 16 atoms appear particularly inert, and relatively large changes in reactivity are observed even for clusters with more than 25 atoms. A strong correlation between cluster ionization potential and reactivity was not found for either vanadium or niobium clusters. Clearly, charge transfer from the metal cluster to the H_2 σ^* orbital is not the whole story.

Chemisorption of hydrogen on Co and Ni clusters has also been studied. For cobalt clusters, a strong size-dependent reactivity was found,[37] although the reactivity pattern is quite different from those found for Fe, V, or Nb clusters discussed above. With Co there is a minimum in the reactivity at 6 and 7 atom clusters and another at 19—22. Rosen and Rantala[49] have performed electronic structure calculations on Co clusters with up to 13 atoms using the local density approximation and assumed geometries for the clusters. They found that the reactivity with hydrogen was correlated with the energy of the HOMO and other electronic properties of the clusters. For Ni clusters, Smalley and co-workers found the reaction rates increased slowly with cluster size. Riley and co-workers, however, found that Ni_9 was particularly inert.[41] The origin of this discrepancy is not clear. Riley and co-workers suggest that it is due to further cluster growth in the reaction region of the pulsed flow reactor of Smalley and co-workers. Another possibility is that in the lower-pressure continuous flow reactor the collision rate is too low to stabilize all the Ni_9H_2 collision complexes. It is therefore not clear whether the apparent unreactivity of Ni_9

[49] A. Rosen and T. T. Rantala, in ref. 43, p. 105.

is due to an activation barrier or a weakly bound Ni_9H_2 adduct. Measurements of absolute rate constants for the larger clusters indicate that the reaction occurs at 60% of the collision rate.[41]

The charge-transfer model for chemisorption discussed above is in essence a purely electronic model. In contrast, Phillips[50] has proposed a model to account for the reactivity of Co and Nb clusters based on geometric structure. It seems reasonable to consider cluster geometry because steps and defects on surfaces often show enhanced reactivity. However, virtually nothing is known about cluster structure, so there are few constraints on the structural model. There is currently no technique which provides detailed structural information on gas-phase metal clusters with more than a few atoms. Spectroscopic methods have been applied to a limited number of trimers (Na_3,[51] Cu_3,[52] and Al_3[53]) and the coulomb explosion technique has recently been applied to C_3^+,[54] but the prospects for extending either of these approaches to larger transition-metal clusters do not appear particularly good. One problem with the spectroscopic techniques is that the lifetimes of excited states of larger metal clusters are exceedingly short. A chemical approach to deducing structural information – saturating the surface of the cluster with ligands – is discussed below.

Recently, Riley and co-workers suggested that geometric structure is also an important factor in determining the reactivity of Fe clusters with H_2.[55] They found a correlation between the reactions of Fe clusters with H_2 and with D_2O and NH_3. Both D_2O and NH_3 are generally less strongly bound to the clusters than H_2 so that desorption in the flow tube is significant and it appears possible to establish an equilibrium (or steady-state). The average number of NH_3 or D_2O molecules that bind to the iron clusters (which is related to the average binding energies) shows a strong correlation with the reactivity of the iron clusters with hydrogen. It was suggested that this correlation between the binding energies for NH_3 and D_2O and the reaction rates with hydrogen (and with the ionization potentials of the bare clusters) are all evidence for structural changes in the clusters, and specific structural models were proposed. For example, it was suggested that Fe_{15} has a body-centred cubic rhombic dodecahedron structure which has six apex sites and can account for the predominance of the $Fe_{15}(NH_3)_6$ species. Similarly, it

[50] J. C. Phillips, *J. Chem. Phys.*, 1986, **84,** 1951.

[51] A. Herrmann, M. Hoffmann, S. Leutwyler, E. Schumacher, and L. Woste, *Chem. Phys. Lett.*, 1979, **62,** 216.

[52] M. D. Morse, J. B. Hopkins, P. R. R. Langridge-Smith, and R. E. Smalley, *J. Chem. Phys.*, 1983, **79,** 5316.

[53] Z. Fu, G. W. Lemire, Y. M. Hamrick, S. Taylor, J. -C. Shui, and M. D. Morse, *J. Chem. Phys.*, 1988, **88,** 3524.

[54] A. Faibis, E. P. Kantor, L. M. Tack, E. Bakke, and B. J. Zabransky, *J. Phys. Chem.*, 1987, **91,** 6445.

[55] E. K. Parks, B. H. Weiller, P. S. Bechthold, W. F. Hoffman, G. C. Nieman, L. G. Pobo, and S. J. Riley, *J. Chem. Phys.*, 1988, **88,** 1622.

was argued that Fe_{19} has a face-centred cubic octahedron structure. Clearly, we have not yet heard the final word on what is responsible for the dramatic changes in reactivity observed for Fe and other transition-metal clusters with hydrogen. It seems likely that geometry and electronic structure (which are of course intimately related) are both important.

The reactivity patterns described above were for the chemisorption of the first hydrogen molecule. After the first hydrogen chemisorbs on to the cluster, further hydrogen molecules adsorb, often more rapidly than the first, up to a limiting saturation value. A detailed study of these saturation values have been performed for Fe[24] and Ni[56] clusters with up to 260 atoms. For both Fe_nH_m and Ni_nH_m, only a narrow range of values for m is observed for each n, suggesting that the clusters have only a few specific geometric configurations. For Fe clusters with up to ~ 30 atoms, m is approximately equal to n; for larger clusters, m approximately follows the number of surface atoms (*i.e.* one H atom per surface atom) assuming a structureless packing model. This result implies that the clusters (or at least the fully hydrogenated clusters) are roughly spherical in shape. For clusters larger than ~90 atoms, m begins to fall below the value expected for the number of surface atoms so that in the 250 atom size range the clusters are only ~80% covered. However, with Ni clusters m closely follows the number of surface atoms all the way up to a 260 atom cluster.

CO *and* N_2. Dissociative and non-dissociative chemisorption of both CO and N_2 occurs on metal surfaces. The chemisorption of CO in particular has been extensively studied. Molecular chemisorption is believed to occur through donation from the CO 5σ orbital to the metal, and back donation from the metal to the 2π orbitals on CO. Compared with CO, chemisorption of molecular N_2 is less common and the heats of non-dissociative chemisorption are generally lower. The reactivity of Co, Fe, and Nb clusters with N_2 and CO has been studied by Smalley and co-workers.[37] The observed reactivity pattern of niobium clusters with N_2 is very similar to that found with H_2. The same sort of correlation was also found for cobalt clusters. Because of this correlation and the small heats of molecular chemisorption of N_2 on bulk surfaces, it seems likely that dissociative chemisorption is occurring in these experiments. Iron clusters were found to be relatively inert towards N_2.

In contrast to the selective reactivity patterns observed with N_2, chemisorption of CO on Co and Nb clusters was found to show only a slow monotonic increase with cluster size. It is likely that the CO is chemisorbing molecularly in these experiments. Cox and co-workers[57]

[56] E. K. Parks, G. C. Nieman, L. G. Pobo, and S. J. Riley, *J. Phys. Chem.*, 1987, **91,** 2671.

[57] D. M. Cox, K. C. Reichmann, D. J. Trevor, and A. Kaldor, *J. Chem. Phys.*, 1988, **88,** 111.

have extended these measurements to clusters of the transition metals V, Fe, Co, Ni, Cu, Nb, Mo, Ru, W, Ir, and Pt containing up to 14 atoms. They found that CO readily chemisorbed on most of the transition-metal clusters containing more than four atoms, and that the reactivity for the larger clusters differed only by a factor of 2—3, depending on the cluster size and the metal. They suggested that the reason the smaller clusters in some cases showed little reactivity was due to a weak M_n—CO bond, the adduct dissociating before collisional stabilization. The results were then used to provide an ordering for M_n—CO bond strengths.

Hydrocarbons. Despite the importance of C—H bond activation, there have been relatively few studies of the reactions of transition-metal clusters with hydrocarbons. Iron clusters have been reported to be unreactive towards methane.[58] Most of the other experiments have been performed with Pt and Nb clusters reacting with C_6 hydrocarbons. Small platinum clusters chemisorb and dehydrogenate cyclohexane to a C/H ratio approaching 1:[27]

$$\mathrm{Pt}_n + \mathrm{C_6H_{12}} \rightarrow \mathrm{Pt}_n\mathrm{C_6H}_{12-m} + \frac{m}{2}\mathrm{H_2} \tag{9}$$

Similar results are found for linear and branched hexanes, although the dehydrogenation does not appear to go quite as far as with cyclohexane. Benzene readily chemisorbs on small platinum clusters.[27] Several molecules adsorb even on the dimer, where the dominant product was found to be $Pt_2(C_6H_6)_2$.

The reactions of niobium clusters with benzene have been studied by El-Sayed and co-workers[59,60] and also by Kaldor and co-workers.[61] Chemisorption of benzene on Nb clusters is apparently an easy process with little cluster-size dependence. However, certain clusters were found to dehydrogenate the benzene completely to yield Nb_nC_6. El-Sayed and co-workers reported that there are maxima in the dehydrogenation probability at $n = 5$, 6, and 11. Kaldor and co-workers found that, under low fluence photoionization conditions, only molecular addition to the clusters occurred for all clusters except those with $n = 5$, 6, and 11. However, with benzene-d_6 only molecular adsorption was found for all cluster sizes. A strong dependence of the degree of dehydrogenation on the photoionizing laser wavelength and fluence was reported by both groups. While it is clear that the ionizing laser plays a role in desorbing the hydrogens from the adduct, it is not clear whether with dehydrogenation is driven by the ionization process, or if dehydrogenation of the benzene occurs before ionization (*i.e.* to give H atoms on the cluster

[58] R. L. Whetten, D. M. Cox, D. J. Trevor, and A. Kaldor, *J. Phys. Chem.*, 1985, **89,** 566.
[59] R. J. St Pierre and M. A. El-Sayed, *J. Phys. Chem.*, 1987, **91,** 763.
[60] R. J. St Pierre, E. L. Chronister, and M. A. El-Sayed, *J. Phys. Chem.*, 1987, **91,** 5228.
[61] M. R. Zakin, D. M. Cox, and A. Kaldor, *J. Phys. Chem.*, 1987, **91,** 5224.

which are then desorbed as H_2 when the cluster is ionized). There is some evidence for dehydrogenation before ionization at least for a few specific cluster sizes.

Other Reactions. The reactions of transition-metal clusters have been studied with various other reagents. In many cases to date, these studies have been performed with one type of transition metal only. A few examples will be mentioned here.

There have been a few studies of the reactions of iron clusters with oxygen. Addition of molecular O_2 was observed to be an easy process for clusters with $n > 2$.[58] With high oxygen concentrations, iron-deficient species with the approximate formula $Fe_{0.90}O$ were found.[62] There may be a relationship between these cluster species and a high-temperature form of iron (II) oxide which exists as $Fe_{1-x}O(x = 0.05—0.15)$. A subsequent study[34] extended this work to include, V, Cr, and Ni clusters. In all cases the metal–oxygen ratio was close to a high-temperature congruently vaporizing phase. The photo-ionization of these oxidized clusters was found to be an extremely high-order multiphoton process (even though two-photon ionization is energetically possible). It was proposed that ionization under these conditions is analogous to thermionic emission, and the extreme heating of the clusters that occurs in the ionization process could account for the observation of the high-temperature oxide phases.

The interaction of Fe clusters with NH_3 was mentioned above in connection with the reactions between Fe clusters and H_2. Here some data on the larger clusters are briefly considered. The reactions were studied using the procedure of following the average number of adsorbates as a function of reagent pressure.[63] In the low-pressure regime, the extent of reaction is determined by kinetics, at higher pressures an equilibrium may be established, and ultimately a saturation coverage is achieved. It appears that ammonia adsorbs on Fe clusters at close to the collision rate. An enthalpy change of 0.72—0.85 eV for chemisorption of NH_3 on Fe_{61} was deduced from an analysis of data from pressure regions where it was believed that an equilibrium was established. The behaviour of Fe_{61} in the saturation regime is unusual.[32,64] There are two main species present: $Fe_{61}(NH_3)_{16}$ and $Fe_{61}(NH_3)_{18}$. Surprisingly, when the NH_3 pressure is increased the $Fe_{61}(NH_3)_{18}$ signal decreases. This result was interpreted as indicating that a structural rearrangement of the cluster skeleton (a phase transition) occurs and that this is promoted by heat generated by the more rapid chemisorption reactions. In general, the maximum number of NH_3 molecules that could be

[62] S. J. Riley, E. K. Parks, G. C. Nieman, L. G. Pobo, and S. Wexler, *J. Chem. Phys.*, 1984, **80,** 1360.

[63] E. K. Parks, G. C. Nieman, L. G. Pobo, and S. J. Riley, *J. Chem. Phys.*, 1987, **86,** 1066.

[64] E. K. Parks, G. C. Nieman, L. G. Pobo, and S. J. Riley, *J. Chem. Phys.*, 1988, **88,** 6260.

adsorbed can be accounted for approximately by a structureless packing model with three Fe atoms per NH_3.

Influence of Ligands. Studying the reactions of a cluster with a pre-adsorbed ligand offers some interesting possibilities. First, the adsorbate is likely to change either locally or globally the electronic properties of the cluster. This is easily seen from the change in ionization potential that is often found to accompany chemisorption on a cluster,[35,55] and from the promotion of bulk surfaces by adsorbed alkali atoms. Second, there is the possibility of performing chemical reactions on the surface of the cluster. So far there have been only a few studies along these lines, at least partly due to the difficulty of interpreting the mass spectral data when there is no selection of the M_nL_m species prior to the reaction.

The reactions of ammoniated iron clusters with hydrogen were found to be an order of magnitude slower than for the bare clusters.[32,64] There were also shifts in the reactivity patterns. For example, the minimum in the reactivity of the bare clusters around Fe_{17} shifts down to Fe_{13} for the fully ammoniated clusters. The shift is apparently a fairly continuous function of the degree of ammoniation. Ammoniated iron clusters will ultimately adsorb as many H_2 molecules as the bare clusters, without loss of ammonia. This result indicates that the NH_3 and H_2 molecules are binding non-competitively at different sites.

Several studies of the influence of oxygen on chemisorption reactions have been performed. Smalley and co-workers noted that for niobium clusters the reactivity of Nb_nO was drastically different from that of the bare clusters for reaction with D_2 and N_2, but about the same for CO chemisorption.[37] Kaldor and co-workers found that the saturation coverage of small vanadium clusters could be accounted for by assuming that each oxygen blocked two surface sites on the cluster.[35] Thus if V_n saturates to give $V_n(H_2)_y$, V_nO_m generally saturates to form $V_nO_m(H_2)_{y-m}$.

The effect of oxygen pre-adsorption on the rates for nitrogen chemisorption on tantalum clusters has been investigated for clusters up to $n = 11$.[65] The presence of oxygen was found to have a dramatic effect on the reaction kinetics. Large changes in the chemisorption rates were found for the small clusters. The differences gradually diminish as the cluster size increases. From the reaction rates as a function of cluster size it appears that the reactivity of Ta_nO tracks that of Ta_{n+1}. It was suggested that the O atom was simply acting as a space filler and preserving the packing arrangement of the clusters.

Reactions of Neutral Main Group Clusters

Carbon Clusters. Theoretical calculations predict that small carbon clusters have the structure of linear chains or monocyclic rings.[66,67]

There is currently a controversy regarding the structures of the larger ($n > 30$) clusters. Rohlfing *et al.*[68] proposed that the structures are related to a high-temperature form of carbon called carbyne, which consists of a series of cross-linked chains. Smalley and co-workers have suggested that the larger carbon clusters form spheroidal shells made up of five- and six-membered rings.[69] These ideas were originally proposed to account for the special nature of C_{60} which, according to this model, would have a very symmetric truncated icosahedron structure.[6] Unfortunately, a discussion of all the data relevant to this interesting question is beyond the scope of this chapter.

In one of the first studies of carbon clusters, Rohlfing *et al.*[68] reacted the clusters with potassium derived from laser vaporization of a KOH saturated graphite target. They found products of the form $C_{2n}K_2$ ($n < 14$) from which they inferred that carbon clusters with up to 24 atoms are linear chains. In similar experiments Heath *et al.*[70] reported that larger clusters (40—70 atoms), and in particular C_{60}, picked up one, and only one, lanthanum atom to give C_nLa. They speculated that the La atom sat in the middle of a spheroidal carbon cluster. This interpretation was challenged by Cox and co-workers,[71] who suggested that the strong signal for $C_{60}La^+$ was due to photophysical effects in the ionization process and was not related to the unique structural stability of C_nLa. The same group have reported the observation of C_{60} with several potassium atoms attached.[72] Recently, Weiss *et al.*[73] have reported more results suggesting that a metal atom can sit inside a carbon cluster shell.

When H_2, H_2O, NH_3, or CH_3CN were added to the buffer gas prior to laser vaporization, the even-numbered clusters were found to react giving products of the form $C_{2n}H_2$ (and $C_{2n}H_6$), $C_{2n}N_2$, and $C_{2n}HCN$ ($n < 12$).[74] These results were also interpreted as indicating that clusters with up to 24 atoms have linear chain structures. However, more recent photoelectron spectroscopy measurements,[75] strongly suggest that clus-

[65] M. R. Zakin, D. M. Cox, and A. Kaldor, *J. Chem. Phys.*, 1987, **87,** 5046.

[66] R. Hoffman, *Tetrahedron*, 1966, **22,** 521.

[67] W. L. Brown, R. R. Freeman, K. Raghavachari, and M. Schluter, *Science*, 1987, **235,** 860.

[68] E. A. Rohlfing, D. M. Cox, and A. Kaldor, *J. Chem. Phys.*, 1984, **81,** 3322.

[69] S. C. O'Brian, J. R. Heath, R. F. Curl, and R. E. Smalley, *J. Chem. Phys.*, 1988, **88,** 220.

[70] J. R. Heath, S. C. O'Brian, Q. Zhang, Y. Liu, R. F. Curl, H. W. Kroto, F. K. Tittel, and R. E. Smalley, *J. Am. Chem. Soc.*, 1985, **107,** 7779.

[71] D. M. Cox, D. J. Trevor, K. C. Reichmann, and A. Kaldor, *J. Am. Chem. Soc.*, 1986, **108,** 2457.

[72] D. M. Cox, K. C. Reichmann, and A. Kaldor, *J. Chem. Phys.*, 1988, **88,** 1588.

[73] F. D. Weiss, J. L. Elkind, S. C. O'Brian, R. F. Curl, and R. E. Smalley, *J. Am. Chem. Soc.*, 1988, **110**, 4464.

[74] J. R. Heath, Q. Zhang, S. C. O'Brian, R. F. Curl, H. W. Kroto, and R. E. Smalley, *J. Am. Chem. Soc.*, 1987, **109,** 358.

[75] S. Yang, K. J. Taylor, M. J. Craycraft, J. Conceicao, C. L. Pettiette, O. Cheshnovsky, and R. E. Smalley, *Chem. Phys. Lett.*, 1988, **144,** 431.

ters with more than 10 atoms are monocyclic rings, so ring opening is presumably induced by the reactions.

In the experiments described above the reactant is present at the point of laser vaporization, so reactions occur as the clusters are growing. There has been only one flow reactor study of the reactivity of carbon clusters where the cluster growth and reaction regions are separated.[76] With small molecules such as NO, CO, SO_2, H_2, O_2, and NH_3 it was found that clusters with <40 atoms were quite reactive and those with 40—80 atoms were unreactive, with C_{60} being particularly inert. It was suggested that these results were further evidence that the larger clusters (with more than 40 atoms) have edgeless spheroidal shell structures.

Aluminium Clusters. The reactions of aluminium clusters with a number of simple molecules have been investigated by Cox and co-workers.[77] With D_2 the smallest cluster that reacts is Al_6, and for clusters larger than Al_6 the reactivity rapidly decreases. Because aluminium has a simple electronic structure it is possible to perform reasonably reliable theoretical calculations on aluminium clusters. Upton has performed some calculations on the chemisorption of hydrogen on small aluminium clusters.[48,78] He found that for clusters with <6 atoms, dissociative chemisorption is endothermic. Dissociative chemisorption on Al_6 was exothermic but, according to his calculations, there is a significant activation barrier associated with this process. Figure 4 shows the interaction geometry and the orbitals of the Al_6 cluster important in the reaction. At large Al_6—H_2 separations, the a_1 orbital is filled and the b_2 orbital is empty. As the H_2 approaches, the a_1 orbital is destabilized by exchange interactions with the filled H_2 σ orbital (and this causes the activation barrier). Eventually, electrons from the a_1 orbital shift into the b_2 orbital. Interaction between the Al_6 a_1 and H_2 σ orbitals and the Al_6 b_2 and H_2 σ^* orbitals is now favourable and ultimately leads to the formation of Al—H bonds. Why the reactivity drops drastically for clusters larger than Al_7 has not yet been explained, but measurements described below suggest that there is a substantial activation barrier for the larger clusters.

A similar reactivity pattern to that found with D_2 (*i.e.* Al_6 most reactive) was also found with CO. With D_2O, clusters up to Al_7 were found to be unreactive, and then oscillations in the reactivity are observed with Al_{10}, Al_{18}, and Al_{24} being particularly reactive. However, with CH_3OH all the clusters appear to be reactive, but Al_7 and Al_{14} are noticeably less reactive than their neighbours. With O_2, the atom and dimer are very reactive, but then the reactivity drops by nearly an order

[76] Q. L. Zhang, S. C. O'Brian, J. R. Heath, Y. Liu, R. F. Curl, H. W. Kroto, and R. E. Smalley, *J. Phys. Chem.*, 1986, **90**, 525.

[77] D. M. Cox, D. J. Trevor, R. L. Whetten, and A. Kaldor, *J. Phys. Chem.*, 1988, **92**, 421.

[78] T. H. Upton, *Phys. Rev. Lett.*, 1986, **56**, 2168.

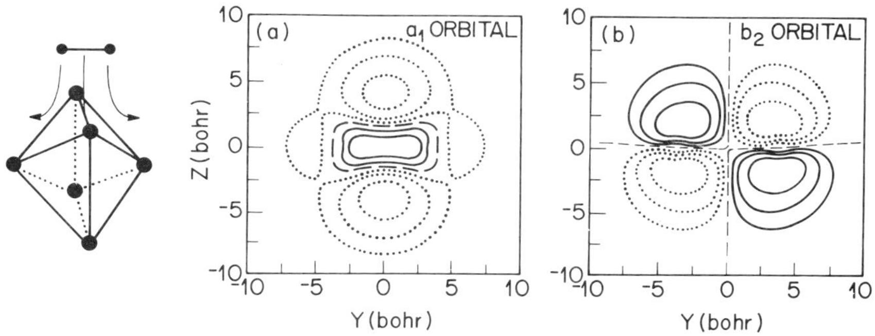

Figure 4 *A diagram showing the interaction geometry between* Al_6 *and* H_2 *and plots of the two highest occupied molecular orbitals. At large* Al_6—H_2 *separations the* Al_6 a_1 *orbital is populated; at the transition state the electrons have shifted into the* b_2 *orbital*

(Reproduced with permission from 'Physics and Chemistry of Small Clusters', Plenum, New York, 1987)

of magnitude and gradually increases again approaching the dimer reactivity at Al_{29}. Overall the reactivity was roughly ordered as $O_2 > CH_3OH > CO > D_2O > D_2 > CH_4$, with CH_4 showing no reaction. It was suggested that this reactivity could be correlated with the HOMO/LUMO energies of the reactants assuming that CH_3OH and D_2O acted as donor molecules and D_2, CO, and CH_4 acted as acceptors. The significance of such a correlation is questionable since different types of orbitals and processes are involved in the various reactions. For example, chemisorption of D_2 is probably dissociative while chemisorption of CO is probably non-dissociative.

Cluster Ions Studied in the Flow Reactor

In addition to studying the reactions of neutral clusters, it is possible to investigate the reactions of cluster ions in the flow reactor. Both positive and negative ions (from ionization caused by the vaporization laser[79]) are present in the flow reactor under normal conditions, so only simple modifications are required to study cluster ions. Unfortunately, there are several potential sources of error with this approach: for example, charge transfer with neutral clusters. As is described below, there are much more elegant ways of studying the reactions of cluster ions. However, this approach has generated some interesting results which are briefly reviewed here.

Smalley and co-workers[80] have investigated the reactions of Nb_n^+ and Co_n^+ with N_2 and CO. They found a strong correlation between the

[79] L. A. Bloomfield, M. E. Geusic, R. R. Freeman, and W. L. Brown, *Chem. Phys. Lett.*, 1985, **121**, 33.

[80] P. J. Brucat, C. L. Pettiette, S. Yang, L. -S. Zheng, M. J. Craycraft, and R. E. Smalley, *J. Chem. Phys.*, 1986, **85**, 4747.

reactivity observed with the ionic clusters and the reactivity of the corresponding neutral species. It was suggested that the positively charged clusters are less likely to donate charge, so charge transfer from the metal cluster is not an important factor in determining reactivity. However, Upton[48] has performed theoretical calculations on the chemisorption of H_2 on Al_6^-, Al_6, and Al_6^+. He found that the activation barriers were only slightly different for the neutral and ionic species. Furthermore, he found that the degree of charge transfer from the cluster to the H_2 σ^* orbital at the transition state was nearly independent of cluster charge. This result was explained by a dipole interaction which was stabilizing for the cation and destabilizing for the anion. Thus, the presence of the charge compensates for the effect that the large ionization potential difference might have been expected to have on the activation barriers.

Kaldor and co-workers[81] have investigated the effect of charge state on the chemisorption of D_2 on Nb clusters. In agreement with the results discussed above, the charge (+ or −) was found to have only a minor effect on the rate for the majority of the clusters. The reactivity of the ions was generally within a factor of 2.5 of the corresponding neutrals. In this study, the kinetic data for Nb_9, Nb_{12}, and Nb_{12}^+ was found to deviate significantly from the expected pseudo first-order behaviour. It was suggested that this may be evidence for different isomeric forms of these clusters.

Bohme and co-workers [82] have used the selected-ion flow tube technique to study some reactions of small carbon cluster cations. The selected-ion flow tube is a modification of conventional flow tube techniques where mass-selected ions are injected into the flowing buffer gas.[83] The ability to study a single cluster size is an obvious advantage over the studies described above using the cluster flow reactor. The disadvantage of this approach appears to be that an intense cluster source is required. In the experiments of Bohme and co-workers the clusters were generated by electron impact on different parent molecules (*e.g.* dicyanodiacetylene was used for C_6^+). They studied the reactions of C_n^+ (n = 1—6) and C_nH^+ (n = 1—5) with CO and found that two CO molecules attached to C_n^+ but only one CO attached to C_nH^+. These results are consistent with linear chain structures. Overall, the rates for the addition reactions were found to increase with cluster size, which was attributed to the increase in the lifetime of the addition complex.

3 Fourier Transform Ion Cyclotron Resonance

Experimental Aspects

Fourier transform ion cyclotron resonance (FT-ICR) is a powerful technique for the investigation of ion–molecule reactions at close to thermal energies.[84] The heart of an ICR experiment is the cell where the ions are trapped. As illustrated schematically in Figure 5, the cell is a rectangular or cubic box inserted into a magnetic field. The magnetic field causes the ions to undergo a cyclotron motion which constrains them in the plane perpendicular to the field. A small voltage on the end plates prevents the ions from escaping parallel to the field direction. The cyclotron frequency is inversely proportional to mass. Ions are detected by applying an r.f. chirp to one of the two sets of plates perpendicular to the magnetic field (see Figure 5). This coherently excites the ions to larger cyclotron orbits and an image current is monitored on the other set of plates. The resulting time-domain signal is Fourier transformed to give the mass spectrum. A mass resolution of 10^8 has been reported for H_2O^+.[85] However, mass resolution is inversely proportional to mass, and under normal operating conditions the upper limit for unit mass resolution is usually ~3000 amu.[84] Ions can be ejected from the cell by irradiating them at their cyclotron frequencies until they collide with the cell walls. Thus it is possible to select a single ion for detailed study. Reactions are followed by introducing gas into the ICR cell and recording mass spectra as a function of time. The reactant-gas pressure in the ICR cell is usually kept low ($<10^{-7}$ Torr) since collisions during the detection process degrade the mass resolution. Reaction times are long (up to several minutes).

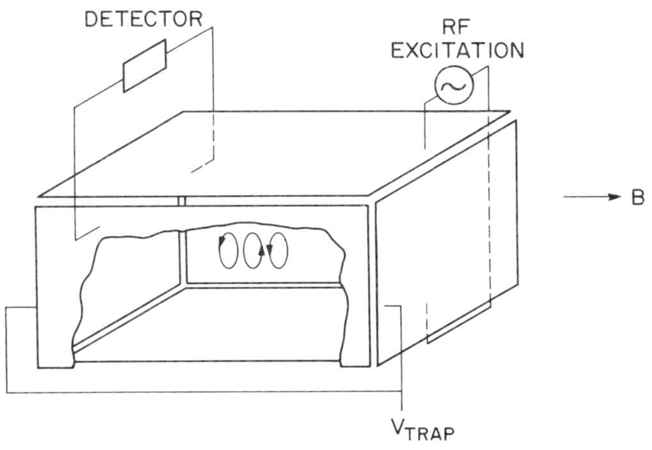

Figure 5 *Schematic diagram of a Fourier transform ion cyclotron resonance experiment*

[81] M. R. Zakin, R. O. Brickman, D. M. Cox, and A. Kaldor, *J. Chem. Phys.*, 1988, **88,** 3555.
[82] D. K. Bohme, S. Wlodek, L. Williams, L. Forte, and A. Fox, *J. Chem. Phys.*, 1987, **87**, 6934.
[83] N. G. Adams and D. Smith, *Int. J. Mass Spectrom. Ion Phys.*, 1976, **21,** 349.
[84] A. G. Marshall, *Acc. Chem. Res.*, 1985, **18,** 316.
[85] M. Allemann, H. Kellerhals, and K. P. Wanczek, *Int. J. Mass Spectrom. Ion Phys.*, 1983, **46,** 139.

Getting cluster ions in an ICR cell is not easy. Three approaches have been employed: injection from an external source,[86] direct laser sputtering,[87] and chemical synthesis of clusters from organometallic species followed by collision-induced dissociation to remove the ligands.[88] Injection from an external source is technically the most difficult, but the other methods are generally limited to quite small clusters. For direct laser sputtering, a pulsed laser is focused on to a target placed just outside the ICR cell. Material ejected from the surface enters the cell where the ions are trapped. For most metals virtually all of the material is removed as atoms with in some cases small amounts of dimer and trimer. With some materials (those whose equilibrium vapour contains molecular species) larger clusters are formed. For example, silicon positive and negative cluster ions with up to seven atoms can be produced in usable quantitites[89] and carbon clusters with over 100 atoms have been observed.[90] Freiser and co-workers[91] have recently reported an enhancement of silver cluster ion production from direct laser sputtering of an AgO–ZnO mixture. Clusters up to Ag_9^+ were observed, with the intensities of the odd-numbered clusters being much greater than the even ones. It is not yet clear whether this will be a generally useful approach for making metal clusters.

Carbon Cluster Ions

O'Keefe, McElvany, and co-workers at the US Naval Research Laboratory (NRL) have studied the reactions of carbon cluster ions, C_n^+, generated by direct laser sputtering.[92,93] With D_2, only clusters up to $n = 9$ were found to react and the main product was C_nD^+. A change in structure from linear chains to monocyclic rings probably occurs at $n = 10$,[75] so that the reactivity of the smaller clusters could be accounted for by the carbene character of the terminal carbon atoms of the linear chains. The reaction probably occurs by insertion of the carbene group into the D_2 bond to give a C_{n-1}—CD_2^+ complex which then dissociates to give the observed products. A drop in reactivity was not found at $n = 10$ for the neutral clusters.[74] However, in the experiments on the neutral clusters the reactant was present at the point of laser vaporization, so the observed products may have been formed from radicals produced in the laser plasma.

[86] J. M. Alford, P. E. Williams, D. J. Trevor, and R. E. Smalley, *Int. J. Mass Spectrom. Ion Processes*, 1986, **72,** 33.
[87] W. D. Reents and V. E. Bondybey, *Chem. Phys. Lett.*, 1986, **125,** 324.
[88] D. B. Jacobson and B. S. Freiser, *J. Am. Chem. Soc.*, 1984, **106,** 4623.
[89] M. L. Mandich, W. D. Reents, and V. E. Bondybey, *J. Phys. Chem.*, 1986, **90,** 2315.
[90] S. W. McElvany, H. H. Nelson, A. P. Baronavski, C. H. Watson, and J. R. Eyler, *Chem. Phys. Lett.*, 1987, **134,** 214.
[91] S. W. Buckner, J. R. Gord, and B. S. Freiser, *J. Chem. Phys.*, 1988, **88,** 3678.
[92] S. W. McElvany, W. R. Creasy, and A. O'Keefe, *J. Chem. Phys.*, 1986, **85,** 632.
[93] S. W. McElvany, B. I. Dunlap, and A. O'Keefe, *J. Chem. Phys.*, 1987, **86,** 715.

The main product observed from the reactions of C_n^+ with O_2 was C_nO^+.[93] The rate constants for the reactions with O_2 show a very similar pattern to those found for D_2. Clusters with more than 9 atoms were found to be unreactive. While performing these reaction studies it was found that there are two components of C_7^+ which react with both D_2 and O_2 at different rates. With D_2 ~ 33% of the C_7^+ reacts and the remaining fraction did not appear to react over times as long as 3 seconds. The most likely explanation for these observations is that there are two structures for C_7^+: possibly a linear chain and a monocyclic ring.

The NRL group have also investigated the collision-induced dissociation of C_n^+.[93] In these experiments the kinetic energy of the ions in the ICR cell is increased by irradiating them at their cyclotron frequency. The cluster ions are then internally excited by collisions with xenon and subsequently dissociate. The main product from cluster ions with $n = 5$—15 was C_{n-3}^+ (loss of C_3). Similar products were observed in the photodissociation of these species.[69,94] Attempts to dissociate larger clusters (C_{50}^+ and C_{60}^+) were not successful.[90] Photodissociation studies have shown that these larger clusters require at least three ArF (6.4 eV) photons to dissociate on a 10^{-5} s timescale[69] and that they are very efficient i.r. emitters.[73]

Marshall and co-workers[95] have investigated the reactions between carbon cluster ions and C_6H_{12}. The main product was $C_6H_{11}^+$. They found that clusters with up to 10 atoms rapidly reacted and those with more than 10 atoms were unreactive. This reactivity pattern is slightly different from those observed with D_2 and O_2, suggesting that it may have a different origin.

Silicon Cluster Ions

The reactions of silicon cluster ions (generated by direct laser sputtering) have been studied by groups at AT&T Bell Laboratories[89,96–98] and the NRL.[99] Si_n^+ (but not Si_n^-) with up to 5 atoms were found to react with CH_3SiH_3.[89] The reaction probabilities (measured rate constant divided by collision rate) are shown plotted in Figure 6. The main products can be accounted for by addition of CH_3SiH_3 to the cluster followed by loss of H or H_2. The silicon cluster has thus gained a silicon atom, and this could be considered a prototypical deposition reaction. It was suggested that the reaction occurred by insertion of a divalent (lone

[94] M. E. Geusic, M. F. Jarrold, T. J. McIlrath, R. R. Freeman, and W. L. Brown, *J. Chem. Phys.*, 1987, **86**, 3862.

[95] R. D. Knight, R. A. Walch, S. C. Foster, T. A. Miller, S. L. Mullen, and A. G. Marshall, *Chem. Phys. Lett.*, 1986, **129**, 331.

[96] M. L. Mandich, V. E. Bondybey, and W. D. Reents, *J. Chem. Phys.*, 1987, **86**, 4245.

[97] W. D. Reents, M. L. Mandich, and V. E. Bondybey, *Chem. Phys. Lett.*, 1986, **131**, 1.

[98] W. D. Reents, A. M. Mujsce, V. E. Bondybey, and M. L. Mandich, *J. Chem. Phys.*, 1987, **86**, 5568.

[99] W. R. Creasy, A. O'Keefe, and J. R. McDonald, *J. Phys. Chem.*, 1987, **91**, 2848.

pair) silicon site on the cluster into a Si—H bond of CH_3SiH_3. This behaviour is less likely for larger clusters which contain predominantly tri- and tetravalent silicon atoms and may explain why the reaction is not observed for clusters with more than 5 atoms (see Figure 6).

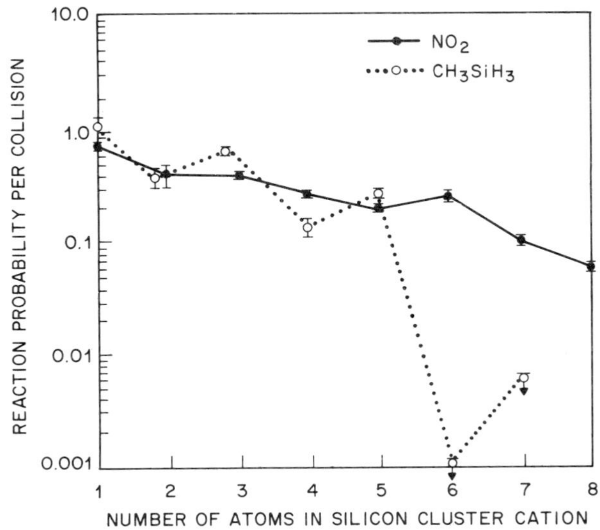

Figure 6 *Plot of the reaction probability (reaction rate constant divided by collision rate) against cluster size for the reactions of silicon cluster ions,* Si_n^+ *(n = 1—8), with* CH_3SiH_3 *and* NO_2
(Adapted from *J. Chem. Phys.*, 1987, **86,** 4245)

NO_2 sequentially etches both Si_n^- and Si_n^+:

$$Si_n^{\pm} + NO_2 \rightarrow Si_{n-1}^{\pm} + NO + SiO \tag{10}$$

all the way down to a monoatomic ion.[96] The other major products were NO^+ for the cations and NO_2^- for the anions. The reaction probabilities for the cations are shown in Figure 6. Unlike the reactions with CH_3SiH_3, the larger clusters are reactive, suggesting that these reactions occur by completely different mechanisms. It was suggested that the reaction with NO_2 occurred by a radical–radical coupling mechanism involving the unpaired electron on NO_2 and an unpaired electron on a silicon atom at a charged or trivalent centre. Both Si_n^+ and Si_n^- react with WF_6.[97] Charge transfer is the major reaction pathway for the anions. For Si^+ to Si_3^+ a range of $Si_nF_m^+$, WF_m^+, and $WSiF_m^+$ species are formed; at Si_4^+ the reaction rate suddenly drops and Si_nF^+ is the only product observed for n = 4—6. Si_n^+ reacts with XeF_2 to give SiF^+, Si_nF^+, and a small amount of Si_{n-1}^+.[98] Since fluorination (to give Si_nF^+) is very exothermic, it is possible that the other products result from the fragmentation of the Si_nF^+. The subsequent reactions of Si_nF^+ with XeF_2 yield Si_{n-1}^+ and SiF^+.

The NRL group has investigated the reactions between Si_n^+ and D_2, H_2O, CH_3OH, C_2H_2, NH_3, O_2, and N_2O.[99] No reaction was observed for D_2. For the other reagents the reaction rates generally show a large drop between $n = 3$ and 4. It was suggested that this may reflect the ease of ring opening to reveal highly reactive terminal silicon atoms. The calculations of Raghavachari and Logovinski[100–102] suggest that the energies for ring opening are ~0.4 eV for Si_3 and 3.4 eV for Si_4. The reactions of Si_n^+ with oxygen are particularly interesting. The clusters are sequentially etched two atoms at a time to give, ultimately, Si^+ or SiO^+. These reactions are similar to the ones mentioned above with NO_2.[96] The only neutral products that are thermodynamically accessible in the reaction with oxygen are two SiO molecules, or Si_2O_2, which is bound by 1.93 eV:[103]

$$Si_n^+ + O_2 \rightarrow Si_{n-2}^+ + 2SiO \text{ or } Si_2O_2 \tag{11}$$

These reactions are exothermic by at least 1.7—2.6 eV, but are very slow, suggesting that there must be an activation barrier. Since the reactions are quite exothermic the activation barrier is most likely to be associated with the entrance channel or the act of breaking the O—O bond and forming new Si—O bonds.

The reactions of larger silicon cluster ions (with up to 65 atoms) have been investigated by injecting clusters into an ICR cell from an external laser vaporization source.[104] The reactions between Si_n^+ and NH_3 were studied. The observed products were $Si_nNH_y^+$. These experiments were not performed with size-selected clusters, so cluster fragmentation could influence the results. Several oscillations in reactivity were observed: clusters with <14, 20, 25, 33, 39, and 45 atoms were relatively unreactive and those with 17, 18, 23, 30, 36, 42–44, and 46 atoms were particularly reactive. Above 47 atoms the reactivity increases slowly with cluster size. Noting that the periodicity in the oscillations is close to 6, Phillips[105] has suggested that the structure of the larger silicon clusters is a stack of six-membered rings.

Transition Metal Cluster Ions

Bare Clusters. The first studies of the chemical reactions of bare metal clusters in the gas phase were performed on clusters such as Mn_2^+ and

[100] K. Raghavachari and V. Logovinsky, *Phys. Rev. Lett.*, 1985, **55**, 2853.
[101] K. Raghavachari, *J. Chem. Phys.*, 1985, **83**, 2320.
[102] K. Raghavachari, *J. Chem. Phys.*, 1986, **84**, 5672.
[103] K. F. Zmbov, L. L. Ames, and J. L. Margrave, *High Temp. Sci.*, 1973, **5**, 235; L. C. Snyder and K. Raghavachari, *J. Chem. Phys.*, 1984, **80**, 5076.
[104] J. L. Elkind, J. M. Alford, F. D. Weiss, R. T. Laaksonen, and R. E. Smalley, *J. Chem. Phys.*, 1987, **87**, 2397.
[105] J. C. Phillips, *J. Chem. Phys.*, 1988, **88**, 2090.

Co_2^+ generated by electron impact on volatile organometallic compounds [*e.g.* $Mn_2(CO)_{10}$ for Mn_2^+]. Using this approach, Ridge and co-workers found that Co_2^+ is unreactive towards alkanes, although the atomic ion reacts.[106] If a CO ligand is attached to the dimer, however, the resulting Co_2CO^+ species does react with alkanes.[107] It was suggested that this enhancement of the reactivity was due to the CO ligand stabilizing an asymmetric charge distribution between the two Co atoms. Ridge and co-workers have also studied the reactions of Mn_2^+ with a range of n-donor bases, B (*e.g.* CH_3OH, H_2O, and H_2S).[108] The main products are MnB^+ species resulting from cleavage of the Mn—Mn bond. The results were used to provide an Mn^+ affinity scale.

Freiser and co-workers[109] have developed methods which increase the variety of small metal clusters that can be studied using ICR techniques. An atomic metal ion generated by laser vaporization of a target just outside the ICR cell is reacted with a volatile metal carbonyl, for example:

$$Co^+ + Fe(CO)_5 \rightarrow CoFe(CO)_3^+ + 2CO \tag{12}$$

The CO ligands are removed by collision-induced dissociation (CID) to yield $CoFe^+$, which is then isolated by ejecting the other ions from the cell. In principle it is possible to repeat this process and build larger clusters, but this is rather tedious. Co_3^+ and $FeCo_2^+$ have been generated by reactions with $Co_2(CO)_8$.[110,111] The same type of approach does not work with anions: a metal atom is lost in the CID step rather than the ligands. Freiser and co-workers have investigated the reactions of a wide range of mixed metal dimers such as $NbFe^+$,[112] VFe^+,[113] $ScFe^+$,[112] $CuFe^+$,[114] $CoFe^+$,[115] $LaFe^+$,[116] $RhFe^+$, and $RhCo^+$,[117] and a few trimers such as Co_3^+ and $FeCo_2^+$.[110,111] Dimers containing a second- or third-row metal atom bound to a first-row metal were generally found to be more reactive than dimers containing two first-row metals. Thus VFe^+, $ScFe^+$, $CuFe^+$, $CoFe^+$, and $NbFe^+$ are unreactive towards alkanes but $LaFe^+$, $RhFe^+$, and $RhCo^+$ dehydrogenate alkanes larger than methane.[117] It was suggested that this could be due to reduced overlap in the metal—metal bonds of dimers composed of metals from

[106] R. B. Freas and D. P. Ridge, *J. Am. Chem. Soc.*, 1980, **102,** 7129.
[107] R. B. Freas and D. P. Ridge, *J. Am. Chem. Soc.*, 1984, **106,** 825.
[108] B. S. Larsen, R. B. Freas, and D. P. Ridge, *J. Phys. Chem.*, 1984, **88,** 6014.
[109] D. B. Jacobson and B. S. Freiser, *J. Am. Chem. Soc.*, 1984, **106,** 4623.
[110] D. B. Jacobson and B. S. Freiser, *J. Am. Chem. Soc.*, 1984, **106,** 5351.
[111] D. B. Jacobson and B. S. Freiser, *J. Am. Chem. Soc.*, 1986, **108,** 27.
[112] S. W. Buckner and B. S. Freiser, in 'Modern Inorganic Chemistry: Gas Phase Inorganic Chemistry' ed. D. H. Russell, Plenum, New York, (in press).
[113] R. L. Hettich and B. S. Freiser, *J. Am. Chem. Soc.*, 1985, **107,** 6222.
[114] E. C. Tews and B. S. Freiser, *J. Am. Chem. Soc.*, 1987, **109,** 4433.
[115] D. B. Jacobson and B. S. Freiser, *J. Am. Chem. Soc.*, 1985, **107,** 1581.
[116] Y. Huang and B. S. Freiser, *J. Am. Chem. Soc.*, 1988, **110,** 387.
[117] Y. Huang, S. W. Buckner, and B. S. Freiser, in ref. 32, p. 891.

different rows. Unlike dimers from the same row, the trinuclear species $FeCo_2^+$ was found to activate C—H bonds in alkanes.[110] With hexane, dehydrocyclization to benzene may occur. However, with oxygen and ethylene oxide the dimers Fe_2^+, $FeCo^+$, and Co_2^+ and the trimers $FeCo_2^+$ and Co_3^+ react in similar ways.[111]

As discussed above, direct laser vaporization of metal targets does not generally yield clusters in usable quantities. Weil and Wilkins have reported that Au and Ag dimer and trimer cations and Au dimer and trimer anions can be formed by laser vaporization of metal salts and oxides,[118] but they did not investigate the chemistry of these species. Moini and Eyler[119] found that it is possible to generate up to Au_6^+ by direct laser vaporization of a gold target. However, the intensities of Au_4^+—Au_6^+ were too small for further study and only a limited investigation of the smaller clusters was performed. Freiser and co-workers[91] have generated Zn_2^+ from laser vaporization of a ZnO pellet. Charge-transfer reactions were used to bracket the ionization potential of Zn_2. With bases, Zn_2^+ was found to undergo displacement reactions, characteristic of a weakly bound cluster, to yield ZnB^+ (B = alkene or alcohol). Freiser and co-workers[91] have also investigated the chemistry of Ag_3^+ and Ag_5^+ clusters generated by direct laser vaporization of an AgO–ZnO mixture. Collision-induced dissociation of Ag_5^+ was found to result exclusively in Ag_3^+. With alkanes, alkenes, and alcohols, the only reaction pathway observed was condensation to give Ag_nB^+ (B = alkane, alkene, or alcohol) products.

Direct synthesis of bare metal clusters in an ICR cell is currently limited to small clusters. An alternative approach is to inject cluster ions into the ICR cell from an external source. This is difficult because the magnetic field acts as a mirror. However, Smalley and co-workers have been successful,[86] having studied hydrogen chemisorption on Nb cluster cations with up to 26 atoms.[120,121] Rate constants for the addition of the first H_2 and the saturation limits were determined. The physical significance of chemisorption rate constants recorded under the low-pressure conditions of an ICR cell is uncertain. The collision complex must survive on average 100 ms before a stabilizing collision occurs, so radiative cooling of the complex could be a factor.[121] A cluster could be unreactive under these conditions either because of an activation barrier or because the collision complex is weakly bound and dissociates.

Rate constants for the addition of the first H_2 on to Nb cluster cations are shown plotted in Figure 7. The values span over 4 orders of magnitude. Two forms of Nb_{19}^+ were found: a reactive form and a non-reactive form. Clusters with 3, 10, and 12 atoms were also found to

[118] D. A. Weil and C. L. Wilkins, *J. Am. Chem. Soc.*, 1985, **107,** 7315.

[119] M. Moini and J. R. Eyler, *Chem. Phys. Lett.*, 1987, **137,** 311.

[120] J. M. Alford, F. D. Weiss, R. T. Laaksonen, and R. E. Smalley, *J. Phys. Chem.*, 1986, **90,** 4480.

[121] J. L. Elkind, F. D. Weiss, J. M. Alford, R. T. Laaksonen, and R. E. Smalley, *J. Chem. Phys.*, 1988, **88,** 5215.

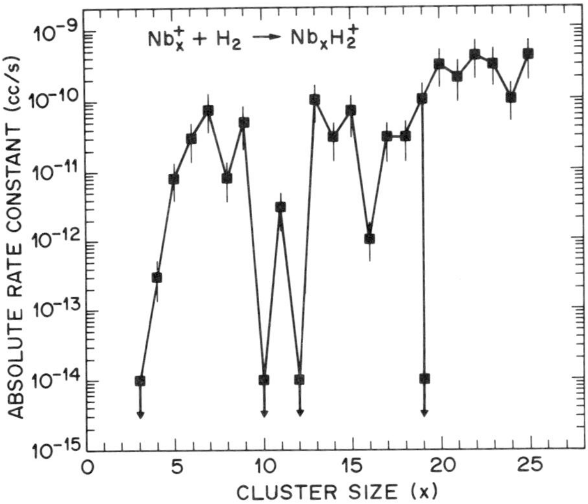

Figure 7 *Rate constants for the chemisorption of the first* H_2 *on to* Nb *cluster cations measured using FT-ICR techniques. Two rates are plotted for* Nb_{19}^+ *for the two different forms of this cluster*
(Reproduced with permission from *J. Chem. Phys.*, 1988, **88,** 5215)

be inert. For neutral Nb clusters studied in the flow reactor, clusters with 8, 10, and 16 atoms are particularly unreactive.[37,42] Cations with 8 and 16 atoms are also somewhat unreactive (see Figure 7). The reactions of niobium cluster cations with H_2 have also been studied in a cluster flow reactor.[81] There appears to be a rough qualitative agreement between these two measurements of chemisorption rates. In the flow reactor the smaller clusters appear more reactive, perhaps because the collision complexes for the smaller clusters dissociate in the low-pressure ICR experiment but are stabilized in the flow reactor.

Organometallic Clusters. The work discussed above dealt primarily with bare metal clusters. Adding ligands to the clusters adds a new dimension, and one enters the field of gas-phase organometallic chemistry. A detailed overview of this subject is beyond the scope of this chapter. However, as an example we will briefly consider one topic: condensation or clustering reactions of organometallic ions. For example, one reaction sequence in $Fe(CO)_5$ is:

$$Fe(CO)_5{}^+ \rightarrow Fe_2(CO)_8{}^+ \rightarrow Fe_3(CO)_{11}{}^+ \quad (13)$$

Following the pioneering work of Foster and Beauchamp,[122] several groups have studied these types of reactions.[123–126] Wronka and Ridge[124]

[122] M. S. Foster and J. L. Beauchamp, *J. Am. Chem. Soc.*, 1975, **97,** 4808.
[123] M. M. Kappes and R. H. Staley, *J. Phys. Chem.*, 1982, **86,** 1332.
[124] J. Wronka and D. P. Ridge, *J. Am. Chem. Soc.*, 1984, **106,** 67.
[125] D. Anderson-Fredeen and D. H. Russell, *J. Am. Chem. Soc.*, 1985, **107,** 3762.
[126] S. L. Mullen and A. G. Marshall, *J. Am. Chem. Soc.*, 1988, **110,** 1766.

investigated the negative ion chemistry of $Fe(CO)_5$ and used an electron counting rule to rationalize the reactivity of the various $Fe_n(CO)_m^-$ species. They proposed that clusters with 18 valence electrons per metal atom were inert and the reactivity increased as the number of valence electrons per atom decreased. This argument usually requires some assumptions about metal–metal binding to bring the experimental data into agreement with the predictions of the 18-electron rule, and so leads to a prediction of the structure of the $M_n(CO)_m$ species. This type of analysis have been extended to several other organometallic systems by Russell and co-workers[125,127] and Ridge and co-workers.[124,128]

4 Low-Energy Ion Beam Techniques

Experimental Aspects

Low-energy ion beam techniques have been used by several groups to investigate the reactions of metal cluster ions.[15,129–132] This experimental approach provides reaction cross-sections as a function of collision energy from below 0.1 eV to above 10 eV. Figure 8 shows a schematic diagram of the experimental apparatus. Two main approaches have been used to generate the clusters: sputtering[129,130,132] and laser vaporization.[17,133] Sputtering of solid targets with a high-energy ion or neutral beam generates clusters by a momentum transfer mechanism.[134] The cluster distribution falls off roughly exponentially with cluster size. The clusters have a wide range of kinetic energies[135] and also contain a substantial amount of internal energy.[129,136] Anderson and co-workers have used an r.f. cooling trap containing a buffer gas to cool the internal and kinetic energies of the clusters to close to room temperature.[18,129]

Cluster-ion sources derived from laser vaporization with continuous buffer-gas flow have also been employed. High pulse-repetition rate lasers (exciplex[17] or copper vapour[133]) are used to provide a reasonable duty factor. Larger clusters are obtainable with these sources compared with sputtering sources, and the clusters are already cooled by the buffer gas. It is possible to use cluster ions originating directly from the laser vaporization plasma or to enhance the degree of ionization by firing a high-energy electron beam into the buffer gas before the expansion.[17]

127 D. Anderson-Fredeen and D. H. Russell, *J. Am. Chem. Soc.*, 1986, **108,** 1860; 1987, **109,** 3903.
128 W. K. Meckstroth and D. P. Ridge, *Int. J. Mass Spectrom. Ion Processes*, 1984, **61,** 149; W. K. Meckstroth, D. P. Ridge, and W. D. Reents, *J. Phys. Chem.*, 1985, **89,** 612.
129 L. Hanley and S. L. Anderson, *Chem. Phys. Lett.*, 1985, **122,** 410.
130 P. Fayet and L. Woste, in ref. 43, p. 77.
131 M. F. Jarrold and J. E. Bower, *J. Chem. Phys.*, 1986, **85,** 5373.
132 T. F. Magnera, D. E. David, and J. Michl, *J. Am. Chem. Soc.*, 1987, **109,** 936.
133 S. K. Loh, D. A. Hales, and P. B. Armentrout, *Chem. Phys. Lett.*, 1986, **129,** 527.
134 P. Sigmund, *Phys. Rev.*, 1969, **184,** 383; 1969, **187,** 768.
135 K. Wittmaack, *Phys. Lett. A*, 1979, **69,** 322.
136 P. Fayet, J. P. Wolf, and L. Woste, *Phys. Rev. B*, 1986, **33,** 6792.

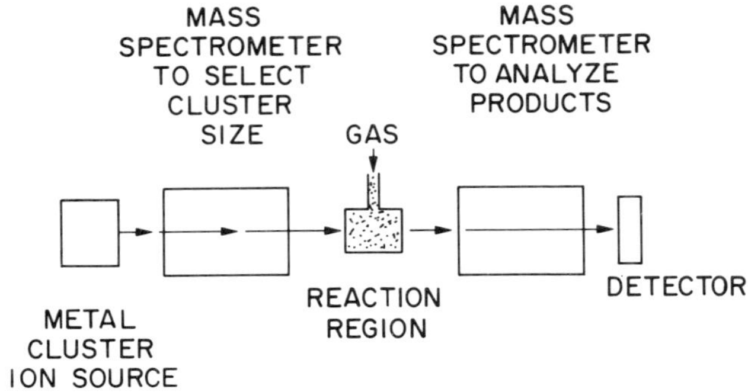

Figure 8 *Schematic diagram of a low-energy ion beam experiment to investigate the reactions of metal cluster ions*

The next step is to isolate a single cluster size. A number of different types of mass spectrometer have been employed: magnetic sector,[133] quadrupole,[17,130,132] and Wien filter.[18] The mass-selected ions are then focused into an ion beam and either injected into an r.f. ion guide or passed through a gas cell where the neutral reagent is introduced. The r.f. ion guide helps to collect the products, particularly at low collision energies. The reagent-gas pressure is kept low to minimize secondary collisions.

After the reaction, the products and unreacted ions are analysed by a second mass spectrometer and generally detected using a high-voltage collision dynode and electron multiplier. Single-ion counting techniques are used to collect the experimental data. A dynamic range in excess of 10^5 is readily attainable.

Mass discrimination is the principal source of error with these experiments, and is most likely to be a problem when the masses of the reactants and products are quite different. Mass discrimination can arise in the second mass spectrometer, which is usually a quadrupole, and at the detector. Often a high-voltage collision dynode is employed to reduce mass discrimination at the detector, and the quadrupole is operated with low resolution to minimize discrimination.

Collision-Induced Dissociation

With this experimental approach it is possible to study cluster interactions at collision energies considerably above room temperature. If the collision energy exceeds the dissociation energy, collision-induced dissociation (CID) can occur. A few years ago Armentrout and coworkers[15] investigated the CID of Mn_2^+ ions derived from $Mn_2(CO)_{10}$. From the collision energy threshold for dissociation they deduced a value for the dissociation energy of Mn_2^+. However, photodissociation experiments[137] suggest that the dissociation energy derived from these

[137] M. F. Jarrold, A. J. Illies, and M. T. Bowers, *J. Am. Chem. Soc.*, 1985, **107**, 7339.

CID measurements is too small, probably because some of the Mn_2^+ clusters were internally excited. Recently, several groups[16—18,138] have been investigating the CID of clusters using various collision energies. A few specific examples of the CID of bare metal clusters are discussed below.

Aluminium Cluster Ions. Meiwes-Broer and co-workers studied the CID of Al_n^+ (n = 2—7) by O_2 and Ar at laboratory energies in the keV range and found that Al^+ was the main product.[16] Subsequently, Jarrold and co-workers[17] studied the CID of aluminium clusters with up to 26 atoms at a collision energy of 5 eV. The main product for the smaller clusters ($n < 15$) was Al^+, and for the larger clusters the main product was Al_{n-1}^+ (loss of an Al atom from the cluster). These results were used to derive estimates of the ionization potentials of the clusters, which were found initially to rise above the atomic value, and then fall (dropping below the atomic value at around Al_{14}) to ultimately approach the bulk work function.[77] Dissociation by loss of a single atom has also been observed in photodissociation studies of metal clusters.[20] This tendency for dissociation by loss of a single atom rather than cluster fission is characteristic of systems where the binding energy per atom increases fairly smoothly with cluster size.

The cross-sections for CID show minima at $n = 7$ and $n = 13,14$. These are presumably particularly stable clusters, and this stability can be accounted for by the electronic shell model of Ekardt[139] and Knight and co-workers.[5] This is an approximate model for metal clusters where the ionic cores are replaced by a uniformly positively charged background ('jellium') and then electronic energies are calculated self-consistently. The electronic energy levels show a shell structure similar to that found for the electronic and nuclear energy levels of atoms. For aluminium, prominent shell closings have been predicted to occur with 20 and 40 valence electrons.[140] Al_7^+ has 20 valence electrons and Al_{13}^+ and Al_{14}^+ are both close to the shell closing with 40 valence electrons.

Anderson and co-workers[18] have investigated the CID of aluminium clusters with up to 7 atoms as a function of collision energy and deduced dissociation energies for the clusters. The dissociation energies are plotted in Figure 9. The dissociation energies generally increase with cluster size and Al_7^+ is particularly strongly bound. Morse and co-workers[53] have deduced a value of 2.4 eV for the dissociation energy of Al_3 from their spectroscopic studies. Combined with estimates of the ionization potentials, this leads to a value of ~2.2 eV for the dissociation energy of Al_3^+. This is considerably above the value obtained by Anderson and co-workers and larger than the theoretical estimates discussed below. The origin of this discrepancy is not clear. The dashed

[138] R. B. Freas, B. I. Dunlap, B. A. Waite, and J. E. Campana, *J. Chem. Phys.*, 1987, **86,** 1276.
[139] W. Ekardt, *Phys. Rev. B.*, 1984, **29,** 1558.
[140] M. Y. Chou and M. L. Cohen, *Phys. Lett. A.*, 1986, **113,** 420.

line in Figure 9 shows dissociation energies calculated by Upton using a generalized valence bond approach.[78,141] The calculated dissociation energies are in reasonable agreement with the experimental values. In Upton's theoretical study, the molecular orbitals were found to be very diffuse and to extend over the whole cluster. Bonding was primarily due to 3*p* atomic orbital combinations, but the 3*s* states significantly affect the 3*p* state occupation and the 3*s* + 3*p* occupation was found to follow approximately the order given by the electronic shell model.

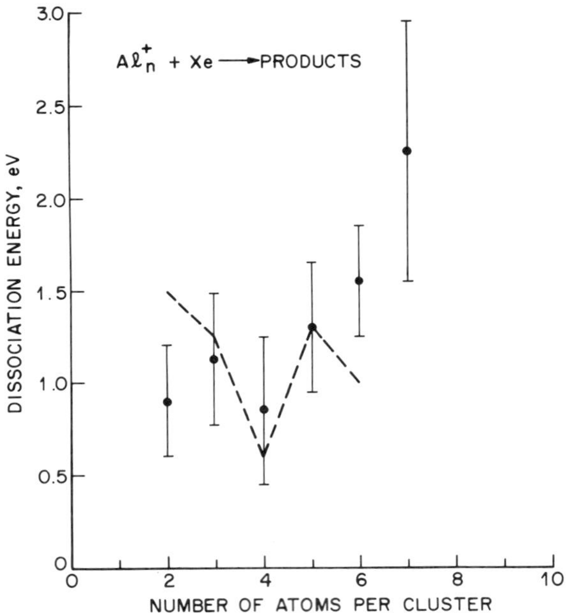

Figure 9 *Plot of the dissociation energies of aluminium cluster ions,* Al_n^+ (n = 2—7), *derived from collision-induced dissociation experiments.*[18] *The dashed line gives calculated dissociation energies taken from ref.* 78.

Silicon Cluster Ions. The CID (and photodissociation) of silicon cluster ions has been investigated on several occasions.[19,142–144] Unlike aluminium, silicon clusters fragment by cluster fission. Generally the same products are observed in both CID and photodissociation. Small clusters (n = 7—14) fragment to give preferentially Si_6^+ and Si_7^+. Clusters with

[141] T. H. Upton, *J. Chem. Phys.*, 1987, **86,** 7054.

[142] Q. -L. Zhang, Y. Liu, R. F. Curl, F. K. Tittel, and R. E. Smalley, *J. Chem. Phys.*, 1988, **88,** 1670.

[143] W. Begemann, S. Dreihofer, G. Gantefor, H. R. Siekmann, K. H. Meiwes-Broer, and H. O. Lutz, 'Elemental and Molecular Clusters', eds G. Benedek, T. P. Martin, and G. Pacchioni, Springer Verlag, Berlin, 1988, p. 230.

[144] M. F. Jarrold and J. E. Bower, *J. Phys. Chem.*, 1988, **92**, 5702.

15—18 atoms dissociate mainly by loss of a Si_6 or Si_7 neutral fragment, and larger clusters preferentially lose a Si_{10} unit. Theoretical studies by Raghavachari and Rohlfing[145] have shown that the observed products in these experiments can be simply accounted for in terms of product stability. It is remarkable that a cluster as large a Si_{20} can break into two equally sized fragments without a significant activation barrier.

Boron Cluster Ions. The CID of boron cluster ions has been studied by Hanley and Anderson.[146] Boron is above aluminium in the periodic table, so a comparison will be interesting. The main product for clusters with 3—8 atoms was found to be B^+ (similar to aluminium). The dissociation energies of the boron clusters were found to be 2—3 times larger than the corresponding aluminium clusters. This is reasonable because the cohesive energy of bulk boron is close to twice that of bulk aluminium.

Chemical Reactions and Reaction Thresholds

Aluminium Cluster Ions. The reactions of aluminium cluster ions have been studied by Anderson and co-workers[147–149] and by Jarrold and Bower.[131,150–152] The reactions with oxygen have been the most extensively studied. The products arising from the reactions of Al_n^+ (n = 3—26) with oxygen at a collision energy of 1.2 eV are shown in Figure 10. Most of the products arise from cluster fragmentation and there is very little oxygen incorporated into the ionic products. For clusters with less than 7 atoms, Al^+ and Al_2O^+ are important products, and loss of 5 atoms to give Al_{n-5}^+ is a major product for clusters with 7—13 atoms. For clusters with more than 13 atoms, Al_{n-4}^+ is the dominant product, except for Al_{19}^+, which loses 5 atoms to give Al_{14}^+. The formation of Al_{14}^+ from Al_{19}^+ is probably related to the extra stability of Al_{14}^+. CID experiments were used to help elucidate the nature of the products formed in these reactions. The main products from the CID of Al_nO^+ and $Al_nO_2^+$ (n = 3—26) were found to arise from loss of Al_2O molecules.[153] Al_2O is very strongly bound, so loss of Al_2O is the lowest-energy dissociation channel. Thus the neutral products associated with the Al_{n-4}^+ ionic product observed in the chemical reaction with oxygen are probably two Al_2O molecules. For the Al_{n-5}^+ product observed with smaller clusters, the neutral products are probably $2Al_2O + Al$. Total reaction cross-sections measured with a collision

145 K. Raghavachari and C. M. Rohlfing, *Chem. Phys. Lett.*, 1988, **143**, 428.
146 L. Hanley and S. L. Anderson, *J. Phys. Chem.*, 1987, **91**, 5161.
147 L. Hanley and S. L. Anderson, *Chem. Phys. Lett.*, 1986, **129**, 429.
148 S. A. Ruatta, L. Hanley, and S. L. Anderson, *Chem. Phys. Lett.*, 1987, **137**, 5.
149 S. A. Ruatta and S. L. Anderson, *J. Chem. Phys.*, 1988, **89**, 273.
150 M. F. Jarrold and J. E. Bower, in ref. 32, p. 801.
151 M. F. Jarrold and J. E. Bower, *J. Chem. Phys.*, 1987, **87**, 5728.
152 M. F. Jarrold and J. E. Bower, *J. Am. Chem. Soc.*, 1988, **110**, 6706.
153 M. F. Jarrold and J. E. Bower, *J. Chem. Phys.*, 1987, **87**, 1610.

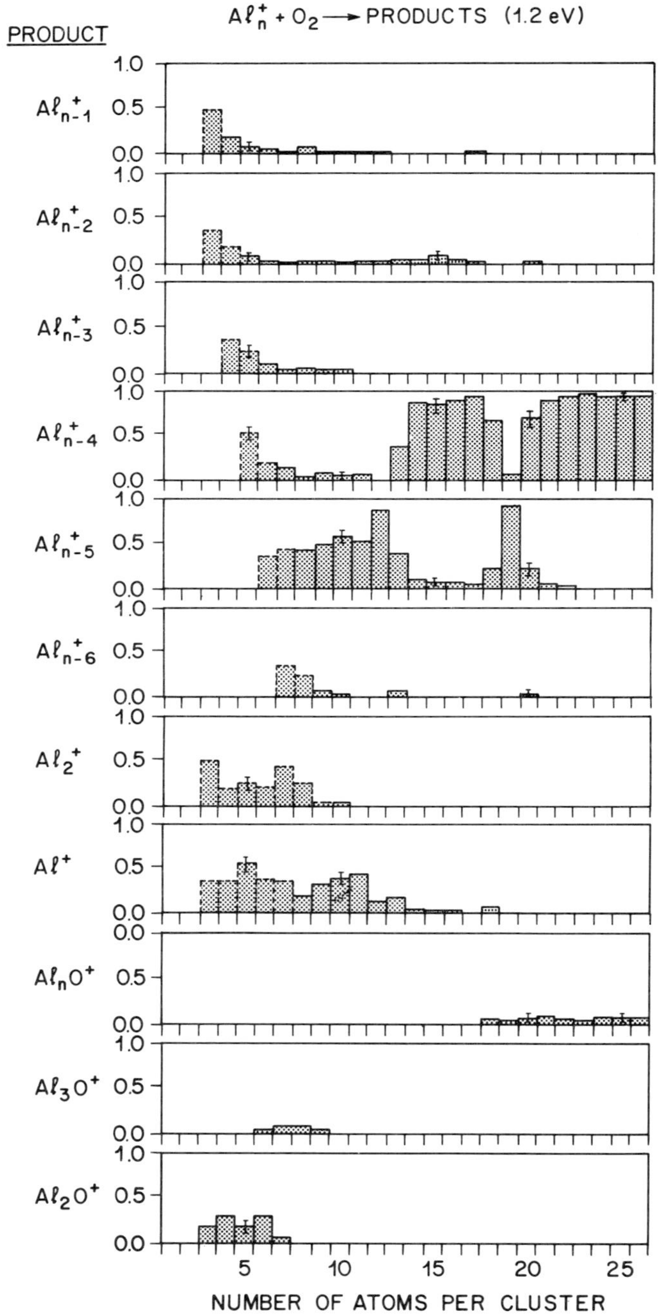

Figure 10 *Histograms showing the products arising from the reactions between* Al_n^+ *(n = 3—26) and oxygen at a collision energy of* 1.2 eV
(Reproduced with permission from *J. Chem. Phys.*, 1987, **87**, 5728)

energy of 4.2 eV were found to be roughly equal to the geometric cross-section of the clusters.

Clean aluminium surfaces rapidly oxidize in air. Aluminium clusters also appear to react readily with oxygen. For the clusters, the simple picture that has emerged from these studies is that the reaction starts by dissociative chemisorption of oxygen on to the clusters. This process is estimated to be exothermic by ~10 eV. This large amount of energy (released on forming the strong Al—O bonds) drives the subsequent fragmentation of the $Al_nO_2{}^{+*}$ adduct. For the larger clusters, two Al_2O molecules evaporate from the adduct; for the smaller clusters (with lower binding energies per atom) there is enough energy left over to fragment the Al_{n-4}^+ species further by loss of an Al atom or formation of Al^+.

Anderson and co-workers[148,149] have investigated the reaction of Al_2^+—Al_9^+ with oxygen as a function of collision energy. They found that Al_2^+ reacted with no activation barrier, while small activation barriers were found for the other clusters. The activation barriers generally increase with cluster size to a maximum value of 0.25 ± 0.31 eV for Al_7^+ and then decrease slightly up to Al_9. Because the overall reaction is so exothermic, the activation barrier must be associated with the first step: dissociative chemisorption of oxygen. As noted above, this is a very exothermic process but, since the oxygen bond is also quite strong (5.12 eV), the presence of activation barrier is not unreasonable. As there is no significant activation barrier for the reaction of oxygen with aluminium surfaces, it would be reasonable to expect the activation barrier to continue to decrease with cluster size for clusters larger than Al_9. However, recent measurements show that this is not the case.[152] The activation barrier for Al_{25}^+ was found to be 0.55 ± 0.3 eV, even larger than the barrier for the smaller clusters. This result suggests that clusters of considerably more than 25 atoms are required before bulk behaviour is approached. As discussed above, neutral aluminium clusters react with oxygen (in the flow tube reactor) to give Al_nO_2 adducts.[77] In view of the activation barriers and the large exothermicity of the oxidation reactions, it is reasonable to assign the Al_nO_2 adducts observed in the flow tube studies to a precursor state with molecular oxygen bound to the cluster surface.

From detailed study of the collision energy dependence of the reactions of the smaller clusters with oxygen, Ruatta and Anderson[149] identified several other processes in addition to the oxidative fragmentation described above. They suggest that aluminium atom abstraction and, at higher collision energies, collision-induced dissociation also contribute to the observed products.

The oxidation of aluminium clusters by N_2O has also been studied.[149] Like the reactions with oxygen, the reactions with N_2O appear to have small activation barriers. No nitrogen-containing product ions were observed and it seems that the N_2O is simply acting as a carrier for the oxygen atom. Most of the products can be accounted for by oxidation of the cluster followed by fragmentation of the Al_nO^+ species.

Nickel Cluster Ions. Woste and co-workers[130,154] and Michl and co-workers[132] have investigated the reactions of Ni cluster ions. Woste and co-workers studied the reactions with CO under multiple-collision conditions in an r.f. quadrupole ion guide. The main products were $Ni_n(CO)_k{}^+$, with small amounts of $Ni_{n-1}(CO)_l{}^+$ and $Ni_nC(CO)_m{}^+$. The observed pressure dependence of the product signals suggests that the $Ni_n(CO)_k{}^+$ species arise from a series of three-body association reactions. With increasing CO pressure, the number of CO ligands increases and then saturates. For example, 10 CO ligands will bind to Ni_4^+ and 22 will bind to Ni_{13}^+. Analogous to the FT-ICR studies of clustering reactions in metal carbonyls,[122–126] it was found that the maximum number of CO ligands bound to the cluster could be accounted for by simple electron counting rules. For example, with Ni_4^+ the number of CO ligands predicted for a tetrahedral metal skeleton is 10, in agreement with the experimental observations. For the larger clusters ($n > 9$), the electron counting rules appear to begin breaking down.

Michl and co-workers[132] have studied the reactions of mass-selected Ni cluster ions with C_4H_{10}. For the atomic Ni ion, the main products result from C—C bond cleavage (or cracking) to give $NiC_2H_4{}^+$, although dehydrogenation to give $NiC_4H_8{}^+$ is also an important process. As the cluster size increases, the cracking reactions are quickly suppressed; some products of C—C bond cleavage are seen for Ni_2^+ but not for larger clusters. For Ni_2^+—Ni_5^+ the main products arise from dehydrogenation to give $Ni_nC_4H_6{}^+$. With larger clusters (Ni_6^+—Ni_{10}^+), more extensive dehydrogenation occurs to give $Ni_nC_4H_4{}^+$. Experiments in which Ni_2^+ and Ni_3^+ were collisionally cooled prior to reaction demonstrated that the product distributions for these reactions were strongly influenced by internal energy in the clusters. Similar reactions were observed for small Pd clusters, although the dehydrogenation activity of Pd clusters appears to be greater than for Ni clusters. Thus Pd_4^+ removes four H_2 molecules from C_4H_{10} to give $Pd_4C_4H_2{}^+$, while Ni_4^+ removes only two.

Measurement of Activation Barriers for Chemisorption

Using low-energy ion beam techniques it is possible to measure directly activation barriers for chemisorption on metal clusters.[155] A low-energy ion beam is passed through a gas cell and under single-collision conditions chemisorption results in an adduct:

$$M_n^+ + AB \rightarrow M_nAB^{+*} \tag{14}$$

[154] P. Fayet, M. J. McGlinchey, and L. H. Woste, *J. Am. Chem. Soc.*, 1987, **109**, 1733.
[155] M. F. Jarrold and J. E. Bower, *J. Am. Chem. Soc.*, 1988, **110**, 70.

This adduct is metastable since it contains enough energy to dissociate back to reactants (or to other products if they are accessible). With large clusters ($n > 7$, for example), the chemisorption energy is distributed among a substantial number of internal degrees of freedom and the metastable adduct can survive long enough to be mass-analysed and detected (if there are no low-energy dissociation channels accessible). If there is an activation barrier for chemisorption then measurements of the cross-sections for adduct formation will show a collision energy threshold. As the collision energy is raised above the threshold for chemisorption, the cross-sections for adduct formation will rise. Because the collision energy is converted into internal energy in the metastable adduct, as the collision energy is raised the lifetime of the metastable adduct decreases and eventually becomes comparable to the time it takes to analyse and observe it. The cross-sections for adduct formation will then begin to decline. This is observed experimentally. Therefore from these studies it is possible to derive a measure of the activation barriers for chemisorption on size-selected clusters. In addition, using an RRKM model for the cross-section behaviour at the higher collision energies, it is also possible to derive activation energies for desorption.

Deuterium on Aluminium Cluster Ions. Activation barriers for the chemisorption of deuterium on aluminium clusters with 10—27 atoms have been measured using the methods described above.[155] The results are shown plotted in Figure 11. The activation barriers increase from a little over 1 eV for the smaller clusters studied up to ~2 eV for clusters with 25 atoms. There appears to be an oscillation in the size of the activation barriers: clusters with an even number of atoms (and an unpaired electron) have lower activation barriers than their neighbours. $Al_nD_2^+$ adduct formation was observed for clusters as small as Al_8^+, although Al_{10}^+ was the smallest cluster for which it was possible to obtain a reliable value for the kinetic energy threshold. No adduct was observed for Al_{13}^+; it is not clear whether this is due to a large activation barrier or a weakly bound $Al_{13}D_2^+$ adduct. The activation barrier for Al_{23}^+ is particularly large. Both Al_{13}^+ and Al_{23}^+ are close to electronic shell closings according to the jellium model, so this could be a factor in the anomalous behaviour of these clusters. The activation energies for desorption of D_2 from the clusters were estimated from an RRKM model. The desorption energies were found to increase from ~1.5 eV up to ~2 eV with increasing cluster size. These values are approximately the same as the activation barriers for chemisorption, so chemisorption of D_2 on to the aluminium clusters is approximately thermoneutral.

As described above, the reactions of neutral aluminium clusters with hydrogen have been studied using the cluster flow reactor technique. Al_6 was the smallest cluster observed to react, and the reactivity decreased rapidly for clusters larger than Al_6, so the data on the ionic and neutral clusters appear to be consistent. The point labelled 'theory'

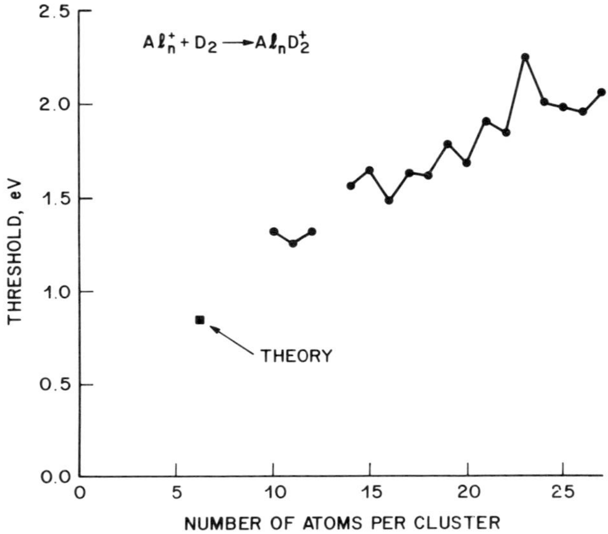

Figure 11 *Plot of the activation barriers for the chemisorption of* D_2 *on to aluminium cluster ions. The absolute uncertainties range from* $\pm$ 0.3 eV *for the smaller clusters to* $\pm$ 0.15 eV *for the larger ones. Relative values are probably more reliable. The point labelled 'theory' is from the work of Upton and co-workers*[48]
(Reproduced with permission from *J. Am. Chem. Soc.*, 1988, **110,** 70)

in Figure 11 is for the activation barrier for chemisorption on Al_6^+ from the work of Upton.[48] Unfortunately, there has not yet been any theoretical study of the chemisorption of hydrogen on larger aluminium clusters. The origin of the increase in the activation barrier with cluster size is currently not well understood. A decrease in activation barrier with cluster size is expected from the charge-transfer model for bond activation. If the activation barrier is mainly due to reorganization of the molecular orbitals to avoid Pauli repulsion, as suggested by Upton's work on Al_6, then the activation barrier would also be expected to decrease with cluster size (as the electronic density of states increases).

Dissociative chemisorption of H_2 on bulk aluminium has not been observed,[156] and therefore the activation barrier for chemisorption on the bulk surface is not known. Jellium model calculations with pseudopotential corrections[157] suggest that the activation barrier for H_2 chemisorption on bulk aluminium is ~1.3 eV. If this is the case then the activation barrier must eventually decrease for clusters with more than 27 atoms. In the jellium model calculations the activation barrier is due to the repulsive interaction between the hydrogen molecule and the electron density that spills over the jellium edge. As the free electron

[156] J. M. Mundenar, R. Murphy, K. D. Tsuei, and E. W. Plummer, *Chem. Phys. Lett.*, 1988, **143,** 593.
[157] P. K. Johansson, *Surf. Sci.*, 1981, **104,** 510.

density increases, the amount of electron density that spills over the jellium edge increases and so does the activation barrier. The increase in the activation barrier with cluster size may then be related to the overlap of the 3*s* and 3*p* orbitals (which results in an increase in the free electron density). However, this does not explain why the activation barriers for the clusters apparently rise above the bulk value.

In addition to metastable adduct formation in $Al_n^+ + D_2$ collisions, the products of chemical reactions were observed at the higher collision energies studied. For clusters with 3–12 atoms, the main products were $Al_{n-1}D^+$ (with presumably AlD) and Al^+. For clusters with more than 12 atoms, the main product was Al_nD^+.

Deuterium on Oxidized Aluminium Cluster Ions. Activation barriers for the chemisorption of deuterium on oxidized aluminium cluster ions have also been measured.[158] Clusters with 10—27 aluminium atoms and 1 or 2 oxygen atoms were studied. Oxygen pre-adsorption was found to have a dramatic effect on the activation barriers for some of the smaller clusters ($n < 18$). For example, with Al_{15}^+, pre-adsorbing one oxygen atom causes the activation barrier to increase by 0.6 eV. Pre-adsorbing a second oxygen atom (to give $Al_{15}O_2^+$) causes the activation barrier to drop slightly. With Al_{16}^+, the first oxygen atom causes a slight drop in the activation barrier and the second causes an increase of 0.6 eV. For most of the smaller clusters, oxygen pre-adsorption causes an increase in the activation barriers. This suggests that the effects of oxygen pre-adsorption on activation barriers are the result of a global perturbation of a cluster's electronic structure rather than the operation of a site-specific mechanism in which the oxygen is directly involved. The activation barriers for the larger clusters ($n > 17$) were only slightly affected by oxygen pre-adsorption. This is easily understood in terms of range and dilution.

CO, N_2, CH_4, *and* C_2H_4 *on* Al_{25}^+. A systematic study of the chemisorption of a number of simple molecules on Al_{25}^+ has been performed by Jarrold and Bower.[152,159] For CO, N_2 and CH_4, metastable adduct formation was observed and activation barriers for chemisorption were determined from the kinetic energy thresholds. In all of these reactions chemisorption is probably dissociative rather than molecular. The activation barriers are shown in Table 1, where the threshold for the reaction with O_2 and the activation barrier for chemisorption of D_2 are included. The activation barriers span a large energy range and there appears to be an approximate correlation between the size of the activation barrier and the LUMO energy of the molecules studied. Such a correlation supports the charge-transfer model for bond activation. While charge transfer to the molecule's LUMO may lead to stabilization of the transition state, other factors are probably important as well. For

158 M. F. Jarrold and J. E. Bower, *Chem. Phys. Lett.*, 1988, **144,** 311.
159 M. F. Jarrold and J. E. Bower, *Chem. Phys. Lett.*, 1988, **149**, 433.

example, with methane, steric factors and distortion of the CH_4 molecule to accommodate the forming M—CH_3 bond may contribute to the activation barrier.

Table 1 *Summary of activation barriers measured for dissociative chemisorption on* Al_{25}^+

Reaction	*Activation barrier* /eV
$Al_{25}^+ + D_2$	1.99
$Al_{25}^+ + CH_4$	3.47
$Al_{25}^+ + O_2$	0.55
$Al_{25}^+ + CO$	1.93
$Al_{25}^+ + N_2$	3.53

At collision energies considerably above the threshold for adduct formation, other products were observed. For example, with CO the product was $Al_{23}C^+$ (with presumably Al_2O), and with CH_4 the main products were $Al_{24}H^+$ (+ $AlCH_3$) and $Al_{24}CH_3{}^+$ (+ AlH). These products can be ascribed to dissociation of the adduct, and since they arise from dissociation of the reactant molecule they are consistent with the idea that chemisorption is dissociative rather than molecular.

In the reactions discussed above, the behaviour of the cross-sections for adduct formation is straightforward: above threshold the cross-sections increase, although if the collision energy is raised sufficiently above threshold the cross-sections will decline as the adduct dissociates. With C_2H_4, the behaviour is considerably more complicated. Cross-sections for the products found in the reaction between Al_{25}^+ and C_2H_4 are shown plotted in Figure 12. C_2H_4 apparently chemisorbs on Al_{25}^+ without a significant activation barrier since the adduct is observed at collision energies as low as 0.2 eV. As the collision energy is raised, the cross-sections for the adduct decline, presumably because this weakly bound adduct (which is formed without an activation barrier) dissociates. However, at collision energies above 0.5 eV the cross-sections for adduct formation begin to increase again. A different type of adduct with an activation barrier for formation of 0.5—1.0 eV is now being made. As the collision energy is raised above 1.5 eV the cross-sections begin to decline as this second adduct dissociates before detection. For collision energies above 3 eV the cross-sections for adduct formation again begin to rise; a third type of adduct is now being made. These experiments show that three different types of $Al_{25}C_2H_4{}^+$ species can be generated in $Al_{25}^+ + C_2H_4$ collisions. These species are probably different forms of C_2H_4 chemisorbed on the cluster surface. C_2H_4 is known to give a range of different products when chemisorbed on metal surfaces. These experiments provide no direct structural information, although it seems likely that the two lower-energy adducts correspond to

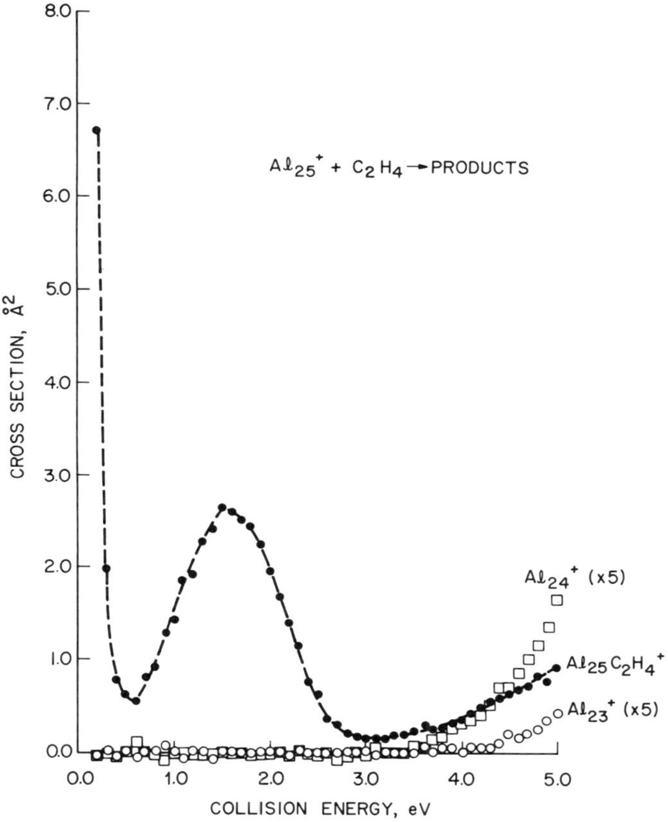

Figure 12 *Plot of the cross-sections of the products observed in the reaction between* Al_{25}^{+} *and* C_2H_4. *The dashed lines drawn through the data are only guides*
(Reproduced with permission from *Chem. Phys. Lett.*, 1988, **149**, 433)

molecular chemisorption. In both cases, as the collision energy is raised and the adduct cross-sections decline (due to adduct dissociation) no other products are observed, indicating that the C_2H_4 desorbs intact. The adduct with an activation barrier of ~3 eV probably arises from dissociative chemisorption. These experiments illustrate the complicated chemistry that can occur on the surface of metal clusters.

5 Concluding Remarks

This chapter has provided an overview of the recent studies of the reactions of bare metal clusters in the gas phase. Virtually all of the work described here has been performed in the past three years. The ultimate goal of these studies is to elucidate the factors that are important in controlling the reactions of metal clusters (for example, the relative importance of electronic factors and geometric structure) and explain why some cluster sizes are particularly reactive or inert. While some progress has been made in this direction, it is clear that considerably more work is required to achieve this goal.

The study of the reactions of metal and semiconductor clusters is but one aspect of cluster science. Studies of the electronic structure and geometry of clusters and the measurement of fundamental quantities such as dissociation energies and ionization potentials are not only important in their own right, but are required in order to improve our understanding of the chemistry of metal clusters. Considerable progress has been made in these areas in recent years, with the possible exception of cluster geometry. There is virtually no experimental data on the geometry of clusters with more than three atoms.

CHAPTER 8

Laser Probing of Ion Collisions in Drift Fields: State Excitation, Velocity Distributions, and Alignment Effects

S. R. LEONE

1 Introduction

The mobility, diffusion, and velocity distributions of ions in gaseous plasmas have been subjects of intense theoretical and experimental investigation for many years.[1–6] Much of this interest comes from a desire to obtain a deeper understanding of the motions of ions and the chemistry that they undergo in discharges, lasers, and in natural phenomena such as the aurora. The recent advance in semiconductor growth using plasma-assisted etching is generating additional interest in the fundamental aspects of ion collisions under the influence of external electric fields.[7] These include such properties as internal excitation, state-to-state chemistry, preparation of novel chemical species, and energy-enhanced surface processes. Such 'low-energy' plasma studies are an important area of concern in both spectroscopy and collision dynamics.

[1] E. W. McDaniel and E. R. Mason, 'The Mobility and Diffusion of Ions in Gases', Wiley, New York, 1973; E. W. McDaniel, 'Collision Phenomena in Ionized Gases', Wiley, New York, 1964.
[2] E. W. McDaniel 'Transport Properties of Ions in Gases', Wiley, New York, 1988.
[3] E. A. Mason and M. A. Biondi, *Comments At. Mol. Phys.*, 1977, **6**, 159.
[4] J. H. Whealton, *Comments At. Mol. Phys.*, 1977, **7**, 35.
[5] D. L. Albritton, in 'Kinetics of Ion-Molecule Reactions', ed. P. Ausloos, Plenum, New York, 1979, p. 119.
[6] Y. Kaneko, *Comments. At. Mol. Phys.*, 1981, **10**, 145.
[7] R. A. Powell, ed. 'Dry Etching for Microelectronics', Elsevier, Amsterdam, 1984.

Many of the studies of ion motion and transport in buffer gases have been successfully carried out with the use of drift tube devices,[1] which have also been combined with the flowing afterglow technique.[8] In a drift tube, ions are created at low densities, separated from electrons and ions of the opposite charge, and drifted through the buffer gas by the application of a uniform external electric field. The motion of the ions is of interest, which on the macroscopic scale is described by the ion transport, and on the microscopic level is related to the individual interaction potentials and the precise trajectories of each elastic or inelastic scattering event. In this chapter, we explore the ramifications of this relationship between ion transport and individual collision events. The discussion considers the connection between microscopic collision processes, the detailed cross-sections for velocity- and state-changing collisions, and macroscopic quantities of ion mobility and diffusion.

In a conventional drift tube apparatus, the velocities of ions drifting in simple neutral buffer gases are measured as a function of applied electric field to obtain velocity distributions and mobilities. The precise measurement of velocity distributions both parallel and perpendicular to the electric field is rendered difficult because of the intrusion of sampling orifices and fields.[9] At high fields, however, some velocity distributions have been reported.[10–15] From drift tube studies, there are many indications of the importance of internal excitation in the collision processes of ions. The distinction between translational kinetic energy and internal vibrational energy is clearly recognized in studies of the rates of reactions.[16] Numerous experiments have also been performed by using the enhanced chemical reactivity of vibrationally excited ions to probe the rates of ion vibrational deactivation[17] and to study ion vibrational excitation and deactivation as a function of velocity in a drift field.[18] Through the clever use of kinetics, a great many state-resolved processes have been investigated by these energy-sensitive methods.

A more detailed investigation of state-resolved properties of ions colliding in drift fields has awaited the introduction of laser probing techniques for ion detection and spectroscopic analysis.[19–28] In addition

[8] M. McFarland, D. L. Albritton, F. C. Fehsenfeld, E. E. Ferguson, and A. L. Schmeltekopf, *J. Chem. Phys.*, 1973, **59**, 6610.

[9] H. R. Skullerud and S. Holmstrøm, *J. Phys. D: Appl. Phys.*, 1985, **18**, 2375.

[10] P. P. Ong, D. Mathur, M. H. Khatri, J. B. Hasted, and M. Hamdan, *J. Phys. D: Appl. Phys.*, 1981, **14**, 633.

[11] H. A. Fhadil, D. Mathur, and J. B. Hasted, *J. Phys. B: At. Mol. Phys.*, 1982, **15**, 1443.

[12] M. H. Khatri, *J. Phys. D: Appl. Phys.*, 1984, **17**, 273.

[13] H. A. Fhadil, A. T. Numan, T. Shuttleworth, and J. B. Hasted, *Int. J. Mass Spectrom. Ion Processes*, 1985, **65**, 307.

[14] P. P. Ong and M. J. Hogan, *J. Phys. B: At. Mol. Phys.*, 1985, **18**, 1897.

[15] T. Makabe and H. Shinada, *J. Phys. D: Appl. Phys.*, 1985, **18**, 2385.

[16] W. Lindinger, M. McFarland, F. C. Fehsenfeld, D. L. Albritton, A. L. Schmeltekopf, and E. E. Ferguson, *J. Chem. Phys.*, 1975, **63**, 2175.

[17] E. E. Ferguson, *J. Phys. Chem.*, 1986, **90**, 731.

[18] M. Kregel, R. Richter, W. Lindinger, L. Barbier, and E. E. Ferguson, *J. Chem. Phys.*, 1988, **88**, 213.

to providing a non-intrusive method for investigating ion collisions in drift fields, laser probing allows the study of the collision dynamics in previously unobtainable detail. Such studies include direct measurements of the mobilities in plasmas,[20,28] velocity distributions of ions both parallel and perpendicular to the electric field,[19,24,27] the alignment of molecular ions by collisions in a drift tube,[26] and the rotational excitation of ions by collisions.[21]

This chapter considers the kinds of measurements that are possible in the study of ion collisions in simple buffer gases using optical detection. The emphasis is on the use of tunable pulsed lasers and high-resolution, sub-Doppler cw (continuous wave) dye lasers. Laser-induced fluorescence (LIF) detection is used to probe both atomic and molecular positively charged ions. A flowing afterglow drift tube device is employed in these studies;[8] the flow tube has been modified and adapted to include numerous ports and special configurations which eliminate scattered light for laser–optical detection.[21]

The flowing afterglow device provides a versatile, well-characterized source for the preparation of many different kinds of ions.[29] It is an extremely effective tool for exploring collision dynamics under well-defined drift field conditions. The uniform drift field is a convenient means of preparing a directed, anisotropic velocity distribution for the study of ion alignment and internal excitation. Different mass combinations and relative kinetic energies can be readily produced. The well-defined flow properties of the flowing afterglow allow the precise calculation of the details of the ion transport. Steady-state kinetics may often be applied to the collision dynamics, resulting in substantial simplifications in the analysis of the measured parameters.

With sub-Doppler lasers, the ion velocity distributions in the drift region of the flowing afterglow may be interrogated directly by the Doppler velocity profiles. Except for the possibility of optical saturation, the method of LIF detection is only a weak perturbation on the ion–buffer gas system. The LIF method is sensitive and wavelength-selective, allowing specificity of detection for the desired ion species. Individual populations of the states of the ions are directly obtained as well.

[19] D. N. Hill, S. Fornaca, and M. G. Wickham, *Rev. Sci. Instrum.*, 1983, **54**, 309.
[20] N. N. Haese, F. -S. Pan, and T. Oka, *Phys. Rev. Lett.*, 1983, **50**, 1575.
[21] M. A. Duncan, V. M. Bierbaum, G. B. Ellison, and S. R. Leone, *J. Chem. Phys.*, 1983, **79**, 5448.
[22] R. A. Gottscho, R. H. Burton, D. L. Flamm, V. M. Donnelly, and G. P. Davis, *J. Appl. Phys.*, 1984, **55**, 2707.
[23] C. S. Gudeman, C. C. Martner, and R. J. Saykally, *Chem. Phys. Lett.*, 1985, **122**, 108.
[24] P. Koslovar and R. McWilliams, *Rev. Sci. Instrum.*, 1986, **57**, 2441.
[25] M. B. Radunsky and R. J. Saykally, *J. Chem. Phys.*, 1987, **87**, 898.
[26] R. A. Dressler, H. Meyer, and S. R. Leone, *J. Chem. Phys.*, 1987, **87**, 6029.
[27] R. A. Dressler, H. Meyer, A. O. Langford, V. M. Bierbaum, and S. R. Leone, *J. Chem. Phys.*, 1987, **87**, 5578.
[28] N. N. Haese and T. Oka, Proc. SPIE – Int. Soc. Opt. Eng., 1983, Vol. 438, p. 165.
[29] E. E. Ferguson, F. C. Fehsenfeld, and A. L. Schmeltekopf, *Adv. At. Mol. Phys.*, 1969, **5**, 1.

In the subsequent sections of this chapter, the relationship between the interaction potentials, internal excitation, and ion velocity and mobility will be explored. A brief description of the theoretical treatment of ion collisions will be given, followed by the experimental details for laser probing of ions in a flowing afterglow apparatus. Specific examples of measurements which have been used to probe population distributions, molecular alignment, and velocity distributions will be discussed.

2 The Coupled Parameters of Ion Drift Motion

The major effect of the drift field is to accelerate the ions to a new velocity, which is higher than the original thermal velocity. A uniform electric field vector, $\boldsymbol{E}$ is assumed. The ions obtain their new component of momentum initially along the direction of the field. Collisions with the buffer gas in the drift tube alter the momentum of the accelerating ions, until a balance is achieved between momentum gained and momentum lost. This results in a phenomenological average drift velocity, $\boldsymbol{v}_d$ of the ion along the direction of the field. At the same time, individual scattering events cause a fraction of the additional momentum to be directed perpendicular to the field direction. Thus the velocity of the ions transverse to the field is also increased to some extent, which causes a spreading of the ion distribution along the velocity vector $\boldsymbol{v}_\mathbf{d}$.

The mobility of the ions is an inverse measure of the effectiveness with which collisions remove the forward momentum that is gained by the ions from the field. Specifically, the mobility can be related to the inverse of the momentum transfer cross-section. Thus, the mobility depends on the attractive and repulsive parts of the ion–buffer gas interaction potential and on the resulting trajectories that transfer momentum from the ion to the buffer gas. An average drift velocity ($\boldsymbol{v}_\mathbf{d}$) can often be ascribed to the ion, and then the phenomenological mobility, K, is obtained from the simple expression:

$$\boldsymbol{v}_\mathbf{d} = K\boldsymbol{E} \tag{1}$$

where $\boldsymbol{E}$ is the field strength. Typically, a reduced mobility is quoted at a particular value of the field strength divided by the number density (E/N), in order to normalize the results of the measurements to a parameter related to the momentum gained from the field per time between collisions. The significance of the additional motivation for the parameter E/N will become clear in the discussion of the free-flight theory below. The reduced mobility is defined as:

$$K_0 = \frac{p}{760}\frac{273.16}{T}K \tag{2}$$

where p is the pressure of the gas in Torr used for the measurements and T is the temperature. Typically, the mobility is only weakly dependent on the relative velocity for small changes in the temperature of the bath-gas.

On a more detailed level, the drifting ions exhibit a range of velocities, *i.e.* a laboratory velocity distribution, $f(v)$. The propensity for momentum transfer out of the forward direction depends on the relative velocity, v_0 of the collision and therefore also on the temperature of the buffer gas. This is clear when one considers that the momentum transfer cross-sections will be a function of the relative velocity between the drifting ion and the thermal buffer gas. Therefore an averaged expression might be more appropriate:

$$\boldsymbol{v}_{\mathrm{d}} = \int K'(v_0) f(v_0) \boldsymbol{E} \, \mathrm{d}v_0 \tag{3}$$

where $f(v_0)$ is the relative velocity distribution and K' describes a quantity that is inversely proportional to an effective cross-section for momentum transfer at a specific velocity. The velocity-dependent quantity defined here can be related to velocity-dependent effective cross-sections for momentum transfer, if they are known. In a simple model (1), K' might be inversely related to the product $v_0 Q(v_0)$, where $Q(v_0)$ is an effective momentum transfer cross-section, which can include more complete information about the differential scattering function. The average drift velocity and v_0 depend on the temperature of the buffer gas, T, and field strength, E/N.

For molecular ions, the microscopic mobility will also be a function of vibrational state V (for which there is already some evidence[20,28]), rotational state J, and the angle of the molecular axis with respect to the relative velocity in the collision, known as alignment, P, with distribution, $f'(P)$. It is known that electronic states also affect the mobility, for example, through the differing ion interaction potentials. One example comes about because of the different values of the charge exchange cross-sections for the spin–orbit states of noble gas ions drifting in their parent buffer gases.[30–32] The drift collisions also readily excite vibrational[18] and rotational[21] states of molecular ions, which will in turn alter the mobility. It has recently been demonstrated that the plane of rotation of a molecular ion, N_2^+, is significantly aligned by He buffer gas collisions in a drift field.[26] The tipping of the plane of rotation of the ion by the collisions ultimately causes a smaller cross-sectional area to be presented to the next oncoming buffer gas atoms. This alignment will undoubtedly affect the mobility as well.

Thus, on the microscopic scale many interrelated state-to-state factors affect the mobility and diffusion of ions, in addition to the fundamental

[30] H. Helm, *Chem. Phys. Lett.*, 1975, **36**, 97.
[31] H. Helm. *J. Phys. B: At. Mol. Phys.*, 1976, **9**, 2931.
[32] H. Helm and M. T. Elford, *J. Phys. B: At. Mol. Phys.*, 1977, **10**, 983.

form of the ground-state interaction potential. The drift equation might be reformulated as:

$$(\boldsymbol{v}_{\mathbf{d}})_{V,J} = \iint K''_{V,J}(v_0, P)f(v_0)f'(P)\boldsymbol{E}\,\mathrm{d}v_0\,\mathrm{d}P \qquad (4)$$

which now allows one to consider the drift velocity of specific vibrational and rotational states at a particular temperature. Such a formulation of the mobility equation has already been noted by Haese and Oka in their pioneering diode laser study of the mobility of ArH^+ as a function of vibrational state.[20,28] It may be possible to ascribe different values of the phenomenological mobility, K, to different vibrational states of ions, largely because of the relatively slow excitation/deactivation of ion vibrations in some buffer gases. It is unlikely, however, that specific values of mobilities can be ascribed to individual rotational states because of the very rapid interchange of population among the rotational levels in collisions. Nonetheless, if detailed state-resolved momentum transfer cross-sections are measured or calculated, these would be very useful in obtaining such state-specific mobility parameters.

When there are many collisions in a steady-state drift experiment, there will be an energy transfer equilibrium between translation, rotation, and vibration. This suggests that not all of the parameters relating to the mobility and drift velocity are independent. A crossed beam experiment with preparation of specific states of a vibrationally and rotationally excited molecule, whose alignment is also defined, would allow the necessary state-specific momentum transfer cross-sections to be measured precisely. In drift experiments, there is a strongly coupled relationship between the drift velocity, the internal excitation, and the mobility. For example, a change in the molecular alignment that occurs in the first collisions of an ion with an anisotropic, directed drift velocity might increase the mobility. This change would further enhance the drift velocity. In turn, this will increase the probability for rotational alignment because of the more directed velocity. In addition, collisions will produce excitation among the rotational levels, which might alter the momentum transfer cross-sections on subsequent collisions by rotation-to-translation (R–T) and translation-to-rotation (T–R) transfer.

Figure 1 shows a qualitative flow chart of some of the major interrelated parameters that affect the state populations, the mobility, and the velocity distributions of a molecular ion drifting in a simple buffer gas. The basic action of the drift field is to produce first an anisotropic drift velocity distribution as a result of collisions which are governed by the zero-order interaction potential and the field strength. The anisotropic velocity distribution excites vibrational and rotational states of the molecular ion (assuming the interaction potential is anisotropic), which can result in an altered interaction potential. The action of the anisotropy of the interaction potential (for non-spherically symmetric systems), when combined with an anisotropy in the velocity distribution, can produce molecular alignment as well. The molecular alignment may influence the mobility and, in turn, affect the drift velocity distribution. In a cyclical fashion, the resulting vibrational and rotational state populations will be altered, which affects the form of the interaction potential. After many collisions, a steady-state results between the alignment, velocity distribution, energy transfer rates, and mobility parameters. It is the goal of this work to measure some of these steady-state parameters by state-resolved optical detection methods and to relate them through the interaction potentials to describe the important physical properties of the collisions which result in the ion motion.

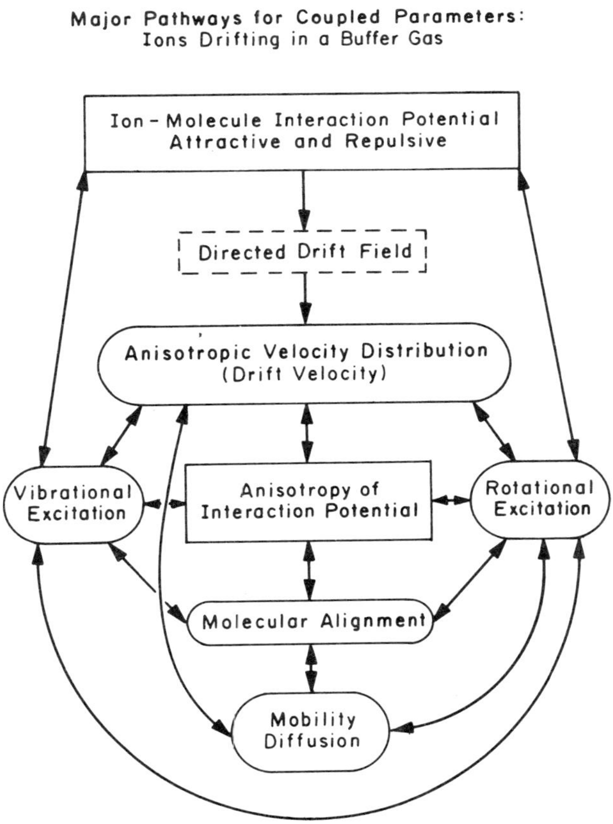

Figure 1 *Flow chart of the major interrelated parameters that occur in the transport of ions in a drift tube*

3 Review of Basic Drift Theories

In principle, the analysis of ion drift motion for purely elastic collisions involves a complete solution of the Boltzmann equation.[1] In practice, this is difficult and requires various approximations. Much of the earlier theoretical literature is thoroughly reviewed in the book by McDaniel and Mason.[1] An important first model of drift motion and mobility was formulated by Wannier in 1953.[33] The results of this model are used as a basis for many more elaborate treatments. Here, a brief discussion of some of the basic principles will be given, with additional emphasis on recent treatments that involve multiple velocity distributions and the internal state details of ions drifting in simple buffer gases.

Most of the theoretical treatments consider only binary collisions between the ion and the buffer-gas species as important. Ion–ion interactions and space charge effects are assumed to be negligible. While the complete ion velocity distribution contains much more detailed information about the interaction potential, it is very useful to consider the general form of the average drift velocity and what information this reveals about the potential of interaction. In the free-flight model, the ion gains an additional velocity, v_a, from the field, which is proportional to eEt/m, where e is the electron charge, E is the field strength, t is the time between collisions, and m is the mass of the ion. The ion loses momentum upon collisions, depending on the magnitudes of the momentum transfer cross-sections. A balance is achieved between the gain and loss of momentum, resulting in the drift velocity, v_d. Recognizing that the frequency of collisions, which limits the acceleration, is proportional to the number density, N, and the cross-section for momentum transfer, Q, then the time, t, will be inversely proportional to the product NQ. Thus, we obtain:

$$v_d \propto \frac{eE}{mNQ} \tag{5}$$

from which it is clear that the drift velocity depends in this simple model on the fundamental parameter E/N. It follows from a dimensional analysis of the Boltzmann equation that all observables depend on the ratio of E/N. In more recent papers, values of E/N are usually given in the unit Townsend (1 Td = 10^{-17} V cm^2).

Since both the ion and the buffer gas may be moving, a transformation to account for the relative velocity between the thermal buffer gas motion and the ion is required. The ion has a thermal component at low fields, but it also has a significant random component of velocity (which persists even at high fields). It has been shown[33] that the average ion velocity, v_i can be obtained from:

$$\langle E_i \rangle = \tfrac{1}{2}mv_i^2 = \tfrac{1}{2}mv_d^2 + \tfrac{1}{2}Mv_d^2 + \tfrac{3}{2}k_BT \tag{6}$$

[33] G. Wannier, *Bell. Syst. Tech. J.*, 1953, **32**, 170.

where $\langle E_i \rangle$ is the average energy of the ion and M is the mass of the buffer gas. The term $\frac{1}{2}Mv_d^2$, which includes the mass of the buffer gas and not the ion, represents random energy of the ion induced by the field upon collisions with the buffer gas. This random component continues to increase with increasing drift velocity. The $\frac{3}{2}k_BT$ term is the thermal energy that the ion has even before the field is applied. Thus, the ion motion is composed of a directed drift velocity and a random motion due to both the initial thermal velocity and the scattering collisions that impart momentum transverse to the field direction.

At low fields, the averaging over the relative velocity results in a drift velocity that is directly proportional to E/N:[1]

$$v_d \propto \left(\frac{1}{m} + \frac{1}{M}\right)^{1/2} \frac{e}{Q(k_Bt)^{1/2}} \frac{E}{N} \tag{7}$$

while at high fields the drift velocity is proportional to $(E/N)^{1/2}$:

$$v_d \propto \left(\frac{1}{m} + \frac{1}{M}\right)^{1/4} \frac{e}{M^{1/2}Q}\left(\frac{E}{N}\right)^{1/2} \tag{8}$$

Thus, assuming that the momentum transfer cross-sections are not strongly dependent on velocity (*e.g.* rigid spheres), we expect as a first approximation that a plot of the drift velocity versus E/N will first increase linearly with E/N and then decline at high E/N. Consequently, a plot of mobility versus E/N ($K = v_d/E$) is almost constant for a range of low E/N and then falls off as $E^{-1/2}$ at high values of E/N (Figure 2).

In the Maxwell model of the ion drift motion, it is assumed that the momentum transfer cross-sections are inversely proportional to the relative velocity, *i.e.* Qv_0 = constant. This would be the case, for example, for the ion-induced dipole Langevin cross-section. Wannier[33] and Kihara[34] recognized that the term Qv_0 could then be removed from the collision operator in the integral of the Boltzmann equation, resulting in a tremendous mathematical simplification. The result, which is also an important basis for many theoretical treatments of the ion drift motion, is equivalent to ascribing a constant mean free time between collision events. Thus, we can see how the phenomenological mobility is qualitatively related to the inverse of the product of Qv_0 for certain realistic functional forms of the cross-section.

The momentum transfer cross-sections may be derived directly from the differential scattering cross-sections $I(\theta)$ for elastic collisions, which can be obtained from beam scattering experiments:[1]

$$Q = \int_0^{\pi}(1 - \cos\theta)I(\theta)\sin\theta\,d\theta \tag{9}$$

[34] T. Kihara, *Rev. Mod. Phys.*, 1953, **25**, 844.

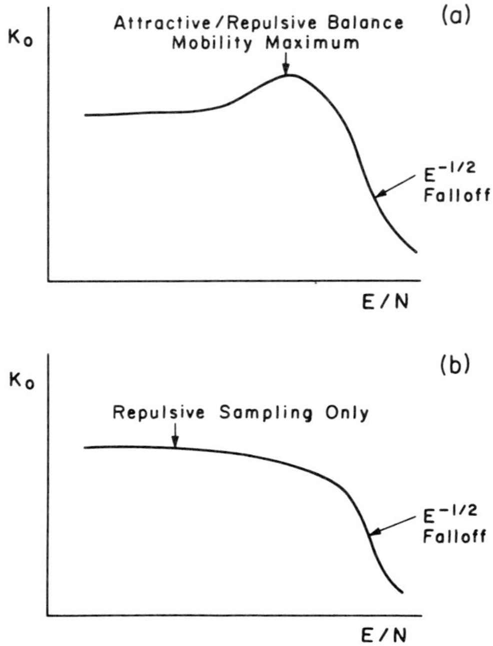

Figure 2 *Typical behaviours of reduced mobility versus E/N for an attractive potential* (a) *and one with repulsion only* (b)

The $(1 - \cos\theta)$ weighting describes predominantly the momentum transfer along the direction of the drift field. Other weighting terms may be used to take into account larger fractions of transverse components and other parameters, such as temperature. Thus, the form of the angular scattering and momentum transfer can be evaluated for various interaction potentials. In the limit of low collision energies, the attractive part of the potential is sampled predominantly; the repulsive part of the potential is sampled at higher energies. A partial cancellation of the attractive and repulsive forces occurs at intermediate ranges of energy, resulting in a value for the momentum transfer cross-section at these energies which can be anomalously low.[35] Thus, the mobility function versus E/N will often exhibit a bump or maximum in its value before the relatively steep decline, if the attractive part of the potential is effectively sampled at the lowest E/N (Figure 2). Qualitatively, a higher percentage of forward glory scattering events is implicated in this region of the mobility maximum.[6]

If the thermal energy is already much greater than the depth of the attractive well, then a flat or monotonically decreasing mobility function is often observed. Helium as a collision partner is especially notable in this regard, because of its small polarizability. At room temperature, the

[35] G. H. Wannier, *Bell. Syst. Tech. J.*, 1970, **49**, 343.

thermal collision energy is already above the limit where the ion-induced polarizability potential well is sampled significantly. The presence or absence of the mobility maximum can readily be affected by changing the gas temperature.[6] At low temperature, the attractive region of the potential well is sampled, resulting in a bump in the mobility plot, while at high temperature, the collisions effectively sample only the repulsive wall, resulting in a flat mobility plot. It is apparent, then, that mobility determinations can be used to gain information about the potential of the interaction. Indeed, there are many calculations to fit mobility data to obtain potential curves.

Even in the early work of Wannier, which used Monte Carlo trajectory simulations among many other analytical techniques, it was appreciated that the distribution of the drift velocity, v_d, would resemble a shifted and 'skewed' Maxwellian function.[33] More precise solutions to the Boltzmann equation have been used by Whealton and Woo[36] to obtain the velocity distributions of the ion as a function of various ion–buffer gas mass combinations; this analysis has been extended by Makabe *et al.*[37] Figure 3 reproduces a set of reduced velocity distribution contour maps calculated for the polarization potential of an ion–atom interaction at $m = M$ for various values of E/N.[36] An

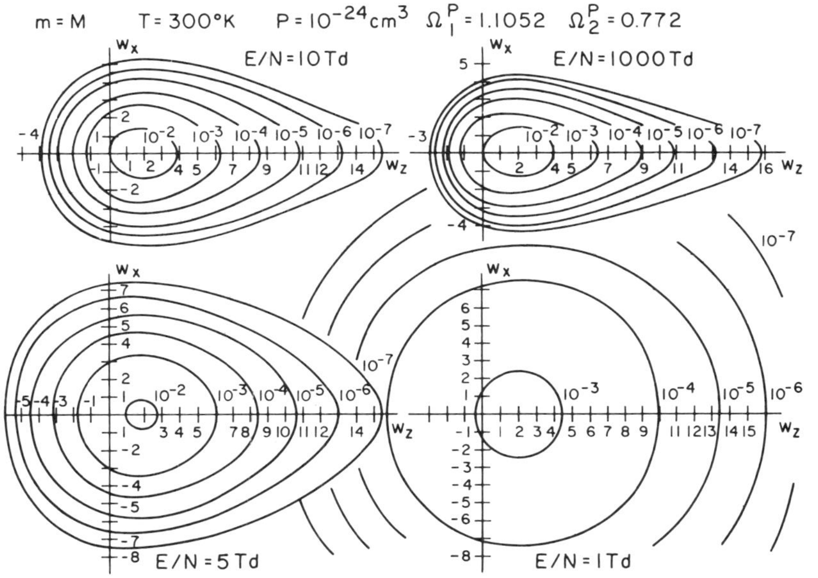

Figure 3 *Velocity contour plots showing the skewing of the distribution* (Reproduced with permission from *Phys. Rev. A*, 1971, **6**, 2319)

[36] J. H. Whealton and S. -B. Woo, *Phys. Rev. A.*, 1971, **6**, 2319.

[37] T. Makabe, K. Misawa, and T. Mori, *J. Phys. D: Appl. Phys.*, 1981, **14**, 199 and 1983, **16**, 1893.

extremely interesting result is the strong 'skewing' of the velocity distributions at higher fields. This skewing comes about from the asymmetry of the differential scattering function, which exhibits a range of impact parameters that produce nearly forward scattering when the kinetic energy is large compared with the well depth.

The skewing of the distribution is more dramatic for collisions with nearly equal masses, as opposed to situations where $m \ll M$ or $m \gg M$.[36] Qualitatively, one expects that the least amount of skewing of the velocity distribution might occur when there is a tendency for attractive orbiting collisions, which would give rise to scattering into all angles. Thus at very low kinetic energies, where the well depth of the attractive part of the potential is sampled significantly, the skewing is very slight. At very high kinetic energies, the high-energy tail of the velocity distribution may also fall off more rapidly than at intermediate energies,[4] but has little effect on the mobility. In this situation, more hard repulsive collisions occur, which take ions away from the high-speed tail of the distribution without replacing them. In an intermediate range of energies, the propensity for forward glory scattering is greatest, which gives rise to the most pronounced high-energy tail in the velocity distribution.

Detailed ion drift velocity distributions have also been calculated using accurate interaction and Monte Carlo trajectory methods.[38–40] At a given mean energy, it is found[39,40] that the velocity distribution of O^+ in Ar has a much higher-energy tail than for O^+ in He. The distribution in He is nearly Maxwellian, but the severe skewing of the distribution in Ar demonstrates that higher-order moments of the velocity distribution must necessarily be included. Measurements that use electrostatic ion sampling methods[13] are not in total agreement with these Monte Carlo calculations, however, possibly due to ion sampling difficulties in the experiments.[9] The anomalous behaviour of the speed distributions also has a profound effect on observed reaction rate constants, in this case determined for various simple atomic ion reactions, $O^+ + O_2$, N_2 and NO.[41] Use of the mean kinetic energy or mean drift velocity from the simple drift equations gives dramatically different rate constants for reactions carried out in different buffer gases as a function of kinetic energy. Thus, it is apparent that more detailed experimental measurements on ion velocity distributions[27] are needed.

It was widely recognized that the ion velocity distribution is dramatically different in the direction parallel to the field compared with the direction transverse to the field and that the parallel distribution is highly skewed in many buffer gases. This stimulated the formulation of

[38] H. R. Skullerud, *J. Phys. B: At. Mol. Phys.*, 1973, **6**, 728.

[39] S. L. Lin and J. N. Bardsley, *J. Phys. B: At. Mol. Phys.*, 1975, **8**, L461.

[40] S. L. Lin and J. N. Bardsley, *J. Chem. Phys.*, 1977, **66**, 435.

[41] D. L. Albritton, I. Dotan, W. Lindinger, M. McFarland, J. Tellinghuisen, and F. C. Fehsenfeld, *J. Chem. Phys.*, 1977, **66**, 410.

a more specific 'three-temperature' theory of ion velocity distributions.[42,43] Wannier[33] and Kihara[34] had already considered the formulation of the Boltzmann equation in terms of moments of the velocity distribution. The contribution of Lin, Viehland, and Mason[42,43] was to incorporate explicit anisotropies into the basis functions used to solve the Boltzmann equation. Qualitatively, the three temperatures are those of the buffer gas, T, the ion velocity components parallel to the field, $T_{\|}$, and the ion velocity components transverse to the field, $T_{\perp}$. Additional flexibility in the basis functions allows the zero-order Maxwellian distributions to be skewed to any degree in the solution of the Boltzmann equation. Waldman and Mason[44] parameterized the equations of Lin, Viehland, and Mason so that simple analytical forms for the parallel and perpendicular ion temperatures can be obtained. They introduced the concept of a differential mobility as a tensor quantity, which allowed for a more rigorous basis for the moment corrections to the velocity distribution.

Recently, Viehland, Lin, and Mason[45] formulated a theory to describe the internal degrees of freedom of polyatomic ions undergoing drift collisions. According to this theory, the final excitation of vibrational and rotational states of an ion in equilibrium with an atomic buffer gas in a drift field is described by an internal temperature, T_{int}, which is given by the centre-of-mass collision energy with the buffer gas:

$$\tfrac{3}{2}k_B T_{\text{int}} = KE_{\text{cm}} = \tfrac{3}{2}k_B T + \tfrac{1}{2}Mv_{\text{d}}^2 \tag{10}$$

This theory does not take into account detailed rate constants for excitation and deactivation of the ion states as a function of kinetic energy, which may be important in a more exact treatment; however, it offers a first approximation to the ion internal excitation and has been verified experimentally in our laboratory for rotational excitation in N_2^+.[21]

Ion 'runaway' is an important concept in the consideration of drift velocities and mobilities.[46–48] Runaway can occur if the ions gain more energy from the field than they lose by collisions. A mobility cannot then be defined. This situation can come about if the potential of interaction is such that the momentum transfer cross-section decreases substantially at some point with increasing relative velocity. Such is the case for H^+ and D^+ collisions with He over a range of higher kinetic energies. Runaway was predicted[46] and experimentally verified[47] for

[42] S. L. Lin, L. A. Viehland, and E. A. Mason, *Chem. Phys.*, 1979, **37**, 411.
[43] L. H. Viehland and S. L. Lin, *Chem. Phys.*, 1979, **43**, 135.
[44] M. Waldman and E. A. Mason, *Chem. Phys.*, 1981, **58**, 121.
[45] L. A. Viehland, S. L. Lin, and E. A. Mason, *Chem. Phys.*, 1981, **54**, 341.
[46] S. L. Lin, I. R. Gatland, and E. A. Mason, *J. Phys. B: At. Mol. Phys.*, 1979, **12**, 4179.
[47] F. Howorka, F. C. Fehsenfeld, and D. L. Albritton, *J. Phys. B: At. Mol. Phys.*, 1979, **12**, 4189.
[48] M. Waldman and E. A. Mason, *Chem. Phys. Lett.*, 1981, **83**, 369.

$H^+(D^+)$ in He. A certain 'softness' exists in the repulsive part of the interaction potential, in which the two species can come into closer proximity with increased kinetic energy. In general, the overlap of two electron clouds will have a somewhat impenetrable repulsive core interaction. However, for the case of protons interacting with a noble gas, when the proton approaches closer and closer owing to higher kinetic energy collisions, a rearrangement of the electron density on the noble gas can occur, which begins to accommodate the formation of a close-range diatomic bond. Thus the repulsive interaction is diminished, leading to a reduction in the effective 'size' of the scattering cross-section. The effect of ion runaway has also been predicted for the system Li^+ in He at $E/N > 200$ Td.[49] Generalizing from these observations, it is unlikely that an ion with a large electron cloud, *e.g.* Ba^+, would exhibit ion runaway. However, other tightly bound or multiply charged ions might exhibit runaway, provided that the ion does not charge transfer with the buffer gas of interest.

So far we have only considered the mobility and drift velocity effects for elastic scattering of ions. Interesting new dynamics occur when reactive processes or symmetric charge exchange channels are open. In symmetric charge transfer, which can occur when an ion is drifted in its parent buffer gas, momentum transfer can be essentially complete on every few collisions. Thus, the mobility is dominated by the charge transfer process rather than momentum transfer in the elastic scattering. Typically the mobility follows $(E/N)^{-1/2}$, since the charge transfer cross-section is very nearly constant over a range of intermediate velocities.

Finally, appropriate links between mobility and diffusion can be made, following from the generalized Einstein relationships.[1] For weak fields:

$$K = \frac{eD}{k_B T} \tag{11}$$

where D is the diffusion coefficient. In more detailed descriptions, the diffusion coefficient has the form of a tensor.[44] While we are not interested in the diffusion coefficients *per se* in this chapter, it should be noted that analogous parallel and perpendicular diffusion coefficients will exist, corresponding to the different ion temperatures, and that a complete solution to the problem of ion drift motion will also yield these coefficients.

Many interesting theoretical challenges remain, including treatment of ion mobility and drift velocity distributions as a function of internal excitation and alignment. Very few microscopic theories exist which include state-to-state cross-sections for inelastic collisions in the treatment of ion motion, and few experimental data are available. With laser

[49] L. A. Viehland, *Chem. Phys.*, 1980, **70**, 149.

probing, however, it is now possible to interrogate ion drift systems with much greater state-resolved detail. Thus, this kind of information will be more readily available in the future. It is also possible to probe ion velocity distributions both parallel and perpendicular to the field with non-intrusive optical methods. Thus, a more detailed level of velocity distribution information will be available to provide accurate tests of theories. The remainder of this chapter considers some of these measurements and the directions that these results will provide for future development in the analysis of ion motion in drift fields.

4 Apparatus for Laser–Optical Probing of Ion Drift Parameters

The flowing afterglow drift tube apparatus[8,29] is ideal for optical probing of ion internal excitation, alignment, and velocity distributions. The basic flowing afterglow apparatus consists of a large-diameter flow tube in which relatively high densities of ions are entrained in a fast flow of a suitable buffer gas, often helium at a pressure of 100 Pa and a flow velocity of 8000 cm s^{-1}. The ions diffuse to the walls on a timescale which causes a factor of two loss in ion density for every 20 cm of flow distance. The expected parabolic velocity distribution of the buffer gas is well documented[29] and results in a well-defined average ion flow velocity.[50,51] This has been crucial for the innumerable kinetic determinations in such devices. A quadrupole mass filter situated at the end of the flow tube allows the ions to be sampled to obtain relative number densities as a function of reagent concentration and thus the branching fractions of products.

Incorporation of a drift region into the flow tube[8] allows many other ingenious experiments to be performed. These include the study of ion mobilities by measuring the flight time of a pulse of ions to arrive at the mass filter,[8] reactivity as a function of translational kinetic energy,[5,8] ion vibrational excitation and deactivation,[18] and enhanced reactivity due to vibrational excitation.[16] The drift region in our apparatus consists of a series of electrically insulated rings, each ~1 cm in length, which are connected together in series with precision resistors. Application of a voltage across the entire drift ring assembly provides a highly uniform electric field with which to accelerate the ions. Typically, a higher voltage is applied to the first section of the drift region to separate the charges completely, so that a low density of ions and ions of only one charge are drifting freely in the buffer gas. In more sophisticated versions, a quadrupole mass filter is used to inject ions of a single mass into the flow tube and drift region.[47,52] This is sometimes necessary to

[50] D. L. Albritton, *At. Data Nucl. Data Tables*, 1978, **22**, 1.
[51] Y. Ikezoe, S. Matsuoka, M. Takebe, and A. Viggiano, 'Gas Phase Ion-Molecule Reaction Rate Constants Through 1986', Maruzen, Tokyo, 1988.
[52] N. G. Adams and D. Smith, *Int. J. Mass Spectrom. Ion Phys.*, 1976, **21**, 349.

avoid deleterious reactions of the ion of interest with its precursor molecules.

The LIF method affords a versatile means of interrogating the number densities of ions in various states. A laser excites the ion on a particular electronic transition, followed by detection of fluorescence from the excited state with a photomultiplier through suitable filters. Tuning of the laser allows different states to be probed. Polarization optics may be used to interrogate alignment effects via the direction of the transition moment of the molecule, as will be discussed in more detail in a section below. Lists of typical ions that can be probed by LIF and the methods of population analysis are discussed in previous reviews of our reaction dynamics work.[53,54]

Modification of a flowing afterglow for laser and infrared optical detection has allowed many new kinds of experiments in reaction dynamics to be carried out.[53,54] Product vibrational states of proton transfer reactions, charge transfer processes, associative detachment events, and heavy-particle exchanges have been investigated with such a device in our laboratory. In this chapter, we are concerned specifically with the coupling of the flowing afterflow/drift tube device with pulsed and cw dye lasers for studies of ion drift velocities, mobilities, internal excitation, and alignment. The flowing afterglow is also ideal for this kind of study because of the very well-characterized parameters in the drift field, the relatively high ion densities that can be achieved, and the great variety of atomic and molecular ions that can be prepared under thermal conditions.

Figure 4 shows a schematic of a version of the flowing afterglow/drift tube when it is coupled to a cw dye laser for optical detection of the Doppler velocity distributions of Ba^+ ions in the direction parallel to the field. The apparatus is composed of various modular sections, which allows it to be assembled in many different configurations with regard to the ion source, laser probe region, photomultiplier, and mass spectrometer detection end. The flow tube incorporates many bends and light baffles to reduce scattered light from both the ion source and the laser. A photomultiplier with suitable filters views the laser excitation zone within the drift field in a direction perpendicular to the laser beam. The laser may also be propagated transverse to the field direction. Pulsed lasers are also used, as well as polarization optics to set the polarization of the exciting laser and the detected light.

The rather bizarre twists and turns in the flow tube device are necessary to obtain the best absolute detection sensitivity for low densities of ions with the LIF method. This arrangement meets the somewhat contradictory goal of achieving almost complete darkness in a

[53] V. M. Bierbaum, G. B. Ellison, and S. R. Leone, in 'Gas Phase Ion Chemistry', ed. M. T. Bowers, Academic Press, New York, 1984, Vol. 3, p. 1.

[54] V. M. Bierbaum and S. R. Leone, in 'Structure/Reactivity and Thermochemistry of Ions', eds. P. Ausloos and S. G. Lias, Reidel, Dordrecht, 1986, p. 23.

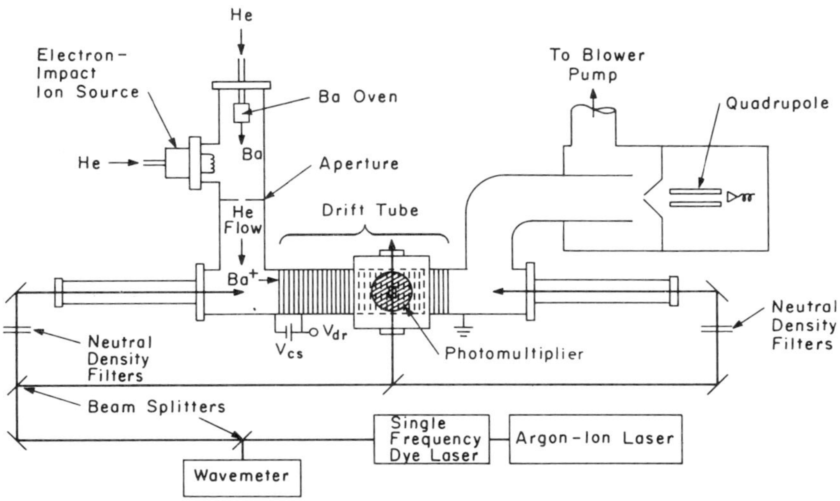

Figure 4 *Schematic of the experimental apparatus used to study ion velocity distributions in a flow drift tube*

flowing afterglow that is often a vigorous light-emitting plasma. While such contortions of the flow tube destroy the well-behaved flow parameters required for kinetics measurements, these modifications do not disturb the optical measurements of product-state distributions or drift parameters. This is because the flow parameters are well established for a sufficient distance both before and after the region of the optical probing.

In the schematic shown in Figure 4, the ion source region consists of either a hollow cathode discharge or an electron impact ionizer. To produce N_2^+ ions, nitrogen gas is ionized in a fast flow of helium. To produce Ba^+, a flow containing neutral Ba atoms is generated by passing a heated flow of He through an oven containing Ba metal, followed by electron impact ionization. Typically, Penning ionization by metastable He produces high densities of Ba^+ ions downstream of the oven. The ion plasma is initially electrically neutral, which accounts to some extent for the high ion densities that are achievable. A significant loss of ion density occurs when the ion flow enters the charge separation field at the beginning of the drift region, due to diffusion to the walls. The ionization region is often operated at higher total pressures than the flow tube itself (~2000 Pa compared with 100 Pa in the flow tube) in order to increase the extent of reaction in the ion source. This is accomplished by the addition of a metal plate in the 7.3 cm diameter flow tube, in which only a 1 cm diameter hole allows the ions and buffer gas to enter the main flow tube. This aperture has the crucial effect of dramatically reducing the background emission from the ionization process that reaches the photomultiplier by allowing all of the metastable He to react before entering the main flow tube.

To further reduce the background light from the ion production region, a blackened 90° bend is inserted between the ion source and the drift region. Baffle arms, each of which incorporates a series of three apertures as well as blackened internal grooves, are used to introduce the laser beam with minimal laser light scattered off the entrance and exit windows. Careful alignment of the laser beam is also crucial to achieve the minimum of scattered light noise from the laser itself. The drift region consists of a series of 1 cm long carbon steel rings that have been treated with an electrically conductive, nickel penetrate blackening agent. The drift rings are separated by thin mylar spacers for electrical isolation and are connected in series by 500 Ω precision resistors. The first five rings are used for charge separation at higher fields and can also be used to stop the ions completely in order to test for background signals. Approximately 40 rings are used in the drift region of our apparatus to achieve the full development of the uniform drift field in the probe zone, which is centred in the middle of these drift rings.

The drift region has been designed to maintain high uniformity of the field by minimizing the number of cutouts for entrance and exit of the perpendicular laser beam and for detection of the fluorescence. In the parallel direction of propagation of the laser, as shown in Figure 4, no cutouts are needed for the passage of the laser beam. Figure 5 shows a detail of the drift region that is currently in use.[21] The rings extend into an evacuated box, so that only small slices of the rings must be removed to pass the laser beam and the fluorescence. The windows, which

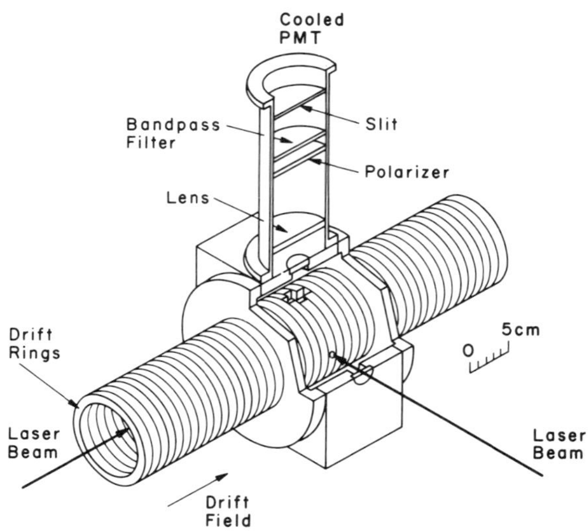

Figure 5 *Close-up view of the ion probe region and drift ring enclosure for laser detection*

occupy much larger areas, are mounted on the outside of the vacuum housing and therefore do not disturb the field uniformity. The cutout for the fluorescence port is covered with a conductive wire mesh to minimize the field non-uniformity of this port. The cutouts for transverse probing with the laser beam are typically 0.7 cm in diameter; in the case of the N_2^+ alignment experiments, they are 2.5 cm × 2.5 cm, in order to minimize scattered laser light. However, tests have been performed using the laser to probe in the parallel direction, during which the cutout rings can be included or removed. No difference has been observed in measurements of sensitive parameters such as molecular alignment.

The lasers used for these studies are a pulsed Nd:YAG pumped dye laser with frequency doubling and mixing capabilities and a single-mode cw ring dye laser, pumped by an argon ion laser. The two lasers allow a wide range of experimental wavelengths and conditions to be achieved for probing of ions in drift fields. The 1 cm^{-1} bandwidth pulsed laser has been used primarily to study N_2^+, where wavelengths at ~390 nm are required to probe the *B–X* transition.[53] The cw laser is used for the velocity distributions of Ba^+, for which single-mode linewidths of 1 MHz at 455 nm are used for accurate determination of the Doppler profiles of the ions. The lasers must frequently be attenuated by neutral density filters (Figure 4) to avoid saturation of the electronic transitions. Even so, mild saturation effects are often observed, manifested by an increase in the observed linewidth in the cw laser experiments. In this case, a Lorentzian component is included in fits to the line profile to account for the saturation. Saturation is also troublesome in the analysis of population distributions.[54,55] When appropriate, an explicit form for the LIF intensity under saturation conditions is used in the population analysis.[54]

Optical characterization of the drift field provides an important new diagnostic to analyse the effectiveness of charge separation and other details of the drift field. Figure 6 shows one set of characterizations, which are used to determine the values of the charge separation field necessary to study the alignment of N_2^+ drifting freely in He.[26] At low voltages on the charge separation field, the LIF signal, which is proportional to the ion density in the drift region, is very high, but the electrons and negative ions have not been fully separated from the plasma. Even at moderately high drift fields (a few Td), the charge separation is incomplete without the application of the higher charge separation field at the beginning of the drift section. At any given drift field, as the charge separation voltage is varied, the ion density dramatically decreases and reaches a plateau. In this plateau region, there is no further dramatic reduction in ion density upon increase of the charge separation field. This plateau region is determined[26] to be

[55] R. Altkorn and R. N. Zare, *Annu. Rev. Phys. Chem.*, 1984, **35**, 265.

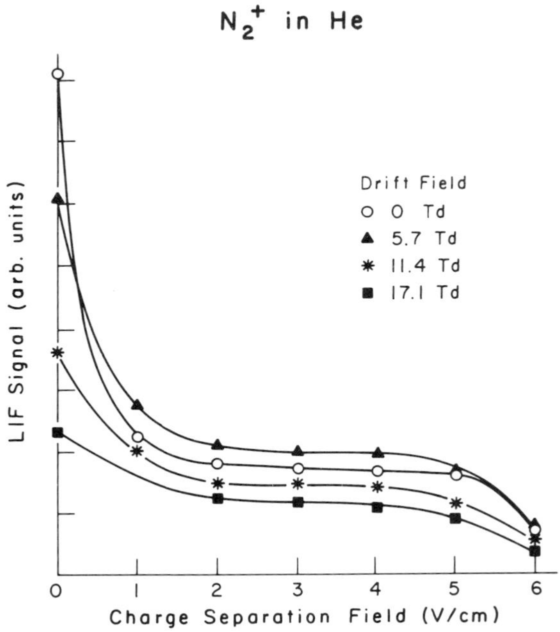

Figure 6 *Characterization of the charge separation using LIF on* N_2^+ *in helium as a function of drift field*

the zone in which the charge separation is complete. The drop-off of signal at high E/N is thought to be due to the effect of collisions on the N_2^+ excited state.

In the following sections, a number of examples are considered in which the flowing afterglow/drift apparatus and laser detection tools are used to probe ion behaviours in drift fields.

5 Ion Internal Excitation in Drift Fields

The enhanced collision energy of ions drifting in a buffer gas can excite vibrational and rotational states of both the ion and the buffer. In the simplest case, a diatomic or polyatomic ion is drifted in an atomic buffer gas, and the internal excitation of the ion may be probed as a function of drift field. The earliest evidence for vibrational excitation of ions in drift fields comes from measurements of anomalously increased reactivity of the ions.[5,16] According to equation (10), the degree of internal excitation of the ions after an equilibrium is fully established between translation and vibration is greater for ions drifting in heavier buffer gases. Thus, it is apparent how the effect of ion internal excitation can be distinguished from that of the kinetic energy in a reaction. Symmetric charge transfer reactions can also be used to quench vibrational excitation efficiently, as a test for the effect of internal excitation.

Recently, 'monitor' ion techniques have been developed to probe for vibrationally excited ions by selective chemical reaction.[17,56–61] Such reactions take advantage of a greatly increased reactivity of a vibrationally excited state, usually in a charge transfer reaction which is endothermic for one vibrational state and exothermic for the next higher vibrational state. The appearance of the newly formed product ion in the monitor reaction is a measure of the concentration of the vibrationally excited ion. In this way, not only have $V = 1$ and $V = 0$ been distinguished, but the presence of higher vibrational states can be ascertained as well. With this type of chemical monitoring, it was possible to determine the branching fraction into the $V = 0$ and $V = 1$ states of the N_2^+ product of the $Ar^+ + N_2$ reaction.[56,57] The monitor ion method has been used to catalogue many rate coefficients for vibrational deactivation of ions with various collision partners in thermal energy collisions.[17,58–61]

Laser probing of the vibrational excitation and deactivation of ions produced by reactions in a flowing afterglow has been a successful technique.[62,63] However, the greatest progress in quantifying vibrationally excited ions in drift fields has been achieved by the monitor ion method. The first evidence for significant vibrational excitation of ions in a drift tube was presented be Federer *et al.*[64] for N_2^+ and O_2^+ ions drifting in He. The vibrational temperature of the ions was estimated to be 3000 K at drift fields (E/N) of 150 Td. Since He is a very inefficient deactivator of vibrationally excited ions, the observed high vibrational temperature is somewhat surprising from detailed balance considerations. However, the result is readily explained by the fact that the rate coefficients for vibrational excitation are a rapidly increasing function of collision energy. In a recent study of the N_2^+–He system,[18] rate coefficients are directly extracted for both vibrational excitation and deactivation as a function of kinetic energy. Figure 7 summarizes these results, which show over three orders of magnitude enhancement in the rate coefficients for both the excitation and deactivation as the drift field is increased. The deactivation of $N_2^+(V = 1)$ by He requires $\sim 10^5$ collisions at thermal energies, but only 10—100 collisions at a high

[56] D. Smith and N. G. Adams, *Phys. Rev. A.*, 1981, **23**, 2327.

[57] W. Lindinger, F. Howorka, P. Lukac, H. Villinger, E. Alge, and H. Ramler, *Phys. Rev. A.*, 1981, **23**, 2319.

[58] H. Bohringer, M. Durup-Ferguson, E. E. Ferguson, and D. W. Fahey, *Planet. Space Sci.*, 1983, **31**, 483.

[59] H. Bohringer, M. Durup-Ferguson, D. W. Fahey, F. C. Fehsenfeld, and E. E. Ferguson, *J. Chem. Phys.*, 1983, **79**, 4201.

[60] W. Dobler, W. Federer, F. Howorka, W. Lindinger, M. Durup-Ferguson, and E. E. Ferguson, *J. Chem. Phys.*, 1980, **79**, 1543.

[61] W. Federer, W. Dobler, F. Howorka, W. Lindinger, M. Durup-Ferguson, and E. E. Ferguson, *J. Chem. Phys.*, 1985, **83**, 1032.

[62] C. E. Hamilton, V. M. Bierbaum, and S. R. Leone, *J. Chem. Phys.*, 1985, **83**, 2284.

[63] C. E. Hamilton, V. M. Bierbaum, and S. R. Leone, *J. Chem. Phys.*, 1985, **83**, 601.

[64] W. Federer, H. Ramler, H. Villinger, and W. Lindinger, *Phys. Rev. Lett.*, 1985, **54**, 540.

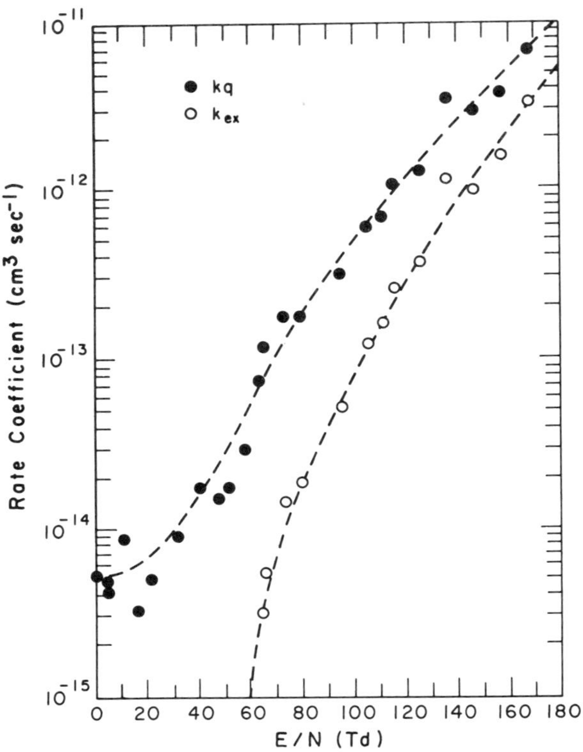

Figure 7 *Excitation* (k_{ex}) *and deactivation* (k_q) *rate coefficients for* $N_2^+(V = 1)$ *in helium as a function of drift field*
(Reproduced with permission from *J. Chem. Phys.*, 1988, **88**, 213)

centre-of-mass collision energy of 0.6 eV. It is clear from the magnitudes of these rate constants that full equilibration of the vibrational temperature may not be achieved for drift fields of finite length.

The rotational excitation of N_2^+ drifted in He has been probed by LIF on the $B^2\Sigma_u^+ \rightarrow X^2\Sigma_g^+$ transition at 390 nm in our laboratory.[21] The drift field was applied continuously for some experiments, and pulsed experiments were carried out as well to generate information about the timescales for rotational equilibration. Figure 8 shows the effect of the application of the drift field on the raw spectra of the *R* branch of the (0,0) band. The *R* branch allows the progression of rotational states to be readily identified, as noted by the quantum numbers shown on the figure. The intensity alternation is 2:1 for even:odd rotational quantum numbers, arising from the nuclear spin statistics. At zero field, the approximate peak in the Boltzmann distribution of rotational states occurs at $N'' = 6$. Upon application of a 5 V cm^{-1} electric field, the maximum in the Boltzmann distribution is seen to shift to $N''= 12$. This corresponds to a rotational temperature change from 300 to 700 K. It is estimated that ~1% of the molecules at this collision energy will also be vibrationally excited if a full equilibrium is developed.

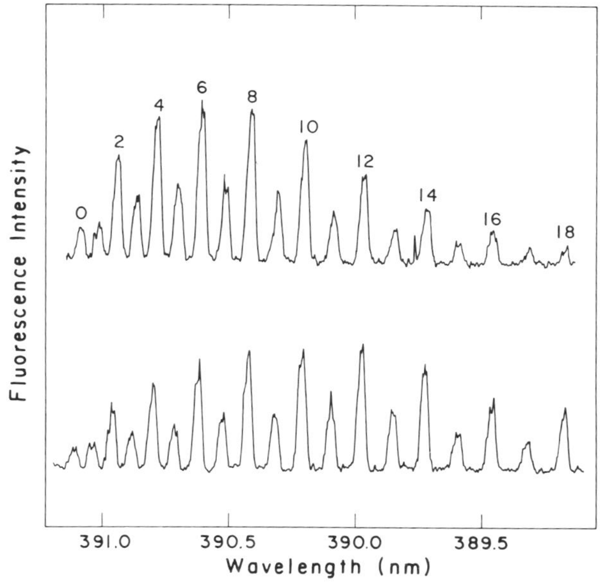

Figure 8 *LIF spectra of the* R *branch of* N_2^+ *at zero field and* 5 V cm^{-1} *drift field*

A fit of the data to the theory of Viehland, Lin, and Mason[45] for the internal temperature of the ion [equation (10)] is shown in Figure 9. These data are obtained by carefully measuring just two lines in the LIF spectra to derive an 'effective' temperature for each field strength. Although the scatter in the data is high, the agreement of the entire data set with the theory is quite good. These measurements confirm the qualitative prediction that the internal energy of the ion drifting in an atomic buffer gas will be given by the centre-of-mass collision energy.

By pulsing the drift field, the timescales required to achieve a particular rotational temperature of the ion are also investigated. Because of the fast rates of rotational excitation, it is not possible to observe the approach to equilibrium directly. The number of collisions required to achieve the full rotational excitation is estimated to be 10 or fewer. This is not surprising given the typically much faster timescales for rotational excitation when compared, for example, with vibration. Collisions with a light buffer gas such as He are also the most favourable for vibrational and rotational excitation in the impulsive limit. Thus, rapid equilibration of rotational degrees of freedom is expected in a very short length of the drift field.

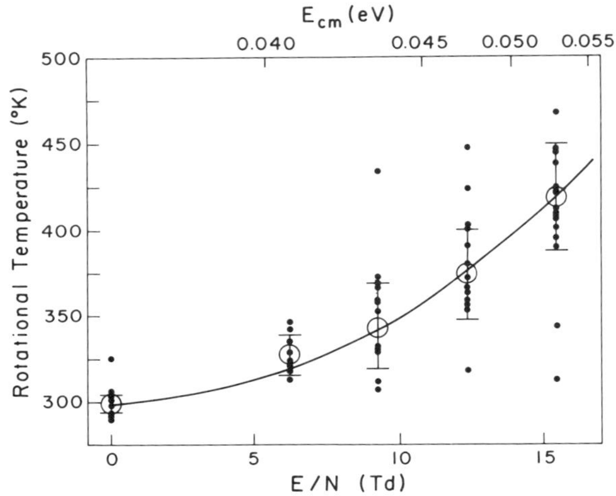

Figure 9 *Comparison of the observed rotational temperatures (●) for* N_2^+ *drifting in helium with the prediction from the theory of Viehland, Lin, and Mason*[45] (——)

Analyses of the full rotational population distributions[21] indicate that the rotational states fit a Boltzmann distribution to within the signal-to-noise. However, this is not necessarily expected or required, especially if the velocity distribution of the N_2^+ ions is non-Maxwellian and rotation is strongly coupled to translation. As noted in the section above on drift theory, significant skewing away from a Maxwell–Boltzmann velocity distribution may occur in the direction of the field, especially for mass combinations where m/M is nearly unity. This skewing is observed in recent experiments[65] using a single-mode dye laser to probe the velocity distribution of Ba^+ colliding with Ar, as discussed below. Thus, in more refined experiments on molecular ions, it may be possible to observe non-Boltzmann rotational distributions and to correlate them with non-Maxwell–Boltzmann centre-of-mass collision energies.

The theory of Viehland, Lin, and Mason[45] accounts for the internal excitation of the ions in a drift field and is completely general. It allows for different ion temperatures in the parallel and perpendicular directions as well as velocity distributions which require higher moments to approximate them. In the simplest approximation, which is embodied in equation (10), spherical basis functions are used for the velocity distribution and an atomic buffer gas is assumed which has no internal degrees of freedom. If the ion is drifting in a molecular buffer gas, additional terms are included[45] to account for the fractional loss of the ion energy to internal modes of the buffer gas. The theory does not treat a case other than steady state, and thus does not contain the details of

[65] S. M. Penn, J. P. M. Beijers, R. A. Dressler, V. M. Bierbaum, and S. R. Leone, *J. Chem. Phys.*, in preparation.

transient excitation (in the application of a pulsed field, for example). Nevertheless, it is clear that many additional experiments can now be performed to verify and quantify the detailed formulation of the problem of internal excitations of ions drifting in a buffer gas.

6 Molecular Ion Alignment in Drift Fields

In 1930, Senftleben observed that the thermal conductivity of neutral O_2 gas changes slightly when a magnetic field is applied.[66] Field effects on other transport coefficients, such as viscosity, were subsequently observed for O_2 and NO. At the time, these effects were assumed to be associated with the paramagnetic nature of these gases. It was found that the transport properties of many other gases exhibit magnetic and electric field effects, generally called Senftleben–Beenakker effects, and these have been reviewed in detail.[67] The fundamental physics were explained by the combination of two important factors:[68] (*a*) a non-equilibrium distribution of velocities, caused by a velocity gradient, which is a necessary component of the transport property itself, and (*b*) a non-equilibrium distribution of rotational angular momentum vectors, called 'alignment', that results from the collisions under the influence of the anisotropic potential and the velocity gradient. The application of an external field influences the rotational alignment by causing a precession of the angular momentum vector and hence a deterioration of the alignment established by the collisions. The change in the transport coefficient with field strength saturates when the precession frequency is high enough to destroy the alignment completely.

In general, a non-spherical potential energy surface for molecular collisions, coupled with an anisotropic spatial distribution of relative velocity vectors, may result in alignment of molecules in collisions. Alignment is a symmetric, anisotropic distribution of the m_J components of the rotational $\boldsymbol{J}$ vectors. Orientation is defined as an asymmetric anisotropy in the m_J components of the $\boldsymbol{J}$ vectors. The cylindrical symmetry of the impact parameters in the collisions leads to an alignment effect, rather than orientation. Aligned molecules can have smaller cross-sections for subsequent collisions, which cause changes in the measured transport coefficients. The application of an external field scrambles the alignment during the time between collisions, thus providing values of the transport coefficients that are more representative of a statistical distribution of m_J states.

The Senftleben–Beenakker effects for *neutral* molecules are typically small, but can be very important. The effect of rotational anisotropy for ionic collisions in a drift field is potentially even larger. In drift field environments, the primary action of the electric field on the ions is not

[66] H. Senftleben, *Z. Phys.*, 1930, **31**, 822 and 961.
[67] J. J. M. Beenakker and F. R. McCourt, *Annu. Rev. Phys. Chem.*, 1970, **21**, 47.
[68] Yu. M. Kagan and L. A. Maksimov, *Sov. Phys. JETP*, 1962, **14**, 604.

to destroy the molecular alignment by precession. Typically, the Senftleben–Beenakker effect requires very high electric fields.[67] However, the primary action of even a weak electric drift field is to induce a strong anisotropy in the velocity. We may then expect dramatic effects on the alignment of molecular ions and the subsequent mobility of these ions due to the combined anisotropies of the velocity and the potential surface.

It was recognized some years ago that a supersonic expansion of gases provides a strong velocity gradient that can align molecules, for example when helium carries a heavy species forward in an expansion. Thus, a number of studies have been undertaken to measure the degree of alignment of neutral Na_2 and I_2 molecules that are expanded with or without He in a jet.[69–72] The experiments have been of two kinds: one type in which the alignment is probed by the different polarizations of the LIF intensity[69,70] and another in which the dependence of the total fluorescence intensity is determined as a function of the strength of an applied magnetic field,[71,72] similar to the Senftleben–Beenakker effect. Only small alignments are reported in some of the measurements, while in others rather large effects are observed. The variation in the results may be caused by the somewhat fragile nature of molecular alignment, which can be easily removed by reactive collisions or isotropic velocity components in the beam expansion. Thus, it is difficult to devise a jet expansion experiment for which reliable alignment measurements can be made under identical conditions in different laboratories.

The generally agreed-upon alignment effect in beam expansions, however, may be modelled on the basis of a classical hard ellipsoid.[69] This potential is representative of the repulsive core of the neutral diatomic molecule in the collision. A velocity gradient occurs in the early part of the expansion before the beam density becomes low enough to freeze out the distribution of m_J states of the diatomic molecule. The large effect reported by Sinha *et al.*[69] suggests that the major axis of the ellipsoid is rotating in a plane parallel to the beam direction, *i.e.* in the direction of the anisotropic velocity. Approximately 1.5 times as many molecules have their angular momentum vector perpendicular to the beam axis rather than parallel to it. Thus, the molecules are found with their planes of rotation preferentially parallel to the beam direction. This makes physical sense for a repulsive ellipsoidal potential, since the average cross-sections for collision with the rigid ellipse will be smaller if the plane of rotation is parallel to the velocity gradient.

While a strongly anisotropic interaction potential is likely for neutral

[69] M. P. Sinha, C. D. Caldwell, and R. N. Zare, *J. Chem. Phys.*, 1974, **61**, 491.
[70] W. R. Sanders and J. B. Anderson, *J. Phys. Chem.*, 1984, **88**, 4479.
[71] A. G. Visser, J. P. Bekooy, L. K. van der Meij, C. de Vreugd, and J. Korving, *Chem. Phys.*, 1977, **20**, 391.
[72] M. A. Treffers and J. Korving, *J. Chem. Phys.*, 1986, **85**, 5071 and 5085.

molecules, it is not as obvious that this will be the case for ions, in which the long-range ion-induced dipole attraction can play an important role in the scattering. Only a small degree of alignment would be expected if the collisions sample primarily the isotropic long-range part of the potential, where orbiting type collisions might occur. This may be the case for low-energy collisions.

In Figure 10 the potential curves calculated for N_2[73] and N_2^+[74] with He are presented. It is readily seen that the long-range part of the ion potential is relatively isotropic. However, at close range, the repulsive interaction is very anisotropic, as it is in the neutral molecule. Thus, higher-energy collisions which sample the repulsive part of the potentials are likely to involve significant m_J alignment for the ion as well as the neutral species. The high-energy limit is very easy to achieve even with

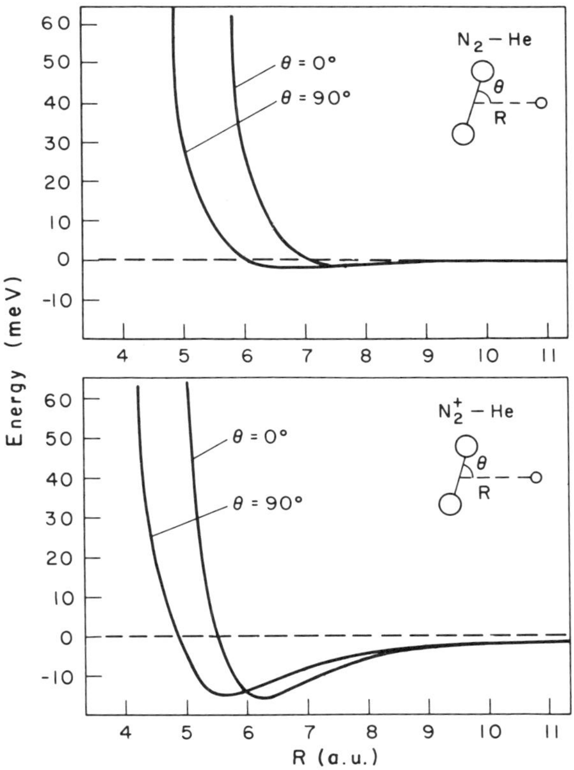

Figure 10 *Potential curves for* N_2 *and* N_2^+ *with* He *at different angles of approach, from refs. 73 and 74, respectively*

[73] A. J. Banks, D. C. Clary, and H. -J. Werner, *J. Chem. Phys.*, 1986, **84**, 3788.

[74] (*a*) S. Miller, J. Tennyson, B. Follmeg, P. Rosmus, and H. -J. Werner, *J. Chem. Phys.*, 1988, **89**, 2178; (*b*) H. -J. Werner, B. Follmeg, and P. Rosmus, to be published.

mild drift fields, since the energy of thermal collisions is approximately twice as great as the well depth for N_2^+ + He. However, this may not be the case for more polarizable collision partners. In addition, the locations of the wells for the two different extreme alignments of the ion shown in Figure 10 are very different. Thus, large impact parameter collisions of N_2^+ in He may also involve a significant anisotropy in the trajectories and state-changing processes.

For a number of reasons, the probing of alignment of ions in a drift field can be a more precise, reproducible measurement than the beam expansion experiments discussed earlier.[26] It should be possible for various groups to perform similar alignment experiments, with a good chance for agreement among the results. The drift experiment may be carefully controlled in terms of the ion and buffer gas densities. The observed alignment effect will be the result of a very large number of collisions under the action of a well-defined anisotropy in the velocity distribution. The drift velocity can be readily controlled by the applied field. The values of the isotropic polarizations can also be exactly determined when the drift field is off. The anisotropy of the velocity distribution may be estimated from the average drift velocity or measured directly via the Doppler velocity shifts and distribution.[27] Finally, a steady-state analysis will be applicable to the description of the observed alignment.[75] It is hoped that the measured alignments will provide definitive information about the degree of anisotropy of the potential surface and the individual state-to-state cross-sections for J- and m_J-changing collisions. It will ultimately be of interest to relate the observed alignment to transport properties such as mobility and diffusion.

The experimental arrangement for the determination of alignment in drift fields is very similar to that for measuring the rotational populations. The only additions are a polarization rotator for selection of the direction of linear polarization of the pulsed laser beam and a polarization analyser for determination of the polarization direction of the detected fluorescence.[26] In these experiments, N_2^+ is drifted in helium buffer gas and the degree of alignment is probed by LIF.

Figure 11 shows a schematic of the propagation directions of the laser beam (x-axis), the He flow and electric field (z-axis), and fluorescence detection (y-axis). The transition moment of N_2^+ is along the direction of the N—N bond. Qualitatively, the linearly polarized laser preferentially excites molecules whose plane of rotation lies along the direction of the laser polarization. The fluorescence also radiates parallel to the internuclear axis. Thus, as shown in Figure 11, some planes of rotation are excited equally with any laser polarization, while other planes are preferentially excited by one polarization or the other. Similarly, the fluorescence of certain planes of rotation are detected with equal

[75] H. Meyer and S. R. Leone, *Mol. Phys.*, 1988, **63**, 705.

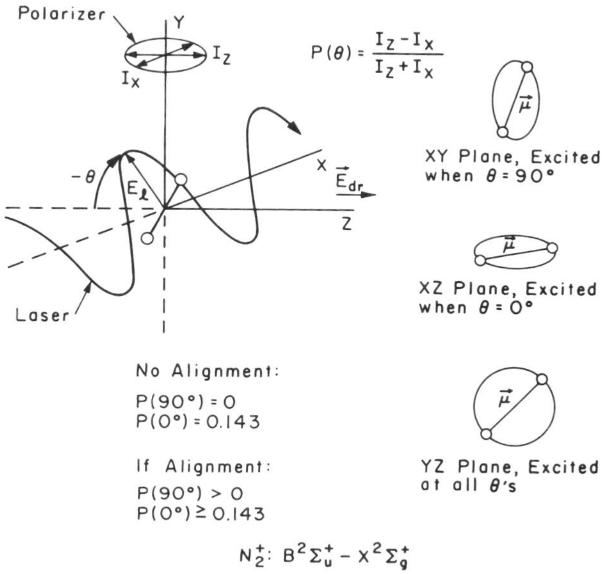

Figure 11 *Schematic of the laser probe and detection geometry for studies of* N_2^+ *alignment in* He

sensitivity through the polarizer in front of the detector, while others are strongly attenuated or transmitted depending on the direction of the polarizer.

The relationships used to determine molecular alignment, as well as other moments of the m_J distribution of rotational angular momentum vectors using linearly polarized LIF, have been thoroughly described.[76,77] For cylindrical symmetry, only the even order moments of the distributions can be obtained. Thus, the alignment and the shape of the aligned distribution might be obtained, for example, indicated by the quadrupole $\mathcal{A}_0^{(2)}$ and hexadecapole $\mathcal{A}_0^{(4)}$ alignment moments, respectively. Circularly or elliptically polarized light may be used to obtain information about asymmetric components, such as the orientation,[78] if they exist.

The molecular polarization, P, is defined as a function of the angle θ that the laser polarization makes with the z-axis (electric field direction):

$$P(\theta) = (I_z - I_x)/(I_z + I_x) \tag{12}$$

where I_z and I_x are the intensities of molecular emission through the polarizer along the z- and x-axes, respectively. Just from the detection geometry itself, the LIF emission is already partially polarized for an

[76] C. H. Greene and R. N. Zare, *J. Chem. Phys.*, 1983, **78**, 6741.
[77] C. H. Greene and R. N. Zare, *Annu. Rev. Phys. Chem.*, 1982, **33**, 119.
[78] A. C. Kummel, G. O. Sitz, and R. N. Zare, *J. Chem. Phys.*, 1988, **88**, 7357.

isotropic sample of N_2^+ when the laser polarization is along the field direction, $\theta = 0°$. Thus, $P = 0.143$ for $\theta = 0°$ and $P = 0$ for $\theta = 90°$. When the electric field is turned on and the N_2^+ molecules may become aligned by collisions, the values of the molecular polarization for either value of the laser polarization can increase or decrease. For example, a positive increase in $P(\theta = 90°)$ indicates that the plane of rotation of the molecule is preferentially aligned parallel to the drift field. A negative change in $P(\theta = 90°)$ indicates that the plane of rotation is perpendicular to the drift field.

In the experiments, the changes in the values for $P(\theta = 0°)$ are too small to be reliably measured. However, the larger changes in $P(\theta = 90°)$ are readily detected for N_2^+ drifting in He. Unfortunately, from the determinations of just the 90° laser polarization, only the first moment of the alignment, the quadrupole alignment term, $\mathcal{A}_0^{(2)}$ can be obtained.

The results of the measurements for the 90° laser polarization[26] are shown in Figure 12. The value of the polarization increases, which indicates that the plane of rotation of the N_2^+ is parallel to the drift field. The observed results are very sensitive to the cleanliness of the system, which once again points out the fragile nature of the molecular alignment with regard to stray fields and collisions with impurity species. The degree of raw polarization varies from 0—5% for the determinations shown, which in this figure are derived from a group of rotational quantum numbers in the P-branch bandhead, $N'' = 8$—12. In separate measurements, the alignment is determined at specific fields for single rotational states.

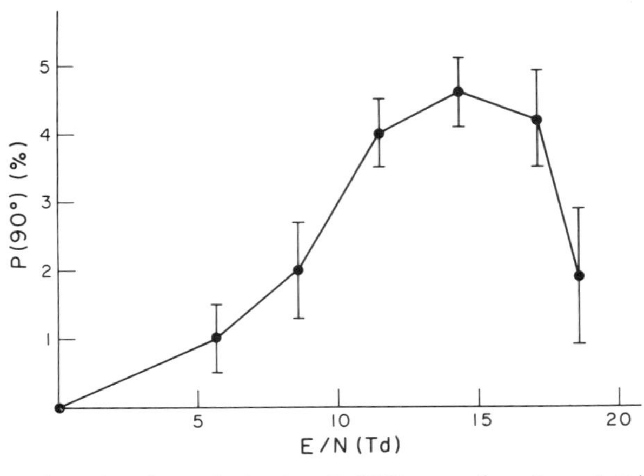

Figure 12 *Raw data for the polarization P* (90°) *as a function of E/N for* N_2^+ *in* He

In Figure 12, the alignment is seen to increase significantly with drift field and then decrease abruptly at an E/N of 18 Td. The initial increase in the alignment is attributed primarily to the increasing anisotropy of the velocity distribution as the drift field is increased. The sharp decrease in the alignment is accompanied by a decrease in total fluorescence intensity, which we attribute to collisions of the upper excited state used to probe the ion by LIF.

While the raw data indicate only a few per cent change in the polarization, the actual changes in the molecular alignment are much greater. Corrections to the data must be made to account for the depolarization of the emission due to the nuclear spin and electron spin, neither of which are aligned by collisions. For N_2^+, these corrections amount to a change of $\mathscr{A}_0^{(2)}$ for the rotational state $N'' = 4$ from -0.04 to -0.10 and for $N'' = 10$ from -0.075 to -0.10. In addition, partial saturation of the electronic transition with the laser deteriorates the observed alignment, resulting in a final corrected value for $\mathscr{A}_0^{(2)}$ of -0.11 for both states.[26]

If the value of the hexadecapole alignment moment $\mathscr{A}_0^{(4)}$ is assumed to be zero, an approximate distribution of m_N values can be plotted, and this is shown in Figure 13 for the $N'' = 4$ state. A statistical distribution of m_N states (given by the dashed line) would appear flat on this figure. The interpretation of this figure is that the plane of rotation parallel to the drift velocity is 1.5 times more probable than perpendicular. This magnitude of alignment is comparable to that reported for neutral Na_2 in the jet expansion of pure sodium.[69]

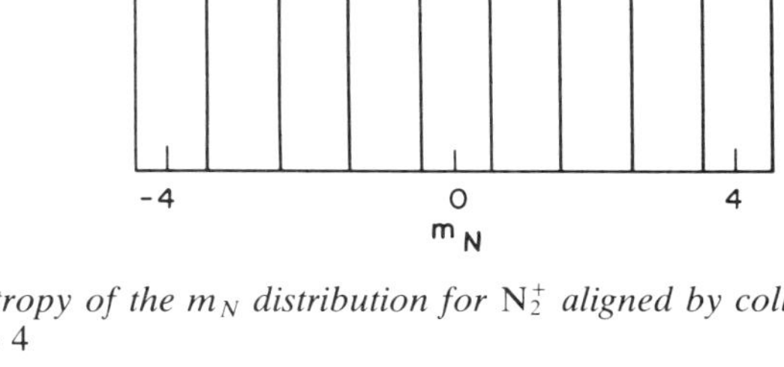

Figure 13 *Anistropy of the m_N distribution for* N_2^+ *aligned by collisions in* He *for* $N'' = 4$

Thus, it appears that the degree of alignment for an ion by collisions in a drift field can be substantial, even at modest drift fields. This result is perhaps of even more consequence when one considers that the abrupt decrease in the alignment observed in these experiments at the higher fields may only be due to an artifact of the method of probing through the electronically excited state. We might expect the alignment of the ground state to increase further with the field until it reaches a plateau. At very high collision energies perhaps the alignment would again diminish if the anisotropy of the potential begins to be negligible compared with the collision energy. N_2^+ is an ion which is very nearly spherically symmetric. Other ions, for example the benzene cation, might be expected to exhibit even stronger alignment effects.

The possible influence of collision-induced alignment on the mobility of ions may be an important effect for further consideration. The substantial alignment caused by the anisotropy of the potential for N_2^+ in He suggests that the difference in mobility for the different molecular alignments may also be large. Collisions with an anisotropic velocity distribution will initially produce aligned molecules, which have concomitantly smaller average momentum transfer cross-sections. Thus the mobility of an ion like N_2^+ in He might be expected to increase anomalously at first, compared with the expectation for a simple isotropic interaction potential. The alignment of molecular ions could also influence reactivity. Further experiments can be carried out to spoil the alignment and determine the mobility and reactivity with and without aligned molecules. Finally, it is important to note that Toschek and co-workers observed a difference in the cross-sections for scattering of different m_J states of Ga atoms in the $^2P_{3/2}$ state.[79,80] Although the effect is small, it suggests a possible dependence of mobility on *p*- or *d*-orbital alignment for atomic ions as well.

7 Theoretical Relationship between the Observed Alignment and Microscopic Tensor Cross-sections

Mobility data have been used with considerable success to determine information about interaction potentials.[81–83] The mobility data alone, in practice, give no direct information on the anisotropy of the potential. If other experimental quantities are obtained which are sensitive to the anisotropy of the potential, such as molecular alignment of ions in drift fields, then it may be possible to use these data to determine the angular dependence of the potential. If accurate potential energy surfaces for ion–noble gas interactions are available, then the detailed microscopic cross-sections for *J*- and m_J-changing collisions can be calculated. The microscopic tensor cross-sections contain all the information which relates to the alignment moments of the molecule after the

[79] K. Berkling, Ch. Schlier, and P. Toschek, *Z. Phys.*, 1962, **168**, 81.
[80] P. Toschek, *Z. Phys.*, 1965, **187**, 56.
[81] L. A. Viehland and E. A. Mason, *Ann. Phys. NY*, 1975, **91**, 499.
[82] I. R. Gatland, W. F. Morrison, H. W. Ellis, M. G. Thackston, E. W. McDaniel, M. H. Alexander, L. A. Viehland, and E. A. Mason, *J. Chem. Phys.*, 1977, **66**, 5121.
[83] D. R. Lamm, M. G. Thackston, F. L. Eisele, H. W. Ellis, J. R. Twist, W. M. Pope, I. R. Gatland, and E. W. McDaniel, *J. Chem. Phys.*, 1981, **74**, 3042.

collision. With a suitable theory, it would be possible to predict the observed degrees of alignment for ions drifting in a simple buffer gas.

The theory to connect the microscopic tensor cross-sections with the observed state multipole moments of the alignment is formulated under steady-state conditions, also using moments to describe the anisotropy of the velocity distribution.[75] As noted above, both an anisotropy in the interaction potential and in the velocity distribution are essential elements to obtain an aligned distribution of m_J states. After application of steady-state kinetics and the transformations required to relate the axis of the new velocity component along the drift field to the centre-of-mass vector in the collisions, a master set of equations is obtained. These equations connect the moments of the angular momentum distribution $\langle T_0^K \rangle$ to the microscopic tensor cross-sections for J- and m_J-changing collisions ${}^{\lambda}\sigma^{K\bar{K}}$ through the moments of the velocity distribution $\tilde{f}_0^{K\lambda}$ and the centre-of-mass velocity g:

$$\sum_{J_f K\lambda} \langle T_0^K(J_f)\rangle \int \mathrm{d}g\, g^3\, \tilde{f}_0^{K\lambda}(J_f,g)\; {}^{\lambda}\sigma^{K\bar{K}}_{J_f\to J_i}(g) =$$

$$\sum_{J_f K\lambda} \langle T_0^K(J_i)\rangle \begin{pmatrix} K & \lambda & \bar{K} \\ 0 & 0 & 0 \end{pmatrix}\begin{pmatrix} K & \lambda & \bar{K} \\ J_i & J_i & J_i \end{pmatrix} \sum \mathrm{d}g\, g^3\, \tilde{f}_0^{K\lambda}(J_i,g)\; {}^{\lambda}\sigma^{\lambda 0}_{J_i\to J_f}(g) \tag{13}$$

The equations determine the Kth order multipole moment. The indices λ, K, and $\bar{K}$ refer to the order of the moments of the velocity distribution, the initial-state multipole, and the final-state multipole, respectively.

Werner, Follmeg, and Rosmus[74b] have recently calculated the detailed tensor cross-sections for collisions of N_2^+ with He. In contrast to cross-sections for just population changes, the tensor cross-sections, take on positive and negative values which contain the detailed alignment information. For example, a cross-section ${}^2\sigma^{02}_{0-2}$ describes the transition from $J = 0$—2 for the second moment of the velocity distribution and characterizes the magnitude of that scattering event when the state multipole moment changes from 0 to 2. The preliminary value of this tensor cross-section[74] is negative, -0.33. This is compared with a positive value of 6.3 when the same $J = 0$—2 change occurs without a change in state multipole for the 0th moment of the velocity distribution. In contrast, the cross-section for $J = 2$—0, when the state multipole also changes from 2 to 0, has a smaller, positive value of 0.13. While a full set of tables of these numbers must be included in a detailed description, the asymmetry in just these few values suggests that alignment will occur as a result of collisions under the influence of an anisotropic velocity. Thus, this information about the interaction potential can soon be employed to connect the anisotropic potential to the observed alignment.

In a simplified form, the set of equations (13) can be used to predict,[75] for example, the expected degree of alignment by: (*a*) measuring a population distribution for the rotational levels; (*b*) obtaining a value for at least the first moment of the anisotropy of the velocity distribution; and (*c*) using the first few terms of the microscopic tensor cross-sections,[75] for example ${}^{0}\sigma^{00}$ and ${}^{2}\sigma^{20}$, which contain information about the cross-sections for population changes for the monopole and quadrupole state moments. It has also been shown[75] how the anisotropy of the velocity distribution can be estimated from the observed rotational temperature of the drifting ions.

8 Laser Probing of Ion Velocity Distributions in Drift Fields

Retarding potential methods[12,14,15] and ion energy analysis techniques[11,13] have been used to determine the velocity distributions of ions in drift fields. However, the intrusion of the extraction fields which are used in these methods can distort the ion motion significantly,[9] and the results of such measurements have been difficult to check accurately against theory. In addition, these methods are applicable primarily to relatively high values of E/N.

Optical probing of ion velocity distributions has the advantage of being non-intrusive, provided that the light field does not saturate or optically pump the transition. Optical probing takes advantage of the Doppler shift with velocity,[84] given by:

$$\omega = \omega_0 \left(1 \pm \frac{v}{c}\right) \tag{14}$$

The plus component is related to molecules moving away from the light source, while the minus component is for molecules moving towards the source. A Maxwell–Boltzmann distribution of velocities at a temperature T for a transition with a negligible natural linewidth produces a Gaussian Doppler line profile described by:

$$I(\omega) = I_0 \exp\left[-\left(\frac{c(\omega - \omega_0)}{\omega_0 v'}\right)^2\right] \tag{15}$$

where the average velocity parameter $v' = (2k_BT/m)^{1/2}$. The Doppler width is given by:

$$\Delta\omega_\mathrm{D} = \frac{2\omega_0}{c}\left(\frac{2k_BT\ln 2}{m}\right)^{1/2} \tag{16}$$

The constants in equation (15) yield $\Delta\omega_\mathrm{D} = 7.16 \times 10^{-7}\ \omega_0(T/m')^{1/2}$, where m' is in units of g mol^{-1}.

[84] W. Demtroder, 'Laser Spectroscopy', Springer Verlag, Berlin, 1982.

Thus, a laser beam can be propagated either parallel or perpendicular to the drift field, and the velocities of the ions can be probed to obtain both the drift velocity shifts as well as the widths, or 'temperatures', of the ion motion. The LIF method has been applied to the study of Ba^+ velocities in discharges and plasmas by a number of investigators.[19,24] Doppler profiles have been measured by diode laser absorption to study the mobility of ArH^+ in different vibrational states in a discharge environment.[20,28] Doppler shift methods are now routinely used to detect ions sensitively in spectroscopy.[22,23] Until recently,[27] however, these optical probe methods have not been applied to the characterization of ions in well-defined drift tube studies.

In the work discussed here,[27,85] a single-mode ring dye laser is used to probe the velocity distributions of Ba^+ drifting in a noble gas buffer at E/N values of ~0—30 Td. The ring dye laser excites the Ba^+ at 455 nm on the $6p\,^2P_{3/2} \leftarrow 6s\,^2S_{1/2}$ transition and fluorescence is detected back to the $5d\,^2D_{5/2}$ state at 614 nm. The lifetime of the excited state is 6.5 ns, which contributes a Lorentzian natural linewidth to the Doppler profiles of 24.5 MHz. At 300 K the Doppler profiles, including the natural linewidth, are observed to be 710 ± 5 MHz.

Saturation by the laser causes an additional breadth to the natural width, which is particularly noticeable if the laser power density is high and the lower ^2D state is not quenched or rapidly removed from the beam. Pappas *et al.*[86] treated the saturation problem of two states in the presence of a third level and showed that the third state accelerates the approach to saturation. Empirically, the saturation problem is observed to be much worse when the laser is used to probe parallel to the drift field than transverse, because the ions stay in the beam interaction region for a longer time.[85] In order to avoid the influence of these saturation effects entirely, laser power densities of less than 60 $\mu W\,cm^{-2}$ are used at all times.

Another interesting effect has been observed in these studies at very high power densities. The lineshapes start to become increasingly rectangular instead of showing larger contributions in the wings, which is expected if an increasing Lorentzian component is added to a Gaussian. A possible explanation for this behaviour is the light-induced drift effect,[87] in which the laser optical pumping in both the red and blue high-velocity wings of the line causes a shift of the ions to lower velocities. This arises from the differing momentum transfer cross-sections in the ground and excited states.

Figure 14 shows a set of Doppler profiles for Ba^+ drifting in 65 Pa of He as function of E/N. The laser is propagated in the direction parallel

[85] R. A. Dressler, J. P. M. Beijers, H. Meyer, S. M. Penn, V. M. Bierbaum, and S. R. Leone, *J. Chem. Phys.*, 1988, **89**, 4707.

[86] P. G. Pappas, M. M. Burns, D. D. Hinshelwood, M. S. Feld, and D. E. Murnick, *Phys. Rev. A.*, 1980, **21**, 1955.

[87] G. Nienhuis, *Phys. Rev. A.*, 1985, **31**, 1636.

to the field, and the ion velocities are shifted to higher frequencies with increasing field strength. The average shift of the Doppler profiles is a direct measure of the average drift velocity, v_d. In addition, a slight broadening of the lines can be noted, which is due to momentum transfer collisions that increase the random component of the ion energy along the drift field. We find that the flow tube system must be kept scrupulously clean to obtain reliable values for the shifts and widths of the lines because of the build-up of strong fields.

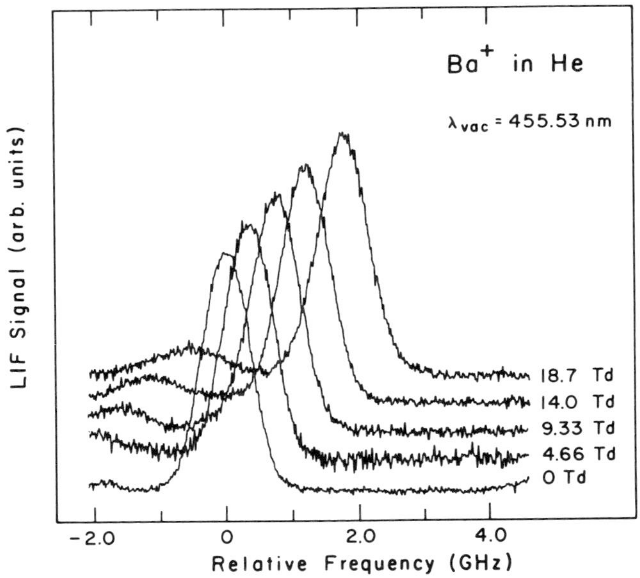

Figure 14 *Doppler profiles for* Ba^+ *in* He *as a function of* E/N *(displaced vertically for clarity) when probing parallel to the field direction*

A series of determinations[85] have been made of the Doppler shifts to obtain the mobility of Ba^+ in He (Figure 15). It can be seen that the reduced mobility decreases monotonically with E/N. This is typical for many ions drifting in He, since the well depth of the interaction potential is smaller than the thermal energy of the system. Thus, a mobility maximum, or bump, in the reduced mobility plot is not expected. The value of the reduced mobility for Ba^+ in He is typical of values obtained for other atomic metal ions.[1] To our knowledge, these are the first absolute mobilities that have been determined optically, and thus it will be important to check the method against other standard techniques in the future.

If the laser is propagated in both directions along the axis of the flow drift tube, then the Doppler shifts in both directions can be observed, as

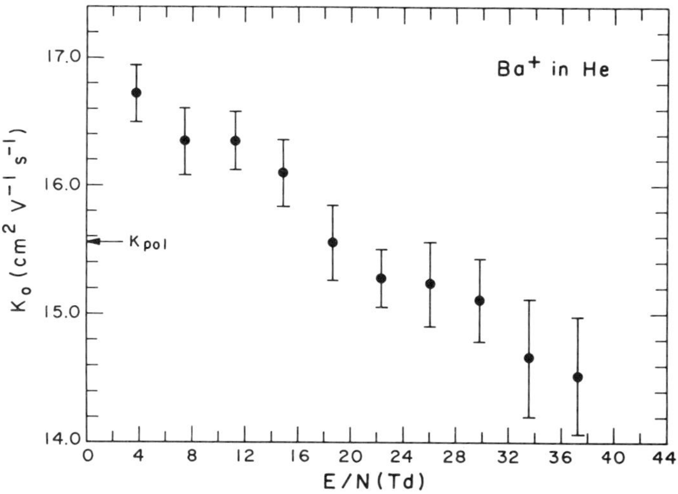

Figure 15 *Reduced mobility of* Ba^+ *in* He *as a function of E/N, determined optically from the Doppler shifts*

in Figure 16. This type of measurement demonstrates the potential use of the laser–optical probing method to obtain the flow velocity of the ions as well. The small shift in the line centres at zero field $\Delta\nu$ is equal to twice the flow velocity of the ions. The average ion velocity is generally between 1.3 and 1.6 times the bulk flow velocity of the carrier gas.[29,88] The laser can determine the exact 'local' velocity of the ions in the centre of the flow tube and can verify the expectations of a parabolic flow profile by measuring the velocity as a function of radial position in the flow tube. The optical probing method may therefore prove to be a valuable new diagnostic for characterization of flow details as well.

The most interesting new results pertain to the measurement of the ion 'temperatures' both parallel and perpendicular to the drift field.[85] Figure 17 shows a set of measurements of the Doppler profiles in the direction transverse to the field. Gaussian profiles are found to give excellent fits to the lineshapes in both the transverse and parallel directions (with the exception of an additional small Lorentzian width to account for the natural lifetime). The Gaussian shape is indicative that a 'temperature' is applicable to these distributions. Such a result is expected since, as noted by Whealton and Woo,[36] the skewness of the velocity distributions is very slight for heavy ions drifting in light buffer gases. One may think of the heavy ion as 'channelling' through the He, with very little momentum transfer. Thus, the original Doppler profile is almost completely preserved, but shifted to higher velocity by the drift motion.

[88] B. L. Upshulte, R. J. Shul, R. Passarella, R. G. Keesee, and A. W. Castleman, jun., *Int. J. Mass Spectrom. Ion Processes*, 1987, **75**, 27.

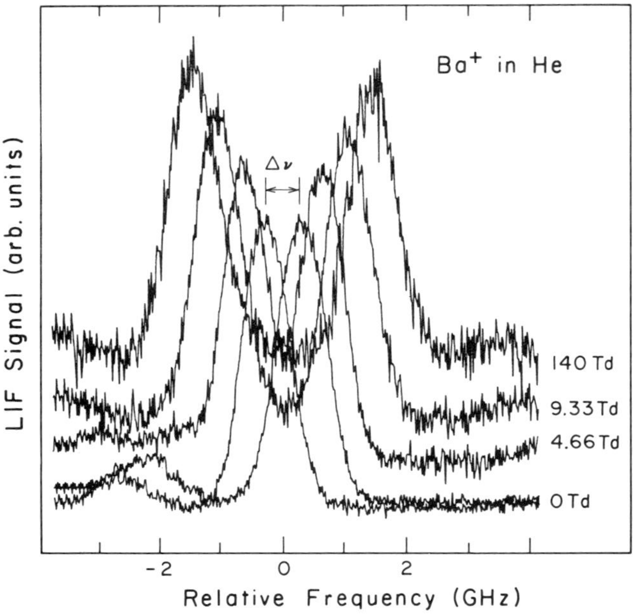

Figure 16 *Doppler profiles of* Ba^+ *in* He *as a function of E/N when probed with the laser propagating in both directions in the parallel geometry (displaced vertically for clarity)*

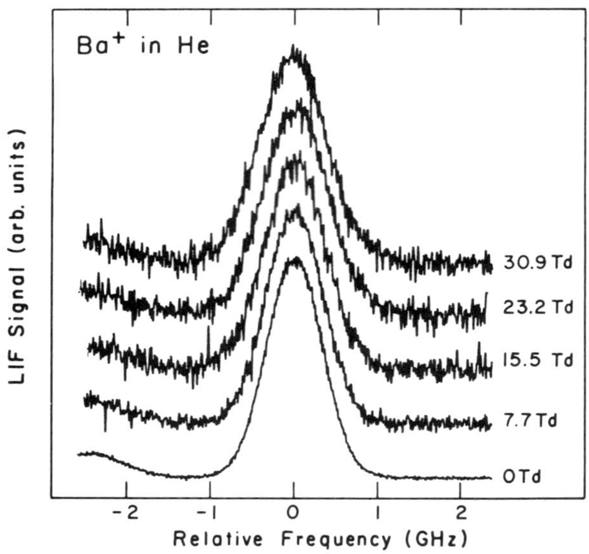

Figure 17 *Doppler profiles of* Ba^+ *in* He *as a function of E/N when probed in the transverse direction to the field (displaced vertically for clarity)*

A slight increase in the Doppler widths in the parallel and perpendicular directions is, nevertheless, measurable. The results of these studies are summarized in Figure 18. Also plotted in the figure is the prediction of the three-temperature theory of Lin, Viehland, and Mason,[42] as parameterized by Waldman and Mason.[48] It can be seen that the data and the predictions are in excellent qualitative and quantitative agreement. The increase in random motion in the parallel direction is greater than the increase in the perpendicular direction, and neither increase is very great for the heavy ion in the light buffer.

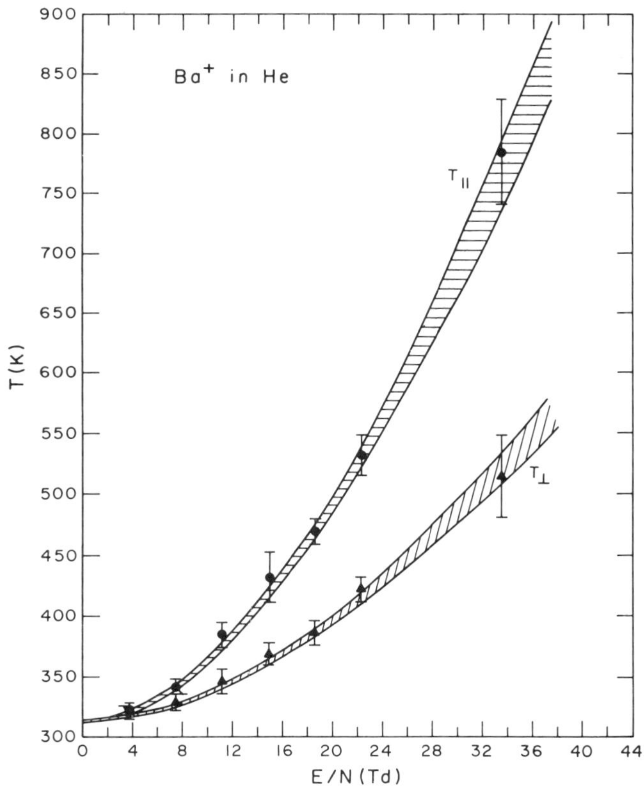

Figure 18 *Plot of the linewidths (temperatures) parallel* (●) *and perpendicular* (■) *to the field for* Ba^+ *in* He *compared with the three-temperature theory*[42] *(shaded regions)*

An even more interesting effect is observed[65] when a small fraction (20%) of argon gas is bled into the He flow (Figure 19). The linewidths are dramatically broader in the parallel direction for a given drift velocity, and a skewing is observed. This is due to the heavier, more polarizable collision partner. Shown in the figure is a Gaussian fit to these preliminary results, which clearly reveals the high-energy tail in the observed distribution of velocities. This exciting result holds promise for many new and interesting measurements of velocity distributions in well-characterized drift fields.

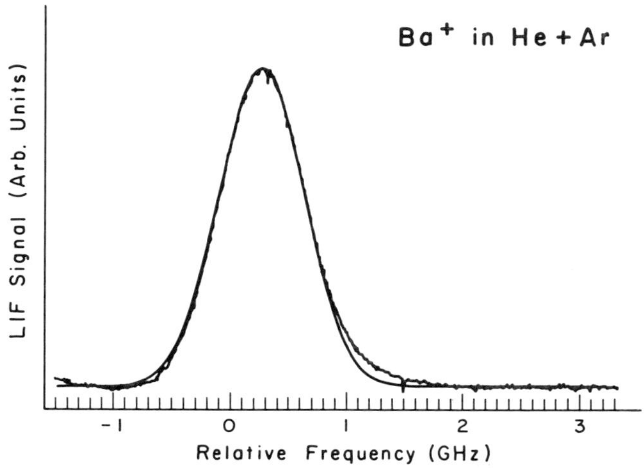

Figure 19 *Doppler velocity in the direction parallel to the field for* Ba^+ *in* He–Ar *mixtures, showing the skewing away from the Gaussian fit (solid line)*

It is hoped that several of the techniques discussed in this chapter can be combined, for example to determine the alignment, population distributions, and velocity distributions of molecular ions simultaneously. The results will provide accurate tests of theoretical predictions for the first time and will stimulate further experimental and theoretical work in the state-resolved description of ion transport phenomena.

Acknowledgement. The author gratefully acknowledges all of the enthusiastic collaborators who participated in these projects, especially Veronica Bierbaum, Michael Duncan, Andrew Langford, Rainer Dressler, Henning Meyer, Stephen Penn, and Hans Beijers. Support for this research was provided by the U.S. Air Force Office of Scientific Research and the U.S. National Science Foundation. This chapter was prepared while the author was a recipient of a John Simon Guggenheim Fellowship.